The 2nd International Electronic Conference on Applied Sciences

The 2nd International Electronic Conference on Applied Sciences

Editors

Takayoshi Kobayashi
Nunzio Cennamo

MDPI • Basel • Beijing • Wuhan • Barcelona • Belgrade • Manchester • Tokyo • Cluj • Tianjin

Editors

Takayoshi Kobayashi
The University of
Electro-Communications
Japan

Nunzio Cennamo
University of Campania Luigi
Vanvitelli
ITALY

Editorial Office
MDPI
St. Alban-Anlage 66
4052 Basel, Switzerland

This is a reprint of articles from the Proceedings published online in the open access journal *Engineering Proceedings* (ISSN 2673-4591) (available at: https://www.mdpi.com/2673-4591/11/1).

For citation purposes, cite each article independently as indicated on the article page online and as indicated below:

LastName, A.A.; LastName, B.B.; LastName, C.C. Article Title. *Journal Name* **Year**, *Volume Number*, Page Range.

ISBN 978-3-0365-4817-3 (Hbk)
ISBN 978-3-0365-4818-0 (PDF)

Contents

engineering proceedings

Proceeding Paper

Overlooked Ionic Phenomena Affecting the Electrical Conductivity of Liquid Crystals †

David Webb and Yuriy Garbovskiy *

Department of Physics and Engineering Physics, Central Connecticut State University, New Britain, CT 06050, USA; davidwebb@my.ccsu.edu

* Correspondence: ygarbovskiy@ccsu.edu

† Presented at the 2nd International Electronic Conference on Applied Sciences, 15–31 October 2021; Available online: https://asec2021.sciforum.net/.

Abstract: Liquid crystal devices, such as displays, various tunable optical components, and sensors, are becoming increasingly ubiquitous. Basic physical properties of liquid crystal materials can be controlled by external physical fields, thus making liquid crystal devices dynamically reconfigurable. The tunability of liquid crystals offers exciting opportunities for the development of new applications, including advanced electronic and photonic devices, by merging the concepts of flat optics, tunable metasurfaces, nanoplasmonics, and soft matter biophotonics. As a rule, the tunability of liquid crystals is achieved by applying an electric field. This field reorients liquid crystals and changes their physical properties. Ions, typically present in liquid crystals in minute quantities, can alter the reorientation of liquid crystals through the well-known screening effect. Because the electrical conductivity of thermotropic liquid crystals is normally caused by ions, an understanding of ion generation processes in liquid crystals is of utmost importance to existing and emerging technologies relying on such materials. That is why measuring of electrical conductivity of liquid crystals is a standard part of their material characterization. Measuring the electrical conductivity of liquid crystals is a very delicate process. In this paper, we discuss overlooked ionic phenomena caused by interactions of ions with substrates of the liquid crystal cells. These interactions affect the measured values of the DC electrical conductivity of liquid crystals and make them dependent on the cell thickness.

Keywords: liquid crystals; liquid crystal devices; ions; ion generation; electrical conductivity

Citation: Webb, D.; Garbovskiy, Y. Overlooked Ionic Phenomena Affecting the Electrical Conductivity of Liquid Crystals. *Eng. Proc.* **2021**, *11*, 1. https://doi.org/10.3390/ASEC2021-11141

Academic Editor: Nicholas Sarlis

Published: 15 October 2021

Publisher's Note: MDPI stays neutral with regard to jurisdictional claims in published maps and institutional affiliations.

1. Introduction

Liquid crystals continue to be at the heart of modern technologies. They include display devices, such as ubiquitous liquid crystal displays (LCD) and miniature liquid crystal on silicon (LCoS) displays for virtual and augmented reality [1]. Active optical components made of liquid crystals can be found in numerous polarization and phase control devices, including liquid crystal variable retarders [2,3], tunable elements of biomedical equipment (dynamic lenses, optical filters for hyperspectral imaging, etc.) [4,5], spatial light modulators, and diffractive optical elements [6,7]. Reconfigurable components of plasmonic [8] and meta-devices [9] also take advantage of the tunability of liquid crystal materials. Switchable liquid crystal light shutters [10–12] and smart windows [13,14] are becoming increasingly ubiquitous. Last but not least, liquid crystals are excellent materials for the fabrication of reconfigurable microwave devices, including phase shifters, tunable antennas, filters, and resonators, to name just a few [15].

As a rule, the tunability of the aforementioned devices is achieved by exploiting the electric-field induced reorientation of liquid crystal materials [7]. This reorientation can be altered by the electric field screening effect caused by ions normally present in liquid crystals in minute quantities [16,17]. In the case of display devices, ions in liquid crystals can lead to many undesirable effects, including image flickering, image sticking, reduced voltage holding ratio, and overall slow response [16,17]. There are also liquid crystal applications relying on ions. For example, liquid crystal shutters [10–12] and smart windows [13,14] are receiving increasing attention these days.

The importance of ions in liquid crystal materials has been recognized since the early 1960s [16,17]. Since that time, numerous reports have been published aimed at broadening our understanding of mechanisms of ion generation in liquid crystal materials ([16–18] and references therein).

Typically, information about ions in liquid crystals is obtained by performing electrical measurements [19–22]. Such measurements utilize sandwich-like liquid crystal cells [19–22]. The obtained experimental results are used to evaluate the value of the DC electrical conductivity, ion mobility, and their bulk concentration [19–22]. In the majority of the reports, electrical measurements are performed at only a single value of cell thickness [23]. At the same time, a very limited number of measurements carried out using several cells with different thicknesses reveal an important experimental fact that the measured electrical parameters of liquid crystals, in general, can depend on the cell gap [20,21,24,25]. Unfortunately, only a very limited set of experimental data is available. An analysis of the dependence of the electrical conductivity of liquid crystals on cell thickness is still missing. In this paper, we discuss how interactions between ions and substrates of the liquid crystal cell make their DC electrical conductivity dependent on the cell thickness.

2. Model

In the case of molecular thermotropic liquid crystals, their finite *DC* electrical conductivity λ_{DC} is caused by ions. It can be written as (1):

$$\lambda_{DC} = \sum_i q_i \mu_i n_i \tag{1}$$

where q_i is the charge of the i-th ion, μ_i is the mobility of the i-th ion, and n_i is its volume concentration [16,17]. In the case of two types of symmetric monovalent ions ($q_i = |e|$ $(i = 1,2)$, $n_1^+ = n_1^- = n_1$, $\mu_1 = \mu_1^+ + \mu_1^-$, $n_2^+ = n_2^- = n_2$, $\mu_2 = \mu_2^+ + \mu_2^-$) the electrical conductivity can be rewritten as (2)

$$\lambda_{DC} = |e|(\mu_1 n_1 + \mu_2 n_2) \tag{2}$$

where $|e| = 1.6 \times 10^{-19}C$.

The bulk concentration of ions can be found by applying a recently developed model [26,27]. This model considers the possibility of both ion capturing and ion releasing regimes in liquid crystal cells. The ion releasing regime takes place if substrates of a liquid crystal cell are contaminated with ions prior to filling the cell with liquid crystals. In this case, contaminated substrates act as sources of ion generation in liquid crystals. The trapping of ions by the substrates (in the simplest case via physical or chemical adsorption) results in the ion capturing regime. The processes of ion capturing and ion generation are described by rate Equation (3):

$$\frac{dn_j}{dt} = -k_{Sj}^{a\pm} n_j \frac{\sigma_{Sj}}{d}\left(1 - \Theta_{S1}^{\pm} - \Theta_{S2}^{\pm}\right) + k_{Sj}^{d\pm} \frac{\sigma_{Sj}}{d} \Theta_{Sj}^{\pm} \tag{3}$$

where j denotes the dominant type of symmetric ions in liquid crystals ($j = 1,2$), $k_{Sj}^{a\pm}$ is the effective rate constant corresponding to the ion capturing process of n_j^+ and n_j^- ions on the surface of substrates, and $k_{Sj}^{d\pm}$ is the effective rate constant characterizing the ion releasing process of n_j^+ and n_j^- ions from the substrates, and $\Theta_{Sj}^{\pm}$ is the fractional surface coverage of substrates corresponding to the j-th ions, σ_{Sj} is the surface density of all surface sites of the liquid crystal substrates, n_j is the concentration (volume density) of mobile ions of the j-th type ($j = 1,2$), d is the thickness of the cell [26,27].

Equation (4) represents the conservation of the total number of ions of the j-th type:

$$n_{0j} + \frac{\sigma_{Sj}}{d}\nu_{Sj} = n_j + \frac{\sigma_{Sj}}{d}\Theta_{Sj}^{\pm} \quad (4)$$

where ν_{Sj} is the contamination factor of substrates [26,27]. Recent papers [26,27] discussed the applicability and limits of the model described by Equations (3) and (4). Equations (3) and (4) are very general and can be applied to a wide range of molecular liquid crystals and substrates. The strength of interactions between ions and substrates determines the value of the parameter $K_i = \frac{k_{Si}^a}{k_{Si}^d}$, the size of ionic contaminants affects the value of the parameter σ_{Si}, and the level of ionic contamination of substrates is quantified by the contamination factor ν_{Si}. The quantification of the ionic contamination of substrates is an important aspect of the proposed model. Depending on the type of materials (liquid crystals and substrates) and ionic contaminants (organic ions, inorganic ions), typical values of basic physical parameters used in this model ($K_i = \frac{k_{Si}^a}{k_{Si}^d}$, σ_{Si}, ν_{Si}, μ_i) can vary over a wide range of values ($K_i = 10^{-25} - 10^{-20}$ m^3, $\sigma_{Si} = 10^{16} - 10^{19}$ m^{-2}, $\nu_{Si} = 0 - 10^{-1}$, additional information can be found in paper [26] and recent review [18]). In the present paper, the values of physical parameters were chosen to represent typical liquid crystal cells, such as cyanobiphenyl nematic liquid crystals sandwiched between polyimide alignment layers.

3. Results

DC electrical conductivity of liquid crystal cells as a function of their thickness was computed using Equations (2)–(4). Table 1 lists values of basic physical parameters used to generate graphs shown in Figures 1 and 2. As a rule, electrical measurements of liquid crystals are carried out using liquid crystal cells. The presence of substrates can lead to several ionic processes. The capturing of ions by substrates results in the ion capturing regime, whereas the use of contaminated substrates leads to the ion releasing regime. Normally, ions that are already present in the liquid crystal bulk prior to filling an empty cell get captured by the substrates once the cell is filled. This scenario is shown in Figures 1a and 2a. At the same time, if substrates of the empty cell are contaminated with ions, these ions contaminate liquid crystal materials upon filling the cell, as can be seen in Figures 1b and 2b. The combination of these two, ion-releasing and ion-capturing, processes can result in a non-trivial dependence of the electrical conductivity on the cell thickness (Figures 1c and 2c).

Table 1. Values of physical parameters used to model DC electrical conductivity.

Physical Parameter	Value
$K_1 = \frac{k_{S1}^a}{k_{S1}^d}$	10^{-21} m^3 (Figures 1 and 3a)
$K_2 = \frac{k_{S2}^a}{k_{S2}^d}$	10^{-22} m^3 (Figures 1 and 3a)
$K_3 = \frac{k_{S3}^a}{k_{S3}^d}$	10^{-23} m^3 (Figures 2 and 3b)
$K_4 = \frac{k_{S4}^a}{k_{S4}^d}$	10^{-21} m^3 (Figures 2 and 3b)
$\sigma_{S1} = \sigma_{S2}$	5×10^{16} m^{-2}
$\mu_1 = \mu_2$	10^{-10} m^2/Vs
ν_{S1}	0
ν_{S2}	10^{-3} (Figures 1 and 2) 10^{-2} (Figure 3a) 10^{-4} (Figure 3b)
n_{01}	6×10^{18} m^{-3}
n_{02}	0

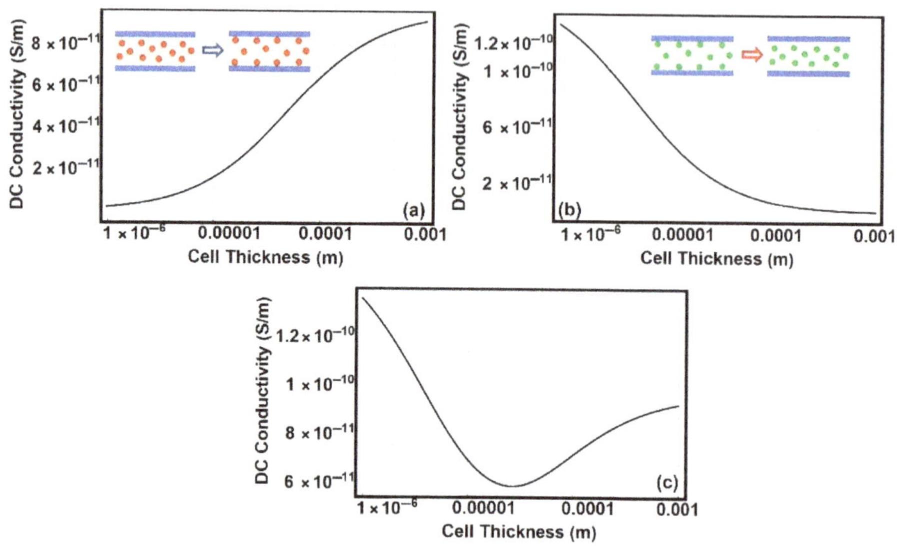

Figure 1. DC conductivity of liquid crystals as a function of the cell thickness. (**a**) DC conductivity caused by ions already present in liquid crystals. Values of parameters used in simulations are listed in Table 1 (n_{01}, ν_{S1}, μ_1, σ_{S1}, K_1). (**b**) DC conductivity due to ions originated from contaminated substrates of the liquid crystal cell. Values of parameters used in simulations are listed in Table 1 (n_{02}, ν_{S2}, μ_2, σ_{S2}, K_2). (**c**) DC conductivity caused by the combination of ion-capturing and ion-releasing processes shown in Figure 1a,b. Values of parameters used in simulations are listed in Table 1 (n_{01}, n_{02}, ν_{S1}, ν_{S2}, $\mu_1 = \mu_2$, $\sigma_{S1} = \sigma_{S2}$, K_1, K_2).

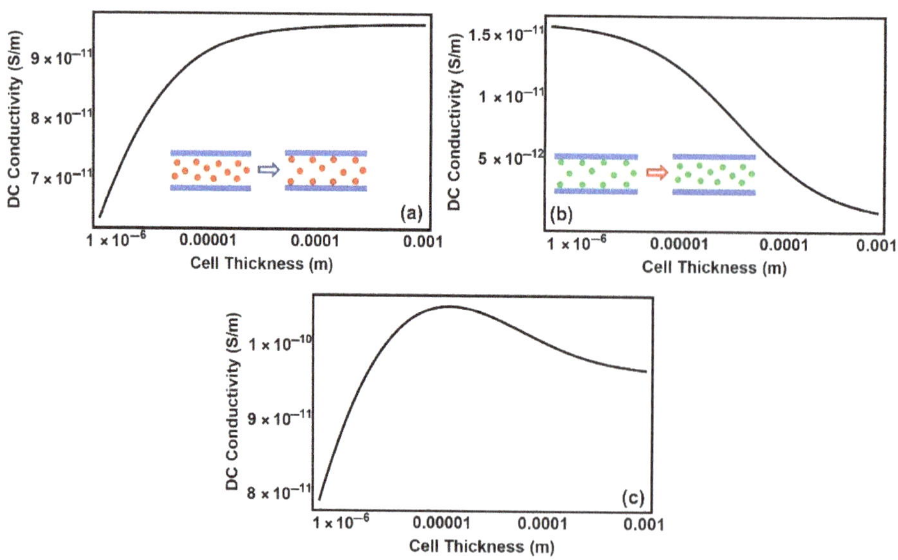

Figure 2. DC conductivity of liquid crystals as a function of the cell thickness. (**a**) DC conductivity caused by ions already present in liquid crystals. Values of parameters used in simulations are listed in Table 1 (n_{01}, ν_{S1}, μ_1, σ_{S1}, K_3). (**b**) DC conductivity due to ions originated from contaminated substrates of the liquid crystal cell. Values of parameters used in simulations are listed in Table 1 (n_{02}, ν_{S2}, μ_2, σ_{S2}, K_4). (**c**) DC conductivity caused by the combination of ion-capturing and ion-releasing processes shown in Figure 1a,b. Values of parameters used in simulations are listed in Table 1 (n_{01}, n_{02}, ν_{S1}, ν_{S2}, $\mu_1 = \mu_2$, $\sigma_{S1} = \sigma_{S2}$, K_3, K_4).

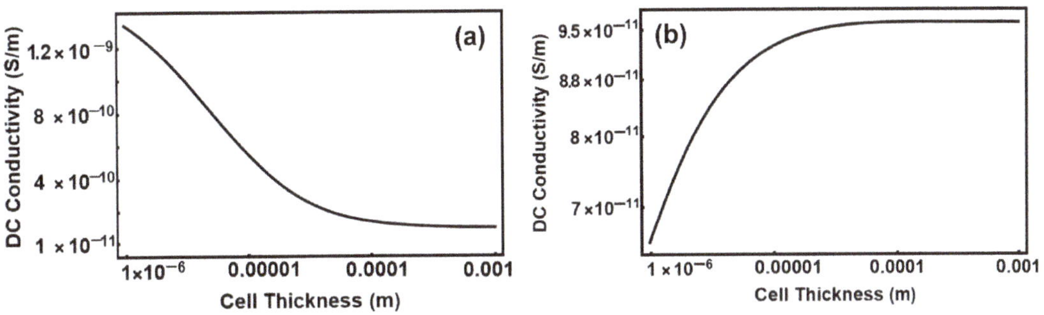

Figure 3. (**a**) Monotonous dependence of DC conductivity on the cell thickness achieved by increasing the value of the contamination factor ν_{S2} by one order of magnitude (Table 1). (**b**) Monotonous dependence of DC conductivity on the cell thickness achieved by decreasing the value of the contamination factor ν_{S2} by one order of magnitude (Table 1).

Interactions between ions and substrates of liquid crystal cells result in the dependence of DC electrical conductivity on the cell thickness. In general, this dependence can be monotonous or non-monotonous. Because monotonous dependence of DC electrical conductivity on the cell thickens is a rather trivial outcome, in the present paper, we decided to discuss some non-trivial cases corresponding to non-monotonous dependencies shown in Figures 1 and 2. At the same time, it is important to show other types of dependencies. For a given set of materials (ionic contaminants, liquid crystals, and substrates), the transition from non-monotonous dependence to monotonous one can be achieved by varying the contamination factor as shown in Figure 3a,b.

4. Conclusions

The dependence of DC electrical conductivity on cell thickness shown in Figures 1–3 has important practical implications. It suggests an importance to consider interactions between ions and substrates of the liquid crystal cell. These interactions can result in a combination of ion capturing and ion releasing effects. A striking manifestation of this interplay between ionic processes is a non-monotonous dependence of the conductivity on the cell thickness exhibiting either minimum (Figure 1c) or maximum (Figure 2c). It should be noted that ionic processes caused by the presence of substrates get weaker and become almost negligible if very thick (>100 μm) cells are used (Figures 1–3). Because typical cells used in experiments are normally much thinner (1–20 μm), the consideration of the dependence of DC electrical conductivity on the cell thickness is very important and should not be ignored. Electrical measurements of this type allow for the correct evaluation of the true values of the electrical parameters of liquid crystal materials. Moreover, they can also reveal possible sources of ionic contamination of substrates of liquid crystal cells.

Author Contributions: Conceptualization, Y.G.; methodology, Y.G.; formal analysis, D.W. and Y.G.; investigation, D.W. and Y.G.; data generation, D.W.; writing—original draft preparation, D.W. and Y.G.; writing—review and editing, D.W. and Y.G.; supervision, Y.G.; project administration, Y.G.; funding acquisition, Y.G. All authors have read and agreed to the published version of the manuscript.

Funding: This research was supported by the CSU—AAUP Faculty Research Grant and by the Faculty—Student Research Grant.

Data Availability Statement: All data that support the findings of this study are included within the article.

Acknowledgments: The authors would like to acknowledge the support provided by the School of Engineering, Science and Technology at Central Connecticut State University.

Conflicts of Interest: The authors declare no conflict of interest.

References

1. Huang, Y.; Liao, E.; Chen, R.; Wu, S.-T. Liquid-Crystal-on-Silicon for Augmented Reality Displays. *Appl. Sci.* **2018**, *8*, 2366. [CrossRef]
2. Otón, J.M.; Otón, E.; Quintana, X.; Geday, M.A. Liquid-crystal phase-only devices. *J. Mol. Liq.* **2018**, *267*, 469–483. [CrossRef]
3. Lazarev, G.; Chen, P.-J.; Strauss, J.; Fontaine, N.; Forbes, A. Beyond the display: Phase-only liquid crystal on Silicon devices and their applications in photonics. *Opt. Express* **2019**, *27*, 16206–16249. [CrossRef] [PubMed]
4. Abdulhalim, I. Non-display bio-optic applications of liquid crystals. *Liq. Cryst. Today* **2011**, *20*, 44–60. [CrossRef]
5. Lin, Y.; Wang, Y.; Reshetnyak, V. Liquid crystal lenses with tunable focal length. *Liq. Cryst. Rev.* **2017**, *5*, 111–143. [CrossRef]
6. De Sio, L.; Roberts, D.E.; Liao, Z.; Hwang, J.; Tabiryan, N.; Steeves, D.M.; Kimball, B.R. Beam shaping diffractive wave plates. *Appl. Opt.* **2018**, *57*, A118–A121. [CrossRef]
7. Chigrinov, V.G. *Liquid Crystal Photonics*; Nova Science Pub Inc.: New York, NY, USA, 2014; 204p.
8. Jeng, S.C. Applications of Tamm plasmon-liquid crystal devices. *Liq. Cryst.* **2020**, *47*, 1223–1231. [CrossRef]
9. Lininger, A.; Zhu, A.Y.; Park, J.S.; Palermo, G.; Chatterjee, S.; Boyd, J.; Capasso, F.; Strangi, G. Optical properties of metasurfaces infiltrated with liquid crystals. *Proc. Natl. Acad. Sci. USA* **2020**, *117*, 20390–20396. [CrossRef]
10. Geis, M.W.; Bos, P.J.; Liberman, V.; Rothschild, M. Broadband optical switch based on liquid crystal dynamic scattering. *Opt. Express* **2016**, *24*, 13812–13823. [CrossRef]
11. Konshina, E.A.; Shcherbinin, D.P. Study of dynamic light scattering in nematic liquid crystal and its optical, electrical and switching characteristics. *Liq Cryst.* **2018**, *45*, 292–302. [CrossRef]
12. Shaban, H.; Wu, P.-C.; Lee, J.-H.; Lee, W. Dielectric and electro-optical responses of a dielectrically negative nematic liquid crystal doped with cationic surfactant. *Opt. Mater. Express* **2021**, *11*, 3208–3222. [CrossRef]
13. Dabrowski, R.; Dziaduszek, J.; Bozetka, J.; Piecek, W.; Mazur, R.; Chrunik, M.; Perkowski, P.; Mrukiewicz, M.; Żurowska, M.; Weglowska, D. Fluorinated smectics—New liquid crystalline medium for smart windows and memory displays. *J. Mol. Liq.* **2017**, *267*, 415–427. [CrossRef]
14. Zhang, Y.; Yang, X.; Zhan, Y.; Zhang, Y.; He, J.; Lv, P.; Yuan, D.; Hu, X.; Liu, D.; Broer, D.J.; et al. Electroconvection in zwitterion-doped nematic liquid crystals and application as smart windows. *Adv. Opt. Mater.* **2020**, *27*, 2001465. [CrossRef]
15. Camley, R.; Celinski, Z.; Garbovskiy, Y.; Glushchenko, A. Liquid crystals for signal processing applications in the microwave and millimeter wave frequency ranges. *Liq. Cryst. Rev.* **2018**, *6*, 17–52. [CrossRef]
16. Blinov, L.M. *Structure and Properties of Liquid Crystals*; Springer: New York, NY, USA, 2010.
17. Neyts, K.; Beunis, F. Ion Transport in Liquid Crystals. In *Handbook of Liquid Crystals: Physical Properties and Phase Behavior of Liquid Crystals*; Wiley-VCH: Weinheim, Germany, 2014; Volume 2, Chapter 11; pp. 357–382.
18. Garbovskiy, Y. Conventional and unconventional ionic phenomena in tunable soft materials made of liquid crystals and nanoparticles. *Nano Ex.* **2021**, *2*, 012004. [CrossRef]
19. Colpaert, C.; Maximus, B.; Meyere, D. Adequate measuring techniques for ions in liquid crystal layers. *Liq. Cryst.* **1996**, *21*, 133–142. [CrossRef]
20. Barbero, G.; Evangelista, L.R. *Adsorption Phenomena and Anchoring Energy in Nematic Liquid Crystals*; Taylor & Francis: Boca Raton, FL, USA, 2006.
21. Khazimullin, M.V.; Lebedev, Y.A. Influence of dielectric layers on estimates of diffusion coefficients and concentrations of ions from impedance spectroscopy. *Phys. Rev. E* **2019**, *100*, 062601. [CrossRef]
22. Karaawi, A.R.; Gavrilyak, M.V.; Boronin, V.A.; Gavrilyak, A.M.; Kazachonok, J.V.; Podgornov, F.V. Direct current electric conductivity of ferroelectric liquid crystals–gold nanoparticles dispersion measured with capacitive current technique. *Liq. Cryst.* **2020**, *47*, 1507–1515. [CrossRef]
23. Garbovskiy, Y. Evaluating the Concentration of Ions in Liquid Crystal Cells: Hidden Factors and Useful Techniques. *Proceedings* **2020**, *62*, 10. [CrossRef]
24. Dhara, S.; Madhusudana, N.V. Ionic contribution to the dielectric properties of a nematic liquid crystal in thin cells. *J. Appl. Phys.* **2001**, *90*, 3483–3488. [CrossRef]
25. Kumar, A.; Varshney, D.; Prakash, J. Role of ionic contribution in dielectric behaviour of a nematic liquid crystal with variable cell thickness. *J. Mol. Liq.* **2020**, *303*, 112520. [CrossRef]
26. Garbovskiy, Y. Ion capturing/ion releasing films and nanoparticles in liquid crystal devices. *Appl. Phys. Lett.* **2017**, *110*, 041103. [CrossRef]
27. Garbovskiy, Y. Ions and size effects in nanoparticle/liquid crystal colloids sandwiched between two substrates. The case of two types of fully ionized species. *Chem. Phys. Lett.* **2017**, *679*, 77–85. [CrossRef]

engineering proceedings

MDPI

Proceeding Paper

Plasma-Induced Fibrillation and Surface Functionalization of Cellulose Microfibrils †

Pieter Samyn

Analytical and Circular Chemistry, Hasselt University, Agoralaan Gebouw D, B-3590 Diepenbeek, Belgium; pieter.samyn@outlook.be

† Presented at the 2nd International Electronic Conference on Applied Sciences, 15–31 October 2021; Available online: https://asec2021.sciforum.net/.

Abstract: The classical production of microfibrillar cellulose involves intensive mechanical processing and discontinuous chemical treatment in solvent-based media in order to introduce additional chemical surface modification. By selecting appropriate conditions of a pulsed plasma reactor, a solvent-free and low-energy input process can be applied with the introduction of microcrystalline cellulose (MCC) and maleic anhydride (MA) powders. The plasma processing results in the progressive fibrillation of the cellulose powder into its elementary fibril structure and in-situ modification of the produced fibrils with more hydrophobic groups that provide good stability against re-agglomeration of the fibrils. The selection of a critical ratio MA/MCC = 2:1 allows separating the single cellulose microfibrils with changeable morphologies depending on the plasma treatment time. Moreover, the density of the hydrophobic surface groups can be changed through a selection of different plasma duty cycle times, while the influence of plasma power and pulse frequency is inferior. The variations in treatment time can be followed along the plasma reactor, as the microfibrils gain smaller diameter and become somewhat longer with increasing time. This can be related to the activation of the hierarchical cellulose structure and progressive diffusion of the MA within the cellulose structure, causing progressive weakening of the hydroxyl bonding. In parallel, the creation of more reactive species with time allows creating active surface sites that allow for interaction between the different fibrils into more complex morphologies. The in-situ surface modification has been demonstrated by XPS and FTIR analysis, indicating the successful esterification between the MA and hydroxyl groups at the cellulose surface. In particular, the crystallinity of the cellulose has been augmented after plasma modification. Furthermore, AFM evaluation of the fibrils shows surface structures with irregular surface roughness patterns that contribute to better interaction of the microfibrils after incorporation in an eventual polymer matrix. In conclusion, the combination of physical and chemical processing of cellulose microfibrils provides a more sustainable approach for the fabrication of advanced nanotechnological materials.

Keywords: cellulose; plasma; microfibrils; surface modification

Citation: Samyn, P. Plasma-Induced Fibrillation and Surface Functionalization of Cellulose Microfibrils. *Eng. Proc.* **2021**, *11*, 2. https://doi.org/10.3390/ASEC2021-11136

Academic Editor: Nicholas Vassiliou Sarlis

Published: 15 October 2021

1. Introduction

The plasma processing of fibers offers an ecologically-friendly method for surface activation and functionalization of the fibers. The plasma-induced surface modification and interface engineering benefits from solvent-free handling of the reactive species and in-situ formation of the coupled materials through a combination of physical processes. The plasma phase is a complex state where particles of a gaseous phase are converted through radical reactions into a variety of active species, i.e., excited atoms and molecules, ionized moieties, electrons, free radicals and ionic complexes. As an advantage, the laboratory setups for plasma modification can be smoothly scaled-up towards (semi-)industrial units for fast and flexible processing of materials. Therefore, plasma treatment has been increasingly used as a green technology to enhance the sustainability of lignocellulosic materials [1].

The plasma modification of cellulose was studied under inert conditions in order to verify the changes in surface conditions. In particular, surface ablation and changes in surface morphology were observed in combination with morphological changes and small reduction in fiber width [2]. The exposure to oxygen plasma induces localized decomposition with the formation of highly functionalized molecules. The bulk of the fibers was selectively changed, where the crystalline zones remained unaffected through the plasma treatment with consequent variations in physical and chemical properties [3]. Ultimately, the depolymerization of cellulose in a non-thermal atmospheric plasma was achieved for the conversion into glucose units [4]. The active species created from a plasma with reactive gasses, e.g., when exposed the oxygen and sulfur hexafluoride, allows creating patterned nanostructures on the cellulose fiber surfaces with both changes in surface chemistry and topography that are needed to change the hydrophilic properties of cellulose into hydrophobic properties [5]. As a result, the plasma (pre-)treatment in combination with gaseous reactants, such as oxygen, ethylene or silane, allows the surface adhesive properties to adapt when used as filler in composite materials and/or changes the hydrophilicity and attachment of bacteria [6]. The hydrophobic modification of more complex cellulose structures can be achieved using CCl_4 as plasma, acting solely on the surface without modifying the bulk structure [7]. Alternatively, the surface modification of cellulose through chemical grafting with maleic anhydride is commonly used as a compatibilization step when mixing cellulose into more hydrophobic matrix materials, but it is usually conducted by wet-end chemical processing [8].

Microfibrillated cellulose (MFC) is a form of nanocellulose that is converted into elementary fibrils, which is usually achieved through an intensive mechanical processing involving high internal shear stresses and energy requirements [9]. The post-step functionalization of cellulose nanomaterials can then be implemented by plasma modification [10]. This has been frequently implemented for cellulose nanocrystalline materials [11,12]. The submerged liquid plasma processing was recently used for the fibrillation of cellulose in combination with ultrasonic treatments in inert argon or reactive oxygen/nitrogen environments to improve the dispersibility in water mixtures [13]. In contrast, the present work starts from dried powders of cellulose and reactive maleic anhydride monomers introduced in a pulsed plasma process, which allows for the simultaneous fibrillation of the cellulose structure and stabilization by chemical grafting of maleic anhydride.

2. Experimental Details

The microcrystalline cellulose (MCC, Merck, Germany) was introduced together with maleic anhydride (MA) as a powdery monomer. The precursors were inserted into a sealed glass tube that was degassed several times and were later connected to the inlet of the plasma reactor. An amount of the monomers was selected according to a ratio MA/MCC = 2:1 (e.g., 2 g of MA and 1 g of MCC) in order to obtain sufficient overload of the polymer relative to the fibrous material.

A cylindrical-shape glass container was used as a vacuum chamber that was connected to a monomer gas inlet and pump and surrounded by a copper coil driven by an RG generator. In a pulsation mode, the power is switched on and off in micro-second intervals, and their ratio becomes important to calculate effective power. The ratio has mainly an effect on the retention time of the functional groups in the plasma. In general, the higher t_{off} time increases deposition thickness, while higher t_{on} times lead to significant monomer fragmentation and loss of functional groups. The plasma conditions were selected for the creation of the MCC and in-situ surface modification with MA according to a pulsed plasma processing. Therefore, the maximum power of 20 W and fixed pulse frequency 820 Hz were applied. The output of the pulsed plasma process is characterized by a selection of the cycling times, including the plasma-on time (i.e., cycling time for creation of active species) t_{on} = 25 µs and the plasma-off-time (i.e., cycling time for reaction and recombination) t_{off} = 1200 µs.

A characterization of the cellulose materials after plasma processing was completed by optical microscopy (LOM), attenuated total reflection Fourier transform infrared spectroscopy (ATR-FTIR), atomic force microscopy (AFM) and X-ray photon spectroscopy (XPS).

3. Test Results

The morphology of pulsed-plasma processed MCC is illustrated in Figure 1, including optimized plasma processing conditions that allow for the in-situ fibrillation of the cellulose structure. The processing of pure MCC resulted in the formation of particles (Figure 1a), resembling the sizes and morphologies of the original powder. It is evident that the plasma processing has no influence on the particle properties, and surface modification is not possible, except for some changes that may be introduced at the surface. However, agglomeration of several particles has not been observed. In the case of a ratio MA/MCC = 2:1 (Figure 1b), an impressive morphology develops where the single particles fibrillate into several separated elementary cellulose fibrils. This morphology was only observed at a critical weight ratio, where the presence of MA is favorable for the stabilization and dispersion of the single fibrils. The fibrillation process obviously proceeds depending on the processing time, where the hierarchical cellulose structure starts to open up after a processing time of 10 min, while the single fibrils become looser and form a more complete fibrillar network structure after longer processing time (Figure 1c). Likely, the presence of MA allows for the diffusion in the gaseous phase within the cellulose structure (more specific the amorphous zones), and the interaction of activated species in the plasma phase allows for simultaneous surface modification and formation of a dense MFC network.

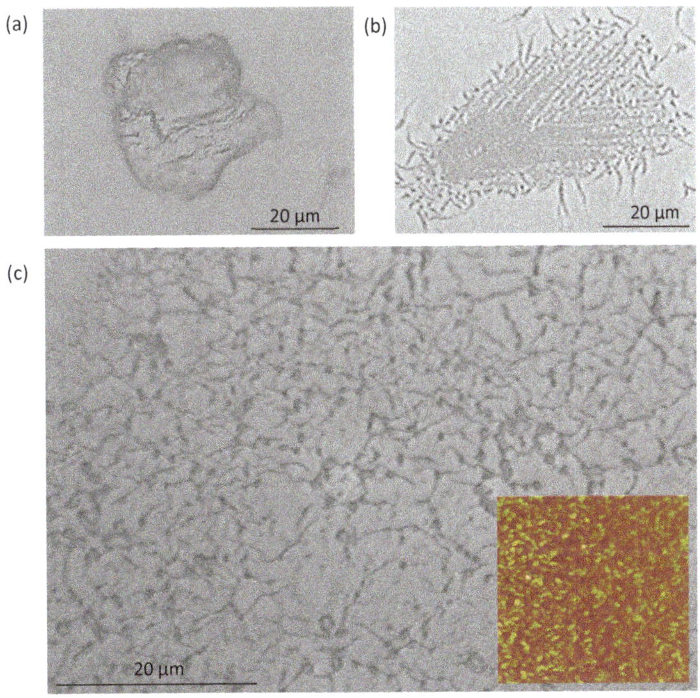

Figure 1. Optical microscopy of (**a**) original MCC, (**b**) starting of the fibrillation of the cellulose particles after plasma processing under optimized conditions with a ratio MA/MCC = 2:1 (processing time t = 10 min), (**c**) formation of and MFC/MA network of surface-modified fibrillated cellulose with AFM inset (10 × 10 μm^2), under optimized conditions with a ratio MA/MCC = 2:1 (processing time t = 30 min).

The variations in chemical structure of the cellulose are confirmed by ATR-FTIR spectra of the processed fibers (Figure 2). The carbonyl region is characterized by a shift in the C=O band of closed-ring maleic anhydride (1800 cm^{-1}) for the original MA, towards an esterified C=O band (1750 cm^{-1}) for the plasma-processed MFC/MA. This confirms the chemical surface modification of the cellulose surface, with an esterification reaction localized near the hydroxyl groups of the cellulose. The structural variations in cellulose are observed near the C–O–C/C–OH band region (1060 cm^{-1}), which suggests variations in crystallinity of the processed MFC. The sharper bands and narrowing of the band are typical for structural variations that include reorientation or confinement of the backbone in cellulose structure into a more ordered phase. The fibrillation of the cellulose during plasma processing may likely result in the removal of more disordered cellulose zones and the creation of fibrils with an enhanced structural organization.

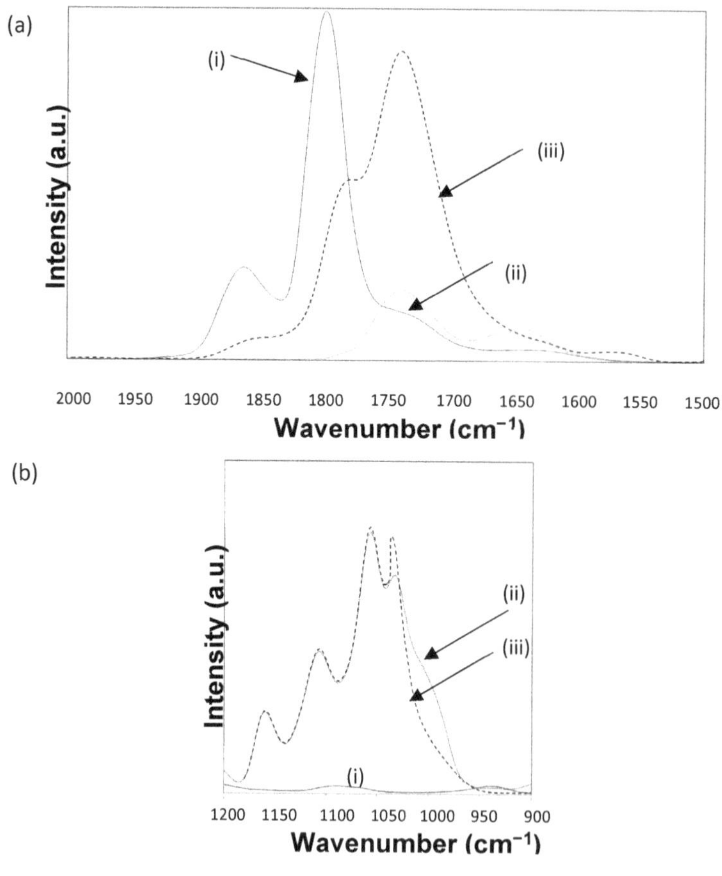

Figure 2. Detail of the ATR-FTIR spectra in different wavenumber regions (**a**) 2000–1500 cm^{-1}, or (**b**) 1200–900 cm^{-1}, for (i) MA, (ii) MCC and (iii) modified MFC/MA after plasma processing.

The XPS results of the high-resolution C1s peak are compared in Figure 3 for the plasma-polymerized MA and MFC/MA. The typical peaks for MA plasma polymers include **1** hydrocarbons (285.0 eV, C–C, CH_x); **2** carbon in alpha position of anhydride/carboxylic acids, esters (285.6 eV, **C**–C=O); **3** carbon single bonds to oxygen (286.6 eV, **C**–O); **4** carbon double bonds to oxygen (287.9 eV, **C**=O); **5** anhydride groups, carboxylic acid, ester groups (289.5 eV, O=**C**–O–**C**=O, O–**C**=O). The calculation of atomic concentrations based on the surface area beneath the respective peaks indicates significant differences for pure

MA compared to MFC/MA. The peak **1** is representative of hydrocarbons formed through monomer fragmentation and is lower for the MFC/MA compared to pure MA, as the likeliness for crosslinking in the pure plasma polymer between single MA moieties reduces in the presence of MFC. Alternatively, the peak **2** indicative for esterification reaction is significantly enhanced for MFC/MA due to favorable chemical surface modification of the cellulose with MA. Finally, the peak **5** representing closed-ring anhydride groups is slightly lower for the MFC/MA than for pure MA as a confirmation for the ring-opening reaction and esterification. However, the intensity ratio of peak ratio 3 (C–OH) relatively to peak 4 (O–C–O) is lower for plasma-modified MFC than for pure cellulose (which should be around 5:1), confirming that an important amount of the cellulose C–OH disappears through ring-opening and esterification with MA.

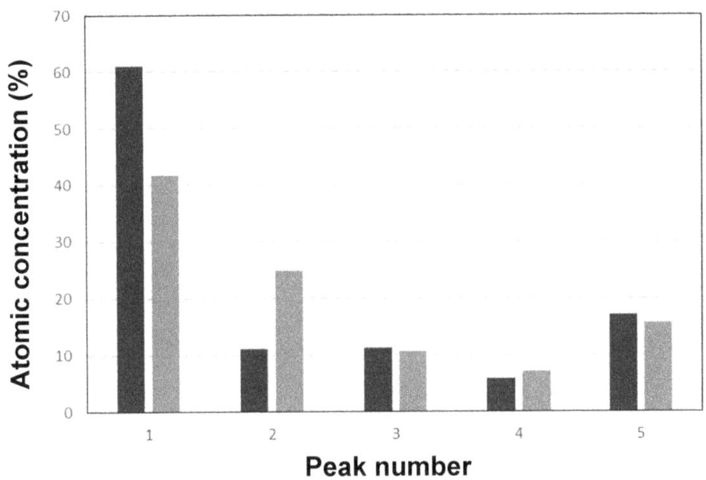

Figure 3. Atomic concentration of different functionalities present in the high-resolution C1s peak obtained by XPS analysis for MA (black bars), and MFC/MA (grey bars). Peak numbers **1**, **2**, **3**, **4**, **5** are explained in the text.

4. Conclusions

The simultaneous fibrillation of cellulose and surface modification with maleic anhydride (MA) has been successfully demonstrated during a pulsed plasma polymerization process of MA, after selecting the optimized plasma processing conditions and a critical ratio of cellulose / MA monomer. The fibrillation is gradually visualized by an opening of the fine fibrillar structure of cellulose and stabilization of the dispersion state of the fibrils. The chemical surface modification of the cellulose fibrils through the esterification reaction with MA has been demonstrated by additional chemical analysis. The characterization of surface modification agrees between XPS and FTIR spectra, while the latter also suggests variations in phase ordering of the cellulose molecules after processing. As such, the feasibility of plasma processing as a sustainable approach for the fabrication of microfibrillated cellulose has been demonstrated.

Funding: This research received no external funding.

Institutional Review Board Statement: Not applicable.

Informed Consent Statement: Not applicable.

Conflicts of Interest: The authors declare no conflict of interest.

References

1. Vanneste, J.; Ennaert, T.; Vanhulsel, A.; Sels, B. Unconventional pretreatment of lignocellulose with low-temperature plasma. *Chem. Sus. Chem.* **2017**, *10*, 14–31. [CrossRef] [PubMed]
2. Kolarova, K.; Vosmanska, V.; Rimpelova, S.; Svorcik, V. Effect of plasma treatment on cellulose fiber. *Cellulose* **2013**, *20*, 953–961. [CrossRef]
3. Calvimontes, A.; Mauersberger, P.; Nitschke, M.; Dutschk, V.; Simon, F. Effects of oxygen plasma on cellulose surface. *Cellulose* **2011**, *18*, 803–809. [CrossRef]
4. Benoit, M.; Rodriguez, A.; Zhang, Q.; Fourré, E.; De Oliveira Viger, K.; Raribouet, J.M.; Jerome, F. Depolymerization of cellulose assisted by a nonthermal atmospheric plasma. *Angew. Chem. Int. Ed.* **2011**, *50*, 8964–8967. [CrossRef]
5. De Camargo, J.S.; de Menezes, A.J.; da Cruz, N.C.; Rangel, E.C.; Delgado-Silva, A. Morphological and chemical effects of plasma treatment with oxygen (O2) and sulfur hexafluoride (SF6) on cellulose surface. *Mat. Res.* **2017**, *20*, 842–850. [CrossRef]
6. Mauger, O.; Westphal, S.; Klopzing, S.; Kruger-Genge, A.; Muller, W.; Storsberg, J.; Bohrish, J. Plasma activation as a powerful tool for selective modification of cellulose fibers towards biomedical applications. *Plasma* **2020**, *3*, 196–203. [CrossRef]
7. Shi, J.; Lu, L.; Guo, W.; Sun, Y.; Cao, Y. An environment-friendly thermal insulation material from cellulose and plasma modification. *J. Appl. Polym. Sci.* **2013**, *130*, 3652–3658. [CrossRef]
8. Cichosz, S.; Masak, A.; Rylski, A. Cellulose modification for improved compatibility with the polymer matrix: Mechanical characterization of the composite material. *Materials* **2020**, *13*, 5519. [CrossRef] [PubMed]
9. Osong, S.H.; Norgren, S.; Engstrand, P. Processing of wood-based microfibrillated cellulose and nanofibrillated cellulose, and applications relating to papermaking: A review. *Cellulose* **2016**, *23*, 93–123. [CrossRef]
10. Chanda, S.; Bajwa, D.S. A review of current physical techniques for dispersion of cellulose nanomaterials in polymer matrices. *Rev. Adv. Mater. Sci.* **2021**, *60*, 325–341. [CrossRef]
11. Alanis, A.; Valdes, J.H.; Guadalupe, N.V.M.; Lopez, R.; Mendoza, R.; Mathew, A.P.; de Leon, R.D.; Valencia, L. Plasma surface-modification of cellulose nanocrystals: A green alternative towards mechanical reinforcement of ABS. *RSC Adv.* **2019**, *9*, 17417–17424. [CrossRef]
12. Matouk, Z.; Rincon, R.; Torriss, B.; Mirzaei, A.; Margot, J.; Chaker, M. Functionalization of cellulose nanocrystals powder by non-thermal atmospheric-pressure plasmas. *Cellulose* **2021**, *28*, 6239–6252. [CrossRef]
13. Vizireanu, S.; Panaitescu, D.M.; Nicolae, C.A.; Frone, A.N.; Chiulan, I.; Ionita, M.D.; Satuly, V.; Carpen, L.G.; Petrescu, S.; Birjega, R.; et al. Cellulose defibrillation and functionalization by plasma in liquid treatment. *Sci. Rep.* **2018**, *8*, 15473. [CrossRef] [PubMed]

engineering proceedings

Proceeding Paper

Developing a Benzimidazole-Silica-Based Hybrid Sol–Gel Coating with Significant Corrosion Protection on Aluminum Alloys 2024-T3 †

Magdi H. Mussa [1,2,3,*], F. Deeba Zahoor [3,4], Oliver Lewis [3] and Nicholas Farmilo [3,5]

1 Mechanical and Energy Department, The Libyan Academy for Graduate Study, Tripoli P.O. Box 79031, Libya
2 Mechanical Engineering Department, Sok Alkhamis Imsehel High Tec. Institute, Tripoli P.O. Box 79031, Libya
3 Materials and Engineering Research institute MERI, Sheffield Hallam University, Howard Street, Sheffield S1 1WB, UK; F.Zahoor@sheffield.ac.uk (F.D.Z.); O.Lewis@shu.ac.uk (O.L.); nfarmilo@gmail.com (N.F.)
4 Department of Chemistry, University of Sheffield, Sheffield S10 2TN, UK
5 Tideswell Business Development Ltd., Ravensfield Sherwood Rd, Buxton SK17 8HH, UK
* Correspondence: magdimosa1976@gmail.com; Tel.: +44-7404496955
† Presented at 8th International Electronic Conference on Sensors and Applications, 1–15 November 2021; Available online: https://ecsa-8.sciforum.net.

Citation: Mussa, M.H.; Zahoor, F.D.; Lewis, O.; Farmilo, N. Developing a Benzimidazole-Silica-Based Hybrid Sol–Gel Coating with Significant Corrosion Protection on Aluminum Alloys 2024-T3. *Eng. Proc.* **2021**, *11*, 3. https://doi.org/10.3390/ASEC2021-11124

Academic Editor: Nicholas Vassiliou Sarlis

Published: 15 October 2021

Abstract: The inherent reactivity of Al–Cu–Mg alloys is such that their use for building structural, maritime, and airplane components with great strength/weight ratios would not be possible without good anti-corrosion systems. These systems could be considered as imitations of the protection mechanism found in the conventional hexavalent chromium-based system, but with additional limited environmental impact, and in particular without toxic or carcinogenic effects. These coatings also are intended to be eco-friendly, using less of the valuable raw materials and energy than more traditional methods. Silica-based hybrid protective coatings have been shown to exhibit excellent chemical stability combined with the ability to reduce the corrosion of metal substrates. However, research shows that sol–gel has some limitations in terms of the period of the anti-corrosive properties. Therefore, this work reports the performance of a silica-based hybrid sol–gel coating encapsulated with benzimidazole (BZI) that can be applied to light alloys to form an inherently inhibited and crack-free coating. This coating was applied on AA 2024-T3 and cured at 80 °C. The high corrosion resistance performance results from the combination of good adhesion, the hydrophobic property of the silica-based hybrid coating, and the presence of the encapsulated (BZI) film-forming volatile corrosion inhibitor, which is released at pores within the coating system, resulting in film-forming, reducing the reaction at cathodic sites. The evaluation of this mechanism is based on using electrochemical testing techniques. The anti-corrosion properties of the coatings were studied when immersed in 3.5% NaCl by using electrochemical impedance spectroscopy (EIS) and potential-dynamic polarization scanning (PDPS). The chemical confirmation was performed by infrared spectroscopy (ATR-FTIR), supported by analyzing the morphology of the surface before and after the immersion testing by using scanning electron microscopy (SEM). The benzimidazole-silica-based hybrid coating exhibited excellent anti-corrosion properties, providing an adherent protective film on the aluminum alloy 2024-T3 samples compared to sol–gel-only and bare metals, as a cost-effective and eco-friendly system.

Keywords: silica-based hybrid sol–gel coating; electrochemical testing; corrosion protection; aluminum alloys

1. Introduction

Silica-based hybrid protective coatings using sol–gel technology have shown an exceptional ability to reduce corrosion on the metal surface combined with high chemical stability. However, using the sol–gel technique alone has some limitations in terms of barrier anti-corrosive properties due to pores in the matrix, which could lead to cracking of the coating. The use of encapsulated corrosion inhibitors will enhance the system's corrosion protection. The systematic development of corrosion inhibitor compounds has produced many effective inhibitors. For example, compounds with heterocyclic organic functional groups consisting of oxygen or nitrogen, with phosphorus or sulfur attached as heteroatoms, are very desirable [1,2].

For instance, benzimidazole (BZI) (structure shown in Figure 1) has been identified as a low-pH film-forming effective corrosion inhibitor used for copper and steel. It possesses a heterocyclic aromatic organic compound structure that is suitable as both an effective volatile or injectable corrosion inhibitor with other soluble carriers. It has a chemical structure containing both a benzene group and the active group imidazole used in oil and gas production [3].

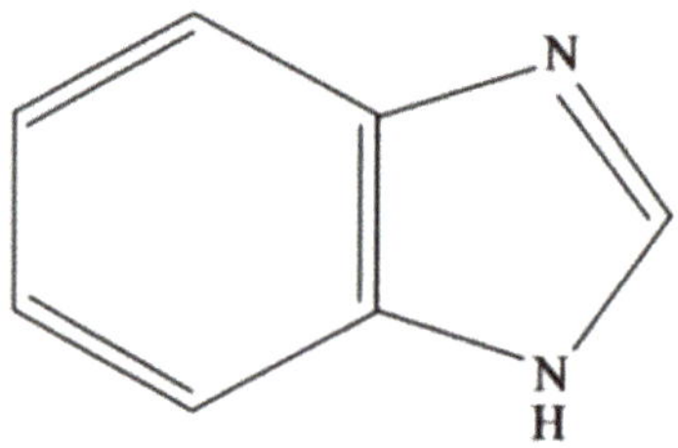

Figure 1. The chemical structure of benzimidazole (BZI).

The main advantages are that BZI is commercially available and cost-effective as a raw material for various uses, the most common of which is as a fungicide in pharmaceutical applications [4]. In addition, BZI derivatives are used in various corrosion applications in applied science. For example, Antonijevic et al. [2] described it as a powerful inhibitor that showed excellent corrosion prevention to protect carbon steel pipelines in corrosive acidic HCl solutions [2,5].

The mechanism of inhibition by BZI or its derivatives on metal has been studied deeply, especially on copper and steel [5–7]. Furthermore, it has been found that BZI and its derivatives may be positioned in a parallel adsorption arrangement, a close joining with the surface, creating a thin layer. The adsorption of BZI in this position may be due to the donation of the π electron on the BZI's nitrogen atoms to the unoccupied D-orbital, balancing the benzimidazole molecules with excited iron or copper atoms [5,6]. These interactions explain the strong adsorption connection to the surface that protects the mild steel or copper from direct corrosion [5,6].

Related research has mentioned the use of BZI chemistries and related derivatives and patents: J. Colreavy et al. and S. Vijaykumar et al. mentioned using the same family of these inhibitors in their patented sol–gel technique on steel or other metals. This work demonstrated the specific application of BZI as a simple silica-based sol–gel additive used as a corrosion inhibitor for aluminum alloy AA2024-T3 [8,9].

2. Experimental Procedures

2.1. Sol–Gel Preparation

In this study, the hybrid silica-based sol–gel was synthesized from tetraethyl orthosilicate silane (TEOS) and trimethoxymethyl silane (MTMS) precursors, purchased from Sigma-Aldrich. The precursors were mixed in isopropyl alcohol by adding deionized (DI) water dropwise in a molar ratio of 18:14:17:220, respectively, until the hydrolyzing and condensation reactions. As mentioned in the previous work, the silica-based sol–gel mixture was then enhanced by adding the poly-siloxane (PSES) solution [10]. The base formula used was labelled as SHX-80.

The benzimidazole-modified hybrid silica-based sol–gel was labelled as BZI-SHX-80 and prepared by encapsulating 3.5 vol.% of a 1:1 solution of ethanol and benzimidazole (BZI) (Sigma-Aldrich) in the base formula SHX-80, added dropwise while stirring. The formulation was then left for 24 h.

2.2. Substrate Preparation and Film Deposition

The aluminum alloy substrates used were purchased from Q-Lab as Standard aluminum alloy AA2024-T3 Q-panels with dimensions of (102 mm × 25 mm × 1.6 mm) [11]. The received Q-panels were washed with a commercial aluminum base surfactant cleaner, then rinsed with DI water, followed by acetone washing to remove the organic residues on the surface.

Then, the sol–gel was sprayed onto the pre-cleaned aluminum alloy substrates. The distance from the spraying gun to the surface was approximately 150 mm. The coating was built up over three passes to keep the thickness standard for all samples at about 15 ± 2 µm. After that, the coated samples were left in the air for 10 min before being annealed at 80 °C for 4 h. Table 1 shows the experiment codes used to identify the samples.

Table 1. Sample identification table.

No.	Identifier	Formula Base Composite	(BZI) *v*/*v*%	Curing Temperature
1-	SHX-80	TEOS + MTMS + PSX	-	80 °C
2-	ZBI-SHX-80	TEOS + MTMS + PSX	3.5%	80 °C
3-	Bare AA2024-T3	-	-	-

2.3. Coating Testing and Characterization

Electrochemical tests were performed on the bare and coated samples to assess their corrosion resistance. Tests were conducted by using a Princeton Applied Research PARSTAT 2273. The corrosion performance of the sol–gel-coated and uncoated aluminum alloy was evaluated using electrochemical impedance spectroscopy (EIS) and potentiodynamic polarization (PDPS) scans. An area of 1.00 mm^2 in the center of the samples in aerated 3.5% NaCl was tested. The tests were carried out at room temperature (20 ± 2 °C). The electrode potential was monitored for approximately 1 h before polarization in the electrolyte solution until stability. The sample was polarized with PDPS at a scanning rate of 1.667 mVs^{-1} from the initial potential of −250 mV vs. OCP to +750 mV vs. SCE. The electrochemical impedance measurements were recorded between 100 kHz and 10 MHz with a sinusoidal AC RMS value of 10 mV [12].

3. Results and Discussion

3.1. ATR-FTIR for the BZI-SBX-80 Sol–Gel Chemical Composition

The organic BZI was successfully incorporated into the SHX sol–gel base formula by comparing the infrared spectrum obtained from the BZI-SHX-80 coating to that of the unmodified SHX-80. This is enlarged in Figure 2. In the spectrum of the BZI-modified coating, several peaks are seen related to the BZI molecule. These peaks can be observed and were retained in the sol–gel-coated BZI-SBX-80 sample. They are as follows: weak imine C=N stretching at 1564.5 cm^{-1}; carbon double-bond C=C stretching peaks at about 1477 cm^{-1}, 1458 cm^{-1}, and 1408 cm^{-1}, respectively. Similarly, the fingerprint of the aromatic amine stretching C–N can be detected at 1364 cm^{-1} and 1300 cm^{-1}, respectively, which can be used to confirm the benzimidazole's presence in the new sol–gel formula. The C–H out-of-plane bending is characterized by the peaks at 768 cm^{-1} and 745 cm^{-1} [13].

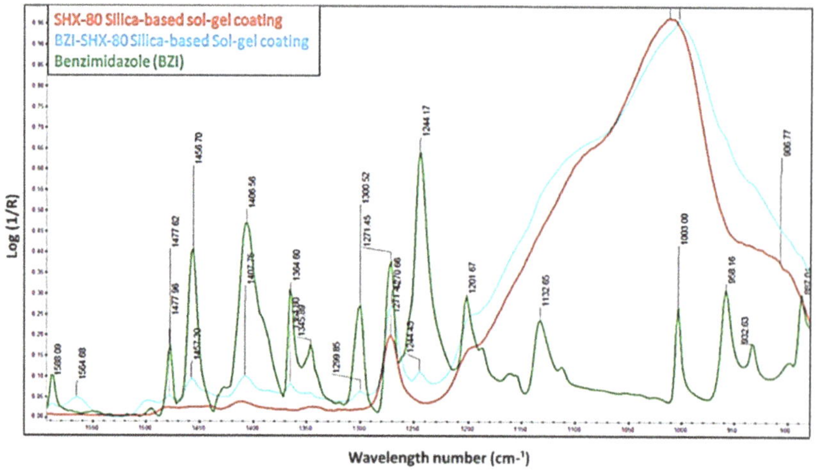

Figure 2. ATR-FTIR spectra, showing the effect of the BZI added to the SHX-80 sol–gel.

3.2. Water Contact Angle of the SBX and BZI-SBX Coatings

As shown in Figure 3, the measured water contact angle (WCA) of the base SHX-80 coating was about $67 \pm 2°$, as shown in Figure 3a, and the measured WCA on the modified BZI-SHX-80 sol-gel coating was $88 \pm 4°$, as shown in Figure 3b; the higher water contact angle recorded for the BZI-SHX-80 shows that its wettability is lower than that of the original SHX-80 [14].

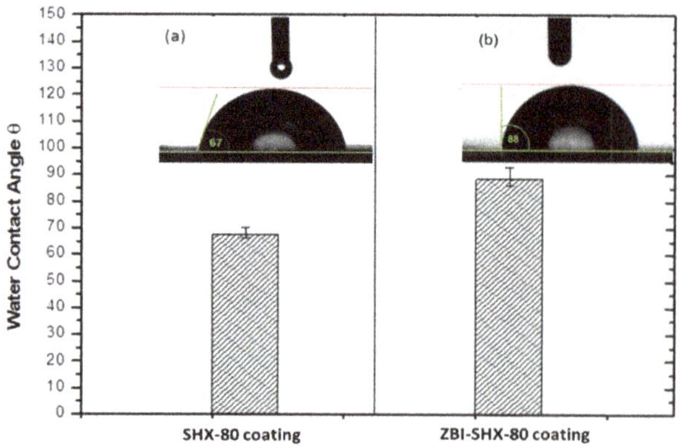

Figure 3. Bar chart showing the mean values of the WCA of the BZI-SHX-80 and SHX-80 coatings. Optical images showing water droplets on the (**a**) SHX-80 and (**b**) modified BZI-SHX-80 coatings.

3.3. Potentiodynamic Polarization Scanning

All coated samples displayed significantly improved performance compared to the bare AA2024-T3 sample. The corrosion potential (E_{corr}) and corrosion current density (I_{corr}) were obtained from PDPS, as shown in Figure 4. The current density on the cathodic branch of the Tafel curve for all coated samples was reduced by more than four magnitudes when compared to the bare AA2024-T3. Nevertheless, the current density of the BZI-SHX-80-coated sol–gel sample was the lowest, as it was reduced by seven orders of magnitude compared to the bare AA2024-T3. This reduction may be attributed to benzimidazole's surface activity and high electronegativity [5]. The SHX-80 sol–gel only showed a reduction in the anodic branch by about four and a half orders of magnitude less than the bare

AA2024-T3. The corrosion current densities of the bare and coated samples were reduced to 5.98×10^{-10} A/cm^2 for BZI-SHX-80 and 1.1×10^{-9} A/cm^2 for SHX-80, respectively, compared to 7.1×10^{-6} A/cm^2 for the bare AA2024-T3 alloy. The shift in E_{corr} indicates that the anode is inhibited to a greater degree than the cathode in the BZI-SHX 80 sol–gel mixture. This could be attributed to the active benzimidazole nitrogen atoms bridging the substrate surface [5,6].

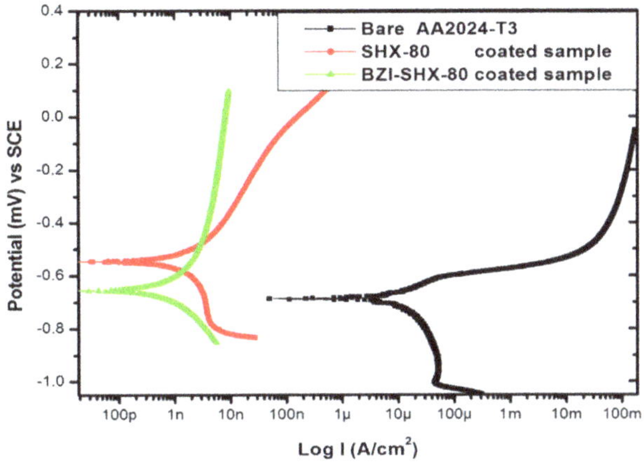

Figure 4. Polarization (PDPS) curves for the bare and sol–gel-coated samples with/without BZI in 3.5% NaCl.

3.4. *Electrochemical Impedance Spectroscopy*

3.4.1. Impedance Magnitude Bode Plots

As shown in Figure 5a,b, after the first hour of immersion, the overall impedance of the BZI-SHX-80 at low frequencies was higher by approximately two orders of magnitude compared to the SHX-80-coated samples, with values of 5.7×10^7 ohms cm^{-2} and 9.1×10^5 ohms cm^{-2} for BZI-SHX-80 and SHX-80, respectively.

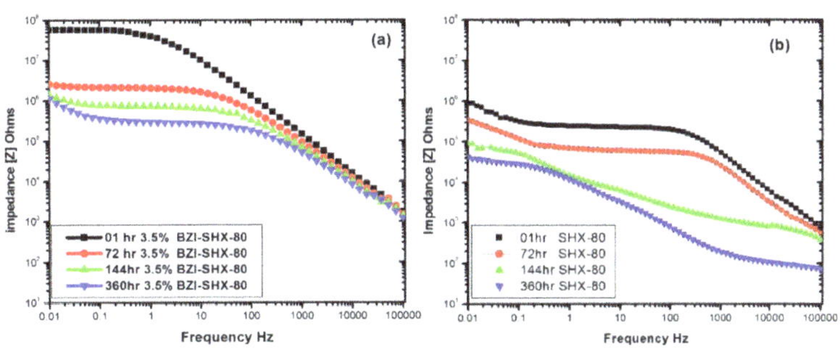

Figure 5. Impedance magnitude Bode plots for (**a**) BZI-SHX-80 and (**b**) SHX-80.

After 360 h, the impedance of the BZI-SHX-80-coated samples dropped by about one and a half orders of magnitude; this could be due to the electrolyte diffusion and expansion of the pores in the sol–gel coating matrix. However, this drop was not reflected in the visible protection in the electrolyte, suggesting that the film of the sol–gel/BZI on the metal surface was stable [10].

A noticeable drop in impedance was observed for the SHX-80-coated sample at about 3.4×10^4 ohms cm^{-2} after 360 h. This might be attributed to the reduction of the coating resistance due to the development of rounded pitting under the coatings. The high-frequency impedance also dropped by about one order of magnitude; this impedance is considered higher than that of the SBX-80 coating in the mid-frequency range between 10^5 Hz and 10^6 Hz. This may be attributed to the coating's pores and the cracking that occurred.

3.4.2. Using Nyquist Plots to Investigate the Corrosion-Protective Behavior

Figure 6 shows the Nyquist plots for both the BZI-SHX-80- (Figure 6a) and the OA-SHX-80- (Figure 6b) coated samples from 1–360 h, respectively. The plots show that the BZI-SHX-80 has the maximum impedance. These plots were used to obtain the equivalent circuit modelling fit using the ZSimpwin electrochemical impedance spectroscopy (EIS) data analysis software. Tables 2 and 3 below demonstrate the fit data for the SHX-80 and the BZI-SHX-80 coatings after various immersion times. The equivalent circuits were used to simulate the corrosion reaction on the surface of the coated samples at 1 h, 48 h and 144 h, respectively. In these circuits, instead of using an ideal capacitor (C), a time-constant element (Q) was used to compare the current leakage in the capacitor and/or frequency dispersion effect of the alternating current signals [12]. The suggested equivalent circuits for each EIS plot after 144 h are provided in Figure 7 for both systems.

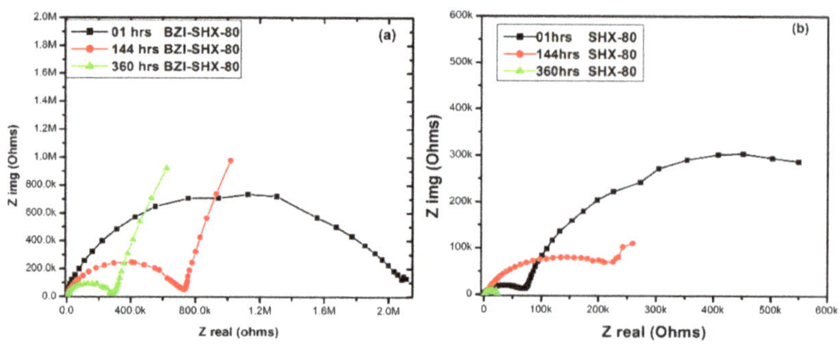

Figure 6. The Nyquist plots for the (**a**) BZI-SHX-80 coating and (**b**) SHX-80 coating systems.

Table 2. The fit data obtained from the EIS spectra for the BZI-SHX-80 sol–gel coating after various immersion times in 3.5 wt. % NaCl solution.

Circuit	Element	Immersion Time (H)		
		01	48	144
		R(Q(R(QR)))	R(Q(R(Q(RW))))	R(Q(R(Q(RW))))
	R_s	105	160	201
	Q_{ct}	1.471×10^{-9}	4.549×10^{-9}	9.148×10^{-9}
	n	0.9626	0.887	0.8406
	R_{ct}	1.326×10^{7}	1.301×10^{6}	6.991×10^{5}
	Q_{iL}	3.453×10^{-9}	6.406×10^{-8}	9.016×10^{-8}
	n	0.800	0.650	0.800
	R_{iL}	4.697×10^{7}	5.885×10^{5}	1.778×10^{3}
	W	-	4.349×10^{-8}	7.044×10^{-10}

The element identifiers used for the equivalent circuits were solution resistance (R_s), coating resistance (R_{ct}), coating constant phase elements (Q_{ct}), interfacial layer resistance (R_{iL}), interfacial layer capacitance (Q_{iL}), oxide layer (pitting) resistance (R_p), oxide layer (pitting) capacitance (Q_p), and the Warburg circuit element (W) [15].

Table 3. The fit data obtained from the EIS spectra for the SHX-80 sol–gel coating after various immersion times in 3.5 wt. % NaCl solution.

Circuit	Element	Immersion Time (H)		
		01	48	144
		R(Q(R(QR)))	R(Q(R(Q(R(QR)))))	R(Q(R(Q(R(QR)))))
	R_s	100	205	225
	Q_{ct}	1.079×10^{-7}	2.850×10^{-7}	4.771×10^{-6}
	n	0.800	0.627	0.490
	R_{ct}	7.287×10^{4}	815	253
	Q_{iL}	4.933×10^{-6}	1.151×10^{-6}	3.912×10^{-6}
	n	0.850	0.694	0.896
	R_{iL}	7.793×10^{5}	4.022×10^{5}	1.221×10^{5}
	Q_p	-	7.364×10^{-5}	4.835×10^{-5}
	n	-	0.900	0.455
	R_p	-	1.369×10^{6}	1×10^{20}

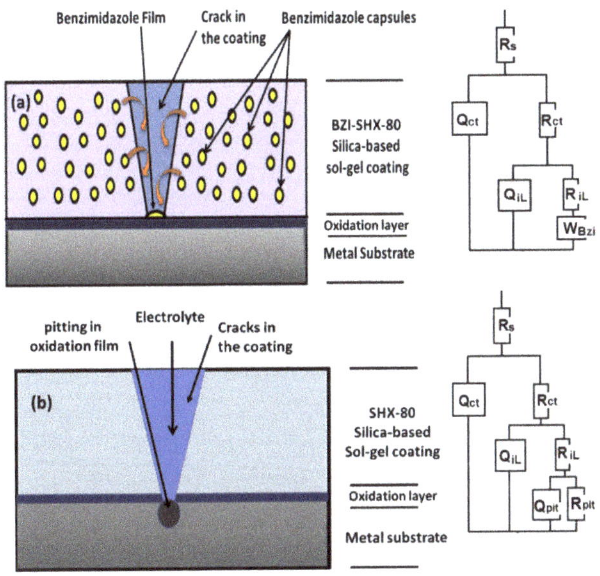

Figure 7. Schematic drawing of the hybrid silica-based system. (**a**) BZI-SHX-80 and (**b**) SHX-80 after 144 h of immersion in 3.5% NaCl solution.

3.5. *Scanning Electron Microscopy Images after Immersion*

Figure 8 shows the surface morphology of the three samples, BZI-SHX-80 (Figure 8a), SHX-80 (Figure 8b), and bare alloy AA2024-T3 (Figure 8c).

As presented in Figure 8a, the BZI-SBX-80 coating showed excellent resistance to corrosion and cracks under similar circumstances. It was expected that the ZBI-SHX-80 would be more stable than the SHX-80, which may prevent the coating system's diffusion, in line with the benzimidazole self-healing inhibition properties.

As shown in Figure 8b, in the SHX-80-coated samples, cracks around 1–6 μm wide were observed on the surface of the coating, with some pitting corrosion under the coating. Due to the cracking of the coating, the cracks could adversely affect the barrier corrosion protection, which has implications for future performance under wet/dry cycling.

On the other hand, the bare sample AA2024-T3 was attacked by aggressive pitting corrosion after 360 h of immersion, as shown in Figure 8c.

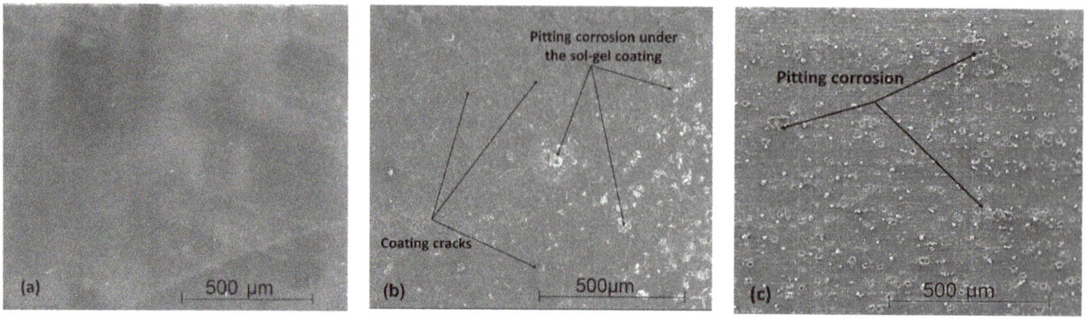

Figure 8. SEM surface images for the (**a**) BZI-SHX-80-coated sample, (**b**) SHX-80-coated sample, and (**c**) bare AA2024-T3 after 360 h of immersion in 3.5% NaCl solution.

4. Conclusions

The base SHX sol-gel formula can provide moderate barrier protection without the presence of any inhibitor. The protection can last for at least 10 d in 3.5% NaCl solution before cracks and pitting corrosion appear visually under the coating surface.

However, by encapsulating the benzimidazole in the silica-based sol–gel, excellent corrosion protection can be achieved, which can provide protection over two weeks without any signs of failure, including cracks and pitting corrosion. Adding benzimidazole as an inhibitor to the sol–gel matrix provides active protection due to the high electronegativity of the active azole group. This was demonstrated by the physical appearance and through AC impedance measurements. In addition, the benzimidazole-sol–gel coating revealed an excellent resistance to post cracking after a long immersion.

Author Contributions: Methodology, M.H.M., O.L., and N.F.; validation and testing M.H.M.; data analysis, M.H.M., N.F., and O.L.; FTIR support, M.H.M. and F.D.Z.; investigation, M.H.M.; resources, M.H.M., O.L., and N.F.; writing, original draft preparation, M.H.M.; writing, review and editing, M.H.M., O.L., F.D.Z., and N.F.; project supervision, N.F. and O.L. All authors have read and agreed to the published version of the manuscript.

Funding: This research was funded by the Libyan scholarship program as part of a Ph.D. project.

Institutional Review Board Statement: Not applicable.

Informed Consent Statement: Not applicable.

Data Availability Statement: The data are not publicly available. The data files are stored on the corresponding instruments and personal computers.

Acknowledgments: The authors would like to acknowledge the support of the Sheffield Hallam University in Material and Engineering Research Institute (MERI), as well as the Libyan Scholarship Program for the financial support.

Conflicts of Interest: The authors declare no conflict of interest.

References

1. Antonijevic, M.M.; Petrovic, M.B. Copper corrosion inhibitors. A review. *Int. J. Electrochem. Sci.* **2008**, *3*, 1–28.
2. Antonijevic, M.; Milic, S.; Petrovic, M.; Radovanovic, M.; Stamenkovic, A. The Influence of pH and Chlorides on Electrochemical Behavior of Copper in the Presence of Benzotriazole. *Int. J. Electrochem. Sci.* **2009**, *4*, 962–979.
3. Wright, J.B. The Chemistry of the Benzimidazoles. *Chem. Rev.* **1951**, *48*, 397–541. [CrossRef] [PubMed]
4. Gupta, P.K. Herbicides and fungicides. In *Biomarkers in Toxicology*; Elsevier Inc.: Amsterdam, The Netherlands, 2014; ISBN 9780124046306.
5. Gutiérrez, E.; Rodríguez, J.A.; Cruz-Borbolla, J.; Alvarado-Rodríguez, J.G.; Thangarasu, P. Development of a predictive model for corrosion inhibition of carbon steel by imidazole and benzimidazole derivatives. *Corros. Sci.* **2016**, *108*, 23–35. [CrossRef]
6. Obot, I.B.; Madhankumar, A.; Umoren, S.A.; Gasem, Z.M. Surface protection of mild steel using benzimidazole derivatives: Experimental and theoretical approach. *J. Adhes. Sci. Technol.* **2015**, *29*, 2130–2152. [CrossRef]

7. Grillo, F.; Tee, D.W.; Francis, S.M.; Früchtl, H.A.; Richardson, N.V. Passivation of copper: Benzotriazole films on Cu(111). *J. Phys. Chem. C* **2014**, *118*, 8667–8675. [CrossRef]
8. Colreavy, J.; Duffy, B.; Varma, P.C.R.; Hayden, H.; Oubaha, M. Organosilane Coating Compositions and Use Thereof. United States US20100330380A1, 30 December 2010.
9. Vijaykumar, S.; Prakash, O.; Raghavan, S.; Ramachandra, K.; Reddy, D. Sol-Gel Coating Compositions Including Corrosion Inhibitor-Encapsulated Layered Metal Phosphates and Related Processes. United States US10421869B2, 24 September 2019.
10. Mussa, M.H.; Rahaq, Y.; Takita, S.; Farmilo, N. Study the Enhancement on Corrosion Protection by Adding PFDTES to Hybrid Sol-Gel on AA2024-T3 Alloy in 3.5% NaCl Solutions. *Albahit. J. Appl. Sci.* **2021**, *2*, 61–68.
11. *ASTM International ASTM Code B209—14 Standard Specification for Aluminum and Aluminum-Alloy Sheet and Plate*; ASTM International: West Conshohocken, PA, USA, 2016; Volume 25, p. 16.
12. Tait, W.S. *Electrochemical Impedance Spectroscopy Fundamentals, an Introduction to Electrochemical Corrosion Testing for Practicing Engineers and Scientists*; Tait, W.S., Ed.; PairODocs Publications: Racine, WI, USA, 1994; ISBN 13-978-0966020700.
13. Mohan, S.; Sundaraganesan, N.; Mink, J. FTIR and Raman studies on benzimidazole. *Spectrochim. Acta Part A Mol. Spectrosc.* **1991**, *47*, 1111–1115. [CrossRef]
14. Kumar, D.; Wu, X.; Fu, Q.; Ho, J.W.C.; Kanhere, P.D.; Li, L.; Chen, Z. Development of durable self-cleaning coatings using organic-inorganic hybrid sol-gel method. *Appl. Surf. Sci.* **2015**, *344*, 205–212. [CrossRef]
15. Yabuki, A.; Yamagami, H.; Noishiki, K. Barrier and self-healing abilities of corrosion protective polymer coatings and metal powders for aluminum alloys. *Mater. Corros.* **2007**, *58*, 497–501. [CrossRef]

engineering proceedings

MDPI

Proceeding Paper

Optical Investigation of PVA/$PbTiO_3$ Composite for UV-Protective Approach Applications †

Asmaa. S. El-Deeb [1], Marwa. M. Abdel Kader [1], Gamal. M. Nasr [2], Mona. A. Ahmed [3] and Eman O. Taha [3,*]

1 Housing and Building National Research Center, Building Physics Institute, Giza 1770, Egypt; asmaphysics@yahoo.com (A.S.E.-D.); marwamahmoud_1211@yahoo.com (M.M.A.K.)
2 Department of Physics, Faculty of Science, Cairo University, Giza 12613, Egypt; rrrrrgmal@yahoo.com
3 Department of Petroleum Applications, Egyptian Petroleum Research Institute (EPRI), Cairo 11727, Egypt; mona_chemist17@yahoo.com
* Correspondence: eman@unm.edu or eman.omar@epri.sci.eg
† Presented at the 2nd International Electronic Conference on Applied Sciences, 15–31 October 2021; Available online: https://asec2021.sciforum.net/.

Abstract: Samples of a new-fangled polymer of poly (Vinyl Alcohol) (PVA) doped with various concentrations of Lead (II) Titanate ($PbTiO_3$, PT) were prepared using the casting method. The prepared samples were identified by Attenuated Total Reflection–Fourier Transform Infrared (ATR-FTIR). Peaks characteristic of PVA at 3280, 2917, 1690, 1425, 1324, 1081, and 839 cm^{-1} appeared; a peak indicating the presence of $PbTiO_3$ also appeared at 713 cm^{-1}. The interaction between PVA and $PbTiO_3$ was confirmed by observing the change in IR absorption intensity. Optical properties in the UV-Vis range were investigated using an Ultraviolet Visible technique (UV-Vis). An enhancement in absorption capacity by the increasing $PbTiO_3$ concentration was observed. Optical properties such as band gap energy, Urbach energy, and extinction coefficient indicate that addition of $PbTiO_3$ into the PVA polymer induced variance in internal states by increasing the ratio of $PbTiO_3$. Obtaining a UV-protective material derived from a PVA/$PbTiO_3$ composite is the aim of this paper.

Keywords: PVA; $PbTiO_3$; optical properties

Citation: El-Deeb, A.S.; Kader, M.M.A.; Nasr, G.M.; Ahmed, M.A.; Taha, E.O. Optical Investigation of PVA/$PbTiO_3$ Composite for UV-Protective Approach Applications. *Eng. Proc.* **2021**, *11*, 4. https://doi.org/10.3390/ASEC2021-11172

Academic Editor: Nicholas Vassiliou Sarlis

Published: 15 October 2021

1. Introduction

Polymeric nanocomposite materials have been receiving a lot of attention lately because of the expanded range of applications that these hybrid materials can be used for [1]. It is widely documented that polymers, as dielectric materials, are good host matrices for nanoparticles [2], and that this is true for both metal and ceramic nanoparticles. While doing so, these embedded particles within the polymer matrix also impact the physical properties of the host [3,4]. Polymer ceramic hybrid composites, in particular, are promising functional materials in a variety of disciplines, demonstrating useful optical, electrical, thermal, mechanical, and antibacterial characteristics [4].

Composites made of piezoelectric ceramics such as barium titanate (BT), lead titanate (PT), triglycine sulphate (TGS), lead zirconate titanate (PZT), etc. have been studied extensively [5]. Lead titanate ($PbTiO_3$) is a ferroelectric ceramic included in the same perovskite family as barium titanate and lead zirconate titanate. All of these materials have one or more phase transitions within a particular temperature range. However, lead titanate exhibits the largest spontaneous polarization in the tetragonal phase (tetragonality factor c/a = 1.064), the lowest dielectric constant (≈200), and a high Curie temperature (≈490 °C) [6].

Polyvinyl Alcohol (PVA) is a polymer with good film-forming and physical properties, high hydrophilicity, processability, and biomaterial and biosensor capabilities [7,8]. PVA is a semicrystalline polymer (Cryo-Amorphous) composed of both crystalline and amorphous phases. When such a polymer is mixed with a suitable ceramic, it interacts either in the amorphous or crystalline fractions, affecting the physical qualities in both cases [9].

It is reported in this study that the preparation and investigation of the optical properties of PVA/$PbTiO_3$ composite films were carried out because these films can combine the advantages of both polymer and ceramic components.

2. Experimental

Polyvinyl Alcohol (PVA) powder and Lead (II) Titanate ($PbTiO_3$) powder were supplied from Sigma-Aldrich (St. Louis, MO, USA). PVA was dissolved in double-distilled water at 80 °C and stirred for 4 h to ensure uniform dispersion. To prepare $PbTiO_3$/PVA composites, different weight percentages (0, 1, 5, and 10 wt.%) were added to the above solution of PVA in water and stirred for 1 h. The mixture was then cast in a glass dish, and the sample was left to dry for a week at room temperature. The dispersion state of the prepared samples was examined using a Field Emission Scanning Electron Microscope (FESEM Sigma 300 VP, Carl Zeiss, GmbH, Jena, Germany). Figure 1 shows the FESEM micrograph for 10 wt.% $PbTiO_3$ in the PVA. The image demonstrates the absence of aggregation of $PbTiO_3$ and its homogenous dispersion.

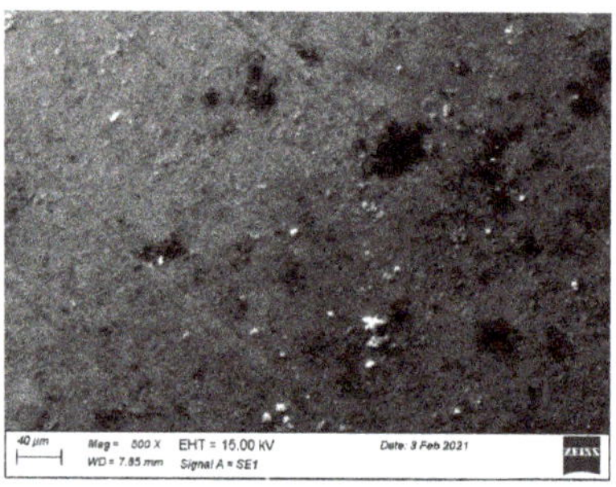

Figure 1. FESEM of the 10 wt.% $PbTiO_3$/PVA composite.

Alpha Bruker platinum Attenuated Total Reflection–Fourier Transform Infrared Spectroscopy (ATR-FTIR) with a wavenumber range of 600–4000 cm^{-1} was used to determine the characterization of the polymer composite. UV-Vis spectra were measured using a Jasco V-630 spectrophotometer.

3. Results and Discussion

3.1. ATR-FTIR

ATR-FTIR spectroscopy represents a key approach for identifying and characterizing polymer composites. The PVA/$PbTiO_3$ composite ATR-FTIR spectra with various $PbTiO_3$ levels are illustrated in Figure 2. The major peaks of PVA were found at 3280 cm^{-1}, indicating the presence of O-H stretching vibration of the hydroxyl group. The peak at 2917 cm^{-1} is linked with the existence of the asymmetrical deep vibration of CH. The peak at 1690 cm^{-1} corresponds to C=O carbonyl stretching. The peak at 1425 cm^{-1} corresponds to the bending vibration of CH_2 [10]. The peaks at 839 cm^{-1}, 1081 cm^{-1}, and 1324 cm^{-1} are associated with C-C stretching vibration, C-O stretching vibration of acetyl groups, and the presence of C-H deformation [11]. For the PVA/$PbTiO_3$ composite, several peaks could be observed, with a belt additive of 713 cm^{-1} indicating metal oxygen stretching, which in turn shows the presence of metal oxygen bonds.

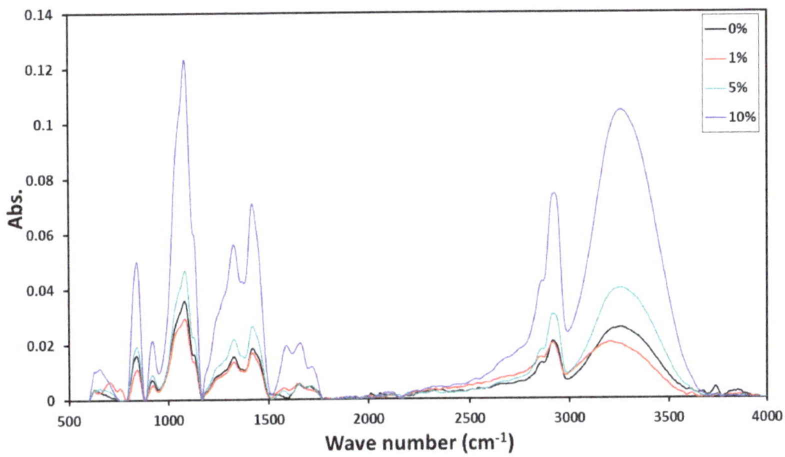

Figure 2. ATR-FTIR spectra for the PVA/$PbTiO_3$ composite.

3.2. Optical Properties

Ultraviolet-visible spectroscopy is one the most prevalent mechanisms that can detect every molecule, identify functional groups, and confirm the concentration of analytes indicated by absorbance using Beer's law. The UV-Vis absorption spectra of PVA/$PbTiO_3$ with various $PbTiO_3$ concentrations (0%, 1%, 5%, and 10%) are depicted in Figure 3. Absorption peaks obtained from 350 to 450 nm for 5% and 10% $PbTiO_3$ can be observed due to the energy of the forbidden band conformable to O-2p→Ti-3d, while no peaks were obtained for neat PVA or a concentration of 1% $PbTiO_3$. The intensity of absorption decreases with increasing wavelength for all samples. High absorption is obtained in the UV region for samples with 5% and 10% concentrations, qualifying them to be promising in UV-protective approaches.

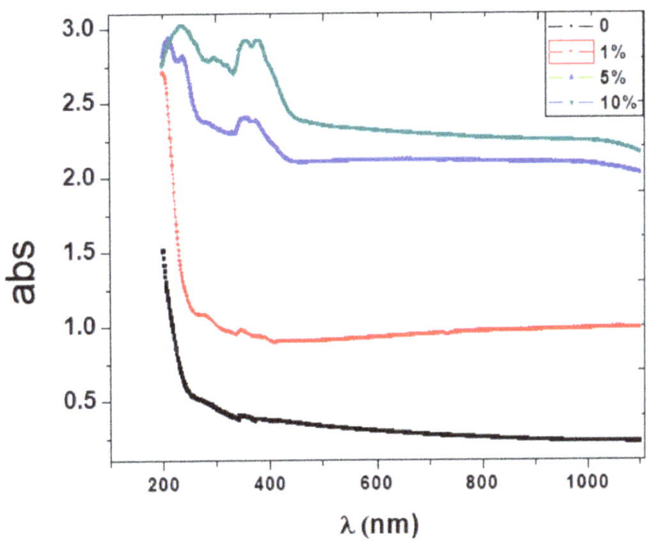

Figure 3. UV-Vis spectra of the PVA/$PbTiO_3$ composite.

In low and poor crystalline materials, disordered and amorphous material exponential tails appear near the optical edge, called Urbach tails, due to the localized states that the materials have. These localized states are prolonged in the band gap and can be described by the Urbach rule [12].

$$\alpha = \alpha_o \ exp\left(\frac{h\gamma}{E_u}\right) \tag{1}$$

where α is the absorption coefficient, $h\gamma$ is the photon energy, α_o is a constant, and E_u is the band tail energy (Urbach energy). The band tail energy is weakly temperature-dependent and interpreted as the band tail width due to the localized states within a bandgap; it is associated with low/poor crystalline materials and amorphous materials. Figure 4 depicts the relation between the absorption coefficient α and the photon energy $E\ (h\gamma)$. It is essential to take the natural logarithm of Equation (1), as the slope of the straight line after taking this logarithm gives the Urbach energy (E_u) [12]:

$$ln\alpha = ln\alpha_o \ + \left(\frac{h\gamma}{E_u}\right) \tag{2}$$

$ln\alpha$ is plotted against the incident photon energy $h\gamma$ in Figure 5.

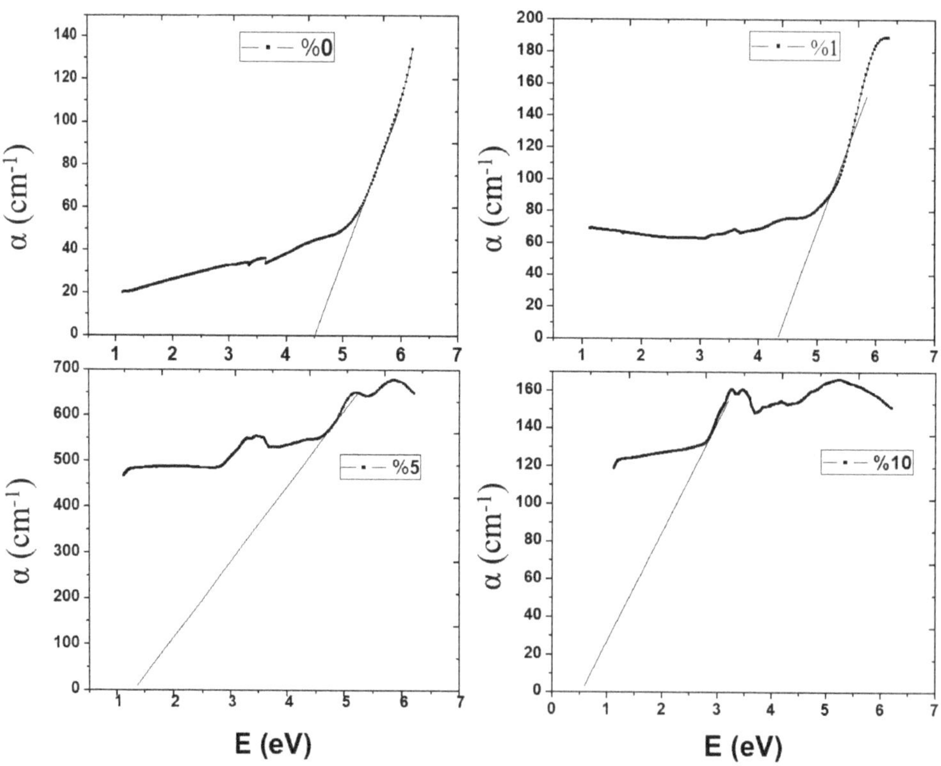

Figure 4. The relation between α and E for the PVA/$PbTiO_3$ composite.

Transition of electrons can take place in semiconductor materials between the valance band and conduction band. Spontaneous emission, absorption, and simulated emission can take place between the conduction band and valance band. Band gap energy can be determined by plotting $(\alpha h\gamma)^{1/k}$ and $h\gamma$, where k depends on the nature of the transition For direct transition (k = 1/2), the electron rising from the valance band to the conduction band changes only its potential, while for indirect transition (k = 2), the electron rising from the valance band to the conduction band changes potential and momentum. The following equation can describe the direct and indirect transition [13]:

$$(\alpha h\gamma) = B\left(h\gamma - E_g\right)^{1/k} \tag{3}$$

B is a constant that depends on the transition probability. Figure 6 shows the plot of $(\alpha h\gamma)^2$ and $(\alpha h\gamma)^{1/2}$ versus E $(h\gamma)$ for the PVA/$PbTiO_3$ composite. The presented values of direct and indirect band gap energies have been determined via the extrapolation of the linear portion of the curve along the x-axis. Table 1 indicates the values of absorption edges, band tail energies, and indirect (E_i) and direct (E_d) optical band gaps for composites. The absorption edge values and the band tail energies decrease with increasing $PbTiO_3$ content. The values of direct and indirect band gaps decrease with increasing $PbTiO_3$ content, indicating a variance in internal states. A composite with 10 wt.% $PbTiO_3$ is promising for UV-protective applications.

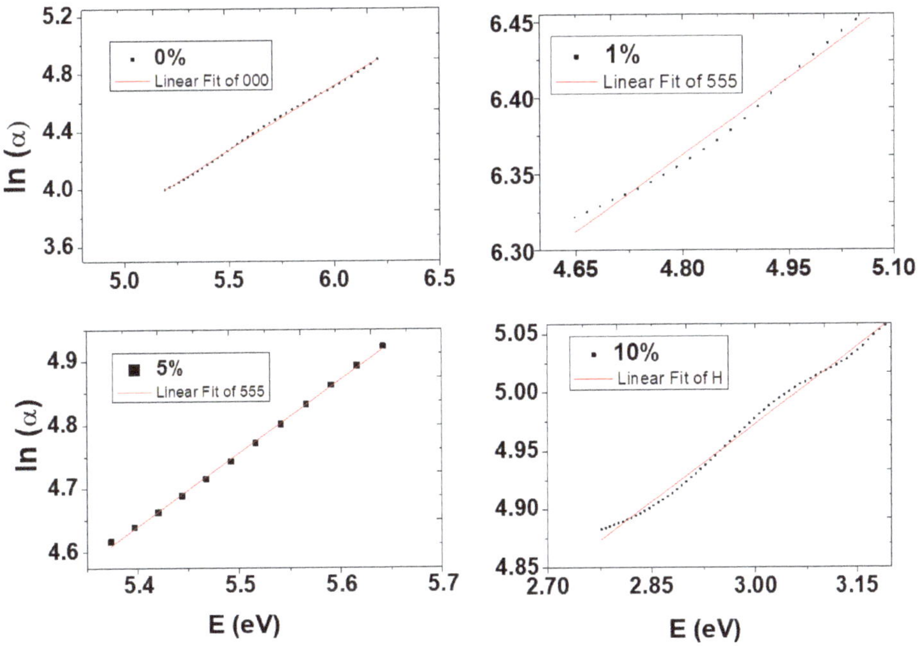

Figure 5. The relation between ln α and E for the PVA/$PbTiO_3$ composite.

Table 1. Values of absorption edge, band tail, and indirect (E_i) and direct (E_g) optical band gap energy for the PVA/$PbTiO_3$ Composite.

PVA/$PbTiO_3$	Absorption Edge (eV)	Band Tail E_u (eV)	Energy Gap (eV)	
			E_i	E_d
0%	4.6	1.12	5.3	5.2
1%	4.46	2	5.8	4.8
5%	1.36	2.34	2.8	2.7
10%	0.56	2.74	2.3	2.17

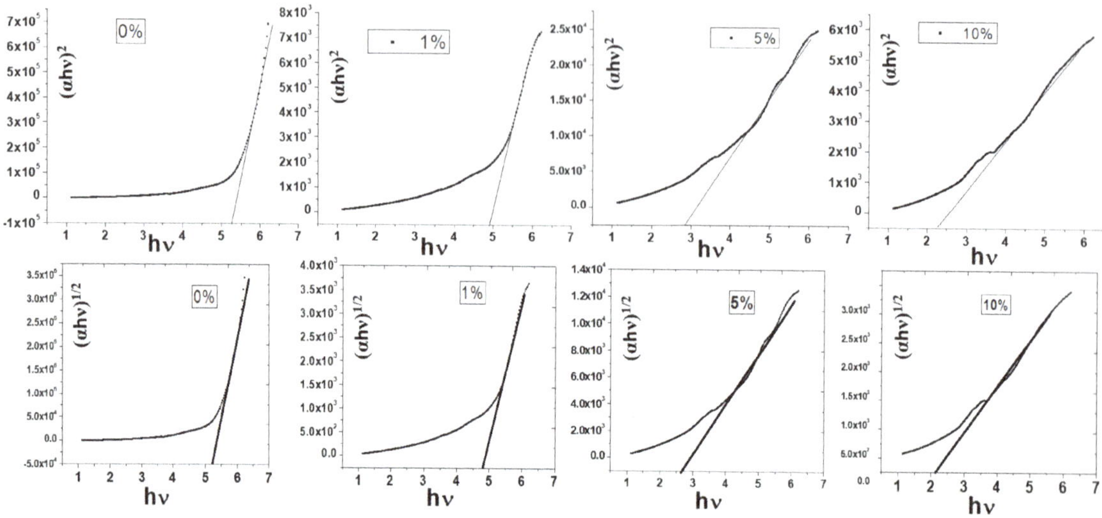

Figure 6. Relation between $(\alpha h\gamma)^2(\text{cm}^{-1}\text{eV})^2$ and $(\alpha h\gamma)^{1/2}(\text{cm}^{-1}\text{eV})^{1/2}$ against $h\gamma$ (eV) for the PVA/$PbTiO_3$ composite.

4. Conclusions

PVA/$PbTiO_3$ composites were prepared using the casting method. The absence of aggregations was examined by using FESEM. The peaks characteristic of PVA appeared when the composites were examined using ATR-FTIR. The absorption edge value and band tail energies decrease with increasing $PbTiO_3$ content. Also, the values of direct and indirect bandgap decrease with increasing $PbTiO_3$ content. A composite with 10 wt.% $PbTiO_3$ is promising for UV-protective applications.

Author Contributions: Conceptualization, G.M.N., E.O.T. and A.S.E.-D.; methodology, M.A.A. and M.M.A.K.; writing—original draft, A.S.E.-D. and E.O.T.; Review and editing, E.O.T. and A.S.E.-D.; supervision, G.M.N. All authors have read and agreed to the published version of the manuscript.

Funding: This research received no external funding.

Conflicts of Interest: The authors declare no conflict of interest.

References

1. Bailey, E.J.; Winey, K.I. Dynamics of polymer segments, polymer chains, and nanoparticles in polymer nanocomposite melts: A review. *Prog. Polym. Sci.* **2020**, *105*, 101242. [CrossRef]
2. Ananth, A.N.; Umapathy, S.; Sophia, J.; Mathavan, T.; Mangalaraj, D. On the optical and thermal properties of in situ/ex situ reduced Ag NP's/PVA composites and its role as a simple SPR-based protein sensor. *Appl. Nanosci.* **2011**, *1*, 87–96. [CrossRef]
3. Ghanipour, M.; Dorranian, D. Effect of ag-nanoparticles doped in polyvinyl alcohol on the structural and optical properties of PVA films. *J. Nanomater.* **2013**, *2013*, 897043. [CrossRef]
4. Sebastian, M.T. Polymer–Ceramic Composites of 0–3 Connectivity for Circuits in Electronics: A Review. *Int. J. Appl. Ceram. Technol.* **2010**, *7*, 415–434. [CrossRef]
5. Uma, S.; Philip, J. Induction of electro-activity in polyvinyl alcohol with addition of nanocrystalline PZT ceramic. *Int. J. Indian J. Pure Appl. Phys.* **2013**, *5*, 717–723.
6. Mir, M.; de Paula, C.C.; Garcia, D.; Kiminami, R.H.G.A.; Eiras, J.A.; Mascarenhas, Y.P. Microstructural characterization using the Rietveld method in lead lanthanum titanate ceramics. *J. Eur. Ceram.* **2007**, *27*, 3719–3721. [CrossRef]
7. Kinadjian, N.; Achard, M.-F.; Julián-López, B.; Maugey, M.; Poulin, P.; Prouzet, E.; Backov, R. ZnO/PVA macroscopic fibers bearing anisotropic photonic properties. *Adv. Funct. Mater* **2012**, *22*, 3994–4003. [CrossRef]
8. Taha, E.O.; Nasr, G.M. Dielectric and dynamic mechanical properties of Rochelle salt/PVA composites. In *IOP Conference Series Materials Science and Engineering*; IOP Publishing: Hurghada, Egypt, 2021; p. 12008.
9. Shehap, A.M.; Akil, D.S. Structural and optical properties of TiO_2 nanoparticles/PVA for different composites thin films. *Int. J. Nanoelectron. Mater.* **2016**, *9*, 17–36.

10. Atta, M.M.; Taha, E.O.; Abdelreheem, A.M. Nitrogen plasma effect on the structural, thermal, and dynamic mechanical properties of PVA/starch/graphene oxide nanocomposite. *Appl. Phys. A* **2021**, *127*, 532. [CrossRef]
11. Afzal, H.M.; Shehzad, F.; Zubair, M.; Bakather, O.Y.; Al-Harthi, M.A. Influence of microwave irradiation on thermal properties of PVA and PVA/graphene nanocomposites. *J. Therm. Anal. Calorim.* **2020**, *139*, 353–365. [CrossRef]
12. Hassaan, M.Y.; Osman, H.M.; Hassan, H.H.; El-Deeb, A.S.; Al-Helal, M.A. Optical and electrical studies of borosilicate glass containing vanadium and cobalt ions for smart windows applications. *Ceram. Int.* **2017**, *43*, 1795–1801. [CrossRef]
13. Aziz, S.B.; Rasheed, M.A.; Ahmed, H.M. Synthesis of polymer nanocomposites based on [methyl cellulose]$_{(1-x)}$:(CuS)$_x$ (0.02 M $\leq$ x $\leq$ 0.08 M) with desired optical band gaps. *Polymers* **2017**, *9*, 194. [CrossRef] [PubMed]

engineering proceedings

MDPI

Proceeding Paper

The Influence of Sample Preparation Techniques on Aluminium Alloy AA2024-T3 Substrates for Sol-Gel Coating †

Magdi H. Mussa [1,2,3,*], Nicholas Farmilo [3,4] and Oliver Lewis [3]

1 Mechanical and Energy Department, The Libyan Academy for Graduate Study, Tripoli, Libya
2 Mechanical Engineering Department, Sok Alkhamis Imsehel High Tec. Institute, Sok Alkhamis, Tripoli, Libya
3 Material and Engineering Research Institute (MERI), Sheffield Hallam University, Sheffield S1 1WB, UK; O.Lewis@shu.ac.uk (N.F.); nfarmilo@gmail.com (O.L.)
4 Tideswell Business Development Ltd., Ravensfield Sherwood Rd., Buxton SK17 8HH, UK
* Correspondence: magdimosa1976@gmail.com; Tel.: +44-740-449-6955

† Presented at the 2nd International Electronic Conference on Applied Sciences, 15–31 October 2021; Available online: https://asec2021.sciforum.net/.

Abstract: Sol-gel coatings provide environmentally friendly surface protection for metals and can replace toxic pre-treatments such as those based on hexavalent chromium on metal alloys. This project ultimately aims to develop silica-based organic–inorganic sol-gel derived thin film coatings possessing anti-corrosion and anti-fouling properties on aluminium alloy substrates. As with any coating, sample preparation plays a significant role in the performance of a sol-gel coating. Therefore, it was necessary to define a preparation method that combines the removal of contaminants and surface roughening to improve adhesion and reproducibility. Four techniques were investigated: fine abrasive sandpaper cleaning, acetone degreasing only and cleaning with an industrial-available alkaline cleaner for 5 min and 30 min.

Keywords: silica-based hybrid sol-gel coating; infrared spectroscopy; electrochemical testing; corrosion protection

Citation: Mussa, M.H.; Farmilo, N.; Lewis, O. The Influence of Sample Preparation Techniques on Aluminium Alloy AA2024-T3 Substrates for Sol-Gel Coating. *Eng. Proc.* **2021**, *11*, 5. https://doi.org/10.3390/ASEC2021-11121

Academic Editor: Filippo Berto

Published: 15 October 2021

Publisher's Note: MDPI stays neutral with regard to jurisdictional claims in published maps and institutional affiliations.

1. Introduction

Aluminium alloys are still considered one of the primary light, high-strength alloys that can be used in aerospace and marine structures with a moderate economic cost. However, depending on the type and grade, aluminium alloys are vulnerable to aggressive environments [1,2]. Surface pollutants, such as organic dirt, grease and lubricants, must be removed to promote ionic or mechanical bonding between coatings and the substrates. Some of the practising procedures focus on degreasing more than removing oxides or cladding film, as it still reduces the corrosion propagation in the surface [3]. The standard aluminium alloys surface preparation in marine application can be achieved using a high alkaline solvent jet to remove the organic and greasing residual from 5 to 30 min, then it could be followed or mixed with sandblasting to remove the oxide film [4].

Hybrid silica-based sol-gel coatings have already been recognised as a potential corrosion mitigation solution for aerospace and marine use as an eco-friendly method [5,6], offering many routes, including using single-or multi-layer coating systems with anti-corrosion and anti-fouling systems. [5,7–9] Additionally, sol-gel coatings can present other desirable properties, such as preventing ice accumulation, oxidation resistance and abrasion resistance [10–12]. However, in sol-gel, the adhesion mechanism with aluminium alloys' surface is different from the other coating types; it is based on (M-O-Si) strong ionic bonding. Therefore, This paper will study the optimisation of sample preparations techniques of the aluminium alloy 2024-t3 for the sol-gel coating used in the previous project [1]. This investigation analyses four types of surface preparation that suit the sol-gel coating technology to provide proper bonding for protection against corrosion.

2. Experimental Work

The testing and characterisation of the techniques will be conducted in two main steps. The first one will examine the treated substrates without the presence of any sol-gel coating by investigating the morphology, appearance, phase construction and XRD analysing techniques. The second step will occur after applying the sol-gel coating on four different treated samples and applying an electrochemical impedance spectroscopy preliminary test to evaluate the corrosion mechanism protection within 6 days in a 3.5% NaCl solution to simulate the aggressive environment.

2.1. Substrate Preparation

Q-panels were supplied by Q-Lab and are made of aluminium alloy AA2024-t3 with dimensions (102 mm × 25 mm × 1.6 mm) for use as test substrates [13]. The Q-panels samples were treated in four different surface preparations as follows:

1. The acetone was applied on the surface followed with DI on the received Q-panels samples to remove the organic contamination and grease from the surface, leaving the oxide film layer on. This sample was labelled as normal cleaning (NC).
2. The second sample followed the same cleaning method by acetone followed by immersing the sample in Super bee® Alkaline etching solution 10% for 5 min as the minimum time in standard and was labelled as (5-AC).
3. The third sample preparation followed the same cleaning method by acetone but was immersed in Super bee® Alkaline etching solution 10% for 30 min as the maximum time in standard and was labelled as (30-AC).
4. The fourth sample preparation was treated with mechanical abrasive sandpaper to reach P1500 directly to remove the top surface and oxides film and was labelled as (SP-C).

2.2. Sol-Gel Preparation

The hybrid silica-based sol-gel was prepared by mixing tetraethylorthosilicate silane (TEOS), trimethoxymethyl silane (MTMS), and isopropanol (all purchased from Sigma-Aldrich, Gillingham, UK) with dropwise additions of DI water at the molar ratio of 18:14:17:220, respectively, as is mentioned in the previous work [1]. The formula was applied by spray coating on the clean substrate and was built up over three passes. After that, the coated samples were left in the atmosphere for 10 min before being thermally annealed at 80 °C for 4 h—the chosen samples with a thickness of 16 ± 2 micrometres were chosen by using an Elcometer 456 Model Coating Thickness Gauge and confirmed with the scanning electron microscope (SEM) via cross-section imaging.

2.3. Testing and Characterisation

The Infinite Focus G5 (IFM), with the capability of surface measurements and reverse engineering modelling, was used for all samples. Tests were performed at SHU labs.

An FEI-QUANTA 650 scanning electron microscope (SEM), with an X-MAX 80 mm^2 energy-dispersive X-ray (EDX) spectrometer (Oxford Instruments, Oxford, UK), was used to analyse the morphology and chemical composition of coated samples.

X-ray diffraction (XRD) measurements of the sample's surface were performed using a Philips X'PERT MPD with operational parameters of 40 kV and 40 mA [13].

The hydrophobicity of the sol-gel coatings was determined by performing water contact angle measurements using a Dataphysics OCA 15EC Goniometer, with deionised water (DI) used as the solvent. A minimum of three analyses was performed across the surface of each sample, and the mean water contact angle value was calculated by the Dataphysics OCA software [14]

Electrochemical tests were performed on the bare and coated samples to assess their corrosion resistance. Tests were conducted using a Princeton Applied Research PARSTAT 2273. The corrosion performance of the sol-gel coated and uncoated aluminium alloy was evaluated using electrochemical impedance spectroscopy (EIS) and potentiodynamic polarisation (PDPS) scans. A tested area of 1.00 mm^2 was created in the centre of the samples in aerated 3.5% NaCl. The tests were carried out at room temperature (20 °C +/− 2 °C). The electrode potential was monitored for approximately 1 h before polarisation in electrolyte solution until stability was obtained. The sample was polarised with PDPS at a scan rate of 1.667 mVs^{-1} from the initial potential of −250 mV vs. OCP to +750 mV vs. SCE. The electrochemical impedance measurements were recorded between 100 kHz to 10 MHz with a sinusoidal AC RMS value of 10 mV [15].

3. Results and Discussion

3.1. Surface Analysis for the Uncoated Substrates

3.1.1. Infinity Focus Microscopy IFM

The IFM images in Figure 1 show the difference between the four sample preparation techniques. This shows that the surface of the 30 min etched sample was affected by aggressive pitting due to the dissolution of the light metals in the alkaline solution. In comparison, the sandpaper P1500 finishing sample preparation shows that the surface was very smooth. The appearance of standard cleaning NC samples and 5 min etching cleaning was smooth, and the preparations did not affect the roughness of the surface Rz, which was approximately 1.2 µm; the highest roughness Rz was 3.3 µm for the 30-AC sample while the smallest was 659 nm for the sandpaper cleaning SP-C.

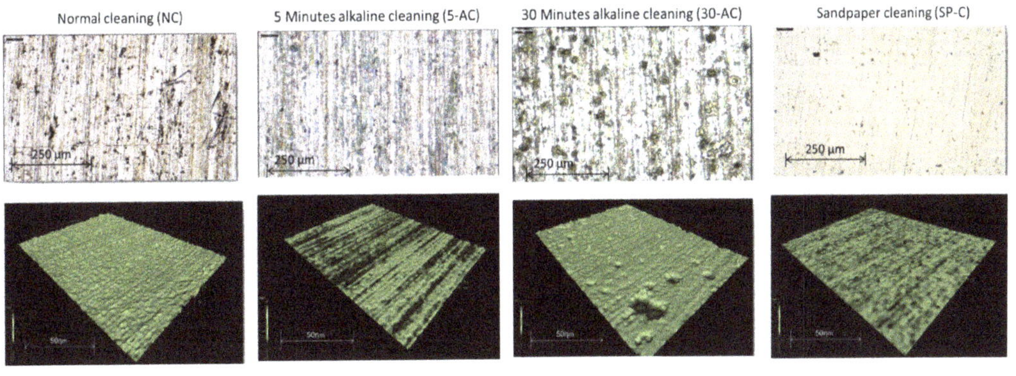

Figure 1. IFM images show the profile of the four surface preparations samples.

3.1.2. Scanning Electron Microscopy (SEM)

The SEM images in Figure 2 show the four sample preparations' morphological surface appearance as follows: the most affected surface was generated by 30 min etching cleaning. The other surface preparation samples show that the surface is not significantly affected. However, the inclusions from the silicate carbides are impeded in the sandpaper sample SP-C preparations' surface as a result of tribology and friction.

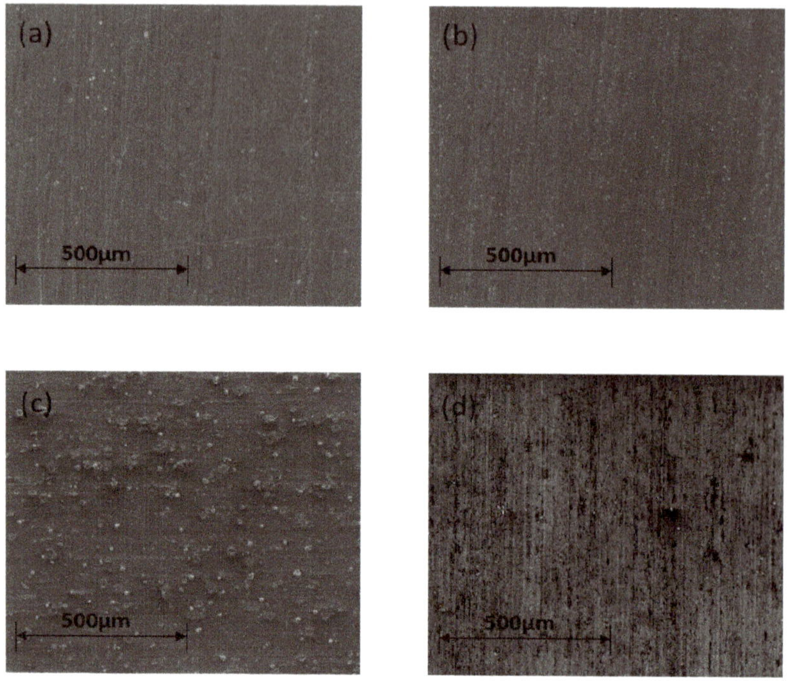

Figure 2. SEM images showing the surface for (**a**) NC, (**b**) 5-AC, (**c**) 30-AC and (**d**) SP-C surface preparation samples.

3.1.3. X-ray Powder Diffraction (XRD)

Generally, Figure 3 show there was no significant phase transformation on the surface of the alloy for the four different sample preparations. However, the sandpaper shows fewer aluminium-oxides Al_2O_3 at 38.4°, which show the level of the oxide on the surface is superficial. At the same time, both alkaline etched samples show a higher level.

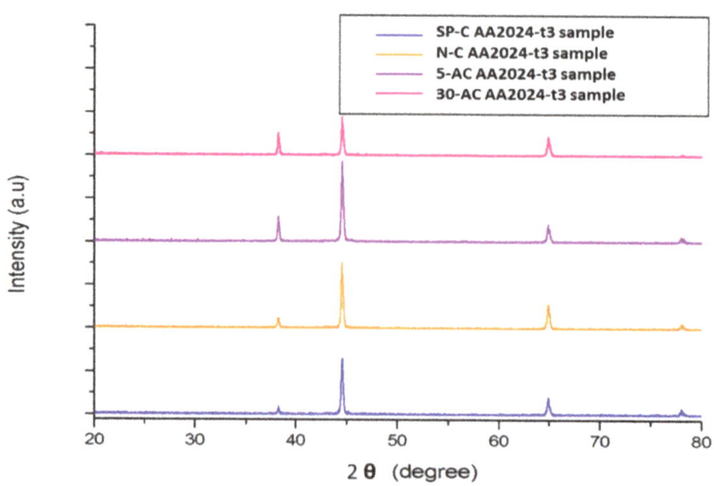

Figure 3. X-ray diffraction (XRD) reading for the four preparation samples.

3.1.4. Wettability and Contact Angle (WCA)

Figure 4 show the droplets of 10 μL of DI water on the surface of the differently treated samples. It is clear that the highest contact angle was for the 30-AC sample while the smallest was on the sandpaper cleaned sample by 92° and 62°, respectively. This confirms that the 30-AC cleaned sample is less-waitable compared to the others. The contact angles of the 5-AC and NC samples were 72° and 78°, respectively.

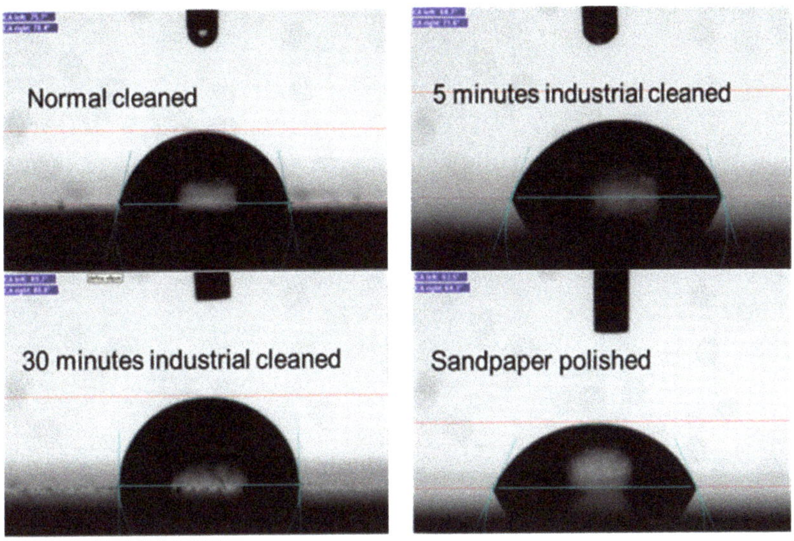

Figure 4. Water contact angle for NC, SP-C, 5-AC and 30-AC sample.

3.2. Testing Characterisations of Sol-Gel Coated Samples

Electrochemical Impedance Spectroscopy (EIS)

Tests were performed over a period of 6 days. Figure 5a,b show the Bode impedance magnitude plots for sol-gel-NC, sol-gel-SP-C, sol-gel-5-AC and sol-gel-30-AC coated samples at the beginning of the investigation and after 144 h. From Figure 6 it can be seen that the overall impedance for the sol-gel coated samples in the first hour lacked variance between 5.2×10^6 to 3.2×10^6 Ohms/cm^2. The maximum reading was with sol-gel-SP-C, and the minimum was for the sol-gel-30-AC; the reading for sol-gel-5-AC and sol-gel-NC was 4.5×10^6 and 4.02×10^6 ohms/cm^2, respectively.

After 144 h, the drop in impedance was clearly apparent by approximately one order of magnitude for the sol-gel-30-AC sample compared to the sol-gel-SP-C sample. This could be due to the microcracks in the sol-gel coating on the 30 min etched samples, also in addition to the oxide film that reduces the adhesion of the sol-gel film on the surface.

On the other hand, the significance reading of the sol-gel-5-AC displayed similar behavior to the sandpaper prepared coated sample but with a slight drop in impedance to reach 1.01×10^6 ohms/cm^2, followed by the standard cleaning sample (degreasing only).

Figure 6a,b show the Nyquist plotting for the four sol-gel coated samples during the first 24 h, and after 144 h. All Nyquist plots support the Bode plots as they display one time-constant capacitive behaviour with the Wurburg diffusion element. However, the maximum impedance was for the sol-gel-SP-C coated sample, while the minimum was for sol-gel-30-AC.

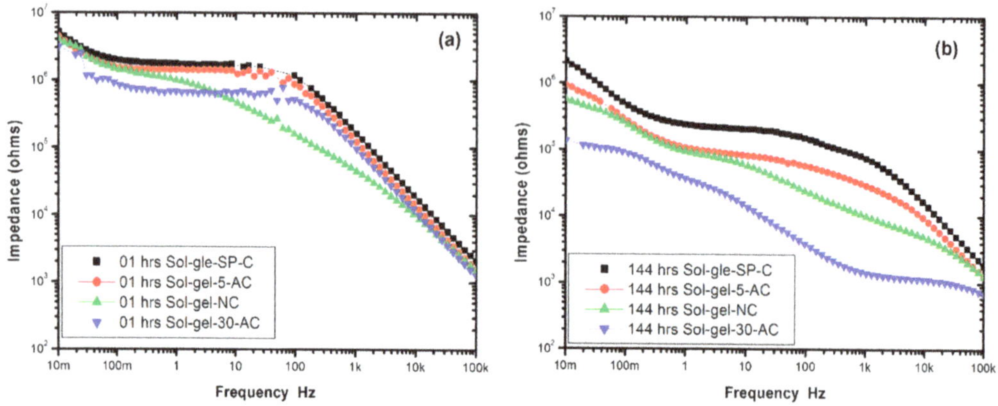

Figure 5. Bode impedance magnitude plots for sol-gel-NC, sol-gel-SP-C, sol-gel-5-AC and sol-gel-30-AC coated samples in the first hour (**a**) and after 144 h (**b**).

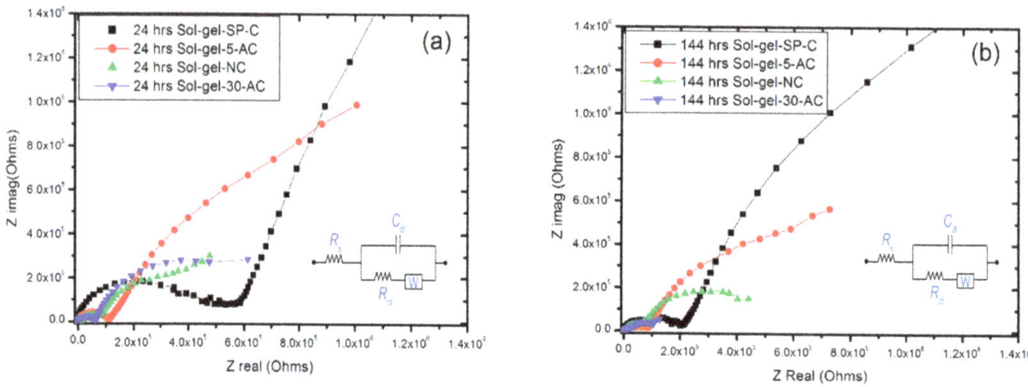

Figure 6. Nyquist plots for sol-gel-NC, sol-gel-SP-C, sol-gel-5-AC and sol-gel-30-AC coated samples in 24 h (**a**) and after 144 h (**b**).

4. Conclusions

Mechanical cleaning using abrasive paper produces a smooth surface with good wettability and adhesion for sol-gel. However, there are probably abrasive particles embedded in the surface, which may cause future corrosion to occur. Longer immersion times (30 min) in 10% Super bee® appear to be less optimal as wettability is decreased, possibly due to excessive surface roughening, while the more active surface results in greater oxidation. On the other hand, 5 min immersion in 10% Super bee® gives good wettability and a degree of surface roughening, which may benefit coating adhesion, as well as the degreasing process. Generally, the sol-gel coating can provide excellent protection against corrosion for the mentioned samples depending on the manner of the surface cleaning process; it is clear from the electrochemical testing that the mechanical sandpaper cleaning was a good combination and offered excellent protection. However, due to the time and cost of this process, the 5 min of etching cleaning was acceptable for sol-gel sample preparation, with the lowest results being provided by the degreasing preparation process.

Author Contributions: Conceptualisation, M.H.M.; methodology, M.H.M., O.L. and N.F.; validation and testing M.H.M.; data analysis, M.H.M., N.F. and O.L.; XRD-support analysis, M.H.M.; investigation and revalidation M.H.M.; resources, M.H.M., O.L. and N.F.; writing—original draft preparation, M.H.M.; writing—review and editing, M.H.M., O.L. and N.F.; project supervision, N.F. and O.L. All authors have read and agreed to the published version of the manuscript.

Funding: This research has no assigned direct fund.

Institutional Review Board Statement: Not applicable.

Informed Consent Statement: Not applicable.

Data Availability Statement: The data are not publicly available; the files are stored on corresponding instruments and personal computers.

Acknowledgments: The authors would like to acknowledge the facilitating support provided by Sheffield Hallam University in the Material and Engineering research institute (MERI) and also to the Libyan Scholarship Program for the financial support.

Conflicts of Interest: The authors declare no conflict of interest.

References

1. Mussa, M. Development of Hybrid Sol-Gel Coatings on AA2024-T3 with Environmentally Benign Corrosion Inhibitors. Ph.D. Thesis, Sheffield Hallam University, Sheffield, UK, 2020.
2. Mussa, M.H.; Rahaq, Y.; Takita, S.; Farmilo, N. Study the Enhancement on Corrosion Protection by Adding PFDTES to Hybrid Sol-Gel on AA2024-T3 Alloy in 3.5% NaCl Solutions. *Albahit J. Appl. Sci.* **2021**, *2*, 61–68.
3. Hughes, A.E.; Taylor, R.J.; Nelson, K.J.H.; Hinton, B.R.W.; Wilson, L. Characterisation of surface preparation of 2024 aluminium alloy for conversion coating. *Mater. Sci. Technol.* **1996**, *12*, 928–936. [CrossRef]
4. ASTM International. *Standard Practice for Inspection of Marine Surface Preparation and Coating Application*; F941–99; American National Standards Institute: Washington, DC, USA, 2009.
5. Detty, M.R.; Ciriminna, R.; Bright, F.V.; Pagliaro, M. Environmentally benign sol-gel antifouling and foul-releasing coatings. *Acc. Chem. Res.* **2014**, *47*, 678–687. [CrossRef] [PubMed]
6. Akid, R.; Wang, H.; Gobara, M.; Smith, T.J.; Gittens, J. Green coatings for industrial applications. *Corros. Manag.* **2011**, *100*, 11–14.
7. Wang, H.; Akid, R.; Gobara, M. Scratch-resistant anticorrosion sol-gel coating for the protection of AZ31 magnesium alloy via a low temperature sol-gel route. *Corros. Sci.* **2010**, *52*, 2565–2570. [CrossRef]
8. Chen, S.; Li, L.; Zhao, C.; Zheng, J. Surface hydration: Principles and applications toward low-fouling/nonfouling biomaterials. *Polymer* **2010**, *51*, 5283–5293. [CrossRef]
9. Eduok, U.; Suleiman, R.; Gittens, J.; Khaled, M.; Smith, T.J.; Akid, R.; El Ali, B.; Khalil, A. Anticorrosion/antifouling properties of bacterial spore-loaded sol-gel type coating for mild steel in saline marine condition: A case of thermophilic strain of Bacillus licheniformis. *RSC Adv.* **2015**, *5*, 93818–93830. [CrossRef]
10. Lev, O.; Wu, Z.; Bharathi, S.; Glezer, V.; Modestov, A.; Gun, J.; Rabinovich, L.; Sampath, S. Sol−Gel Materials in Electrochemistry. *Chem. Mater.* **1997**, *9*, 2354–2375. [CrossRef]
11. Wang, D.; Bierwagen, G.P. Sol-gel coatings on metals for corrosion protection. *Prog. Org. Coatings* **2009**, *64*, 327–338. [CrossRef]
12. Pathak, S.S.; Khanna, S. Sol-gel nano coatings for corrosion protection. *Met. Surf. Eng.* **2012**, *12*, 304–329.
13. Rahaq, Y.; Mussa, M.; Mohammad, A.; Wang, H.; Hassan, A. Highly reproducible perovskite solar cells via controlling the morphologies of the perovskite thin films by the solution-processed two-step method. *J. Mater. Sci. Mater. Electron.* **2018**, *29*, 16426–16436. [CrossRef]
14. Bracco, G.; Holst, B. Contact Angle andWetting Properties. In *Springer Series in Surface Sciences*; Springer: New York, NY, USA, 2013; Volume 51, ISBN 978-3-642-34242-4.
15. Tait, W.S. *Electrochemical Impedance Spectroscopy Fundamentals, an Introduction to Electrochemical Corrosion Testing for Practicing Engineers and Scientists*; Tait, W.S., Ed.; PairODocs Publications: Racine, WI, USA, 1994; ISBN 13-978-0966020700.

engineering proceedings

MDPI

Proceeding Paper

Changes in Anthocyanin and Antioxidant Contents during Maturation of Australian Highbush Blueberry (*Vaccinium corymbosum* L.) Cultivars †

Joel B. Johnson [1,*], Michelle Steicke [2], Janice S. Mani [1], Shiwangni Rao [2], Scott Anderson [3], Lara Wakeling [2] and Mani Naiker [1]

1 School of Health, Medical & Applied Sciences, Central Queensland University, Rockhampton, QLD 4701, Australia; janice.mani@gmail.com (J.S.M.); m.naiker@cqu.edu.au (M.N.)
2 School of Health and Life Sciences, Federation University Australia, Ballarat, VIC 3350, Australia; michellesteicke@hotmail.com (M.S.); wani.rao@gmail.com (S.R.); l.wakeling@federation.edu.au (L.W.)
3 Faculty of Health, Deakin University, Waurn Ponds, Geelong, VIC 3216, Australia; scott@sda0.com
* Correspondence: joel.johnson@cqumail.com

† Presented at the 2nd International Electronic Conference on Applied Sciences, 15–31 October 2021; Available online: https://asec2021.sciforum.net/.

Citation: Johnson, J.B.; Steicke, M.; Mani, J.S.; Rao, S.; Anderson, S.; Wakeling, L.; Naiker, M. Changes in Anthocyanin and Antioxidant Contents during Maturation of Australian Highbush Blueberry (*Vaccinium corymbosum* L.) Cultivars. *Eng. Proc.* **2021**, *11*, 6. https://doi.org/10.3390/ASEC2021-11155

Academic Editor: Nunzio Cennamo

Published: 15 October 2021

Abstract: The Australian blueberry industry is worth over $300 million, but there is limited information on factors influencing their chemical composition, particularly their ripeness and harvest stage. This pilot study investigated changes in total monomeric anthocyanin content (TMAC; measured using the pH-differential method) and total antioxidant capacity (TAC; measured with the cupric reducing antioxidant capacity assay) of four Australian highbush blueberry cultivars (Denise, Blue Rose, Brigitta and Bluecrop) at four time points and three maturity stages (unripe, moderately ripe and fully ripe). The TAC of most cultivars decreased by 8–18% during ripening, although that of the Blue Rose cultivar increased markedly. However, the TAC of ripe fruit from this cultivar also fluctuated markedly throughout the harvest season (between 1168–2171 mg Trolox equivalents 100 g^{-1}). The TMAC increased sharply between the medium-ripe and fully ripe maturity stages, with the Blue Rose cultivar showing the highest TMAC values (211 mg 100 g^{-1}, compared to 107–143 mg 100 g^{-1} for the remaining varieties). The TMAC of ripe fruit from this cultivar also rose steadily throughout the harvest season, in contrast to most other cultivars where the TMAC fell slightly over time. These results indicate that the levels of health-benefitting compounds in Australian-grown highbush blueberries may depend not only on the cultivar, but also upon the time of harvest.

Keywords: ripening; phytochemical composition; functional food; blueberry

1. Introduction

Highbush blueberries (*Vaccinium corymbosum* L.) are the second-most grown berry crop in Australia, second to strawberries. After being commercially established in Victoria in 1974, rapid growth in the past 15 years has seen a 10-fold expansion in the blueberry industry value to reach $300 million farmgate value in 2019 [1]. Most of the crop (75%) is consumed fresh by the domestic market, with 15% used in domestic processing [1].

Blueberries are a well-known functional food, with purported health benefits including antioxidative, anti-inflammatory, neuroprotective, anti-obesity, anti-diabetic and cardioprotective effects [2]. The majority of these health benefits are derived from their high levels of anthocyanins and polyphenols [3]. At least 25 different anthocyanins have been identified in highbush blueberries, with malvidin, delphinidin and peonidin being the predominant aglycones (anthocyanidins) present [4,5]. The phenolic acids present are similarly diverse, with hydroxycinnamic acid esters (in particular chlorogenic acid) found to be the most abundant polyphenols [6]. Both the anthocyanins and polyphenols present in blueberries contribute to the exceptional antioxidant capacity of these matrices.

Previous studies have investigated changes in anthocyanin and phenolic content throughout the ripening process in highbush blueberries [7–9], generally finding a marked increase in anthocyanin content during maturation, accompanied by decreasing total phenolic content and antioxidant capacity. Both genotype and environment influence the accumulation process and final content of anthocyanins and phenolic compounds in blueberries [4,9,10]. However, there is limited information available on the effect of growing conditions and other physiological factors on anthocyanin content [11].

Furthermore, there is little published literature available on the phytochemical composition of Australian-grown blueberries, with previous studies comparing the anthocyanin content of ripe fruit between different cultivars [5] or studying the effects of varying food preservation techniques on anthocyanin content [12]. Furthermore, there does not appear to be any previous work investigating the changes in anthocyanin and antioxidant capacity during the ripening process of Australian-grown blueberry cultivars. Consequently, the aim of this study was to undertake a one-year pilot study to investigate the changes in anthocyanin content and antioxidant capacity in Australian highbush blueberries throughout different stages of berry development.

2. Methods

2.1. Blueberry Sample Preparation

Four northern highbush blueberry cultivars were included in this study (Denise, Blue Rose, Brigitta and Bluecrop). Brigitta was originally developed in Australia and has now become popular worldwide due to its excellent storage and shipping characteristics. Blueberry samples were collected from a farm in Buninyong, western Victoria (Buninyong Blueberry Farm) during the 2015 summer harvest season. The sampling time points were at approximately weekly intervals for four weeks, on the 16, 23 and 30 January, and 12 February. At each sampling time point, ripe, medium-ripe and unripe blueberries were collected (where available for each variety), based on the appearance, colour and hardness of the fruit. Ripeness was qualitatively determined, with dark purple berries classified as ripe, reddish berries classified as medium-ripe, and green berries classified as unripe. For each sample, approximately 200 g of berries were collected across the rows for each cultivar, ensuring that all positions on the plants were sampled. The samples were stored at −20 °C prior to extraction.

2.2. Extraction of Anthocyanins and Phenolics

For each sample, approximately 20 g of frozen berries were subsampled and homogenised in a mortar and pestle. Extractions were performed in triplicate, using around 5 g of the homogenate in 15 mL of extraction solvent (95% methanol; 5% glacial acetic acid). The extracts were shaken at 250 rpm for 10 min (Ratek orbital shaker), followed by centrifugation (10,000 rpm; 10 min) and collection of the supernatant. The extraction was repeated twice more on the sample pellet, with the combined supernatant made up to 50 mL, vacuum filtered (Whatman No. 1) and stored at −20 °C.

2.3. Measurement of Total Anthocyanin Content and Antioxidant Capacity

Total monomeric anthocyanin content (TMAC) was measured on the triplicate extracts using the pH-differential method, as previously described [13]. Results were expressed as equivalents of cyanidin-3-glucoside. The total antioxidant capacity (TAC) of the extracts was determined on the triplicate extracts using the previously described CUPRAC protocol [13]. From the absorbance at 450 nm, TAC results were quantified as a function of the equivalent absorbance of Trolox standards (R^2 = 0.99).

3. Results and Discussion

3.1. Changes in Anthocyanin Content and Antioxidant Capacity during Maturation

The first aim of this study was to determine the changes in anthocyanin content and antioxidant capacity at different ripeness stages. In order to do this, ripe, medium-ripe and unripe blueberry samples for each cultivar were harvested on the same date (16 January 2015) and subsequently analysed.

The range of TAC and TMAC values found across all cultivars and maturity stages generally agreed with the range of results reported by Connor et al. [10] in 16 varieties of highbush blueberries grown in the United States. As shown in Figure 1a, the total antioxidant capacity of the blueberry samples generally decreased throughout the ripening process, as previously reported by several authors [7–9]. However, not every cultivar followed this trend, with the TAC of the Brigitta cultivar increasing between the medium-ripe and ripe stages (Figure 1a). The TAC of the Blue Rose cultivar showed the greatest deviation, increasing markedly between the unripe and ripe stages. However, as no medium-ripe fruit could be obtained for this variety, further investigation is required to confirm this trend.

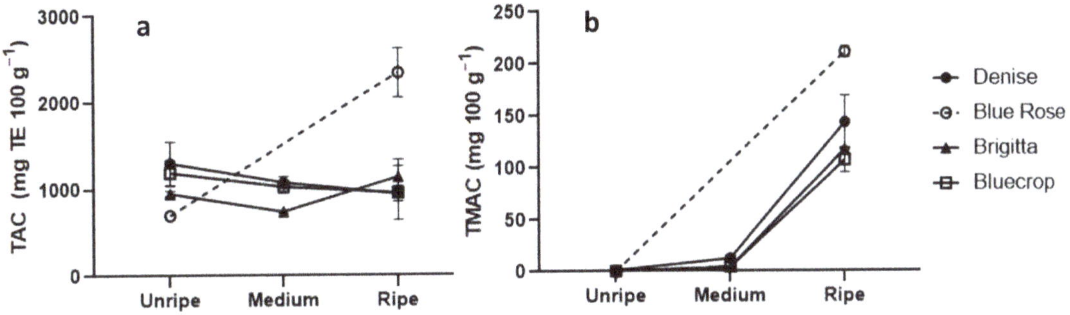

Figure 1. (**a**) There was no clear trend visible in the total antioxidant capacity (TAC) of the four blueberry varieties at different stages of ripeness (unripe, medium-ripe and ripe). All samples were collected on the same date (16 January 2015) to avoid potential effects of temporal variation. Note that no sample of "medium" ripeness was available for the Blue Rose cultivar. (**b**) Total monomeric anthocyanin content (TMAC) increased markedly in the four blueberry cultivars at different stages of ripeness. All samples were collected on the same date (16 January 2015). Note that no sample of "medium" ripeness was available for the Blue Rose cultivar.

The total anthocyanin content increased in a non-linear fashion throughout the maturation process, with a sharp increase between the medium and ripe stages (Figure 1b), as previously documented in other cultivars [7,9]. However, some previous researchers only recorded the development of anthocyanin content in already ripened fruit [8], rather than the changes from unripe to ripe fruits, as presented here. This development of anthocyanin content during the ripening process occurs as a temporally-dependent extension of the flavonoid synthesis pathway, primarily controlled by the transcription factor *MYB1* [14].

As observed with the TAC, the final TMAC found in ripe fruits from the Blue Rose cultivar was considerably higher (mean of 211 mg 100 g^{-1}) compared to the three remaining cultivars (means ranging between 107–143 mg 100 g^{-1}), highlighting the opportunity for further investigation of the phytochemical constituents and potential health benefits of this specific cultivar. Overall, the anthocyanin content of all cultivars fell within the average range reported by Stevenson and Scalzo [4] for 80 different blueberry genotypes.

3.2. *Anthocyanin Content and Antioxidant Capacity in Ripe Fruit at Different Timepoints during the Season*

The second aim of this study was to investigate if there is temporal variation in the anthocyanin content and antioxidant capacity of blueberry fruit at different times within the harvest season. In order to investigate this possibility, ripe fruit from each cultivar were collected at four different sampling timepoints (mid-Jan to mid-Feb) and analysed.

Neither the TAC or TMAC showed any clear inter-varietal trends throughout the harvest season (Figure 2); however, there were significant changes associated with specific varieties. Both Denise and Bluecrop showed a slight reduction in TAC throughout the season, while the TAC of Blue Rose fell sharply in late January before increasing again. The TAC of Brigitta increased around the end of January, before falling back to its original levels by mid-February.

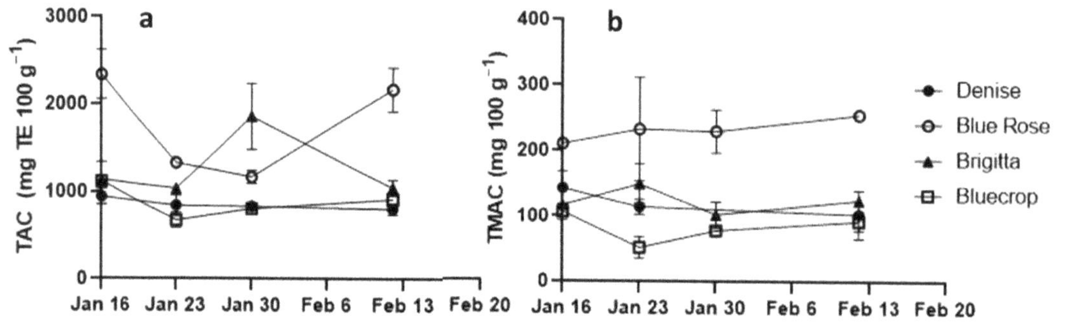

Figure 2. (**a**) Variation in the total antioxidant capacity (TAC) of ripe blueberries from four different cultivars throughout the harvest season. (**b**) Variation in the total monomeric anthocyanin content (TMAC) of ripe blueberries from four different cultivars throughout the harvest season.

The anthocyanin content of the Blue Rose cultivar increased steadily throughout the harvest season (Figure 2b), while that of Denise showed a slight fall. The anthocyanin content of Brigitta and Bluecrop fluctuated during the season, with little net trend in TMAC between mid-January and mid-February for these two cultivars. In nearly all cultivars, there was a small increase in anthocyanin content between the end of January and the middle of February. Viewed holistically, these results appear to show that the time of picking within the blueberry season may have a significant impact on the chemical composition of Australian-grown blueberries (in terms of both anthocyanin and antioxidant content); however, the specific impact of harvest time depends on the cultivar in question. Given that these compounds are largely responsible for the well-known health benefits of blueberries, this suggests that the potential health benefits associated with the consumption of these berries could also vary throughout the growing season.

4. Conclusions

In this pilot study, we profiled the changes in total monomeric anthocyanin content and total antioxidant capacity in four highbush blueberry cultivars during three maturation stages. While the TAC of most cultivars decreased with increasing ripeness, that of Blue Rose increased markedly. The TMAC increased sharply between the medium-ripe and fully ripe maturity stages in all cultivars. Throughout the harvest season, the TAC and TMAC of ripe fruit generally fluctuated over time, with the exact trends appearing to be cultivar-specific. This suggests that the levels of health-benefitting compounds in Australian-grown highbush blueberries will depend not only on the cultivar, but also upon the time of harvest. Although not explored in the present study, agronomic conditions are also likely to have a considerable impact on these compounds.

The spectrophotometric methods used for the measurement of TAC and TMAC in this study benefit from their speed and ease of use. This makes them suited to the rapid analysis of phytochemical contents in a large number of food samples, such as those included in this study. However, they are likely to be less specific compared to separation-based methods such as high-performance liquid chromatography (HPLC). Hence future work could focus on comparing the accuracy and precision of spectrophotometric and HPLC-based methods for the analysis of TMAC and specific antioxidant compounds (e.g., phenolic acids). Future studies could also investigate the temporal variation of TAC and TMAC over longer time periods than those included in the present study.

Author Contributions: Conceptualization, L.W. and M.N.; methodology, M.S. and M.N.; software, J.B.J.; validation, M.S.; formal analysis, J.B.J.; investigation, M.S., S.R. and S.A.; resources, M.N.; data curation, S.A. and J.B.J.; writing—original draft preparation, J.B.J.; writing—review and editing, J.B.J., M.S., J.S.M., S.R., S.A., L.W. and M.N.; visualization, J.B.J.; supervision, L.W. and M.N.; project administration, L.W. and M.N.; funding acquisition, L.W. and M.N. All authors have read and agreed to the published version of the manuscript.

Funding: This research received no specific grant from any funding agency in the public, commercial, or not-for-profit sectors.

Institutional Review Board Statement: Not applicable.

Informed Consent Statement: Not applicable.

Data Availability Statement: The data presented in this study are available on request from the corresponding author.

Acknowledgments: We thank Buninyong Blueberry Farm (https://www.buninyongblueberries.com.au/) for providing the blueberry samples used in this study. The assistance of the Biomedical Sciences team at Federation University is also gratefully acknowledged.

Conflicts of Interest: The authors declare no conflict of interest.

References

1. Australian Blueberries. Blueberry Facts. Available online: https://australianblueberries.com.au/is-good/berry-facts/ (accessed on 24 December 2020).
2. Kalt, W.; Cassidy, A.; Howard, L.R.; Krikorian, R.; Stull, A.J.; Tremblay, F.; Zamora-Ros, R. Recent Research on the Health Benefits of Blueberries and Their Anthocyanins. *Adv. Nutr.* **2019**, *11*, 224–236. [CrossRef] [PubMed]
3. Rutledge, G.A.; Sandhu, A.K.; Miller, M.G.; Edirisinghe, I.; Burton-Freeman, B.B.; Shukitt-Hale, B. Blueberry phenolics are associated with cognitive enhancement in supplemented healthy older adults. *Food Funct.* **2021**, *12*, 107–118. [CrossRef] [PubMed]
4. Stevenson, D.; Scalzo, J. Anthocyanin composition and content of blueberries from around the world. *J. Berry Res.* **2012**, *2*, 179–189. [CrossRef]
5. Lohachoompol, V.; Mulholland, M.; Srzednicki, G.; Craske, J. Determination of anthocyanins in various cultivars of highbush and rabbiteye blueberries. *Food Chem.* **2008**, *111*, 249–254. [CrossRef]
6. Grace, M.H.; Xiong, J.; Esposito, D.; Ehlenfeldt, M.; Lila, M.A. Simultaneous LC-MS quantification of anthocyanins and non-anthocyanin phenolics from blueberries with widely divergent profiles and biological activities. *Food Chem.* **2019**, *277*, 336–346. [CrossRef] [PubMed]
7. Kalt, W.; Lawand, C.; Ryan, D.A.J.; McDonald, J.E.; Donner, H.; Forney, C.F. Oxygen Radical Absorbing Capacity, Anthocyanin and Phenolic Content of Highbush Blueberries (*Vaccinium corymbosum* L.) during Ripening and Storage. *J. Am. Soc. Hortic. Sci.* **2003**, *128*, 917. [CrossRef]
8. Castrejón, A.D.R.; Eichholz, I.; Rohn, S.; Kroh, L.W.; Huyskens-Keil, S. Phenolic profile and antioxidant activity of highbush blueberry (*Vaccinium corymbosum* L.) during fruit maturation and ripening. *Food Chem.* **2008**, *109*, 564–572. [CrossRef]
9. Spinardi, A.; Cola, G.; Gardana, C.S.; Mignani, I. Variation of Anthocyanin Content and Profile Throughout Fruit Development and Ripening of Highbush Blueberry Cultivars Grown at Two Different Altitudes. *Front. Plant Sci.* **2019**, *10*, 1045. [CrossRef] [PubMed]
10. Connor, A.M.; Luby, J.J.; Tong, C.B.S.; Finn, C.E.; Hancock, J.F. Genotypic and Environmental Variation in Antioxidant Activity, Total Phenolic Content, and Anthocyanin Content among Blueberry Cultivars. *J. Am. Soc. Hortic. Sci.* **2002**, *127*, 89. [CrossRef]
11. Routray, W.; Orsat, V. Blueberries and Their Anthocyanins: Factors Affecting Biosynthesis and Properties. *Compr. Rev. Food Sci. Food Saf.* **2011**, *10*, 303–320. [CrossRef]
12. Lohachoompol, V.; Srzednicki, G.; Craske, J. The change of total anthocyanins in blueberries and their antioxidant effect after drying and freezing. *J. Biomed. Biotechnol.* **2004**, *2004*, 248. [CrossRef] [PubMed]
13. Johnson, J.; Collins, T.; Walsh, K.; Naiker, M. Solvent extractions and spectrophotometric protocols for measuring the total anthocyanin, phenols and antioxidant content in plums. *Chem. Pap.* **2020**, *74*, 4481–4492. [CrossRef]
14. Die, J.V.; Jones, R.W.; Ogden, E.L.; Ehlenfeldt, M.K.; Rowland, L.J. Characterization and Analysis of Anthocyanin-Related Genes in Wild-Type Blueberry and the Pink-Fruited Mutant Cultivar 'Pink Lemonade': New Insights into Anthocyanin Biosynthesis. *Agronomy* **2020**, *10*, 1296. [CrossRef]

engineering proceedings

MDPI

Abstract

Idea of Rapid Preparation of Fatty Acid Methyl Ester Using In Situ Derivatization from Fresh Horse Mussel †

Siriluck Pojjanapornpun [1,*], Apinya Cheewaphan [2], Akkaradech Nakornsadet [3], Salisa Chumsantea [3], Piraporn Sombutsuwan [3], Kanit Krisnangkura [3] and Kornkanok Aryusuk [3]

1 Lipid Technology Research Laboratory, Pilot Plant Development and Training Institute, King Mongkut's University of Technology Thonburi, Bangkok 10150, Thailand
2 Department of Basic Medical Science, Faculty of Medicine Vajira Hospital, Navamindradhiraj University, Bangkok 10300, Thailand; apinya@nmu.ac.th
3 Division of Biochemical Technology, School of Bioresources and Technology, King Mongkut's University of Technology Thonburi, Bangkhuntien, Bangkok 10150, Thailand; xtra_54@hotmail.com (A.N.); s.chumsantea@gmail.com (S.C.); pirapon_sw@yahoo.com (P.S.); kanit.kri@kmutt.ac.th (K.K.); kornkanok.ary@kmutt.ac.th (K.A.)
* Correspondence: siriluck.poj@mail.kmutt.ac.th
† Presented at the 2nd International Electronic Conference on Applied Sciences, 15–31 October 2021; Available online: https://asec2021.sciforum.net/.

Abstract: The analysis of the fatty acid (FA) profile requires multiple preparation steps, which are lipid extraction followed by derivatization of the FA into a fatty acid methyl ester (FAME). The procedures are time-consuming, and generally require large volumes of sample sizes and solvents. This report proposes a technique for the preparation of FAME from fresh horse mussels without a step of lipid extraction. A rapid in situ derivatization using *N*,*N*-dimethylformamide dimethyl acetal (DMF-DMA) methylation followed by alkali-transesterification was examined. In this method, acylglycerols and free fatty acids (medium to long-chain FA) of the sample are targeted to convert into FAME. Direct alkali-transesterification of the fresh sample gave only 58.7% FAME with 12.4% triglyceride and 21.1% FFA. The alkali in situ method showed low conversion efficiency due to the initial sample containing high contents of moisture and FFA (75.11% of the fresh sample and 14.3% of total oil, respectively). The reaction was developed by using two steps in situ derivatization. A 50 mg sample was methylated with 1 mL of DMF-DMA (100 °C, 15 min), followed by transesterified with 10 mL of 1% (*w*/*v*) NaOH in methanol (60 °C, 3 min). The conversion into FAME was monitored using size-exclusion HPLC with evaporative light-scattering detection. The column was a 100 Å Phenogel with toluene and 0.25% acetic acid as a mobile phase. The FAME yield of 79.9% with 7.8% triglyceride and 8.5% FFA was obtained. The two steps in situ derivatization gave a promising result with the higher conversion with lower FFA. It is a simple and rapid (less than 20 min) method that requires a low volume of sample and solvent for FAME preparation. However, increasing the conversion efficiency as well as the variety of samples should be further studied.

Keywords: fatty acids; fresh sample; in situ transesterification; methylation; *N*,*N*-dimethylformamide dimethyl acetal

Citation: Pojjanapornpun, S.; Cheewaphan, A.; Nakornsadet, A.; Chumsantea, S.; Sombutsuwan, P.; Krisnangkura, K.; Aryusuk, K. Idea of Rapid Preparation of Fatty Acid Methyl Ester Using In Situ Derivatization from Fresh Horse Mussel. *Eng. Proc.* **2021**, *11*, 7. https://doi.org/10.3390/ASEC2021-11142

Academic Editor: Samuel Adeloju

Published: 15 December 2021

Supplementary Materials: The following are available online at https://www.mdpi.com/article/10.3390/ASEC2021-11142/s1. S1: Idea of rapid preparation of fatty acid methyl ester using in situ derivatization from fresh horse mussel.

Funding: The authors acknowledge King Mongkut's University of Technology Thonburi for funding the KMUTT Research Center of Excellence Project to Lipid Technology Research Group.

Institutional Review Board Statement: Not applicable, as commercial samples were studied.

Informed Consent Statement: Not applicable.

Data Availability Statement: The data presented in this study are available on request from the corresponding author.

Conflicts of Interest: The authors declare no conflict of interest.

engineering proceedings

MDPI

Proceeding Paper

A Real-Time Snore Detector Using Neural Networks and Selected Sound Features †

Stelios A. Mitilineos, Nicolas-Alexander Tatlas, Georgia Korompili, Lampros Kokkalas and Stelios M. Potirakis *

Department of Electrical and Electronic Engineering, University of West Attica, 12243 Aigaleo, Greece; smitil@uniwa.gr (S.A.M.); ntatlas@uniwa.gr (N.-A.T.); gkorompili@uniwa.gr (G.K.); lkokkalas@uniwa.gr (L.K.)
* Correspondence: spoti@uniwa.gr

† Presented at 2nd International Electronic Conference on Applied Sciences, 15–31 October 2021; Available online: https://asec2021.sciforum.net/.

Abstract: Obstructive sleep apnea hypopnea syndrome (OSAHS) is a widespread chronic disease that mostly remains undetected, mainly due to the fact that it is diagnosed via polysomnography, which is a time and resource-intensive procedure. Screening the disease's symptoms at home could be used as an alternative approach in order to alert individuals that potentially suffer from OSAHS without compromising their everyday routine. Since snoring is usually linked to OSAHS, developing a snore detector is appealing as an enabling technology for screening OSAHS at home using ubiquitous equipment like commodity microphones (included in, e.g., smartphones). In this context, we developed a snore detection tool and herein present our approach and selection of specific sound features that discriminate snoring vs. environmental sounds, as well as the performance of the proposed tool. Furthermore, a real-time snore detector (RTSD) is built upon the snore detection tool and employed in whole-night sleep sound recordings, resulting in a large dataset of snoring sound excerpts that are made freely available to the public. The RTSD may be used either as a stand-alone tool that offers insight concerning an individual's sleep quality or as an independent component of OSAHS screening applications in future developments.

Keywords: obstructive sleep apnea hypopnea syndrome; apnea screening; snoring detection; machine learning; neural networks

Citation: Mitilineos, S.A.; Tatlas, N.-A.; Korompili, G.; Kokkalas, L.; Potirakis, S.M. A Real-Time Snore Detector Using Neural Networks and Selected Sound Features. *Eng. Proc.* **2021**, *11*, 8. https://doi.org/10.3390/ASEC2021-11176

Academic Editor: Roger Narayan

Published: 15 October 2021

1. Introduction

Obstructive sleep apnea-hypopnea syndrome (OSAHS) is a chronic condition held responsible for a number of well-documented effects on patients' health. It is linked to increased cardiovascular morbidity and mortality, including sudden heart death [1], while an estimated 4% and 2% of the male and female population respectively suffer from OSAHS. Interestingly enough, an estimated 85% of patients remain undiagnosed [2]. This underestimation poses an increased risk for individuals and society as a whole and is mainly due to polysomnography being the only method for OSAHS diagnosis currently trusted by doctors. Polysomnography is a time and resource-consuming procedure that monitors sleep with a multitude of specialized sensors and equipment and is performed in dedicated sleep laboratories or hospital care clinics. As such, most of the suffering population remains unscreened and, hence, undiagnosed.

The APNEA research project aims at accurately and cost-efficiently screening patients at home, using sound recordings via the users' smartphone during sleep [3]. In an ongoing measurement campaign, the APNEA project is collecting polysomnography data together with time-synchronized and high quality tracheal and ambient microphone recordings. The data are collected during sleep studies that are performed by project partners following the relevant medical protocols and are of a duration of about 8-h each. The acquired database consists of more than 200 complete polysomnography studies and our respective findings are reported in [4]. In parallel, and inspired by literature findings linking snoring to OSAHS episodes (e.g., see Refs. [5–7]), APNEA aims at developing a real-time snore detector (RTSD) in order to use it for the pre-screening of microphone recordings at home. The RTSD is intended to be either used as a stand-alone tool for apnea screening or integrated within more sophisticated apnea detection solutions by allowing to the latter to focus on timeslots of increased OSAHS probability.

As long as snore classifiers are concerned, we have focused on neural networks. They have been used in the literature for snoring detection with substantial classification accuracy, usually in the order of 90% or larger [8–10]. However, neural networks are usually trained using a relatively small dataset or a fragment of whole night sleep sound recordings. On the contrary, RTSDs, neural based or otherwise, are meant to be employed in much larger datasets (see whole night recordings of multiple patients), while larger datasets are typically related to reduced accuracy performance. In this respect, our contribution lies in (i) our approach and findings about which sound features are more promising and should be used for snoring classification, (ii) the training of a successful neural network for snoring detection with superior classification accuracy despite been trained using a much larger dataset compared to those used in the literature, (iii) the development of a RTSD tool, and (iv) the availability of a large body of annotated snoring sound excerpts (upon which the neural network training was implemented) together with an extremely large body of snoring sound excerpts that correspond to the output of the RTSD upon a large subset of whole-night sleep sound recordings. We present the architecture of the proposed classification tool in Section 2, while we report our findings regarding feature selection in Section 3.1. Numerical results on the performance of the proposed neural network and RTSD are demonstrated in Section 3.2. Section 4 concludes the paper and includes a discussion regarding future work for RTSD improvement.

2. Architecture of the Proposed Classification Tool and Real-Time Snore Detector

The architecture of the proposed classification tool is illustrated in Figure 1. Sound excerpts are used as input to the classification tool. Each sound excerpt is de-noised using wavelet filtering and then normalized with respect to its average energy. Selected features are calculated for each sound excerpt (sampled at 48 kHz, 24-bit), including temporal (time-domain), spectral (frequency-domain) and time-frequency features (for a complete discussion please refer to the subsequent Sections of this paper). Due to the high dimensionality of the available features, a Gaussian mixture model (GMM) is calculated for the time-frequency features. Afterwards, a neural-network classifier is employed in order to infer whether the input sound excerpt is a snore or not based on the calculated features and GMM models. Keeping in mind the big picture of a RTSD that will ultimately run in smartphones at home, we selected the implementation of a shallow neural network classifier with one hidden layer. After extensive trial-and-error, the number of nodes of the network hidden layer was selected to be equal to (rounded) 2.5 times the number of nodes of the network input layer. A detailed discussion on the features that we implemented and used for this work is provided in Sections 3 and 4, while details on the architecture of the neural network per se as well as the implementation of wavelet de-noising, energy normalization, GMM, and neural network training are provided in our previous work [11,12].

As far as the neural network classifier is concerned, a schematic of the internal architecture of the neural network that we used is illustrated in Figure 2. We consider only feed-forward artificial neural networks (FANNs) with the training function being an error back-propagation variant. The input layer of the network is used for data entry and weighting and consists of a number of nodes that is equal to the number of the independent feature inputs. The weights that multiply each data entry are subject to the network's training that is performed off-line and prior to classification. The weighted input features are then forwarded to an intermediate layer of neurons. The middle layer's number of nodes is tuned around the empirical rule-of-thumb value of two and a half times the number of input layer nodes. These neurons sum up all the weighted features and, essentially, configure all possible convex classes of data in the feature space. The output of the intermediate layer is then forwarded to two output neurons. These neurons at the output layer are essentially combining convex classes in order to configure non-convex classes to classify the input data. Each output is taking a value of "+1" or "−1" that corresponds to "true" or "false" state respectively. The usage of two output nodes instead of one with a true/false implementation is empirically proven to add robustness to the network and improve performance. As such, a decision for snoring is formed in the case where $(Z_1, Z_2) = (1, -1)$ while a decision for non-snoring is formed in the case where $(Z_1, Z_2) = (-1, 1)$.

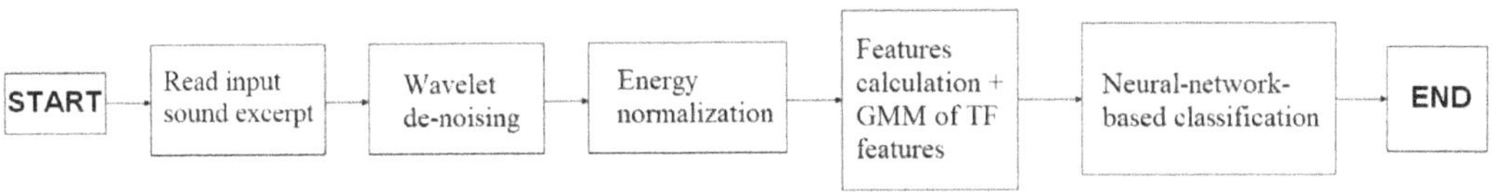

Figure 1. Architecture of the proposed classification tool and neural network.

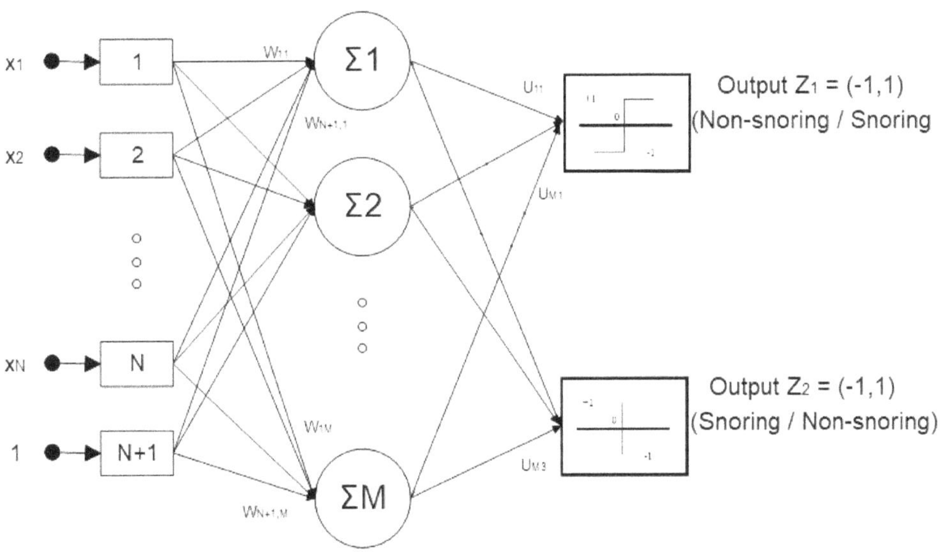

Figure 2. Neural network classifier for high−level snoring/non-snoring classification.

Furthermore, the architecture of the proposed RTSD is illustrated in Figure 3. The RTSD is designed to be used in real-time, but its operation is herein emulated using whole-night sleep recordings as its input. As such, the input sound recording is parsed with a sliding window of duration 6 s and a sliding step of 2 s (i.e., there is an overlap of around 66.7% between adjacent windows). The window duration of 6 s was selected because we have observed that a typical breathe-in-breathe-out cycle is about 4 s, so we opted for a guard interval of 1 s before and after. The sliding length is then equal to the sum of these guard intervals. The sound within each window is captured and the proposed classification tool of Figure 1 is employed in order to infer whether the specific time window corresponds to snoring or not. If this is the case, then we record the sound excerpt within the specific window to a separate .wav file for further processing or else we proceed to the next time window according to the predefined time-step. The procedure is repeated until the end of the whole-night sound recording or, in a real-life scenario, until the user aborts the application in her/his smartphone.

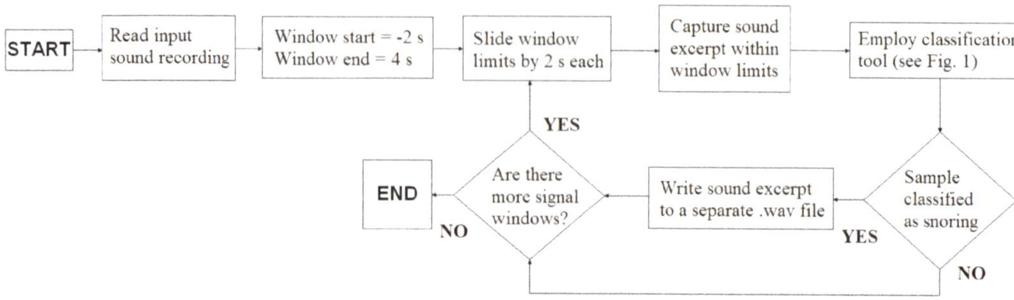

Figure 3. Architecture of the proposed Real-Time Snore Detector.

3. Numerical Results

3.1. Features Selection and Performance of the Proposed Neural Network Snore Detection Tool

In the literature, sound classification is performed using carefully selected features that are broadly categorized in time-domain (such as zero-crossing-rate (ZCR), energy, volume, etc.) and frequency-domain features (pitch, bandwidth, mel-frequency cepstral coefficients (MFCCs), etc.). However, such "static" features fail to capture the time evolution of the signal. Time-frequency (TF) features are therefore proposed and consist in crafting a sequence of static features calculated on a time window that is sliding over the entire sound signal. With such an approach, the temporal evolution of the signal is captured. However, the resulting feature space is usually huge and therefore needs to be reduced by the means of, e.g., a Gaussian mixture modeling in the case of shallow neural networks [11,12] (or repeated convolutional layers in the case of convolutional or deep neural networks).

Most of the aforementioned sound features are also used for snore detection in the literature [13]. On top of these, features that are used for snore detection include low-level descriptors and functional-based features that are reported in [14], positive/negative amplitude ratio, sampling entropy, and 500 Hz power ratio reported in [15], local dual octat pattern reported in [16], and many more. Nonetheless, there is not yet a clear consensus on what should be considered an appropriate feature selection when it comes to snore detection in whole night sleep studies [13]. In this respect, we performed a preliminary study about selecting a set of well-performing sound features.

The first features subset that we opted to compare consists of scalar features, including (i) the ZCR, pitch, bandwidth, volume, and intensity of the signal, (ii) a set of entropy metrics, specifically the Shannon, Tsallis, wavelet, and permutation entropy, and (iii) a few statistical metrics, namely the median, average, variance, skewness, and kurtosis of the signal amplitude. The second features' subset includes the MFCCs of the sound signal; more specifically, 13 MFCCs are calculated over the frequency range between 20 Hz and 6 kHz of the recorded signal. Implementation details for scalar features and MFCCs are provided in [11,12], while both are calculated over the entire signal portion that corresponds to the relative position of the sliding window described in Section 2.

Furthermore, inspired by studies reporting that snoring frequencies are mostly centered on specific and narrow ranges [17,18], we developed a modified spectrogram of the input signal to be used as a sound feature suitable for snore detection. More specifically, we calculate the spectrogram of each sliding window. Each sound excerpt is down-sampled to 12 kHz. Hence the resulting spectrogram ranges from 0 up to 6 kHz. Then, we calculate the average spectral coefficients in adjacent, non-overlapping windows of length 100 Hz each, resulting to the so-called modified spectral coefficients (MSC). Finally, we extract the normalized MSC values in order to capture the energy concentration within specific frequency ranges. As an example, Figure 4 compares the normalized MSC between a snoring and a non-snoring sound excerpt. In this case, snoring sound energy exhibits a peak at around 170 Hz that complies with the snoring frequencies reported in [17]. On the contrary, the non-snoring excerpt exhibits a smoother distribution of energy vs. frequency. Following multiple similar by-visual-inspection comparisons, we considered that the normalized MSC can be successful in discriminating snoring events and we herein report numerical results that justify this approach.

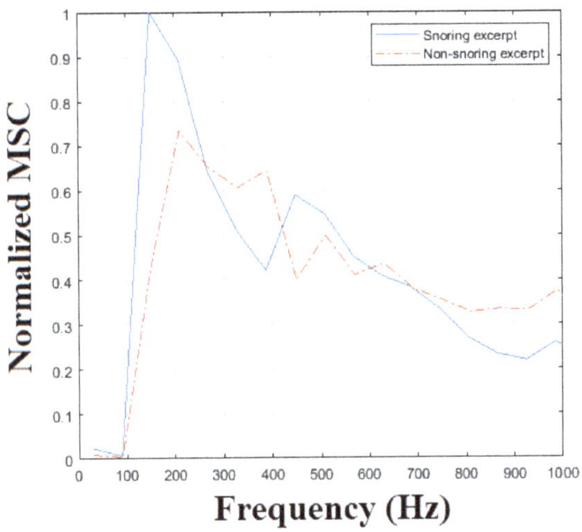

Figure 4. Modified spectral coefficients for a snoring and a non−snoring sound excerpt (solid and dash-dotted curve, respectively).

Then, in order to infer which combination of the implemented features provides the best accuracy with respect to snoring classification, we executed multiple training sessions of the proposed neural network using different features and feature combinations. More specifically, we selected fifty different whole-night sound recordings from fifty different patients. For each one of them, we manually selected and isolated 50 snoring sound excerpts and 50 non-snoring sound excerpts, of 6 s duration each. This results constitute a snoring and non-snoring database of 2500 + 2500 sound excerpts respectively (a total of 5000 excerpts), with a total duration of about 30,000 s (15,000 s of snoring and 15,000 s of non-snoring). This database of manually annotated sound excerpts is freely available upon request and the interested reader is referred to the Data Availability section below.

We then selected 70% of this dataset for network training and the remaining 30% for testing. The resulting classification accuracy of the test set vs. selected features combinations is tabulated in Tables 1 and 2. Classification accuracy is defined as the ratio of correct classification events (snoring excerpt classified as snoring plus non-snoring excerpt classified as non-snoring) vs. the total number of classification attempts (which is equal to the number of available excerpts, i.e., 5000 sound excerpts). According to Table 1, the normalized MSC exhibit better accuracy compared to MFCC or the set of scalar features. On the other hand, the normalized MSC exhibit similar accuracy when combined with either MFCC or the set of scalar features. Given that scalar features are computationally less intensive than MFCC to calculate, and that the proposed tool is envisioned to be used in portable devices, we selected the combination of normalized MSC plus scalar features to be used in our experiments.

Table 1. Test set classification accuracy per feature class.

Scalar Features	MFCC	Normalized MSC
93.4%	95.7%	97.7%

Table 2. Test set classification accuracy per feature classes' combination.

	Scalar Features	MFCC	Normalized MSC
Scalar features	-	96.0%	98.6%
MFCC	-	-	98.7%
Normalized MSC	-	-	-
All feature classes		97.3%	

Furthermore, the *Precision* and *Recall* of the selected combination of normalized MSC and scalar features are calculated. Precision corresponds to the proportion of positive identifications that was actually correct and is calculated as the ratio of true positives vs. the sum of true positives plus false positives. Recall corresponds to the proportion of actual positives that were identified as such and is calculated as the ratio of true positives vs. the sum of true positives plus false negatives. For the aforementioned features combination and test set, the resulting precision is equal to 99.59% while the recall is equal to 98.32%. Taking into account these performance metrics together with the reported overall accuracy of 98.6%, we consider that the proposed classification tool is eligible to be used as a building block of a RTSD.

3.2. Application of the Real-Time Snore Detector

Following the training and testing of the proposed classification tool, we employed it within the proposed RTSD scheme illustrated by Figure 1. Then, we used a set of twenty-five whole-night sound recordings, other than those used for the training and testing of the classification tool, in order to test the proposed RTSD. The total duration of the whole-night recordings that were tested is equal to 51 h, 45 min and 13 s. A total of 12,090 different sound excerpts of duration 6 s each are classified as snoring by the RTSD, corresponding to a total duration of 20 h and 9 min. These snoring sound excerpts are freely available to the interested reader upon request (please see the Data Availability section below).

In order to assess the performance of the proposed RTSD, a scoring campaign of experts listening to the extracted excerpts is required. For a complete evaluation, experts should also listen to the parts of the recordings that were classified as non-snoring, in order to evaluate with respect to false negatives. In order to provide a quick performance estimate, we opted to score 100 excerpts that were classified as snoring from each whole-night recording. Statistics of the resulting sound excerpts classified as snoring are tabulated in Table 3. Out of 2500 tested excerpts, a total of 2032 (or about 80%) were scored as snoring. However, this figure should not be translated to accuracy or TPR, since accuracy and TPR assume that the denominator (i.e., total number of true snoring excerpts and true non-snoring excerpts) of the ratios is known. Notwithstanding, 80% is an encouraging ratio in the case where we are solely interested in providing a database of actual snoring excerpts. Furthermore, it is also expected that during a whole-night recording there are large periods of non-snoring (actually, this should be the default state). As such, even a small percentage of FPR shall produce a large number of non-snoring excerpts that are classified as snoring.

Table 3. Performance results of the proposed real-time snore detector.

Number of Sound Excerpts Classified as Snoring	Subset of Sound Excerpts That Were Manually Scored by Experts	True Positive Ratio within Subset of Sound Excerpts (Number of True Positive Excerpts)
12,090 (total duration equal to 20 h, 9 min)	2500 (100 excerpts from each whole-night sound recording)	~80% (2032 excerpts out of 2500)

4. Conclusions and Future Work

We report herein a snoring classification tool with substantial performance (estimated accuracy equal to 98.6%), as well as the availability of a small dataset of annotated snoring and non-snoring excerpts together with a large dataset of non-annotated excerpts classified as snoring. In the immediate future, we intend to fully annotate the latter and offer a large, freely available database of annotated snoring excerpts. We also intend to use this full annotation in order to train a cascaded neural network that will have as input only the positive output of the classification tool proposed herein, as shown in Figure 5. The cascaded neural network will be trained with the aim of discriminating between true and false positives, thus providing a new classification output that is expected to be much more accurate than that of the first neural network alone.

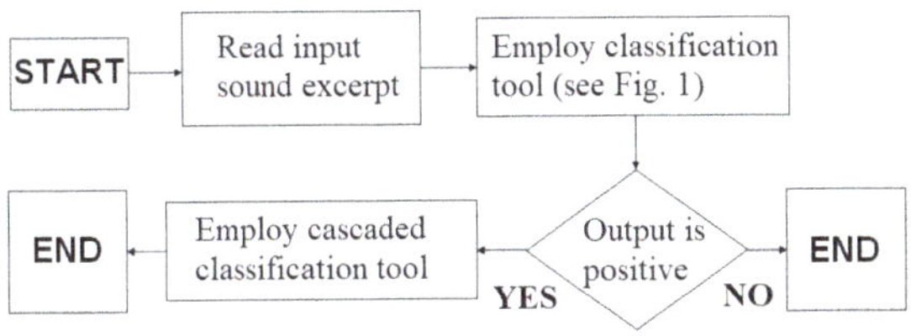

Figure 5. Architecture of a cascaded neural-network classification tool.

Supplementary Materials: The following are available online at https://www.mdpi.com/article/10.3390/ASEC2021-11176/s1. S1: A Real-Time Snore Detector Using Neural Networks and Selected Sound Features.

Author Contributions: S.A.M.: conceptualization, methodology, software, validation, writing—original draft preparation, N.-A.T., G.K., L.K. and S.M.P.: writing—review and editing, S.M.P.: conceptualization and supervision. All authors have read and agreed to the published version of the manuscript.

Funding: This research was co-financed by the European Union and Greek national funds through the Operational Program Competitiveness, Entrepreneurship and Innovation, under the call RESEARCH—CREATE—INNOVATE (project code: T1EDK-03957_Automatic Pre-Hospital, In-Home, Sleep Apnea Examination).

Institutional Review Board Statement: The study was conducted according to the guidelines of the Declaration of Helsinki, and approved by the Ethics Committee of Sismanoglio—Amalia Fleming General Hospital of Athens on 16 March 2017, with protocol number 05/16.03.2017.

Informed Consent Statement: Informed consent was obtained from all subjects involved in the study.

Data Availability Statement: Two sets of data samples are provided free of charge to the interested reader(s). These include (i) the database of manually annotated sound excerpts upon which the proposed neural network is trained, including snoring and non-snoring sounds (2500 + 2500 sound excerpts respectively, summing up to a total of 5000 excerpts and a total duration of about 30,000 s), and (ii) the database the RTSD output, including 12,090 snoring sound samples of duration 6 s each, corresponding to a total duration of 20 h and 9 min of snoring sounds. These data are provided upon request at spoti@uniwa.gr.

Acknowledgments: The authors acknowledge M. Kouvaras for his help during the manual annotation of the neural network training dataset.

Conflicts of Interest: The authors declare no conflict of interest. The funders had no role in the design of the study; in the collection, analyses, or interpretation of data; in the writing of the manuscript, or in the decision to publish the results.

References

1. Qaseem, A.; Holty, J.-E.C.; Owens, D.K.; Dallas, P.; Starkey, M.; Shekelle, P. Management of Obstructive Sleep Apnea in Adults: A Clinical Practice Guideline From the American College of Physicians. *Ann. Intern. Med.* **2013**, *159*, 471–483. [CrossRef] [PubMed]
2. Pack, A.I. Advances in Sleep-disordered Breathing. *Am. J. Respir. Crit. Care Med.* **2006**, *173*, 7–15. [CrossRef] [PubMed]
3. APNEA Project: "Automatic Pre-Hospital and In-Home Screening of Sleep Apnea", Operational Programme "Competitiveness, Entrepreneurship and Innovation. Available online: http://apnoia-project.gr/ (accessed on 1 April 2021).
4. Korompili, G.; Amfilochiou, A.; Kokkalas, L.; Mitilineos, S.A.; Tatlas, N.-A.; Kouvaras, M.; Kastanakis, E.; Maniou, C.; Potirakis, S.M. Poly-SleepRec: A scored polysomnography dataset with simultaneous audio recordings for sleep apnea studies. *Nat. Sci. Data* **2021**, *8*, 1–13.
5. Karunajeewa, A.S.; Abeyratne, U.; Hukins, C. Multi-feature snore sound analysis in obstructive sleep apnea–hypopnea syndrome. *Physiol. Meas.* **2010**, *32*, 83–97. [CrossRef] [PubMed]
6. Ben-Israel, N.; Tarasiuk, A.; Zigel, Y. Obstructive Apnea Hypopnea Index Estimation by Analysis of Nocturnal Snoring Signals in Adults. *Sleep* **2012**, *35*, 1299–1305. [CrossRef] [PubMed]
7. Azarbarzin, A.; Moussavi, Z. Snoring sounds variability as a signature of obstructive sleep apnea. *Med. Eng. Phys.* **2013**, *35*, 479–485. [CrossRef] [PubMed]
8. Arsenali, B.; van Dijk, J.; Ouweltjes, O.; Brinker, B.D.; Pevernagie, D.; Krijn, R.; Van Gilst, M.; Overeem, S. Recurrent Neural Network for Classification of Snoring and Non-Snoring Sound Events. In Proceedings of the 2018 40th Annual International Conference of the IEEE Engineering in Medicine and Biology Society (EMBC), Honolulu, HI, USA, 18–21 July 2018. [CrossRef]
9. Emoto, T.; Abeyratne, U.R.; Akutagawa, M.; Nagashino, H.; Kinouchi, Y.; Karunajeewa, S. Neural networks for snore sound modeling in sleep apnea. In Proceedings of the 2005 IEEE International Conference on Computational Intelligence for Measurement Systems and Applications, Messian, Italy, 20–22 July 2005. [CrossRef]
10. Kang, B.; Dang, X.; Wei, R. Snoring and apnea detection based on hybrid neural networks. In Proceedings of the 2017 International Conference on Orange Technologies (ICOT), Singapore, 8–10 December 2017. [CrossRef]
11. Mitilineos, S.A.; Tatlas, N.-A.; Potirakis, S.M.; Rangoussi, M. Neural Network Fusion and Selection Techniques for Noise-Efficient Sound Classification. *J. Audio Eng. Soc.* **2019**, *67*, 27–37. [CrossRef]
12. Mitilineos, S.A.; Potirakis, S.M.; Tatlas, N.-A.; Rangoussi, M. A Two-Level Sound Classification Platform for Environmental Monitoring. *J. Sens.* **2018**, *2018*, 1–13. [CrossRef]
13. Dafna, E.; Tarasiuk, A.; Zigel, Y. Automatic Detection of Whole Night Snoring Events Using Non-Contact Microphone. *PLoS ONE* **2013**, *8*, e84139. [CrossRef] [PubMed]
14. Zhang, Z.; Han, J.; Qian, K.; Janott, C.; Guo, Y.; Schuller, B. Snore-GANs: Improving Automatic Snore Sound Classification with Synthesized Data. *IEEE J. Biomed. Health Inform.* **2019**, *24*, 300–310. [CrossRef] [PubMed]
15. Wang, C.; Peng, J.; Song, L.; Zhang, X. Automatic snoring sounds detection from sleep sounds via multi-features analysis. *Australas. Phys. Eng. Sci. Med.* **2016**, *40*, 127–135. [CrossRef] [PubMed]
16. Tuncer, T.; Akbal, E.; Dogan, S. An automated snoring sound classification method based on local dual octal pattern and iterative hybrid feature selector. *Biomed. Signal Process. Control.* **2020**, *63*, 102173. [CrossRef] [PubMed]
17. Agrawal, S.; Stone, P.; McGuinness, K.; Morris, J.; Camilleri, A. Sound frequency analysis and the site of snoring in natural and induced sleep. *Clin. Otolaryngol.* **2002**, *27*, 162–166. [CrossRef] [PubMed]
18. Saunders, N.; Tassone, P.; Wood, G.; Norris, A.; Harries, M.; Kotecha, B. Is acoustic analysis of snoring an alternative to sleep nasendoscopy? *Clin. Otolaryngol.* **2004**, *29*, 242–246. [CrossRef] [PubMed]

engineering proceedings

MDPI

Proceeding Paper

A Case Study for Implementing a Plant Corrosion Inspection and Maintenance Anomaly and Integrity Management System on the Sabratha Gas Production Offshore Platform in the Mediterranean Sea †

Magdi H. Mussa [1,2,3,*], Mahmoud Dukali [4] and Yaqub Rahaq [5]

1 Mechanical and Energy Department, The Libyan Academy of Graduate Study, Tripoli P.O. Box 79031, Libya
2 Mechanical Engineering Department, Sok Alkhamis Imsehel High Tec. Institute, Tripoli P.O. Box 79031, Libya
3 The Institute of Marine Engineering, Science And Technology, London SW1H 9J, UK
4 Libyan Experts Platform (LEP), Ministry of Higher Education and Scientific Research, Tripoli P.O. Box 80045, Libya; mahmoud.dukali@gmail.com
5 Materials and Engineering Research Institute (MERI), Sheffield Hallam University, Sheffield S1 1WB, UK; yr1181@exchange.shu.ac.uk
* Correspondence: magdimosa1976@gmail.com; Tel.: +44-7404496955
† Presented at the 2nd International Electronic Conference on Applied Sciences, 15–31 October 2021; Available online: https://asec2021.sciforum.net/.

Abstract: Corrosion and maintenance anomaly and integrity management systems (AIMS) are now the foundation of many industrial and engineering systems regarding sustainability and long-lasting assets. The oil and gas industry has started developing new, integrated management systems to keep its assets safe from real external failures. However, the overlapping of assets' integrity responsibilities occur when a conflict of interests, such as production, safety, environmental, and financial interfacing, are inaccurately weighed against each other. This paper will review the case study of the anomaly and integrity management systems implemented on the Sabratha offshore platform. In order to achieve sustainable asset implementation, it is essential to identify the different weights given to the critical factors controlling the operational anomaly and integrity of facilities on offshore platforms and re-classify the potential failures. Therefore, design practices are reviewed. Moreover, inspection techniques and strategies are re-assessed and used to describe the consistent integrity assessment techniques linked to anomaly monitoring and maintenance criteria. Finally, the anomaly and integrity management system design use activity, process models, structures, and flow diagrams are reviewed. This work will be helpful for the further enhancement of a new machine learning system to support this approach.

Keywords: failure analysis; oil and gas offshore platform; risk-based inspection; corrosion; maintenance assets integrity management

Citation: Mussa, M.H.; Dukali, M.; Rahaq, Y. A Case Study for Implementing a Plant Corrosion Inspection and Maintenance Anomaly and Integrity Management System on the Sabratha Gas Production Offshore Platform in the Mediterranean Sea. *Eng. Proc.* **2021**, *11*, 9. https://doi.org/10.3390/ASEC2021-11153

Academic Editor: Dibyendu Sarkar

Published: 15 October 2021

1. Introduction

Offshore structures such as fixed offshore platforms, compliant towers, semi-submersible platforms, jack-up drilling rigs, drillships, floating and production systems, and the subsea system, as shown in Figure 1, are considered costly capital assets in the offshore oil and gas industry [1,2]. Therefore, they need exceptional management to mitigate and avoid disasters, shutdowns, corrosion and unnecessary emergency failures by using a reasonable inspection and anomaly-reducing control system [3].

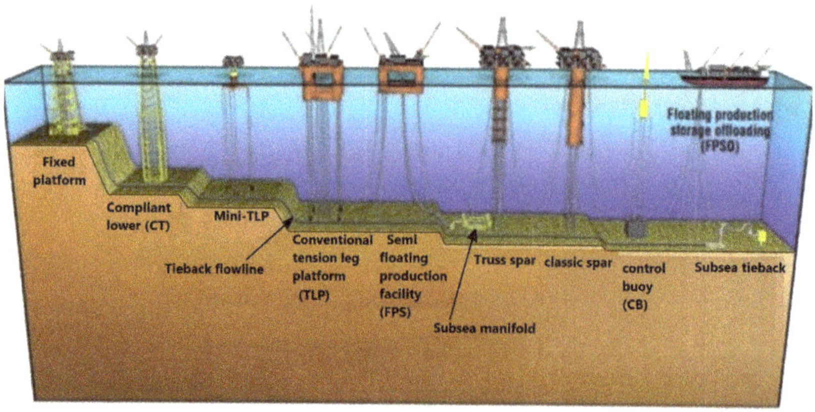

Figure 1. Shows offshore rigs and marine structure platforms [4].

As of now, there are many mitigation management systems designed on inspection-based time (TBI) or risk-based inspection (RBI). Therefore, risk identification and evaluation techniques (RIET) are critical for designing any assets of anomaly and integrity management systems. Many of these techniques, including hazard and operability study (HAZOP), hazard identification study (HAZID), fault tree analysis (FTA), human reliability analysis (HRA) and failure mode effect and criticality analysis (FMECA), can be used to design decent mitigation systems for offshore facilities [5,6].

In the past decades, world statistic failure reports from marine, oil and gas industries exhibited that the cause of significant accidents and explosions on offshore structure systems were the result of equipment and facilities failures, operation errors, artificial damage, natural disasters, and other unknown reasons, as shown in Table 1 [6].

Table 1. Percentage of causes of major accidents and explosions on marine and offshore rigs and platforms.

Equipment and Facilities	Operation Errors	Artificial Damage	Natural Disaster	Unknown Reasons
40%	20%	12%	10%	18%

Since the offshore platform contains complex integrated systems arranged by hundreds of pipes, pressure vessels, auxiliary equipment, and machines, it could also contain a process production system with specific series characteristics. Notably, the living quarters are located in the top of the platform, make it a riskier place than onshore production units [7]. The offshore platform production system has some characteristics that could increase onboard risks, including dealing with inflammable and explosive materials, high temperatures and high pressures, aggressive corrosion, and rough working conditions [8,9]. Half of these major accidents can be avoided if equipment and facilities are inspected, maintained and managed systematically and strategically [6]. Therefore, looking to reduce the impact of failure could cause catastrophic disasters. Scheduled maintenance is not enough to avoid unexpected circumstances. The strategical risk-based inspection of critical facilities, pressure vessels and the high-risk piping system could reduce the likelihood of accidents [10]. However, the overlapping in responsibility between the operation sections on the offshore platform could reduce the effectiveness of the mitigation action. That is why we shall clearly define the responsibility and action plan with an easy understanding of the process flowchart [11]. This case study will review the anomaly and integrity assets management system procedure, implemented on the Sabratha platform. It will cover the period from 2010 to 2014 of implementing the procedure in the offshore field; Sabratha

Platform is located 110 km from the Libyan coast in the Bahr Essalam offshore field. The platform, considered one of the biggest Libyan Offshore condensate and gas production units, is fixed to the seabed in a water depth of 190 m, as shown in Figure 2. The platform consists of all of the facilities required for preliminary separation and treatment of the gas produced from the Bahr Essalam field, as well as a fast-moving workover rig (FMWR), a helideck and living quarters for 120 persons [12].

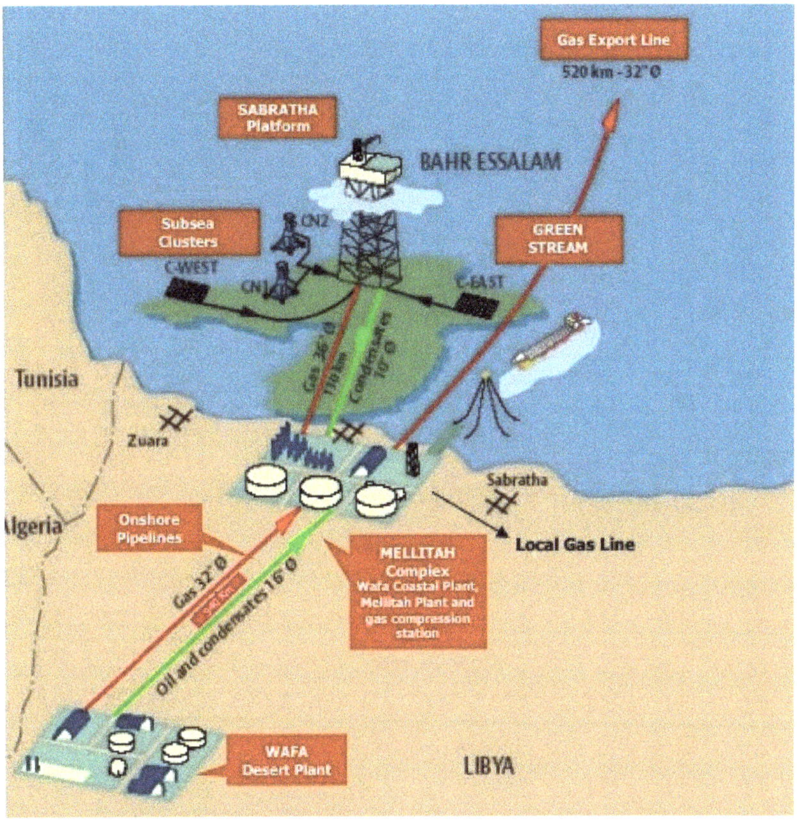

Figure 2. Location of the Sabratha platform in front of the Libyan coast in the Mediterranean Sea [12].

2. Methodology and Procedure

2.1. AIMS System Overview

Figure 3 shows the flowchart procedure that was implemented onboard [5,13]. The corrosion inspection and maintenance anomaly and integrity management system (AIMS) objective is to identify the risk of an anomaly in the early stages, so that mitigation plans can be developed and appropriately implemented, with clear responsibility and actions. The purpose of this management system is:

- to brief the requirements to record anomalies occurring in the pressure system.
- to inspect the critical facilities following up the mitigation (including maintenance and replacement),
- to resolve any anomaly that might threaten the integrity of the offshore facility.

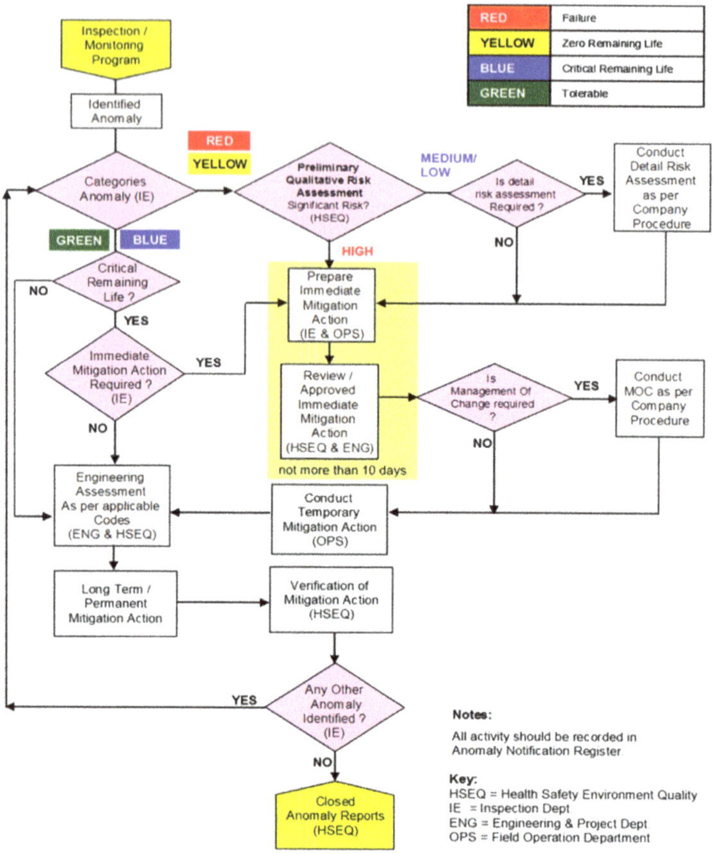

Figure 3. System flowchart procedure, shows the action and decision making based on the corrosion inspection and maintenance anomaly and integrity management system.

2.2. *Responsibilities of Departments*

Table 2 summarizes the responsibilities in the procedure according to the flow of the required work, based on the risk identification of the platform equipment, including inspection, planning, maintenance, and continued improvements.

Table 2. Shows the responsibilities of departments in the implemented system.

HSEQ Department	• Ensure compliance with this procedure. • Report to chairman regarding the anomaly status. • Notify all related parties to take immediate action if there is a critical anomaly that may affect the integrity of the plant and lead to unsafe operating conditions. • Acknowledge and bring to the attention of the individual department the anomaly status of their operation. • Perform detailed risk assessments with the engineering department. • Perform integrity analysis for existing anomalies. • Register the received transmittal. • Monitor the status of anomalies. • Review and approve short-term/temporary mitigation action recommendations. • Close long-term/permanent mitigation action recommendations. • Review and approve company procedures as required.

Table 2. *Cont.*

Engineering Department	• Perform detailed risk assessment with the HSEQ department. • Perform support in engineering analysis for existing anomalies. • Issue recommendations for permanent mitigation actions. • Issue recommendations for any plant modification requests as a result of the engineering assessment. • Provide necessary documents, calculations, etc., for detailed execution of long-term mitigation actions. • Review and approve company procedures as required.
Inspection Section	• Defect reporting. • Perform preliminary qualitative risk assessment. • Raise an anomaly notification. • Distribute the transmittal to the relevant parties. • Verify that any mitigation actions in the maintenance management software MMS (MAXIMO) have been appropriately taken. • Verify any field procedures available as required, according to the approved procedures. • Discuss the short-term mitigation action with the field operation department.
Operation, maintenances, and production Department	• Discuss the short-term mitigation action with the Inspection Engineer. • Follow the short-term mitigation action as agreed. • Issue a work order for short-term/temporary mitigation action recommendations. • Issue a work order for mitigation action recommendations. • Enter mitigation action to the maintenance management software MMS (Maximo) for follow up and monitoring. • Issue a plant modification request as necessary or follow the engineering department recommendation.

2.3. Evaluation of Using the Database Results

Using the system from the period 2010 to 2014, showed that the platform contained over 163 pressure vessels of equipment from the process and auxiliary facilities. These pressure vessels have been identified as one of two categories according to the risk of safety-critical equipment (SCE) and non-safety critical equipment (non-SCE). Figure 4 shows the pie chart of the inspected pressure vessels within the four years.

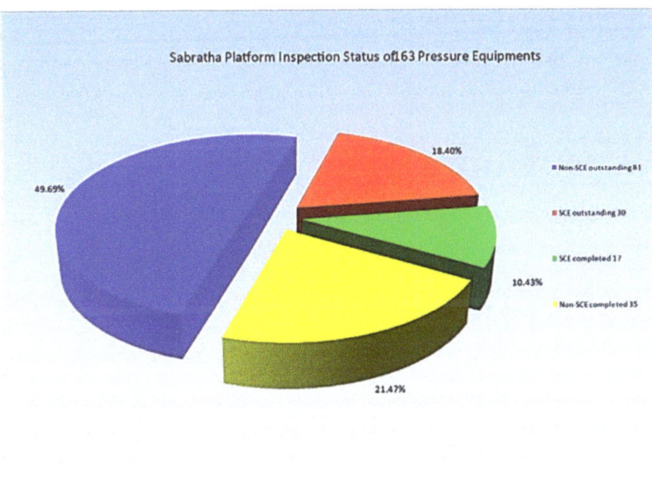

Figure 4. Sabratha Platform inspection status of pressure vessels and equipment.

As the design requires, the pressure vessels and processing pipes system shall be fitted with safety pressure valves (PSV). These PSVs on the platform should be inspected and recalibrated a minimum of once every two years; the total number of PSVs onboard is 198 PSV. Figure 5 shows the total inspected and calibrated PSV valves per the reviewing period.

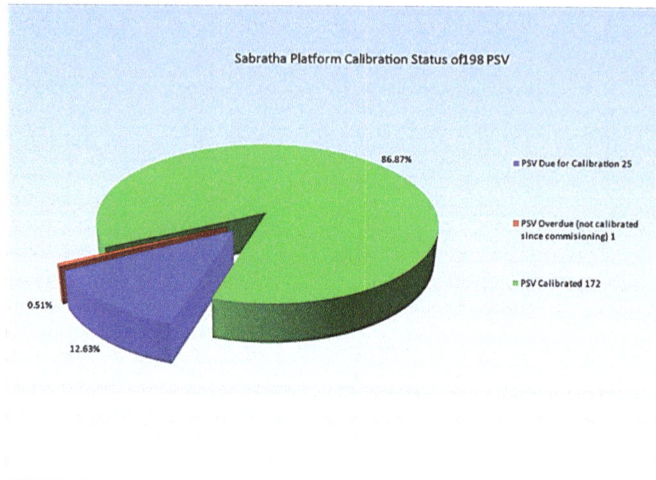

Figure 5. Sabratha Platform inspection status of pressure safety valves.

From the implemented system, this should obtain the inspected and calibrated data for the safety-critical equipment and other equipment, and show the trend of required maintenance action that should be taken, according to the periodic inspection and maintenance plan, as shown in Figure 6.

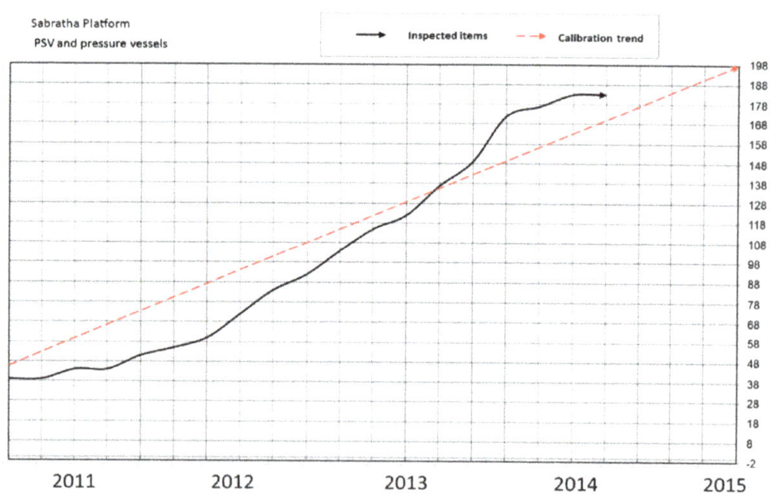

Figure 6. The real statues of inspected PSV and pressure equipment over four years.

3. Conclusions

Offshore corrosion inspection and maintenance anomaly and integrity management systems (AIMS) have become the mainstream system for safe gas operation management of the equipment and facilities on the Sabratha offshore platform. With the implementation of this management on the offshore platform, it is clear that inspection priorities and mitigation actions are well maintained. To avoid any shutdown or accident, the AIMS system has changed from qualitative to quantitative when dealing with an inspection. This system could be applied in other fields, including onshore and desert oil and gas fields.

Funding: This research received no external funding.

Institutional Review Board Statement: Not applicable.

Informed Consent Statement: Not applicable.

Data Availability Statement: The data are not publicly available; The data files are stored on corresponding instruments and personal computers.

Acknowledgments: The authors would like to acknowledge the facilitating support by Mellitah Oil and gas BV for data resources.

Conflicts of Interest: The authors declare no conflict of interest.

References

1. Tan, Y.; Li, H.X.; Cheng, J.C.P.; Wang, J.; Jiang, B.; Song, Y.; Wang, X. Cost and environmental impact estimation methodology and potential impact factors in offshore oil and gas platform decommissioning: A review. *Environ. Impact Assess. Rev.* **2021**, *87*, 106536. [CrossRef]
2. Kaiser, M.J. Offshore oil and gas records circa 2020. *Ships Offshore Struct.* **2020**, 1–37. [CrossRef]
3. Guédé, F. Risk-based structural integrity management for offshore jacket platforms. *Mar. Struct.* **2019**, *63*, 444–461. [CrossRef]
4. NOIA. The Basics of Offshore Oil & Gas—NOIA. Available online: https://www.noia.org/basics-offshore-oil-gas/ (accessed on 20 July 2021).
5. American Petroleum Institute. *API- RECOMMENDED PRACTICE 581: Risk-Based Inspection Technology*; American Petroleum Institute: Washington, DC, USA, 2016; Volume 7056.
6. Tang, Y.; Yao, J.; Wang, G.; Zhang, Z.; He, Y.; Jing, J. Risk Identification and Quantitative Evaluation Method for Asset Integrity Management of Offshore Platform Equipment and Facilities. *Math. Probl. Eng.* **2019**, *19*, 1–14. [CrossRef]
7. Olsen, E.; Næss, S.; Høyland, S. Exploring relationships between organizational factors and hydrocarbon leaks on offshore platform. *Saf. Sci.* **2015**, *80*, 301–309. [CrossRef]
8. Mussa, M. Development of Hybrid Sol-Gel Coatings on AA2024-T3 with Environmentally Benign Corrosion Inhibitors. Ph.D. Thesis, Sheffield Hallam University, Sheffield, UK, 2020.
9. Mussa, M.H.; Rahaq, Y.; Takita, S.; Farmilo, N. Study the Enhancement on Corrosion Protection by Adding PFDTES to Hybrid Sol-Gel on AA2024-T3 Alloy in 3.5% NaCl Solutions. *Albahit J. Appl. Sci.* **2021**, *2*, 61–68.
10. El-Reedy, M.A. Practical risk management for offshore projects. In *Offshore Projects and Engineering Management*; Elsevier: Amsterdam, The Netherlands, 2021; pp. 215–238. ISBN 9780323857956.
11. Hogenboom, S.; Vinnem, J.E.; Utne, I.B.; Kongsvik, T. Risk-based decision-making support model for offshore dynamic positioning operations. *Saf. Sci.* **2021**, *140*, 105280. [CrossRef]
12. Mellitah Oil & Gas B.V. Sabratha Platform—Mellitah Oil & Gas. Available online: https://mellitahog.ly/en/sites/sabratha-platform/ (accessed on 2 July 2021).
13. Mellitah Oil & Gas B.V. *Management Corporate Procedure Anomaly Management System MG-HSEQ-P-035*; Mellitah Oil & Gas B.V.: Tripoli, Libya, 2010; Volume 5, pp. 1–17.

engineering proceedings

MDPI

Proceeding Paper

Studies on the Preparation of Nanoparticles from Betulin-Based Polyanhydrides †

Daria Niewolik [1,*], Grzegorz Dzido [2] and Katarzyna Jaszcz [1]

1 Department of Physical Chemistry and Technology of Polymers, Silesian University of Technology, M. Strzody 9, 44-100 Gliwice, Poland; Katarzyna.Jaszcz@polsl.pl
2 Department of Chemical Engineering and Process Design, Silesian University of Technology, M. Strzody 7, 44-100 Gliwice, Poland; Grzegorz.Dzido@polsl.pl
* Correspondence: daria.niewolik@polsl.pl
† Presented at the 2nd International Electronic Conference on Applied Sciences, 15–31 October 2021; Available online: https://asec2021.sciforum.net/.

Abstract: Nanoparticles were obtained by nanoprecipitation and by emulsion solvent evaporation (ESE) method. In the ESE method, the size of the particles depended on the type and concentration of surfactant (in the water phase) and the polymer concentration (in the organic phase). The best results were obtained with ionic surfactants, however, the use of such compounds may accelerate the degradation process of polymers. In the nanoprecipitation method, the ratio of solvent (methylene chloride) to non-solvent (hexane) has a significant influence on the particle size. The smallest particles were obtained with a solvent to non-solvent ratio of 1:150.

Keywords: betulin; polyanhydrides; biodegradable polymers; nanoparticles

Citation: Niewolik, D.; Dzido, G.; Jaszcz, K. Studies on the Preparation of Nanoparticles from Betulin-Based Polyanhydrides. *Eng. Proc.* **2021**, *11*, 10. https://doi.org/10.3390/ASEC2021-11160

Academic Editor: Samuel Adeloju

Published: 15 October 2021

1. Introduction

Polymeric nanoparticles are considered as useful carriers in controlled drug delivery systems. Due to their small sizes and high surface area, they have many advantages. Polymer-based nanocarriers may improve solubility and bioavailability of highly insoluble hydrophobic drugs, increase the stability of volatile pharmaceutical agents, or deliver a higher concentration of drugs to a desired location [1,2]. They can be used for treatment of various diseases, including bacterial and fungal infections, hypertension, asthma, or cancer [3]. Polymeric nanoparticles can be prepared from both synthetic polymers, e.g., polycaprolactone, poly(methyl methacrylate) and poly(lactide-co-glycolide), or natural polymers, such as gelatin, chitosan, sodium alginate, or albumin [2,3]. Polymers used in nanoparticle formulation should be biocompatible, non-toxic, and non-antigenic [4].

Polyanhydrides are FDA-approved, biodegradable polymers that possess favorable properties as the material for drug carriers, such as biocompatibility and lack of toxicity. They easily undergo hydrolytic degradation to their respective diacids, which are completely eliminated from the body within a short period of time. Polyanhydride nanoparticles can be successfully used as drug carriers [5–7].

The aim of this work was to obtain nanoparticles from polyanhydride based on betulin disuccinate (polyDBB) and to evaluate the influence of the preparation conditions, including: Homogenization time, type of surfactant and concentration of the polymer solution, on the shape, morphology and size of particles. Betulin and its derivatives (e.g., betulin disuccinate) have a broad spectrum of biological relevance, including anticancer activity, thus these compounds are promising as new, potentially therapeutic agents [8,9]. The major problem, which limits their potential pharmaceutical uses, is the poor aqueous solubility of lupane triterpenes faced when trying to formulate pharmaceutical compounds from betulin [1,9]. However, this problem can be solved by obtaining polymeric form of betulin and forming it into nanospheres, thus nanoparticles prepared from betulin-based polyanhydrides may have significant applications in drug delivery systems.

2. Materials and Methods

2.1. Materials

Betulin disuccinate (obtained in the laboratory according to the procedure described earlier [9]), acetic anhydride (POCh S.A., Gliwice, Poland), poly(vinyl alcohol) (M_w = 88,000 g/mol, 88% hydrolyzed) (ACROS Organics, Geel, Belgium), sodium dodecyl sulfate (SDS, Fisher Scientific, Hampton, NH, USA), cetyltrimethylammonium bromide (CTAB, ACROS Organics, Geel, Belgium), methylene chloride (Chempur, Piekary Śl., Poland), hexane (Chempur, Piekary Śl., Poland), diethyl ether (Chempur, Piekary Śl., Poland) were used as supplied.

2.2. Polyanhydride Synthesis

Polyanhydride was obtained by two-step melt polycondensation of betulin disuccinate (DBB) according to the procedure described earlier [9]. DBB (5 g) was refluxed in acetic anhydride (1:10, *w*/*v*) under nitrogen flow for 40 min to get prepolymer. Excess of acetic anhydride and acetic acid formed in reaction were removed under vacuum. In the next step, prepolymer was heated at 150 °C for 2 h with constant stirring under vacuum (0.1 mm Hg) and nitrogen. The obtained polyanhydride (polyDBB) was stored in a freezer prior to being used for nanoparticle formulation.

2.3. Formulation of Nanoparticles Using ESE Method

Nanoparticles were prepared by solvent evaporation from O/W polymer emulsion. The polymer solution in methylene chloride (20 mL, concentration 10, 30, or 50 mg/mL) was emulsified in 400 mL of aqueous solution (1% *w*/*w*) of surfactant (PVAl, SDS, or CTAB), using ULTRA-TURRAX T18 homogenizer, for 60 s. The speed of homogenizer was 18,000 rpm. Afterwards, the emulsion was subjected to ultrasonication under ice cooling for 180 s in a pulse regime (50% pulses, 50% power) using an Omni Sonic Ruptor 250 W and 3/4″tip. The emulsion was then stirred with a magnetic stirrer at 1100 rpm at room temperature for 3 h to evaporate the organic solvent. After that, nanoparticles were collected by centrifugation at 5000 rpm for 5 min, washed 3 times with distilled water, lyophilized, and stored in a freezer.

2.4. Formulation of Nanoparticles by Nanoprecipitation

The polymer solution in methylene chloride (concentration 50 mg/mL) was added dropwise via a syringe into non-solvent bath of hexane at a solvent to non-solvent ratio 1:50, 1:100, or 1:150, which was continually stirred with a magnetic stirrer at 1100 rpm at room temperature. Upon contact with the non-solvent, the polymer spontaneously precipitated from the solution, forming polymeric nanospheres. The suspension was stirred for 2 h. After that, hexane was decanted and nanoparticles were filtered, dried under vacuum, and stored in a freezer.

2.5. Characterization of the Particles

The average size and size distribution of polymer particles were determined by dynamic light scattering (DLS) using a Zetasizer Nano S90. Before measurements, particles were dispersed in distilled water.

The morphological characterization of particles was carried out using a Phenom ProX scanning electron microscope (SEM) using an accelerating voltage of 10 kV. Samples were coated with a 10 nm gold layer under vacuum using sputter coater Quorum Q150R ES. The particle size was measured by imaging software (ImageJ).

The content of anhydride groups in polyanhydride after the nanoparticle fabrication process was monitored by ^{1}H NMR. ^{1}H NMR spectra of polyDBB nanoparticles in $CDCl_3$ were recorded on a Varian 600 MHz spectrometer with TMS as an internal standard.

3. Results and Discussion

Polyanhydride (polyDBB) was obtained by melt polycondensation of betulin disuccinate. The use of DBB was intended to obtain polyanhydride with potential antitumor activity. The chemical structure of polyDBB (Figure 1) was confirmed by IR, ^{1}H NMR, and ^{13}C NMR spectroscopy. The molecular weight of polyDBB was calculated from ^{1}H NMR and determined by GPC. The molecular weight was approx. 8500 g/mol. The detailed characteristics of polyDBB were described previously [9].

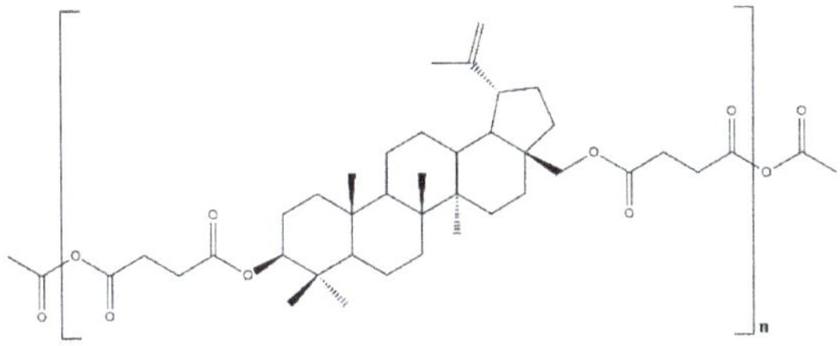

Figure 1. Chemical structure of polyDBB.

Nanoparticles were prepared using two methods: Solvent evaporation (ESE) method and nanoprecipitation.

3.1. Nanoparticles Obtained by ESE

Particles obtained by the solvent evaporation method were in the range of 250–2400 nm for different formulation conditions. The nanoparticles were spherical in shape and had a smooth surface. The morphology of the polyDBB particles is shown in Figure 2.

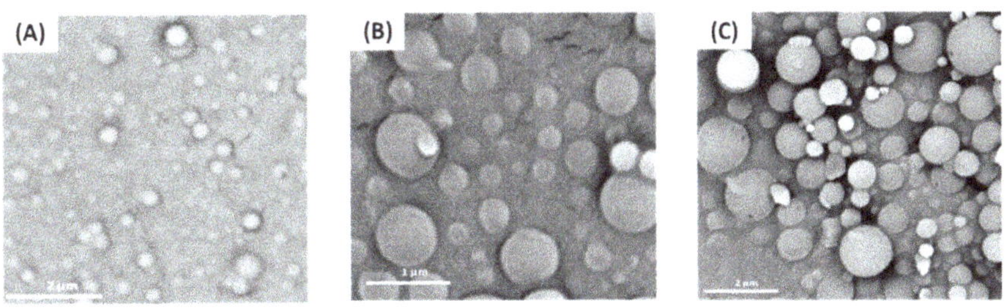

Figure 2. SEM images of polyDBB nanoparticles obtained by the ESE method with the use of: (**A**) SDS, (**B**) CTAB, and (**C**) PVAl as surfactant.

Different process parameters, including homogenization time, type of surfactant, and concentration of the polymer solution have a significant influence on the particle size. Changing the homogenization time from 30 s to 180 s reduced the diameter of particles. The obtained results showed that extending the homogenization time above 60 s slightly reduces the particle size, therefore in further experiments, the homogenization time was 60 s.

3.1.1. Effect of the Polymer Concentration and Surfactant Type

During the preparation of nanoparticles such parameters as concentration of polymer solution and type of surfactant were changed. These parameters play an important role in the final size of particles, but they did not affect the surface morphology of nanoparticles. The particle sizes were determined by DLS and also calculated from the SEM images. Due to the tendency of nanoparticles to agglomeration, the sizes determined by DLS were larger than sizes calculated from SEM (except for particles obtained using PVAl as surfactant). The diameters of particles obtained by ESE in different conditions are listed in Table 1.

Table 1. The sizes of polyDBB nanoparticles with different polymer concentration and type of surfactant.

Surfactant	Polymer Concentration in Organic Phase								
	50 mg/mL			30 mg/mL			10 mg/mL		
	Particle Size [nm]		PDI [1]	Particle Size [nm]		PDI [1]	Particle Size [nm]		PDI [1]
	SEM	DLS		SEM	DLS		SEM	DLS	
1% SDS	820.0	802.9	0.212	770.0	1008.0	0.296	560.0	732.8	0.294
1% CTAB	504.2	859.8	0.434	330.0	1000.0	0.445	250.0	1268.0	0.386
1% PVAl	2420.0	2436.0	0.261	1870.0	1548.0	0.427	1420.0	896.8	0.356

[1] PDI = polydispersity index determined by DLS. Homogenization = 18,000 rpm.

The influence of concentration of polymer solution and surfactant type on the characteristics of nanoparticles was determined using homogenization speed at 18,000 rpm. Nanoparticles with the smallest diameters were obtained with the use of a polymer solution of concentration 10 mg/mL. Increasing the polymer concentration increases the diameters of particles. Additionally, when the smaller particles were formed, the tendency to agglomeration was higher. A comparison of the data in Table 1 shows that the use of ionic surfactants (anionic SDS and cationic CTAB) leads to smaller particles compared to the non-ionic surfactant (PVAl). However, the use of such compounds may accelerate the degradation process of polymers.

3.1.2. Effect of the Use of Ultrasound

Another parameter influencing the size of the obtained nanoparticles is the use of ultrasound. The diameters of particles obtained with the use of ultrasonication are listed in Table 2.

Table 2. The sizes of polyDBB nanoparticles after ultrasonication.

Surfactant	Ultrasonication	Particle Size [nm]		PDI [1]
		SEM	DLS	
1% SDS	—	820.0	802.9	0.212
	before stirring	330.0	288.3	0.225
	after 1.5 h of stirring	930.0	1218.0	0.325
1% CTAB	—	504.2	859.8	0.434
	before stirring	440.0	480.9	0.278
	after 1.5 h of stirring	549.5	1064.0	0.237
1% PVAl	—	2420.0	2436.0	0.261
	before stirring	918.4	1552.0	0.305

[1] PDI = polydispersity index determined by DLS. Polymer concentration in organic phase = 50 mg/mL, ultrasonication time = 3 min.

Ultrasonication reduces the size of the obtained nanoparticles. However, it is important at which point of the formulation of nanoparticles the ultrasound will be used. Nanoparticles with the smallest diameters were obtained when ultrasonication was used directly after the homogenization process. Ultrasonication used at a later stage results in the formation of particles with a larger size compared to the particles obtained without ultrasound.

3.2. Nanoparticles Obtained by Nanoprecipitation

Nanoprecipitation was used for the preparation of polyDBB nanoparticles, varying the solvent to non-solvent ratio. Obtained particles were in the range of 370–950 nm. Nanoparticles with the smallest diameters were obtained with the solvent to non-solvent ratio of 1:150, thus it can be concluded that increasing the hexane to methylene chloride ratio reduces the particle diameter. The diameters of particles obtained by nanoprecipitation are listed in Table 3.

Table 3. The sizes of polyDBB nanoparticles obtained by nanoprecipitation.

Sample	Solvent to Non-Solvent Ratio	Homogenization	Ultrasonication	Particle Size [nm]		PDI [1]
				SEM	DLS	
1	1:50	—	—	—	953.2	0.354
2	1:100	—	—	—	867.4	0.416
3	1:150	—	—	440.0	682.0	0.240
4	1:100	—	3 min.	370.0	703.9	0.262
5	1:100	18,000 rpm/1 min.	—	—	689.5	0.191

[1] PDI = polydispersity index determined by DLS.

Taking into account the hydrolytic instability of the polyanhydrides, nanoprecipitation is the more suitable method because it does not require the use of water. However, obtained nanoparticles indicated a very high tendency to agglomeration, which can be observed at the SEM image.

In this method, it is also difficult to control the shape of the particles. The morphology of the nanoparticles is shown in Figure 3. Among the spherical particles, many irregular ones were also observed (Figure 3).

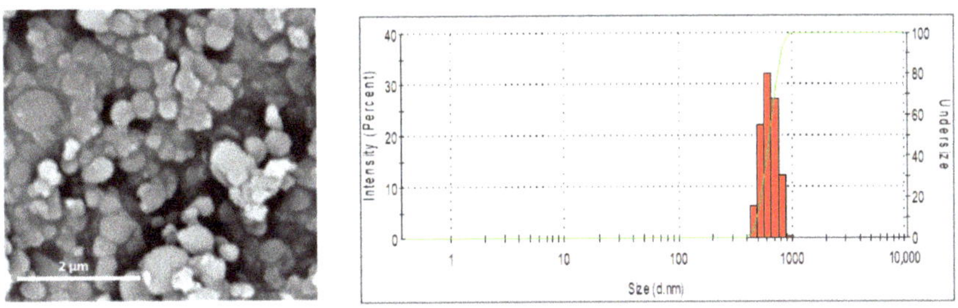

Figure 3. SEM image (**left**) and DLS size distribution profile (**right**) of polyDBB nanoparticles obtained by nanoprecipitation with the solvent to non-solvent ratio of 1:150.

As in the case of the ESE method, the use of additional homogenization and/or ultrasonication during the particle formation process reduces the particle size.

3.3. Stability of Polyanhydride in Nanoparticle Formulation Process

Because the polyDBB is hydrolytically degradable, it is necessary to quantify the degree of hydrolytic degradation that occurs during the nanoparticle formulation. For this purpose, the anhydride bond content was determined by ^{1}H NMR spectroscopy for neat polymer (before nanosphere formulation) and polymer after nanosphere preparation, according to the formula described earlier [9]. The anhydride bonds content was calculated using the intensity of the signals at δ = 2.74–2.64 ppm and at δ = 2.82–2.77 ppm, which are assigned to methylene protons close to ester and anhydride groups, respectively. When the nanospheres were produced by the ESE method, the disappearance of anhydride bonds was less than 10%, when using PVAL as a surfactant, but almost 20% when SDS or CTAB was used. The observed results confirmed that the presence of ionic groups on the particle surface accelerates the polyanhydride degradation rate due to better water access. When the nanoprecipitation was used to produce the nanoparticles, the degradation of polyDBB also occurred but was less significant than in the case of the ESE method (the disappearance of anhydride bonds was less than 5%).

4. Conclusions

In this work, polyanhydrides based on disuccinate betulin were successfully fabricated into nanoparticles. Nanoparticles were obtained by two methods. The process parameters influence the characteristics of obtained particles. The ESE method provided regularly spherical particles, while among the particles obtained from the nanoprecipitation method, many irregular ones were observed. In the ESE method, the size of the particles depended on the type of surfactant (in the water phase) and the polymer concentration (in the organic phase). The smallest particles were obtained with ionic surfactants, however, the use of such compounds accelerated the degradation process of polymers. In the nanoprecipitation method, the ratio of solvent (methylene chloride) to non-solvent (hexane) had a significant influence on the particle size. The smallest particles were obtained with a solvent to non-solvent ratio of 1:150.

Author Contributions: Conceptualization, D.N. and K.J.; methodology, D.N. and K.J.; investigation, D.N., G.D.; resources, D.N.; data curation, D.N. and K.J.; writing—original draft preparation, D.N.; writing—review and editing, D.N. and K.J.; visualization, D.N.; supervision, K.J. All authors have read and agreed to the published version of the manuscript.

Funding: This research was funded by the Polish Budget Founds for Scientific Research in 2021 as core funding for R&D activities in the Silesian University of Technology—funding for young scientists, grant number 04/040/BKM21/0181.

Institutional Review Board Statement: Not applicable.

Informed Consent Statement: Not applicable.

Data Availability Statement: The data presented in this study are available on request from the corresponding author.

Conflicts of Interest: The authors declare no conflict of interest.

References

1. Rao, J.P.; Geckeler, K.E. Polymer nanoparticles: Preparation techniques and size-control parameters. *Prog. Polym. Sci.* **2011**, *36*, 887–913. [CrossRef]
2. Nagavarma, B.V.N.; Yadav, H.K.; Ayaz, A.V.L.S.; Vasudha, L.S.; Shivakumar, H.G. Different techniques for preparation of polymeric nanoparticles—A review. *Asian J. Pharm. Clin. Res.* **2012**, *5*, 16–23.
3. Moritz, M.; Geszke-Moritz, M. Recent developments in the application of polymeric nanoparticles as drug carriers. *Adv. Clin. Exp. Med.* **2015**, *24*, 749–758. [CrossRef]
4. Michalak, G.; Głuszek, K.; Piktel, E.; Deptuła, P.; Puszkarz, I.; Niemirowicz, K.; Bucki, R. Polymeric nanoparticles—A novel solution for delivery of antimicrobial agents. *Med. Stud.* **2016**, *32*, 56–62. [CrossRef]
5. Jain, J.P.; Modi, S.; Domb, A.J.; Kumar, N. Role of polyanhydrides as localized drug carriers. *J. Control. Release* **2005**, *103*, 541–563. [CrossRef] [PubMed]
6. Wafa, E.I.; Geary, S.M.; Goodman, J.T.; Narasimhan, B.; Salem, A.K. The effect of polyanhydride chemistry in particle-based cancer vaccines on the magnitude of the anti-tumor immune response. *Acta Biomater.* **2017**, *50*, 417–427. [CrossRef] [PubMed]
7. Lee, W.C.; Chu, I.M. Preparation and degradation behavior of polyanhydrides nanoparticles. *J. Biomed. Mater. Res. Part B Appl. Biomater.* **2008**, *84*, 138–146. [CrossRef] [PubMed]
8. Alakurtti, S.; Mäkelä, T.; Koskimies, S.; Yli-Kauhaluoma, J. Pharmacological properties of the ubiquitous natural product betulin. *Eur. J. Pharm. Sci.* **2006**, *29*, 1–13. [CrossRef] [PubMed]
9. Niewolik, D.; Krukiewicz, K.; Bednarczyk-Cwynar, B.; Ruszkowski, P.; Jaszcz, K. Novel polymeric derivatives of betulin with anticancer activity. *RSC Adv.* **2019**, *9*, 20892–20900. [CrossRef]

engineering proceedings

MDPI

Proceeding Paper

Preliminary Design and Validation of a 3D-Printed Continuously Variable Transmission for an Electric Vehicle Prototype †

Marcos R. C. Coimbra [1], Társis P. Barbosa [1] and César M. A. Vasques [2,*]

1 Department of Telecommunications Engineering and Mechatronics, Federal University of São João del-Rei, Alto Paraopeba Campus, Rodovia MG 443, km 7, Ouro Branco 36420-00, MG, Brazil; mrafaelcc@live.com (M.R.C.C.); tarsisbarbosa@ufsj.edu.br (T.P.B.)

2 proMetheus, Escola Superior de Tecnologia e Gestão, Instituto Politécnico de Viana do Castelo, Rua Escola Industrial e Comercial de Nun'Álvares, 4900-347 Viana do Castelo, Portugal

* Correspondence: cmavasques@gmail.com

† Presented at the 2nd International Electronic Conference on Applied Sciences, 15–31 October 2021; Available online: https://asec2021.sciforum.net/.

Abstract: This article discusses the progress made in developing a new 3D-printed continuously variable transmission (CVT) for an electric vehicle (EV) prototype competing in the Shell Eco-marathon electric battery category, a global energy efficiency competition sponsored by Shell. The proposed system is composed of a polymeric conic gear assembled in the motor axle and directly coupled to the rear tire of the vehicle. The conical shape allows to implement a continuous variation of the gear diameter in contact with the tire. The motor with the gear was mounted over a board with linear bearings, allowing the speed ratio to change by moving the board laterally. A 3D-printing slicing software with an optimization algorithm plug-in was used to determine the best printing parameters for the conic gear based on the tangential force, maximum displacement and safety factor. When compared to the original part with a 100% infill density, the optimized solution reduced the component mass by about 12% while maintaining safe mechanical resistance and stiffness.

Keywords: continuously variable transmission; powertrain design; electric vehicle; 3D printing; Shell Eco-marathon

Citation: Coimbra, M.R.C.; Barbosa, T.P.; Vasques, C.M.A. Preliminary Design and Validation of a 3D-Printed Continuously Variable Transmission for an Electric Vehicle Prototype. *Eng. Proc.* **2021**, *11*, 11. https://doi.org/10.3390/ASEC2021-11178

Academic Editor: Filippo Berto

Published: 15 October 2021

1. Introduction

The automotive industry has been heavily reliant on the use of fossil fuels for decades. However, due to rising diesel and gasoline prices, as well as environmental pollution and fossil fuel depletion, it was concluded that alternate vehicle propulsion methods were required. Electric vehicles (EVs) have been intensively studied and appear as a possible solution for reducing global warming emission gases in this scenario. To put it in perspective, the transportation industry alone accounts for over 30% of global warming emissions in the United States; thus, switching to electric vehicles would be a very responsible and environmentally sound option [1]. International conferences, such as the United Nations Framework Convention on Climate Change (UNFCCC), also show international pressure for the development of sustainable technologies. According to Miyamoto [2], the Kyoto Protocol, which was signed in 1997, had an indirect impact on climate change discussion by boosting the number of worldwide patent applications for renewable energy, emphasizing the need for international cooperation on the subject.

The Shell Group sponsors an international competition called the *Shell Eco-marathon*, which focuses on lowering vehicle fuel consumption and is competed in by university groups of enthusiastic automobile teachers and students all over the world [3–8]. This and similar events foster several automotive technology developments, including the demonstration of distinctive car designs, energy management systems, powertrain innovations, advanced use of materials and manufacturing methods, mechanical design and the development of new 3D-printed automotive parts [9–14]. A continuously variable transmission (CVT) provides a continuous range of gear ratios between predetermined limits, improving a vehicle's fuel economy and dynamic performance by better matching engine operating conditions to variable driving circumstances [15].

Additive manufacturing and 3D printing technologies are rapidly changing machinery engineering as we know it, allowing for more creativity and freedom in machine design innovation, with numerous advantages over traditional manufacturing [16]. Free and widely distributed computer-aided design (CAD) and manufacturing (CAM) software tools, as well as well-established computer numerical control (CNC) standards, combined with readily available and affordable commercial 3D printers and materials, are fueling an increase in the number of users and potential applications of these technologies, including powertrain and other automotive components [17,18]. While almost anyone with a basic understanding of 3D modeling and printing can create their own self-made parts and devices, the wide design freedom in terms of geometry and materials, as well as professional quality assurance, raises a slew of new complex engineering questions and scientific uncertainties that go beyond basic design and functionality. Due to the design and manufacturing versatility of 3D printing, as well as continuous change and evolution in the range of available materials, the development and application of 3D-printed powertrain components is still in its infancy, but research and development activities underway promise alternative feasible and breakthrough solutions.

In response to society's desire for efficient and environmentally friendly small automotive vehicles, this study proposes to design and validate a new 3D-printed CVT enabling improved powertrain efficiency with a simpler, lighter and cheaper design than existing options for EV prototypes. The competition and EV prototype's essential specifications are presented first, with the main requirements for transmission design identified, followed by the required engineering calculations and analysis. Following that, information about the 3D-printing technologies used is presented, as well as the analysis and results obtained from the optimized printing parameters for CVT gear manufacturing. Finally, the most important findings and conclusions are summarized.

2. Materials and Methods

At the Shell Eco-marathon, prototype vehicles must complete a 12 km circuit in less than 28 min, which equates to 10 turns on the circuit depicted in Figure 1. As a result, an average speed of 25.71 km/h is required to complete the circuit without exceeding the time limit. Normally, the circuit's speed limit is set between 0 and 35 km/h.

Figure 1. Circuit diagram and image of the Brazilian start line for the Shell Eco-marathon competition (5 m wide rectangular circuit with 90° curves with an 8 m radius).

The energy consumption unit, commonly used for conventional vehicles with internal combustion engines (ICE), is a physical unit of fuel volume per unit distance, such as liters per 100 km (l/100 km). In the case of EVs, energy consumption is evaluated in kWh per unit distance in km (kWh/km) [19]. At the Shell Eco-marathon, the most energy-efficient vehicle that wins the competition is the one that completes the entire circuit in the allotted time while consuming the least amount of energy—fuel, in the case of ICE vehicles, or battery, in the case of EVs. The vehicle's output speed is affected by the motor's input speed, the transmission speed ratio and the tire diameter. For simplicity, the prototype EV transmission system was first idealized with a fixed speed ratio in which a gear was directly connected to the tire of the rear wheel tire, eliminating the need for gear shifting and the construction and implementation of a more complex driveline system. In this architecture, when the throttle is pressed by the pilot, signals are sent to the motor controller to change the motor speed and, consequently, the wheel rotation. In the case of DC electric motors, the speed range can vary from zero to thousands of rotations per minute (rpm) depending on the voltage received from the motor controller. The first gear prototype design is shown in Figure 2a. This component was 3D printed and assembled in a prototype vehicle in order to validate the reliability of a plastic gear in the vehicle's powertrain. However, this solution harmed the performance and efficiency of the EV's powertrain, affecting acceleration time, top speed, available torque at the wheels and energy consumption. Indeed, the narrow output speed range obtained with this design in comparison to the torque required to start the vehicle demonstrated the importance of a variable transmission.

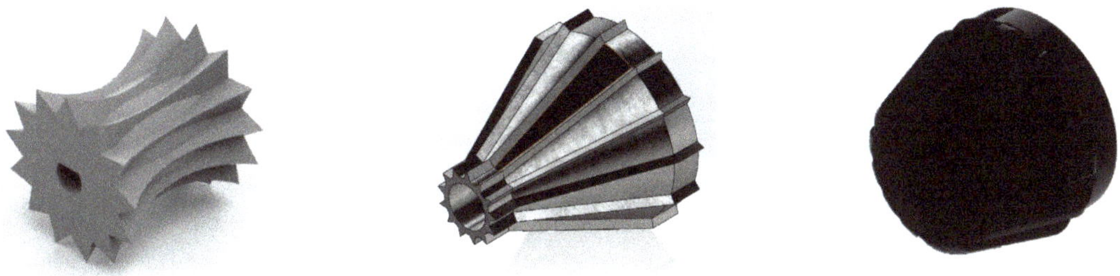

(a) (b) (c)

Figure 2. Gear design concepts: (**a**) first gear attempt, (**b**) idealized conic gear and (**c**) final gear.

The simplest implementation of a variable transmission system began with an idealized conic gear, as illustrated in Figure 2b. This component was originally believed to be assembled in the motor axle and directly coupled to the vehicle's rear tire. If the electric motor was assembled with the proper inclination angle and on a movable platform, the conical shape would allow the gear diameter in contact with the tire to gradually increase. As a result, the solution was to mount the motor on a board with linear bearings, allowing the speed ratio to be varied by lateral movement of the board. To move the motor base and thus the gear, a system of strings and pulleys connected to a lever system can be attached to the board. When necessary, the pilot may pull or push this lever, thereby changing the speed ratio. The first conic gear designed, shown in Figure 2b, was able to achieve the proposed speed range. However, its geometric format was not properly designed to handle the contact force with the vehicle rear tire, resulting in teeth fractures, wear and loss of performance. The EV prototype and idealized variable transmission concept employing the final gear design in Figure 2c is shown in Figure 3.

The maximum and minimum gear speed ratios u from the variable transmission can be obtained considering the wheel tire diameter D and the maximum and minimum values of the conic gear diameter d, given by

$$\max u = \frac{D}{\min d} \quad \text{and} \quad \min u = \frac{D}{\max d}. \tag{1}$$

The vehicle speed V_x as a function of the gear ratio u and angular velocity of the electric motor $\omega = 2\pi n/60$, where n is the rotating velocity of the electric motor in rpm, from which the limiting EV speeds can be determined, is given by

$$V_x = \frac{\omega D}{2u}. \tag{2}$$

Similar relations can be established to determine torque magnification at the wheel. The corresponding torques at the wheel and gear, T_D and T_d, respectively, can be determined according to the rated power of the motor P and the transmission efficiency η, yielding

$$T_d = \frac{P}{\omega} \quad \text{and} \quad T_D = u\eta T_d. \tag{3}$$

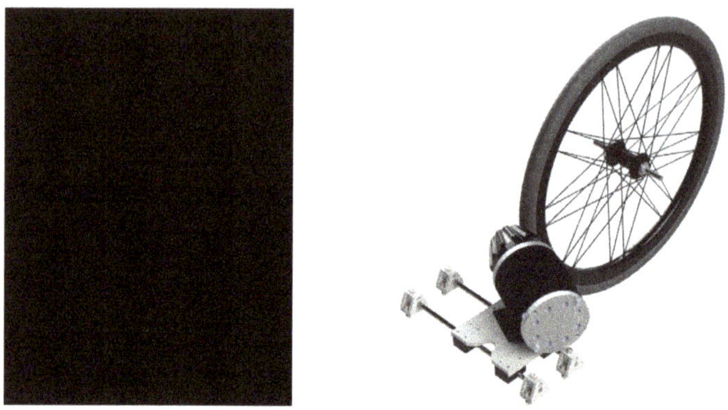

Figure 3. EV prototype (**left**) and virtual representation of the idealized CVT concept (**right**).

As previously stated, the final gear design depicted in Figure 2c was prototyped using polymer additive manufacturing (AM) technologies. These involve layering materials to create objects from 3D model data via material extrusion in which the material is selectively dispensed through a nozzle or orifice. The process is called FDM (fused deposition modeling; a trademark of the company Stratasys), and it is a material extrusion process that uses heated extrusion and layer deposition of materials to create thermoplastic parts. In its untrademarked form, the FDM process is also referred to as FFF (fused filament fabrication), and is typically associated with a less industrial grade technology (desktop, hobbyist level 3D printing technology). The 3D printer model used is the widely accessible, low-cost and hobbyist grade technology, Anet A8. To determine the optimal printing parameters for the STL part file generated from the 3D geometric model of the gear, a commercially available plug-in and tool called SmartSlice was used. It is designed to be used with Ultimaker's Cura slicer software and incorporates an optimization algorithm. To ensure structural requirements are met, the software is fed with the force acting on the component, the component's maximum displacement and the desired safety factor. Additionally, this plug-in makes use of modifier meshes to reinforce infill density in critical regions without requiring the entire body to be reinforced with additional material.

3. Results and Discussion

Considering an intermediary value of motor speed $n = 1900$ rpm and the wheel diameter $D = 480$ mm, it is possible to estimate the conic gear's maximum and minimum diameters according to Equations (1)–(3). The minimum and maximum gear diameters chosen were min $d = 68.71$ mm and max $d = 96.19$ mm and the corresponding gear speed ratios found were max $u = 6.99{:}1$ and min $u = 4.99{:}1$, respectively. For simplicity, adopting 100% transmission efficiency, the variation of the vehicle's speed and wheel torque as a function of the gear diameter can be estimated. To gain a better understanding of the forces and torques acting on the vehicle powertrain and to evaluate the proposed CVT, a computational dynamic vehicle model of the EV prototype was implemented in MATLAB/Simulink but is not shown here due to space constraints.

PETG (polyethylene terephthalate glycol) was chosen as the material for the second version of the gear. Table 1 compares the mechanical properties of PETG to those of some of the most commonly used filaments in 3D printing from the 3Dlab brand. Although PLA (polylactic acid) is stronger than PETG, PETG is better suited for outdoor use due to its resistance to weather and sunlight. PETG has a temperature tolerance of approximately 75 °C. On the other hand, 3D printed parts made of PLA must be kept below 55 °C to avoid thermal deformation and stiffness relaxation.

Table 1. Typical material properties values of common filament materials for 3D printing.

Properties	ABS	PLA	PETG
Tensile modulus (MPa)	1335	1896	1067
Yield strength (MPa)	14.7	24.8	18.6
Ultimate strength (MPa)	29	46	32.6
Strain at break (%)	7.08	3.69	7.74
Melting point (°C)	220	185	240
Specific weight (g/cm^3)	1.04	1.24	1.27

The input data used to analyze the gear printing process are shown in Table 2. The safety factor was defined as 1.5 and maximum displacement on the gear teeth flanks to an applied distributed force was considered equal to 0.3 mm. The total load force distributed on the teeth flanks was equal to 621.78 N and was obtained considering the maximum traction force and the respective gear ratio discussed in the previous section. As shown in Table 2, by adjusting the printing parameters to its optimized setting it was possible to reduce the mass from 437.2 to 386.7 g, maintaining an acceptable maximum displacement and guaranteeing the safety factor previously stipulated. Figure 4 illustrates the gear slicing process before and after optimization. In Figure 4, it is possible to verify the mesh reinforcement close to the teeth surface. Moreover, local reinforcements were used in the part holes. Due to local reinforcement complexity, the printing time in the optimized configuration was approximately 40 min higher; however, it is not critical in this project since the main concern is about improving the gear mechanical strength and stiffness.

Table 2. Three-dimensional printing optimization parameters and comparison of results.

Parameters	Initial Settings		Optimized Settings	
	Extruder	Local Reinforcements	Extruder	Local Reinforcements
Top and bottom layers	3	-	4	3
Infill density (%)	100	-	35	85
Wall thickness (mm)	0.6	-	0.6	0.6
Wall line counting	3	-	2	3
Results				
Mass (g)	437.2		386.7	
Max displacement (mm)	0.24		0.28	
Printing time	22 h 57 min		23 h 36 min	
Factor of safety	3.92		3.24	

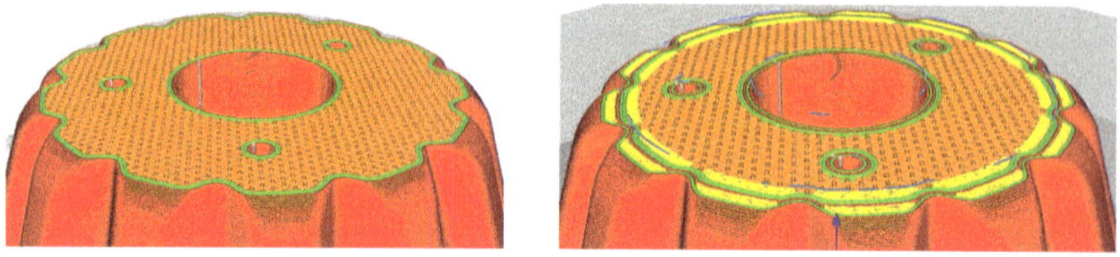

Figure 4. A 3D printed gear slicing comparison: initial (**left**) and optimized slicing (**right**) performed with the 3D printing software Cura slicer from Ultimaker. The red lines denote external shell surface; orange, the infill region; and green, the inner wall.

4. Conclusions

The present work proposes a new CVT architecture utilizing a conic gear. This gear engages with the rear wheel's tire and is coupled to the electric motor axle. The speed ratio change is obtained moving the motor, which is mounted over a movable board with linear bearings. The proposed variable transmission is unprecedented and its constructive simplicity suggests opportunities for application in small vehicles. A computational model was implemented to simulate a prototype three-wheeled vehicle with this designed transmission. Furthermore, an optimization algorithm was used to obtain optimized printing parameters for manufacturing the gear in PETG. The obtained set of parameters demonstrated possibilities to reduce the gear mass about 12% in relation to the 3D-printed gear with 100% of material infill without losing mechanical resistance and stiffness. In future work, the EV prototype will be tuned and tested in a real-driving cycle. To determine the powertrain's efficiency, data on speed, time, motor current, battery voltage and battery consumption will be collected. Additional testing will be conducted to determine the transmission's efficiency and durability.

Author Contributions: Conceptualization, all; methodology, all; software, M.R.C.C. and T.P.B.; validation, M.R.C.C. and T.P.B.; formal analysis, all; investigation, all; writing—original draft preparation, M.R.C.C. and C.M.A.V.; writing—review and editing, T.P.B. and C.M.A.V.; funding acquisition, T.P.B. and C.M.A.V. All authors have read and agreed to the published version of the manuscript.

Funding: The first two authors gratefully acknowledge that this study was funded, in part, by the National Council for Scientific and Technological Development (CNPq), as well as the support provided by the Center of Innovation, Research and Teaching of Mechatronics (NIPEM) and by the Federal University of São João del-Rei (UFSJ). The third author gratefully acknowledges the support provided by the Foundation for Science and Technology (FCT) of Portugal, within the scope of the project of the Research Unit on Materials, Energy and Environment for Sustainability (proMetheus), Ref. UID/05975/2020, financed by national funds through the FCT/MCTES.

Institutional Review Board Statement: Not applicable.

Informed Consent Statement: Not applicable.

Data Availability Statement: Not applicable.

Conflicts of Interest: The authors declare no conflict of interest.

References

1. Reddy, K.J.; Natarajan, S. Energy sources and multi-input DC-DC converters used in hybrid electric vehicle applications—A review. *Int. J. Hydrogen Energy* **2018**, *43*, 17387–17408. [CrossRef]
2. Miyamoto, M.; Takeuchi, K. Climate agreement and technology diffusion: Impact of the Kyoto Protocol on international patent applications for renewable energy technologies. *Energy Policy* **2019**, *129*, 1331–1338. [CrossRef]
3. Alnunu, N.; Said, S.; Al-Sharman, S.; Al-Ibrahimi, A.; AbdulAziz, A.; Hellabi, M.A.; Touati, F.; Ghani, S.; Mahdi, E.S.; Benammar, M. Design of Qatar University's first solar car for Shell Eco-Marathon competition. In Proceedings of the 2012 First International Conference on Renewable Energies and Vehicular Technology, Nabeul, Tunisia, 26–28 March 2012; pp. 49–54. [CrossRef]
4. Baldissera, P.; Delprete, C. Human powered vehicle design: A challenge for engineering Education. In Proceedings of the ASME 2014 12th Biennial Conference on Engineering Systems Design and Analysis (ESDA2014), Copenhagen, Denmark, 25–27 July 2014. [CrossRef]

5. Buck, L.; Mclening, C.; Burgess, J. Eco-car: A perfect vehicle for technical design teaching? In Proceedings of the 16th International Conference on Engineering and Product Design Education (E&PDE14), Design Education and Human Technology Relations, University of Twente, Enschede, The Netherlands, 4–5 September 2014.
6. Abdulwahed, M.; Ahmad, S.; Hasna, M.O.; Ghani, S.; Benammar, M. Contribution of Shell Eco-Marathon engineering design experience to soft skills development: A qualitative analysis in the Asian context. In Proceedings of the 2014 International Conference on Interactive Collaborative Learning (ICL), Dubai, United Arab Emirates, 3–6 December 2014. [CrossRef]
7. Von Solms, S.; Nel, H. Reflective learning in engineering education: A case study of shell Eco-Marathon. In Proceedings of the 2017 IEEE International Conference on Industrial Engineering and Engineering Management (IEEM), Singapore, 10–13 December 2017. [CrossRef]
8. Verma, A.R.; Chaurasia, A.; Jaiswal, S.S.; Bhonde, L.; Guha, R.; Sahu, H.; Patel, S.; Banthiya, S.; Maddeshiya, S.; Mirzapure, S.; et al. Team AVERERA's Alterno V4.0—A hyper energy-efficient electric prototype vehicle for Shell Eco-Marathon. *SAE Tech. Pap.* **2021**. [CrossRef]
9. Ary, A.K.; Sanjaya, Y.; Prabowo, A.R.; Imaduddin, F.; Nordin, N.A.B.; Istanto, I.; Cho, J.H. Numerical estimation of the torsional stiffness characteristics on urban Shell Eco-Marathon (SEM) vehicle design. *Curved Layer Struct.* **2021**, *8*, 167–180. [CrossRef]
10. Stabile, P.; Ballo, F.; Mastinu, G.; Gobbi, M. An ultra-efficient lightweight electric vehicle—Power demand analysis to enable lightweight construction. *Energies* **2021**, *14*, 766. [CrossRef]
11. Carello, M.; Pinheiro, H.C.; Longega, L.; Di Napoli, L. Design and modelling of the powertrain of a hybrid fuel cell electric vehicle. *SAE Tech. Pap.* **2021**. [CrossRef]
12. Sethi, N.; Chauhan, P.; Bansal, S.; Singari, R.M. Robust vehicle development for student competitions using fiber-reinforced composites. In *Lecture Notes in Mechanical Engineering*; Springer: Singapore, 2021; pp. 61–76. [CrossRef]
13. Gilewski, M.; Czarnigowski, J.; Górski, W.; Mitrus, K.; Różyło, P.; Trocha, S.; Wypychowski, M. Strength analysis of the drive wheel hub of a hydrogen-powered prototype hyper-light vehicle. *J. Phys. Conf. Ser.* **2021**, *1736*, 012056. [CrossRef]
14. Kılıç, A.E. Redesign of Drivetrain Component of a Shell Eco-Marathon Vehicle for Additive Manufacturing via Topology Optimization. Master's Thesis, Piri Reis Üniversitesi, Istanbul, Turkey, 2020.
15. Vaughan, N. Transmission and Driveline: Introduction. In *Encyclopedia of Automotive Engineering*; Crolla, D., Foster, D., Kobayashi, T., Vaughan, N., Eds.; Wiley: Hoboken, NJ, USA, 2014; p. 11. [CrossRef]
16. Attaran, M. The rise of 3-D printing: The advantages of additive manufacturing over traditional manufacturing. *Bus. Horizons* **2017**, *60*, 677–688. [CrossRef]
17. Agapovichev, A.V.; Balaykin, A.V.; Smelov, V.G. Production technology of the internal combustion engine crankcase using additive technologies. *Mod. Appl. Sci.* **2015**, *9*, 335–343. [CrossRef]
18. Gray, J.; Depcik, C. Review of additive manufacturing for internal combustion engine components. *SAE Int. J. Eng.* **2020**, *13*, 617–632. [CrossRef]
19. Ehsani, M.; Gao, Y.; Longo, S.; Ebrahimi, K.M. *Modern Electric, Hybrid Electric, and Fuel Cell Vehicles*, 3rd ed.; CRC Press: Boca Raton, FL, USA, 2018. [CrossRef]

engineering proceedings

MDPI

Proceeding Paper

Bangladesh Agriculture: A Review of Modern Practices and Proposal of a Sustainable Method †

Nazifa Tabassum * and Fatema Rezwana *

Department of Environmental Sciences, Jahangirnagar University, Savar, Dhaka 1342, Bangladesh
* Correspondence: tabassumnazifa09@gmail.com (N.T.); rezwanaepty07@gmail.com (F.R.)

† Presented at the 2nd International Electronic Conference on Applied Sciences, 15–31 October 2021; Available online: https://asec2021.sciforum.net/.

Abstract: Agriculture is the largest economic and employment sector in Bangladesh, accounting for 23 percent of its gross domestic product and 65 percent of its labour force. It has a total land area of 14,570 km^2, with cultivated land accounting for 60% of the total land area. The population is still growing at a 1.37 percent annual rate, but cultivated land is shrinking at the same time. The agricultural land is being converted as a result of uncontrolled urbanization, industrialization, and an increase in human activity. Modern sustainable agricultural methods approach to agricultural innovations and farming practices that increase farmers' efficiency and reduce the use of natural resources. This study uses secondary information to provide a general view on modern practices used in Bangladesh for sustainable agriculture (i.e., crop diversification, change in cropping pattern and rotations, integrated farming systems, etc.) and suggests a sustainable method (polyculture and crop rotation) based on SWOT and PESTEL analysis. This paper recommends that Bangladesh should adopt polyculture and crop rotation more to improve soil health and for higher crop resistance to plant pests that will hedge against a loss in agricultural sustainability.

Keywords: sustainable agriculture; environment; polyculture and crop rotation; modern practices

Citation: Tabassum, N.; Rezwana, F. Bangladesh Agriculture: A Review of Modern Practices and Proposal of a Sustainable Method. *Eng. Proc.* **2021**, *11*, 12. https://doi.org/10.3390/ASEC2021-11190

Academic Editor: Dibyendu Sarkar

Published: 15 October 2021

1. Introduction

An integrated system of plant and animal production practices, known as sustainable agriculture, produces sufficient quantities of high-quality food while protecting natural resources in an environmentally safe and profitable manner [1]. Bangladesh agriculture is deteriorating due to degradation of land and water resources and the excess use of toxic chemicals. This is the reason the government of Bangladesh has adopted sustainable agricultural methods to protect environment stability and economic profitability. Crop diversification, integrated farming systems, changes in cropping pattern and rotations, adoption of new crop varieties, modifying sowing dates, polyculture and crop rotations, and other modern cropping practices are all being employed in Bangladesh to ensure food self-sufficiency and environmental sustainability.

A conventional farming technique where crops are cultivated together or separately (at either short or long intervals) on the same field is known as crop rotation. Crops are grown using this method because of its high productivity, high revenue level, and adaptability for combination management practices. Crop rotation is currently the foundation of agriculture systems around the world, allowing for the efficient use of environmental resources [2]. Crop rotation provides various advantages over monoculture planting in terms of agronomy, economics, and the environment. In the long run, crop rotation can increase soil organic matter and soil structure, as well as minimize soil degradation. Water and nutrients are better retained when there is more soil organic matter. Using crop rotation, weeds and diseases can be controlled, as well as insect and other pest infestations, reducing pesticide consumption [3]. Another possibility is for the farmer to cultivate two or more crops at the same time in a same place. This system is known as polyculture. Increased diversity in cultivated species, but mostly in the soil, promotes the possibility of ecological equilibrium. As opposed to monocultures, polycultures offer better protection against soil deterioration and the expansion of pests, pathogenic bacteria, and fungus than monocultures do. In addition, this system has a favorable economic impact on the size of inputs and outputs and the stabilization of the food chain. Even though polyculture is not often practiced in developed countries, it has a major impact on food security in the developing world [2].

Most of the studies reviewed in this paper discussed the sustainability, challenges, acceptance, advantages and disadvantages, and profitability of polyculture and crop rotation. Very few attempts have so far been made to analyze the environmental impacts and external–internal aspects of this process. The overall goal of this research paper is to review modern agricultural practices used in Bangladesh and suggest polyculture and crop rotation as a comparatively effective agricultural practice for sustainability that reduces environmental consequences.

2. Methodology

The data for this review were gathered from a variety of secondary sources, including national or international peer reviewed journals, relevant books, articles, and other sources. As the topic of this paper has rarely been linked and considered, a broad search query was used with limited constraints on years or titles. With the search string TITLE-ABS-KEY ("sustainable agriculture", "polyculture and crop rotation"), academic publications were searched in Google Scholar and PubMed, databases of abstracts and citations of peer reviewed scientific journal articles. The review of modern practices in Bangladesh is based on literature search and the proposed method mainly relies on SWOT and PESTEL analysis. Environmental implications have been studied utilizing the SWOT analysis process in order to achieve a systematic approach to decision making in the case of polyculture and crop rotation. The impacts will be classified using the SWOT analysis framework, which stands for strengths, weaknesses, opportunities, and threats. The PESTEL model will provide an external assessment of polyculture and crop rotation in terms of political, economic, social, technological, environmental, and legal aspects.

3. Review of Modern Practices Used in Bangladesh

Strengthening agricultural production systems for greater sustainability and higher economic returns is a crucial strategy for enhancing income and food and nutrition security in underdeveloped countries [4]. In Bangladesh, there is no way to expand agricultural land. Furthermore, due to increased population pressure, net crop acreage is shrinking at an alarming rate. It is crucial to identify and investigate major cropping trends [5]. To enrich and sustain soil fertility, increase crop productivity, and improve crop sequences, Bangladesh has adapted some modern agricultural practices such as:

- Crop diversification: Crop diversification is a cropping method that adds new crops to agricultural production on a specific farm in order to assist the farming community to thrive economically. This agricultural approach lessens farmers' reliance on a single crop and mitigates unforeseeable climate catastrophes such as the appearance of pests and the unexpected advent of frost or drought [6]. Lack of market access, established soil conditions, flood depth levels, and decreased rainfall are some of the obstacles to crop diversification [7]
- Change in cropping pattern and rotations: The distribution of a farm's land to various crops cultivated over the course of a year is referred to as cropping pattern. It entails allocating land to different crops at different times of the year [8]. Cropping pattern is essentially a yearly strategy for maximizing agronomic and economic production while being sustainable and also an essential measure of a locality's land use, environment, and socioeconomic elements of its farmers [5].

- Integrated farming system: The integration of various agricultural enterprises, such as cropping, animal husbandry, fishery, forestry, and so on, into the farming system has great potential in the agricultural economy [9]. Integrated farming with crops, livestock, and aquaculture has the potential to increase yields and provide financial benefits, extend the harvest period and alleviate seasonal food shortages, improve the stability of household food access, and reduce erosion risks [10]. Since the final customers and processing industries are so far apart, a large upfront investment is required [11].

4. Analysis of Polyculture and Crop Rotation

4.1. SWOT

Polyculture and crop rotation have a significant impact on soil, water, climate, and pest and disease control, hence those are the sectors that have been considered for the SWOT analysis (Figure 1) and discussed below:

	Strengths	Weaknesses
Soil	1.Prevent soil biodiversity loss 2.Variety of plants grow in the fields and increase crop productivity. 3.Provide habitat and food for more species 4.Higher soil enzyme and microbial activity 5.Lower risk of soil erosion.	1.Requiring tillage and ploughing has a negative impact on soil biodiversity. 2.Pesticide-intensive crop rotations have a negative influence on wildlife above the soil.
Water	1.Increase soil nitrogen availability by reducing N leaching. 2.Reduce water erosion depending on soil cover provided by crops. 3.Reduce ground water contamination.	1.Crops that are heavily reliant on pesticides during wet seasons may leach more residues into groundwater and rivers.
Climate	1.Conserve water in water-stressed regions to prevent excessive evaporation. 2.Rotations that do well without irrigation help conserve water. 3.Reduces nitrogen loss through leaching and reduces environmental pollution.	1. Influence the emission of greenhouse gases.
P&D control	1. Disrupt pest biology and control damage and disease. 2. Increased plant diversity benefit pest management. 3.Entire farm is not at danger if a disease strikes.	1. fungal and bacterial soil-borne disease pathogens can live on crop leftovers are difficult to control. 2. Poor management skill
	Opportunities	**Threats**
Soil	1.Rotations that maintain a cover throughout the year benefit soil microbes and earthworms. 2.Planting crops on the margins of fields and cutting them with grasses and flowers benefit biodiversity.	1. As a result of pesticide use, genetic diversity is reduced. 2. Short rotations reduce the ability of rotations with many years to build a good structure.
Water	1. crops with high soil cover index reduce losses of water. 2.Crops with a long vegetation period develop large mass of roots.	1. Crop with short vegetation period have a negative impact.
Climate	1.Beneficial if it contains legumes and crops that require low or no N fertilization.	
P&D control	1.Encourage biological control agents manage crop pests. 2.Pests and weeds are managed with fewer chemical pesticides using multiyear, multi crop rotations.	

Figure 1. SWOT Analysis of environmental impacts of polyculture and crop rotation [12–14].

4.2. PESTEL

4.2.1. Political Aspects

Food security is a huge concern in a world where the agricultural area is shrinking, soil fertility is declining, and the population is growing. As a result, it is critical to place a special emphasis on productivity increase. The National Agriculture Policy 2018 is developed from all relevant laws, policies, development plans, and perspective plans for the agricultural sector's development. Its fundamental goal is to ensure sustained food and nutrition security while also ensuring safe and profitable agriculture. For sustainable production and conservation of natural resources, this policy promotes and inspires ecologically friendly technology, consistent pest management, soil microorganism protection, and minimum tillage methods [15]. Development of the agricultural sector in Bangladesh has somehow been hindered by corruption activities like bribery or unauthorized payments, embezzlement, etc. According to a state-wide survey on corruption by Transparency International Bangladesh (TIB) in Bangladesh, 20.4 percent of service recipients in the agricultural sector reported corruption [16].

4.2.2. Economic Aspects

Profitability is used to compare the economic performance of crop rotation systems on a micro-level from the farmer's perspective. Pricing and profitability are interdependent and are calculated by subtracting the total cost of production from the value of production [12]. Profitability is not the only factor that risk-averse farmers consider. Agricultural prices will not be significantly affected by individual farmers because agriculture resembles a perfectly competitive sector [17]. Crop rotation contributes to the economic and social consequences of farming systems in a variety of ways, including uniformity of crop planting areas, a reduction in equipment costs and different skill sets, and increased interaction with the local community for labor [18]. In order to reduce agriculture's dependence on external inputs, crop rotation systems are considered to be one of the most important cropping system alternatives. They do this by recycling nutrients internally, maintaining the long-term productivity of the land, and disrupting weed and disease cycle [17].

4.2.3. Social Aspects

In many agriculturally based developing countries, cultivation systems and farmers' income are the most important issues. Polyculture farming systems increase farmers' income from a financial standpoint and generate a higher income than monoculture. Decreased economic losses due to fluctuating product prices are another benefit of polyculture. Agricultural production cost efficiency can be improved by reducing operational costs, such as labor and plant maintenance in a polyculture system. Thus, farmers' losses due to price fluctuations would be reduced, while the potential benefit from production cost efficiency would be increased [19]. Because the market determines the price of monoculture crops, farmers' earnings are solely based on yields. A crop grown in rotation will have an average output price that is equal to the average yield-weighted price of all crops combined. A farmer can alter the average output price of their production by changing the crop structure in rotational cropping [12].

4.2.4. Technological Aspects

Crop rotation and polyculture approaches can benefit from technological advancements. In this process, advances in sensing devices and embedded systems can help to enhance tillage and crop yields. Farmers may find that resource management efficiency and autonomous data collecting are essential. Field measurements of crop requirements, as well as planned improvements in crop productivity and resource consumption, are all moving towards more sustainable ideas and lowering environmental consequences [20]. In terms of researcher numbers and qualifications, research and development efforts in Bangladesh's agriculture sector are worsening. In Bangladesh, further research and investment in polyculture and crop rotation studies will help this system thrive and expand. As a result, the environment will be more sustainable [21].

4.2.5. Environmental Aspects

Polyculture and crop rotations could be a useful tool for climate change adaptation and mitigation. High biomass production crops contribute to CO_2 sequestration, whereas low biomass production crops, such as legumes, are good for reducing N_2O emissions since they do not require N fertilization, which adds to N_2O losses [12]. Bangladesh's food security depends on sustainable agricultural development and management, as the country faces a gradual loss in the productivity and quantity of agricultural land, as well as the negative consequences of climate change. Some policies are incorporated in the National Environmental Policy 2018 to ensure food security and sustainable agricultural management. As per the policies, technologies and development activities for agricultural development must be implemented in an environmentally friendly manner. Farmers should be encouraged to diversify their crops, employ green manure derived from legumes, and use legumes in crop rotation in their farms in order to increase food output [22]. These policies provide a clear concept for sustainable agriculture. In Bangladesh, NGOs such as PROSHIKA-MUK, Friends in Village Development Bangladesh (FIVDB), CARE International, and Rangpur Dinajpur Rural Service (RDRS) promote environmentally friendly farming system in which community members maximize the utilization of their resources through traditional, indigenous, as well as modern science [23].

4.2.6. Legal Aspects

Bangladesh is a Least Developed Country that is heavily reliant on agriculture. Plant breeding, seed commercialization and exchange, and farmers' traditional knowledge (TK) are central to the country's agricultural development. Among the laws, the Patents and Designs Act of 1911 contains provisions for patentable inventions involving plant varieties and seeds, while the Trademarks Act of 2009, the Geographical Indication of Goods Act of 2013, and the Seeds Ordinance of 1977 are considered to be only relevant for seed trading. Furthermore, these laws are not initially regarded as significant in terms of farmers' rights, food security, sustainable agriculture, and plant genetic resources (PGRs). As a member of the TRIPS, the country acknowledges these issues with the adoption of the WTO agreement on Trade-Related Aspects of Intellectual Property Rights (TRIPS) providing intellectual property rights (IPRs). Because these laws are treaty specific, they frequently fail to address local needs and situations, impeding farmers' ability to make a living and causing problems for sustainable agriculture [24]. By implementing the Plant Varieties Protection (PVP) Act Bangladesh can promote plant varieties and sustainable use of plant genetic resources.

5. Recommendation

SWOT analysis indicates that, as a sustainable agricultural method with some limitations, polyculture and crop rotation has numerous advantages and future prospects. Most of the benefits of polyculture and crop rotation include increased productivity, improved soil health, reduced pest and disease problems, increased soil N availability, and reduced water erosion. The PESTEL study indicates that the National Agriculture Policy 2018, the National Environmental Policy 2018, and the Plant Varieties Protection Act are the policies and act that encourage and support Bangladesh's sustainable agriculture. In terms of economic profitability, social status, and environmental quality improvements, polyculture and crop rotation are valuable practices to implement. Polyculture and crop rotation are effective methods for achieving more sustainable crop production from an environmental and economic standpoint. However, combining these methods can make them more long lasting. Changing crop rotations quickly and drastically disrupts the natural balance of weeds, insect pests, and disease. Consequently, a polyculture that incorporates crop rotation will have greater crop productivity and soil health. In other words, by combining crop rotation with polyculture, Bangladesh can achieve a sustainable agricultural goal over the long term.

6. Conclusions

Polyculture and crop rotation are excellent techniques to make the most of natural resources while also protecting the environment. Alone or in combination, this procedure will boost environmental elements and strengthen sustainability. Based on internal–external analysis, this research suggests polyculture and crop rotation as a sustainable method. Through the transformation of agricultural production into environmental advantages, this technology will contribute to green agriculture and promote sustainable agriculture.

Supplementary Materials: The following are available online at https://www.mdpi.com/article/10.3390/ASEC2021-11190/s1.

Author Contributions: All authors were directly involved and contributed equally in reviewing, writing and editing the paper. All authors have read and agreed to the published version of the manuscript.

Funding: This research received no external funding.

Institutional Review Board Statement: Not applicable.

Informed Consent Statement: Not applicable.

Data Availability Statement: Not applicable.

Conflicts of Interest: The authors declare no conflict of interest.

References

1. Velten, S.; Leventon, J.; Jager, N.; Newig, J. What is sustainable agriculture? A systematic review. *Sustainability* **2015**, *7*, 7833–7865. [CrossRef]
2. Meena, R.S. *Soil Health Restoration and Management*; Springer: Berlin/Heidelberg, Germany, 2020.
3. Friends of the Earth Europe. *Benefiting Farmers, the Environment and the Economy*; Friends of the Earth Europe: Brussels, Belgium, 2012.
4. Soni, R.P.; Katoch, M.; Ladohia, R. Integrated Farming Systems—A Review. *IOSR J. Agric. Vet. Sci.* **2014**, *7*, 36–42. [CrossRef]
5. Nasim, M.; Shahidullah, S.M.; Saha, A.; Muttaleb, M.A.; Aditya, T.L.; Ali, M.A.; Kabir, M.S. Distribution of Crops and Cropping Patterns in Bangladesh. *Bangladesh Rice J.* **2018**, *21*, 1–55. [CrossRef]
6. Khanam, R.; Bhaduri, D.; Nayak, A.K. Crop diversification: An important way-out for doubling farmers' income. *Indian Farming* **2018**, *68*, 31–32.
7. Islam, M.M.; Hossain, M.E. Crop Diversification in Bangladesh: Constraints and Potentials. 2011. Available online: https://bea-bd.org/site/images/pdf/057.pdf (accessed on 10 September 2021).
8. Pervez, A.S.M.F.; Rahman, M.S.; Al-amin, A.K.M.A. Change in Cropping Patterns and Its Impacts on Farmers' Livelihood in Some Selected Areas of Mymensingh District. *Bangladesh J. Agric. Econ.* **2015**, *36*, 1–13.
9. Al Mamun, S.; Nasrat, F.; Debi, M.R. Integrated Farming System: Prospects in Bangladesh. *J. Environ. Sci. Nat. Resour.* **2012**, *4*, 127–136. [CrossRef]
10. Uddin, M.T.; Takaya, H. Comparative Study on Integrated Farming in Bangladesh and other countries. *Bangladesh J. Agric. Econ.* **2006**, *14*, 81–92.
11. Balbino, L.C.; Kichel, A.N.; Bungenstab, D.J.; de Almeida, R.G. Integrated systems: What they are, their advantages and limitations. In *Integrated Crop-Livestock-Forestry Systems, a Brazilian Experience for Sustainable Farming*; Embrapa: Brasília, Brazil, 2014; pp. 11–18.
12. Mudgal, S.; Lavelle, P.; Cachia, F.; Somogyi, D.; Majewski, L.; Fontaine, E.; Bechini, L.; Debaeke, P. *Environmental Impacts of Different Crop Rotations in the European Union*; European Commission: Brussels, Belgium, 2010; Volume 33, p. 149.
13. Reddy, P.P. *Agro-Ecological Approaches to Pest Management for Sustainable Agriculture*; Springer: Singapore, 2017; pp. 1–339. [CrossRef]
14. Barus, C. Alpheus spring packard. *Science* **1905**, *21*, 404–406. [CrossRef] [PubMed]
15. Xudayberdievich, X.X. National policy. *Acad. Int. Multidiscip. Res. J.* **2020**, *10*, 442. [CrossRef]
16. Anik, A.R.; Bauer, S. *Corruption in the Agriculture Sector–Micro Level Evidence from Bangladesh*; Emerald Publishing Limited: Bingley, UK, 2017; pp. 363–389. [CrossRef]
17. Gebremedhin, B.; Schwab, G. *The Economic Importance of Crop Rotation Systems: Evidence from the Literature*; Michigan State University: East Lansing, MI, USA, 1998. [CrossRef]
18. Francis, C.A. Crop Rotations. *Encycl. Soils Environ.* **1990**, *4*, 318–322. [CrossRef]
19. Arsyad, M.; Sabang, Y.; Agus, N.; Bulkis, S.; Kawamura, Y. Intercropping farming system and farmers income. *Agrivita* **2020**, *42*, 360–366. [CrossRef]

20. Miranda, B.S.; Yamakami, A.; Rampazzo, P.C.B. A New Approach for Crop Rotation Problem in Farming 4.0. *IFIP Adv. Inf. Commun. Technol.* **2019**, *553*, 99–111. [CrossRef]
21. International Food Policy Research Institute (IFPRI). *ASTI (Agricultural Science and Technology Indicators)*; International Food Policy Research Institute (IFPRI): Washington, DC, USA, 2015.
22. MoEF. *National Environmental Policy 2018*; Ministry of Environment, Forest and Climate Change, Ed.; Government of People's Republic of Bangladesh: Dhaka, Bangladesh, 2018.
23. Vaxelaire, J. *A Country Paper of Bangladesh on Promoting Dialogue and Collaboration in Sustainable Agriculture & Rural Development (SARD) between NGOs/RPOs and Government*; Food and Agriculture Organization of the United Nations (FAO-UN): Rome, Italy, 1994.
24. Arner, R.P.B.D.W.; Barberis, J. The Legal Regime of Plant Varieties and Farmers' Rights Protection in Bangladesh: Options and Challenges. *Grou* **2008**, *23529*, 1–45.

engineering proceedings

MDPI

Proceeding Paper

Optical Chemo-Sensors for Specific Markers in Transformer Insulating Oil Exploiting Molecularly Imprinted Polymers and Plasmonic Optical Fibers †

Letizia De Maria [1,*], Francesco Arcadio [2], Maria Pesavento [3], Antonella Profumo [3], Nunzio Cennamo [2] and Luigi Zeni [2]

1 RSE S.p.A, Technologies for Transmission and Distribution, Via Rubattino 54, 20134 Milan, Italy
2 Department of Engineering, University of Campania "L. Vanvitelli", Via Roma 29, 81031 Aversa, Italy; Francesco.Arcardio@unicampania.it (F.A.); nunzio.cennamo@unicampania.it (N.C.); luigi.zeni@unicampania.it (L.Z.)
3 Department of Chemistry, University of Pavia, Via Taramelli 12, 27100 Pavia, Italy; maria.pesavento@unipv.it (M.P.); antonella.profumo@unipv.it (A.P.)
* Correspondence: letizia.demaria@rse-web.it

† Presented at the 2nd International Electronic Conference on Applied Sciences, 15–31 October 2021; Available online: https://asec2021.sciforum.net/.

Abstract: 2-FAL (2-furaldehyde) and furanic derivatives are the main by-products of the thermal degradation of cellulose paper insulation of power transformers' windings. The detection of these compounds in the insulating oil of transformers is essential to investigate the ageing of the oil-paper system in order to avoid failures. To this aim, a non-conventional surface plasmon resonance (SPR) platform in plastic optical fiber (POF) was proposed for the monitoring of a biomimetic receptor specific to detect 2-FAL in transformer oil. In particular, the investigation was performed in mineral oil, which is currently the main insulating liquid for power transformers. A molecularly imprinted polymer (MIP) receptor was used, giving the sensor device a noticeable selectivity and many advantages with respect to the biological counterparts. Furthermore, the study was extended to safer and more environmentally acceptable insulating fluids representing an alternative to mineral oil (i.e., esters). To this aim, the principle and limitations of the SPR chemo-sensor performances have been discussed in this work.

Keywords: surface plasmon resonance (SPR); MIP; plastic optical fiber; power transformer

Citation: De Maria, L.; Arcadio, F.; Pesavento, M.; Profumo, A.; Cennamo, N.; Zeni, L. Optical Chemo-Sensors for Specific Markers in Transformer Insulating Oil Exploiting Molecularly Imprinted Polymers and Plasmonic Optical Fibers. *Eng. Proc.* **2021**, *11*, 13. https://doi.org/10.3390/ASEC2021-11180

Academic Editor: Nicholas Vassiliou Sarlis

Published: 15 October 2021

1. Introduction

The power transformer is one the most expensive components in the electricity grid. It is an essential asset of the transmission and distribution network, since its reliable operation makes trustworthy the delivery of electricity in the network. Nowadays, power transformers can be exposed to severe operating conditions like irregular stresses or overloads, which can dramatically shorten their useful life. Monitoring power transformer health is a way to enhance their service life and increase power availability [1,2]. Some improvements in predictive maintenance and cost-saving, for this electrical equipment, are achievable by the widespread integration of new sensors and devices, especially those based on optical fibres-based technologies. Due to their low invasiveness and the dielectric nature of the optical fibers, these sensors could be effective for monitoring different physical and chemical parameters in the presence of high electromagnetic interferences as in the transmission and distribution network. An extensive review of the potential applications of fiber optic sensors in power transformers is presented in [2]. In particular, for oil-filled transformers, the frequent control of chemical markers in the insulating oil could provide an early warning of incipient failures or accelerated ageing of dielectric parts. In this framework, a surface plasmon resonance plastic optical fiber sensor, combined with specific molecularly imprinted polymers (MIPs), was reported by the authors in [3,4] as a diagnostic tool for chemical markers detection in transformers. In particular, dibenzyl disulfide (DBDS), which is correlated to the presence of corrosive sulfur in insulating mineral oil and to the formation of conductive deposits inside the insulating paper of the windings (copper sulfide) [5], and furfuraldehyde (2-FAL), which is a by-product of the thermal degradation of the paper insulation (paper windings, pressboards) in transformers [6,7] have been considered as significant markers. Because of the particular nature of the liquid in which the two markers must be determined, i.e., the transformer oil, bioreceptors as antibodies or aptamers were not considered due to the well-known characteristic of the biomolecules to better perform in aqueous media. Instead, a synthetic receptor as an MIP was applied, which has the additional advantages of being more easily developed and much less expensive than bioreceptors. For these reasons, MIPs are particularly suitable for implementation of low-cost sensing devices. In our case, they have been implemented in a plastic optical fibre, which acts as an optical platform for the synthetic receptor and makes signal transduction based on the surface plasmon resonance phenomenon possible.

As a proof of concept, this SPR sensor, combined with MIPs (POF-SPR-MIP), was used for the detection of two important markers of degradation of the solid insulating system of transformers by analysing the mineral oil, which is currently the main insulating liquid for power transformer.

In the last few decades, the new requirements of oil filled transformers with safer performances and with lower impact on the environment are pushing for the use of fluids alternative to mineral oil, like synthetic and natural esters. Esters, which consist of triglycerides or esters of fatty acids, present excellent performances in terms of biodegradability and fire points [8] and some benefits in terms of prolonging the life of the transformers' solid insulating system (that is, the oil-paper insulation). According to the state-of-art on oil-paper degradation phenomena, some physical and chemical properties of esters, like hygroscopic features, acidity, etc., could influence the nature or concentration of the chemical markers representing the by-products of transformers' oil-paper system ageing [9]. From a diagnostic point of view, the use of a new matrix in which the detection of the chemical markers should be performed can influence the optical response of the POF-SPR-MIP sensors. In this work, we focus on some features of the optical platform and on the MIP layer of the proposed POF-SPR-MIP sensor, in view of an extension of the chemical markers detection in the new ester-based insulating oil matrix. We focalized on the 2-FAL detection since the presence of antioxidants, as DBDS, is highly unlikely in ester oils [10].

2. SPR Chemo-Sensor System Based on MIP Receptor

As reported in previous works of the same authors [4,5], the optical chemo-sensor named POF-SPR-MIP is achieved through a combination of the SPR platform, manufactured in a plastic optical fiber, and a synthetic receptor, MIP, synthesized by molecular imprinting methods. An outline of the experimental configuration used for the measurements is reported in Figure 1.

The measurement apparatus consists of a halogen lamp and a spectrometer connected to a PC. The white light source presents an emission range from 360 nm to 1700 nm, whereas the spectrometer has a detection range from 350 nm to 1000 nm. The transmission spectra and data values are displayed online on a computer screen and saved by proprietary software.

The use of a relatively thick MIP layer provides several advantages with respect to other bio- and chemical receptors, as for example antibodies, guaranteeing high stability in an aggressive matrix, like the mineral oil, while maintaining high selectivity and sensitivity. Figure 1 shows a cross-section of the POF-SPR-MIP sensor.

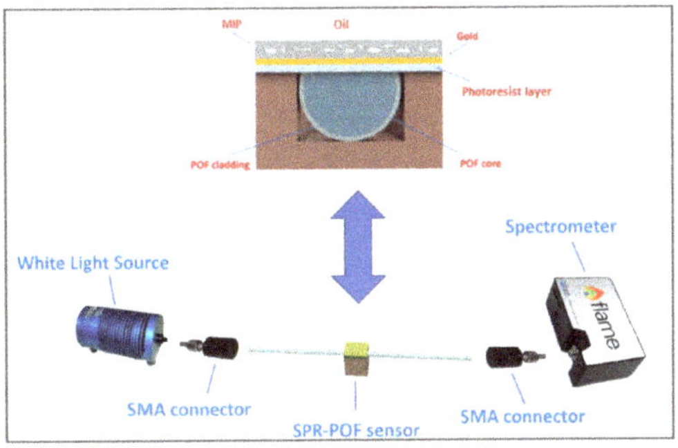

Figure 1. Scheme of the experimental setup and cross-section of the POF-SPR-MIP sensor.

3. MIP Layer

With respect to more standard molecular recognition elements (MRE), MIPs represent a new emerging class increasingly employed to realize biosensors. Up to now, very few examples have been reported in the literature of the coupling between MIPs and POFs to realize biomimetic optical sensors [11–13], a number of which came from our group. MIPs are synthetic solids with many favorable characteristics with respect to bio-receptors, such as an easier and faster preparation, the possibility of application outside the laboratory, for example in real environment conditions, and in non-aqueous solutions, and a longer duration [14]. They are porous solids containing specific sites interacting with the molecule of interest, according to a "key and lock" model. In the case of MIPs, the receptor site is a "cavity" or "pocket" which is formed in the polymeric structure by using the molecular imprinting techniques. They consist of the polymerization of the aggregates formed between the molecule of interest (which is the template) and a suitable functional monomer. The principle of the preparation of a specific MIP is briefly reported in the Scheme of Figure 2.

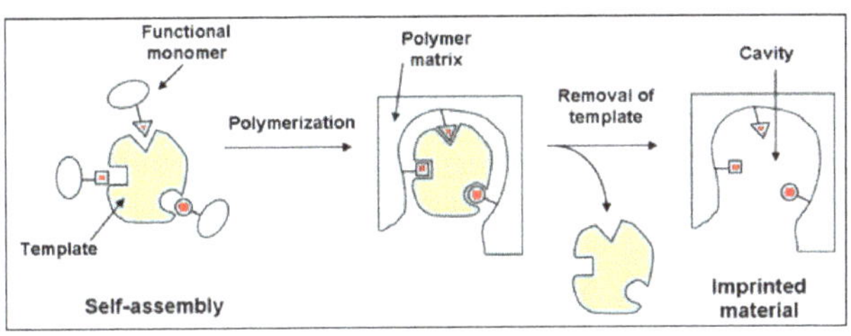

Figure 2. Principle of the preparation of a specific MIP.

In the sensors here examined, the MIP is obtained by radical polymerization of a mixture of template (2-FAL, 1 mole): functional monomer (methacrylic acid, 4 moles): cross-linker (divinylbenzene, 40 moles). This liquid is called the pre-polymeric mixture. The mixture is simply deposited over gold, and spun in order to reduce and homogenize the amount of polymer, which will be successively formed by radical polymerization at high temperatures (72 °C).

In this last decade, different MIPs have been specialized for proving the high versatility of this kind of platform in matrices with a refractive index higher than water. Water is a well-suited solvent for measurements with the platforms here described since its refractive index (RI) matches the operative range of the proposed SPR sensors (1.33–1.42) [14]. Problems can arise when liquid samples with higher RI are considered. In that case, an elegant solution to the problem can be the use of an MIP layer with RI matching the useful range, and sufficiently thick as to shield the effect of the overlaying liquid, as explained in the following section.

4. Plasmonic Response of the POF-SPR-MIP Sensor System

As an example, typical spectra of a POF-SPR-MIP sensor in different media (bulk) are reported in Figure 3. The pristine transmission spectrum is normalized to the corresponding spectrum of the platform without an MIP layer (bare surface of the sensor) in air.

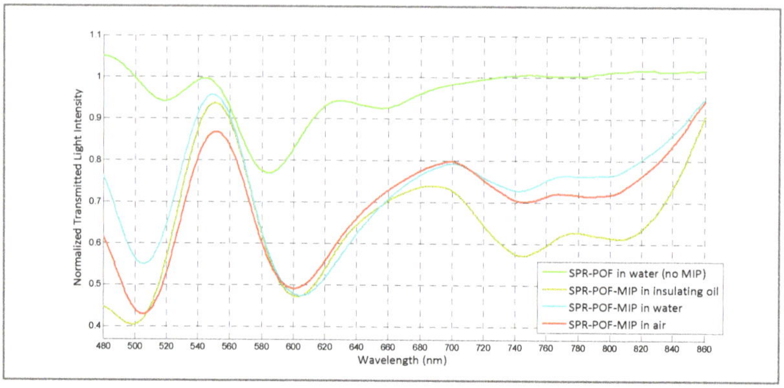

Figure 3. SPR spectra of a POF-SPR-MIP sensor in different media: air, water, and mineral oil. For comparison, the SPR spectrum of the bare surface in contact with water (without an MIP layer) is also reported.

The spectra of the POF-SPR-MIP in media with largely different refractive index, i.e., air, water, and mineral oil are very similar, different from that of the bare surface (without MIP) in water. As shown in Figure 3, the peak at about 740 nm is similar in air and in insulating oil, indicating that the plasmonic resonance does not depend on the RI of the overlaying medium (bulk solution) but only on the presence of the specific chemical marker.

The resonance, in fact, shifts to higher wavelengths when the sensor is contacted with bulk solutions containing different concentrations of 2-FAL [4]. This gives information about the affinity of the MIP for the molecule of interest. In the considered case, the affinity, measured as the equilibrium constant of the adsorption equilibrium K_{aff}, is at about 10^6 M^{-1}, corresponding to a lower detection range of ~10^{-6} M. This lower detection limit could be suitable for the detection of the 2-FAL marker in a different insulating oil matrix (ester) for a warning on accelerated ageing of the transformer.

With previous investigations, we have demonstrated the optical response of the POF-SPR-MIP for 2-FAL in different case studies, i.e., in exhausted oil samples spilled from two ex-serviced current transformers used as a bench test [4,15], in mineral oil samples, representative of the solid insulating system of transformers, exposed to controlled thermal treatment to simulate aging of the insulating paper [16,17]. In these cases, the likely presence of impurities did not induce any shift of resonance wavelengths in the SPR spectra, indicating that they are not adsorbed at the imprinted sites. This is another favorable aspects for extending this application to a new insulating matrix.

5. Conclusions

Molecularly imprinted polymers represent a key solution for the development of SPR optical fiber chemo-sensor for detecting transformers' oil degradation. Results of previous investigations confirmed that the high sensitivity of the developed POF-SPR-MIP sensors and the high specificity of the MIP layer are the main strengths allowing to face the new challenging application of these sensors to a different insulating matrix (i.e., ester oil).

Author Contributions: Conceptualization, validation, writing—original draft, L.D.M., M.P., F.A. and N.C.; writing—review & editing, L.D.M., L.Z., M.P. and A.P. All authors have read and agreed to the published version of the manuscript.

Funding: This work was financed by the Research Fund for the Italian Electrical System under the Contract Agreement between RSE and the Ministry of Economic Development-General Directorate for Energy and Mining Resources stipulated on 29 July 2009 in compliance with the Decree of 19 March 2009.

Institutional Review Board Statement: Not applicable.

Informed Consent Statement: Not applicable.

Data Availability Statement: Not applicable.

Conflicts of Interest: The authors declare no conflict of interest.

References

1. Chakravorti, S.; Dey, D.; Chatterjee, B. *Recent Trends in the Condition Monitoring of Transformers: Theory, Implementation and Analysis;* Springer: London, UK, 2013.
2. N'cho, J.S.; Fofana, I. Review of fiber optic diagnostic techniques used in power transformers. *Energies* **2020**, *13*, 1789. [CrossRef]
3. Cennamo, N.; De Maria, L.; D'Agostino, G.; Zeni, L.; Pesavento, M. Monitoring of Low Levels of Furfural in Power Transformer Oil with a Sensor System Based on a POF-MIP Platform. *Sensors* **2015**, *15*, 8499–8511. [CrossRef] [PubMed]
4. Cennamo, N.; De Maria, L.; Chemelli, C.; Profumo, A.; Zeni, L.; Pesavento, M. Markers detection in transformer oil by plasmonic chemical sensor system based on POF and MIPs. *IEEE Sens. J.* **2016**, *16*, 7663–7670. [CrossRef]
5. Scatiggio, F.; Tumiatti, V.; Maina, R.; Pompili, M.; Bartnikas, R. Corrosive Sulfur in nsulating Oils: Its Detection and Correlated Power Apparatus Failures. *IEEE Trans. Power Deliv.* **2008**, *23*, 508–509. [CrossRef]
6. Lundgaard, L.E.; Hansen, W.; Linhjell, D.; Painter, T.J. Aging of oil-impregnated paper in power transformers. *IEEE Trans. Power Deliv.* **2004**, *19*, 230–239. [CrossRef]
7. N'cho, J.S.; Fofana, I.; Hadjadj, Y.; Beroual, A. Review of Physicochemical-Based Diagnostic Techniques for Assessing Insulation Condition in Aged Transformers. *Energies* **2016**, *9*, 367. [CrossRef]
8. Scatiggio, F.; Pompili, M. Evaluation of vegetable ester for filling large power transformers. In Proceedings of the 2016 IEEE International Conference on Dielectrics (ICD), Montpellier, France, 3–7 July 2016; pp. 1052–1056. [CrossRef]
9. Rozga, P.; Beroual, A.; Przybylek, P.; Jaroszewski, M.; Strzelecki, K. A Review on Synthetic Ester Liquids for Transformer Applications. *Energies* **2020**, *13*, 6429. [CrossRef]
10. Scatiggio, F.; Rebolini, M. *Preliminary Study for Use of Vegetable Esters in Big Power Transformers;* Paper D1-202; CIGRE: Paris, France, 2014.
11. Pesavento, M.; Marchetti, S.; De Maria, L.; Zeni, L.; Cennamo, N. Sensing by Molecularly Imprinted Polymer: Evaluation of the Binding Properties with Different Techniques. *Sensors* **2019**, *19*, 1344. [CrossRef] [PubMed]
12. Uzun, L.; Turner, A.P.F. Molecularly-imprinted polymers sensors: Realising their potential. *Biosens. Bioelectron.* **2016**, *76*, 131–144. [CrossRef] [PubMed]
13. Haupt, K.; Mosbach, K. Molecularly imprinted polymers and their use in biomimetic sensors. *Chem. Rev.* **2000**, *100*, 2495–2504. [CrossRef] [PubMed]
14. Cennamo, N.; Pesavento, M.; Zeni, L. A review on simple and highly sensitive plastic optical fiber probes for bio-chemical sensing. *Sens. Actuators B Chem.* **2021**, *331*, 129393. [CrossRef]
15. De Maria, L.; Scatiggio, F.; Pesavento, M.; Cennamo, N.; Zeni, L. Toward an optical monitoring of chemical markers in transformers insulating oil. In Proceedings of the 2019 IEEE 20th International Conference on Dielectric Liquids (ICDL), Roma, Italy, 23–27 June 2019; pp. 1–4. [CrossRef]
16. De Maria, L.; Borghetto, J.; Cennamo, N.; Scatiggio, F.; Pesavento, M.; Zeni, L. Frequency dielectric spectroscopy and an innovative optical sensor to assess oil-paper degradation. *IEEE Trans. Dielectr. Electr. Insul.* **2020**, *27*, 1728–1735. [CrossRef]
17. De Maria, L.; Cice, L.; Bartalesi, D.; Borghetto, J.; Tavakoli, A.; Scatiggio, F.; Gasparini, D.; Pesavento, M. An optical SPR sensor for monitoring accelerated ageing of oil-paper insulation of transformers. In *Sensors and Microsystems; Lecture Notes in Electrical Engineering;* Springer: Cham, Switzerland, 2019; Volume 629.

engineering proceedings

Proceeding Paper

Effects of MgO-Based Expansive Agent on the Characteristics of Expansive Concrete †

Abir Mahmood [1,*], A. B. M. Amrul Kaish [1], Nor Farhana Binti Ab Gulam [1], Sudharshan N. Raman [2], Maslina Jamil [3] and Roszilah Hamid [1]

1 Department of Civil Engineering, Faculty of Engineering and Built Environment, Universiti Kebangsaan Malaysia, Bangi 43600, Selangor, Malaysia; amrul.kaish@ukm.edu.my (A.B.M.A.K.); P112072@siswa.ukm.edu.my (N.F.B.A.G.); roszilah@ukm.edu.my (R.H.)
2 Civil Engineering Discipline, School of Engineering Monash University Malaysia, Jalan Lagoon Selatan, Bandar Sunway 47500, Selangor, Malaysia; sudharshan.raman@monash.edu
3 Department of Architecture and Built Environment, Universiti Kebangsaan Malaysia, Bangi 43600, Selangor, Malaysia; maslinajamil@ukm.edu.my
* Correspondence: P108539@siswa.ukm.edu.my; Tel.: +60-172294341
† Presented at the 2nd International Electronic Conference on Applied Sciences, 15–31 October 2021; Available online: https://asec2021.sciforum.net/.

Abstract: Expansive concrete are used to reduce cracking caused by drying shrinkage in concrete structures such as slabs, beams, columns, and pavement constructions. Although CaO and Sulphoaluminate based expansive agents have been used for decades, MgO-based expansive agents have demonstrated superior performance since 1970, especially for concrete dam structures. It has been proven that compensating shrinkage with MgO expansion efficiently prevents thermal cracking of mass concrete, reduces the expense of temperature control systems, and speeds up the construction process. This paper reviews several parameters such as reactivity, thickness of water film, curing condition, additive ratio, and calcination condition that affects expansibility, strength, soundness, durability, flowability, pore structures, crystal size, and hydration activity. The review indicates that the expansion characteristics of MgO may be designed flexibly by altering the calcination conditions (calcining temperature and residence time), maintaining a certain curing temperature, and tweaking its microstructure.

Keywords: expansive agent; MgO-based expansive agent; curing conditions; calcining temperature

Citation: Mahmood, A.; Kaish, A.B.M.A.; Gulam, N.F.B.A.; Raman, S.N.; Jamil, M.; Hamid, R. Effects of MgO-Based Expansive Agent on the Characteristics of Expansive Concrete. *Eng. Proc.* **2021**, *11*, 14. https://doi.org/10.3390/ASEC2021-11165

Academic Editor: Samuel Adeloju

Published: 15 October 2021

Publisher's Note: MDPI stays neutral with regard to jurisdictional claims in published maps and institutional affiliations.

1. Introduction

Buildings, bridges, tunnels, dams, and highways are examples of civil infrastructure that may be damaged in many ways across the world. Cracks are one of the types of damages that can severely reduce the service life of reinforced concrete structural parts [1]. Concrete constructions can crack as a result of structural design flaws, poor construction techniques, or unfavourable interactions with the environment. Concrete fractures might start off as tiny microcracks that link throughout the microstructure, eventually producing significant internal damage before becoming evident on the outer surfaces [2]. One of the most common causes of cracking is the shrinkage of cement-based materials, which has a detrimental impact on building durability [3]. It often generates cracks which are produced by heat contraction or drying shrinkage when exposed to climatic conditions and restricted internally or externally [4]. To cater the shrinkage cracking of concrete, for more than 50 years, researchers and practitioners have investigated and applied expansive hydraulic cements, including expansive agents, which accommodates the shrinkage characteristics of concrete and further enhances strength and durability [4]. The three common expansive agents are based on calcium sulfoaluminate, calcium oxide (CaO), and magnesium oxide (MgO). When it comes to preparing shrinkage-compensating concrete, sulfoaluminate-type and calcium oxide-type expansive agents are the most often employed expansive agents. Some drawbacks, such as poor temperature stability and an unpredictable hydration pace, have limited their application. A potential expansive agent, MgO-based expansive agent, has recently been discovered. It is typically generated by calcining magnesite, and the reactivity of expansive agents may be readily regulated by the calcining conditions used to produce it. Years of study have resulted in the successful use of MgO-type expansive agents in a variety of applications. Because of having chemically stable hydration characteristics, relatively low water requirement for hydration, designable expansion property, and lack of obvious shrinkage at a late age, MgO-based expansive agents have been used in various parts of more than 60 dams in China. The poor thermal conductivity of mass concrete allows for considerable temperature increases caused by the heat of cement hydration. This produces significant thermal shrinkage throughout the cooling stage, which is a key contributing factor to shrinkage cracking at later ages. Conventional methods of preventing the thermal shrinkage cracking in mass concrete, such as using cement with low hydration heat, mixing with a high volume of supplementary cementitious materials as cement replacements, pre-cooling the raw materials and/or post-cooling the concrete by embedding cooling pipes, are widely used to reduce the maximum temperature rise and, consequently, the thermal shrinkage cracking. Nonetheless, some of these conventional processes are expensive, and due to the extended duration of the process, thermal shrinkage of super big mass concrete, such as dam concrete, cannot always be fully avoided.

Engineers have been looking for an expansive agent that can provide the necessary expansion to compensate for the particular shrinkage of concrete. MgO-based expansive agents surpass conventional expansion agents in terms of their designable expansion properties, which are very important. As a result, this paper will review the factors that effect the characteristics of an expansive concrete that contains MgO as the expansive agent. The characteristics will be defined by an assessment of the expansibility, durability, volume deformation, thermal cracking and so on.

2. MgO-Based Expansive Agent: A Brief History and Application

It all started in 1884, when periclase (magnesia) was used to develop expansive behavior in concrete. However, due to delayed excessive expansion and the hydration process, some bridges in France and the city hall of Kassel in Germany failed. As a result, the usage of magnesia was limited to 6% in America and 5% for the rest of the world. Later, in 1970, a concrete dam named Baishan dam was constructed using 5% MgO as an expansive agent. The results were surprisingly good: no cracks developed in the concrete foundation, although no measures were taken regarding temperature control. Its success attracted attention from many researchers, who began developing MgO-based expansive agents and, thus, several dams were constructed in China using the same expansive cement. Dam construction technology has demonstrated that delayed micro-expansion of MgO concrete can be used to allow the concrete to produce appropriate pre-compressive stress, reducing or balancing the tensile stress generated during concrete temperature drop, which improves the crack-resisting performance of concrete, and fundamentally resolves the crack control problem. Longer blocks, thicker layers, non-installed construction joints from the left to the right bank, and all-weather continuous concrete placement may also be achieved, as well as a reduction in construction time and project cost. This is a major advancement in concrete dam building techniques [5].

MgO offers numerous benefits over typical expansive cements or expansive additives such as sulfoaluminate, aluminate clinker, or CaO-based expansive additives, which are currently widely utilised in Japan, the United States, and China [6]. This comprehends $Mg(OH)_2$, its chemically stable hydration characteristics, relatively low water requirement for the hydration of MgO, and specifically its designable expansion properties [7]. The benefits of MgO in correcting concrete shrinkage and improving concrete durability have been demonstrated in both experimental and industrial applications.

Coming forward to 2021, the demand for dams to produce hydropower, reservoir of fresh water etc. is enlarging day by day. As a result, more and more large concrete dams with heights over 200 m will be constructed worldwide, which provides a potentially huge market for the application of MgO concrete. The mix proportions of MgO concrete must be well-designed, and the expansion must be carefully controlled to meet the shrinkage of the concrete. Thus, for dam concrete, temperature rise, thermal stress, and concrete shrinkage in various structural elements of the dam must first be anticipated using simulation calculations, after which the optimum expansion process of MgO concrete required for complete shrinkage compensation can be projected.

3. Factors Influencing the Characteristics of Expansive Concrete

MgO-based expansive agents manipulate the characteristics of expansive concrete in various ways. In this paper, the factors are categorized into five different areas in which several researchers have conducted experiments and have reached conclusions, as shown in Table 1.

Table 1. Factors affecting the characteristics of expansive concrete.

Factors	Authors
Hydration Process	[8–11]
Calcining Temperature	[1,12–16]
Curing Conditions	[11,17–19]
Autogeneous Volume Deformation	[20–22]
Mechanical and Self-healing Properties	[7,23]

3.1. Hydration Process

Desiccation, chemical reactivity, and temperature change can produce volumetric shrinkage of cementitious materials, which can create tensile stress and cracking of concrete that is confined in a structure [8]. Expansive agents were added which react with water and cement in order to develop volume expansion and decrease concrete shrinkage. Due to its controlled reactivity speed and expansive capacity, as well as its extremely stable hydration characteristics, MgO is an excellent expansive agent. Furthermore, MgO's hydration characteristics is just $Mg(OH)_2$, a very simple and stable crystal that does not dehydrate until temperatures reach 350 °C, making MgO a safe expanding agent, whereas other expansive agents such as Sulphoaluminate-based ones start to dehydrate from 70 °C [9]. An experiment conducted by Lu et al. [10] found that the conductivity of MgO-based expansive agents at 50 °C are double as of at 25 °C, proving that curing temperature influences the hydration characteristics of Mgo-based expansive agents in concrete. Supporting this, Mo et al. [11] also found out that if the type of EA and curing age is kept constant, the temperature is proportional to the hydration rate.

3.2. Calcination Temperature

MgO particles may be employed as expanding agents to compensate for thermal shrinkage and therefore avoid cracks in mass concrete caused by thermal stresses [12]. MgO is typically produced by calcining magnesite at a temperature ranging from 900 °C to 1450 °C, which creates differences in the reactivity and expansion rate [13]. Sherir et al. [1] found that when MgO particles were burned at 900 °C, they produced a more reactive MgO-type expanding agent than when they were burned at 1300 °C. Moreover, Mo et al. [14] investigated that higher calcination temperatures and longer residence times promote MgO grain development, resulting in a reduction in inner pore volume and specific surface area, and consequently a reduction in hydration activity. Early hydration causes shrinkage cracking at a later age. Magnesium oxide crystals become increasingly perfect as the calcining temperature rises, lowering reaction activity [15]. Chen et al. [16] maintained the residence time constant at 1 h and found out that MgO-based expansive agents calcined at 900 °C reacted with water quickly and accomplished 85% of hydration within 2 days.

When the temperature was raised to 1100 °C, it took 30 days to hydrate 83% and only 10% after increasing the temperature to 1300 °C. With this in mind, the adjustment of magnesite calcination temperature and residency duration could be achieved by various types of expansive agents designed to compensate for dam concrete needs.

3.3. Curing Conditions

Li et al. [17] studied the temperature history of MgO-based expansive agents, finding it very important in regard to hydration and expansive characteristics. Standard water curing occurs in between 20 °C to 80 °C. Fang and Yongchao [18] experimented the effects by keeping the curing duration to 90 days with 10% of MgO-based expansive agent mixed with the cement. He found out that the range of 20 °C–40 °C has very little effect on the expansion but the range of 40 °C to 80 °C increases the expansion rate exponentially as the temperature increases. Findings of Mo et al. [11] also shows the increment in expansion rate as the temperature raises, observed after 360 days. Because the expansion of cement paste is determined by the conflicting effects of the expansive stress generated by MgO hydration and the resistance of the hydrated cement matrix to expansion, the influence of curing temperature on cement paste expansions may be attributed to the increment of the hydration of MgO by increasing the temperature. The MgO hydrated quicker and generated more hydration products at a higher curing temperature, inducing a faster and bigger expansion, especially at an early age. Qian et al. [19] investigated the mechanical effects of different curing temperatures. He found that at later ages, 20 °C–60 °C curing temperature results in decrement of compressive strength, while cured below 20 °C provides better compressive strength values. Two well-known processes have been proposed to explain the crossover effect observed in MgO-based expansive agents. The first, and probably most important, is that the quick hydration rate at a younger age would result in inhomogeneous and porous microstructures, decreasing the final strength of cementitious materials. The second point to note is that the response rate is significantly quicker than the diffusion rate. As a result, the majority of the hydration products remain near the unhydrated grains. As a result, these densely deposited hydration products will form barriers to ionic diffusion, limiting hydration rate and compressive strength increase at later ages.

3.4. Autogenous Volume Deformation

Autogenous volume deformation of the concrete is an essential index often employed in building design and control design of mass concrete projects when calculating a temperature stress. An experiment conducted by Gao et al. [20] showed that when the concrete specimens had been cured for 180 days, the autogenous volume expansion of conventional concrete with MgO-based expansive substance increased with the amount of MgO and curing age and tended to remain constant. The shrinkages of concrete were not found when the MgO-based expansive agent were blended in the concrete, which were compensated by the hydration of the expansive materials in concrete. Nguyen et al. [21] researched on the effect of different curing temperature and percentage of MgO content on autogenous volume deformation. He found that the hydration of the expanded elements used in concrete compensated for concrete shrinkage, whereas MgO mixed concrete tends to expand fast at a young age but remains constant at older ages. Moreover, Chen et al. [22] studied on autoclave expansion deformation and found that with the rise of magnesium oxide mixture concentration, the autoclave expansion rate for cement-based materials increases. This applies regardless of the kind or the size of the material. When the amount of combined magnesium oxide reaches a particular threshold, the autoclave expansion of the specimen increases significantly.

3.5. Mechanical and Self-Healing Properties

Proper expansion of concrete with the appropriate amount of MgO-based expansive agent addition can enhance mechanical characteristics. Mo et al. [7] stated that the effect of MgO-based expansive agents on the mechanical performance or durability of concrete is determined by the impact of expansion on the microstructure. The expansion helps to improve the microstructure of concrete by reducing pore size and total pore volumes, for example. It has the potential to improve both the mechanical strength and the durability of concrete. Qureshi et al. [23] also found out that stiffness improvements were found to increase with the addition of higher proportions of MgO. While treated in an immersed state for self-healing, samples were discovered to have expanded. The crack points might become closer as a result of this expansion, which speeds up the self-healing mechanisms. Stronger amounts of expansive MgO in the samples revealed higher expansive characteristics during healing, resulting in faster healing of larger cracks. Overall, the addition of MgO enhanced the durability by increasing crack sealing and strength recovery (load recovery and stiffness improvement). However, the quantity of expanding MgO that causes the best self-healing performance has a limit.

4. Conclusions

MgO-based expansive agents have vastly improved since their original applications, and are now used industrially as a reliable expansive agent. The volume stability, mechanical strength, and durability of MgO concrete are all directly connected to the raw ingredients, mix percentage, curing conditions, calcination temperature etc. The effect of MgO's expansion on concrete performance is determined by its influence on the microstructure of the concrete. Appropriate expansion helps to densify concrete and improves its performance, whereas excessive expansion can damage the microstructure of concrete and degrade its performance. However, the manufacture of MgO with adequate properties on a large industrial scale, precise control of the expansion process of concrete containing the appropriate type and addition content, and evaluation of MgO concrete soundness are some of the challenges faced in the application of MgO concrete in constructing super large concrete dams. The expansion of MgO concrete must be carefully planned and precisely regulated, taking into consideration the impacts of temperature, restraint conditions in various structural components, and the thermal stress field in the dam structure. Furthermore, simulation technology for predicting MgO concrete temperature, thermal deformation, and expansion is critical and should be enhanced.

Institutional Review Board Statement: Not Applicable.

Informed Consent Statement: Not Applicable.

Data Availability Statement: Not Applicable.

Acknowledgments: The authors acknowledge Universiti Kebangsaan Malaysia for providing the necessary opportunities & funding for this research through "Research Graduate Assistance" scheme (FRGS/1/2019/TK01/UKM/02/2) and all my family members for encouraging me throughout the entire journey.

Conflicts of Interest: The authors declare that there is no conflict of interest.

References

1. Sherir, M.A.; Hossain, K.M.; Lachemi, M. The influence of MgO-type expansive agent incorporated in self-healing system of Engineered cementitious Composites. *Constr. Build. Mater.* **2017**, *149*, 164–185. [CrossRef]
2. Sherir, M.A.; Hossain, K.M.; Lachemi, M. Development and recovery of mechanical properties of self-healing cementitious composites with MgO expansive agent. *Constr. Build. Mater.* **2017**, *148*, 789–810. [CrossRef]
3. Mo, L.W.; Deng, Y.; Lu, A.Q.; Deng, M. Preparation of MgO-and CaO-bearing expansive agent used for cement-based materials. *Key Eng. Mater.* **2013**, *539*, 211–214. [CrossRef]
4. Li, S.; Feng, Y.; Yang, J. Expansion Mechanism and Properties of Magnesium Oxide Expansive Hydraulic Cement for Engineering Applications. *Adv. Mater. Sci. Eng.* **2021**, *2021*, 5542072. [CrossRef]
5. Chen, C.L.; Tang, C.S.; Zhao, Z.H. Application of MgO concrete in China Dongfeng arch dam foundation. *Adv. Mater. Res.* **2011**, *168*, 1953–1956. [CrossRef]
6. Mather, B.; Mehta, P.K. Committee; Discussion of the report by ACI Committee 223, "Expansive Cement Concretes–Present State of Knowledge". *J. Am. Concr. Inst. Proc.* **1971**, *68*, 293–296.

7. Mo, L.; Deng, M.; Tang, M.; Al-Tabbaa, A. MgO expansive cement and concrete in China: Past, present and future. *Cem. Concr. Res.* **2014**, *57*, 1–12. [CrossRef]
8. Cao, F.; Yan, P. The influence of the hydration procedure of MgO expansive agent on the expansive behavior of shrinkage-compensating mortar. *Constr. Build. Mater.* **2019**, *202*, 162–168. [CrossRef]
9. Yan, P.; Qin, X. The effect of expansive agent and possibility of delayed ettringite formation in shrinkage-compensating massive concrete. *Cem. Concr. Res.* **2001**, *31*, 335–337. [CrossRef]
10. Lu, X.; Geng, F.; Zhang, H.; Chen, X. Influence of MgO-type expansive agent hydration behaviors on expansive properties of concrete. *J. Wuhan Univ. Technol.-Mater. Sci. Ed.* **2011**, *26*, 344–346. [CrossRef]
11. Mo, L.; Fang, J.; Hou, W.; Ji, X.; Yang, J.; Fan, T.; Wang, H. Synergetic effects of curing temperature and hydration reactivity of MgO expansive agents on their hydration and expansion behaviours in cement pastes. *Constr. Build. Mater.* **2019**, *207*, 206–217. [CrossRef]
12. Mahmood, A.; Kaish, A.B.M.; Raman, S.; Jamil, M.; Hamid, R. Effects of Different Expansive Agents on the Properties of Expansive Cementitious Materials. In Proceedings of the 3rd Global Congress on Construction, Material and Structural Engineering, online, 23–24 August 2021.
13. Mehta, P.K.; Pirtz, D. Magnesium oxide additive for producing self-stress in mass concrete. In Proceedings of the 7th International Congress on the Chemistry of Cement, Paris, France, 1 January 1980; Volume 3.
14. Mo, L.; Deng, M.; Tang, M. Effects of calcination condition on expansion property of MgO-type expansive agent used in cement-based materials. *Cem. Concr. Res.* **2010**, *40*, 437–446. [CrossRef]
15. Gao, P.W.; Geng, F.; Lu, X.L. Hydration and expansion properties of novel concrete expansive agent. *Key Eng. Mater.* **2009**, *405*, 267–271. [CrossRef]
16. Chen, X.; Yang, H.Q.; Li, W.W. Factors analysis on autogenous volume deformation of MgO concrete and early thermal cracking evaluation. *Constr. Build. Mater.* **2016**, *118*, 276–285. [CrossRef]
17. Li, H.; Tian, Q.; Zhao, H.; Lu, A.; Liu, J. Temperature sensitivity of MgO expansive agent and its application in temperature crack mitigation in shiplock mass concrete. *Constr. Build. Mater.* **2018**, *170*, 613–618. [CrossRef]
18. Yu, F.; Li, Y. The effect of curing conditions on the expansion efficiency of MgO expansion agent. In *E3S Web of Conferences*; EDP Sciences: Les Ulis, France, 2013; Volume 233.
19. Qian, X.; Jiang, L.; Song, Z.; Chu, H.; Yan, X.; Zhang, Y. Impact of elevated curing temperature on mechanical properties and microstructure of MgO-based expansive additive cement mortars. *Struct. Concr.* **2020**, *21*, 1082–1092. [CrossRef]
20. Gao, P.W.; Xu, S.Y.; Chen, X.; Li, J.; Lu, X.L. Research on autogenous volume deformation of concrete with MgO. *Constr. Build. Mater.* **2013**, *40*, 998–1001. [CrossRef]
21. Nguyen, V.C.; Zhang, X.S.; Nguyen, V.N.; Phan, D.T.; Liu, G. Experimental study on autogenous volume deformation of RCC mixed with MgO. *IOP Conf. Ser. Mater. Sci. Eng.* **2020**, *794*, 012047. [CrossRef]
22. Chen, C.; Li, W.; Yang, D.; Zhao, Z.; Yang, H. Study on autoclave expansion deformation of MgO-admixed cement-based materials. *Emerg. Mater. Res.* **2017**, *6*, 422–428. [CrossRef]
23. Qureshi, T.S.; Al-Tabbaa, A. Self-healing of drying shrinkage cracks in cement-based materials incorporating reactive MgO. *Smart Mater. Struct.* **2016**, *25*, 084004. [CrossRef]

engineering proceedings

MDPI

Proceeding Paper

Development of a Curd Cake for Patients with Type 2 Diabetes: Influence of Replacing Sugar with Sweetener on Nutritional Value †

Elena Aksonova, Victoria Evlash and Sergey Gubsky *

Depertment of Chemistry, Biochemistry, Microbiology and Hygiene of Nutrition, State Biotechnological University, 61051 Kharkiv, Ukraine; eaksonova@gmail.com (E.A.); evlashvv@gmail.com (V.E.)

* Correspondence: sergey.m.gubsky@gmail.com

† Presented at the 2nd International Electronic Conference on Applied Sciences, 15–31 October 2021; Available online: https://asec2021.sciforum.net/.

Abstract: Sugar is one of the main ingredients of baked products, so reduction of sucrose negatively affects product appearance, texture, and mouthfeel. However, by replacing it with sweeteners such as sucralose, it is possible to create desserts with a low glycemic index and low calorie content. The aim of this study was to investigate the physico-chemical and nutritional properties, and sensory characteristics of sugar replaced with curd cake made using sucralose. Optimal results were obtained with replacement up to 50%, leading curd cake similar to the control products and having a high level of acceptance in sensory evaluation. A vibrational analysis using FTIR spectroscopy showed in these samples the presence of hydrogen bonds between water molecules and a dense packing of texture. The study of adsorption/desorption isotherms indicates the predominance of the mechanism of monomolecular physical adsorption with a weak interaction of adsorbent-adsorbate in samples with sugar replacement up to 50%. A further increase in the sugar substitute content increases the contribution from polymolecular adsorption and capillary water, which is reflected in the deterioration of sensory characteristics. The developed technology made it possible to obtain a curd cake with a caloric content of 14% lower than that of an analogue, which in its organoleptic characteristics and structure does not differ from the analogue product and contains 26% more protein.

Keywords: diabetes; sweetener; sucralose; curd cake; nutritional value

fv **Citation:** Development of a Curd Cake for Patients with Type 2 Diabetes: Influence of Replacing Sugar with Sweetener on Nutritional Value. *Eng. Proc.* **2021**, *11*, 15. https://doi.org/10.3390/ASEC2021-11123

Academic Editor: Nunzio Cennamo

Published: 15 October 2021

Publisher's Note: MDPI stays neutral with regard to jurisdictional claims in published maps and institutional affiliations.

1. Introduction

The rise in diabetes and obesity is the driving force behind the development of the food industry towards the development of low-calorie products by reducing sugar content. One of the ways to implement this trend is to use natural and synthetic sweeteners [1]. Substances with virtually "zero calories" and a very intense sweet taste that are used in small quantities to replace much higher amounts of sugar in products and are approved for use in the food industry include sucralose. However, in the process of optimizing such technologies, it becomes necessary to solve the problem of compensating for the negative impact on texture due to partial or complete replacement of sugar with a sweetener in the product formulation. In such cases, it is necessary to correct the texture of the final product by adding other components to the recipe [2]. Earlier in [3], a curd cake technology with partial replacement of sugar with sucralose was proposed. The problem of compensating for the negative effect on the consistency of the product with a low sugar content was solved by introducing additional amounts of cottage cheese and melange into the recipe (cottage cheese and melange were introduced to replenish the mass of removed sugar). This technique, of course, did not allow to restore the dry matter content; however, it made it possible to increase the amount of protein in the finished product. As is known, proteins (both egg white and casein proteins) are involved in protein network formation during thermosetting [4]. However, given the fact that cottage cheese and melange contain about 70% water, significant weight loss can be predicted during heat treatment. However, the hydrophilic properties of proteins and the ability to retain moisture should also be considered.

The aim of this study was to produce high quality curd cake using different combinations from sucrose and sucralose in order to reduce the calorie content without affecting the physico-chemical characteristics and nutrition value (Figure 1).

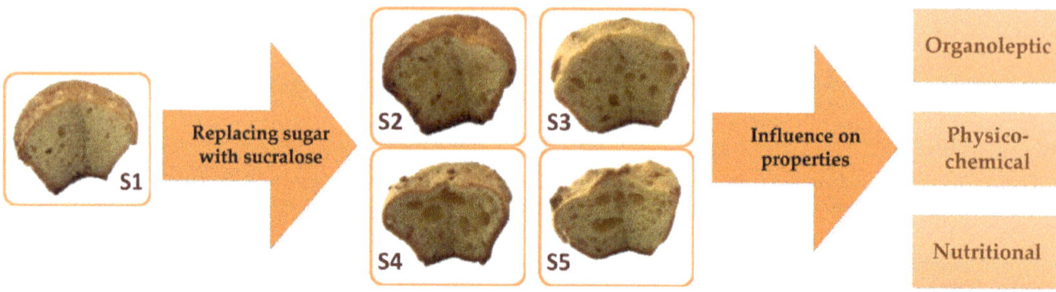

Figure 1. A graphical scheme of study approach with appearance of the curd cake samples with the replacement of 0% (**S1**), 30% (**S2**), 50% (**S3**), 70% (**S4**), and 100% (**S5**) sugar with a sucralose.

2. Materials and Methods

2.1. Materials and Samples Preparation

The ingredients used in this study were obtained from local stores in city Kharkiv, Ukraine. Sucralose sold as SPLENDA™ was purchased from Tate & Lyle (PLC, UK). This sweetener contains maltodextrin and sucralose (the amount of sucralose is 1%).

A sample S1 as control curd cake was manufactured according to [5]. Other samples were as in [3] with partial or complete substitution of sugar for sucralose (Figure 1).

2.2. Methods

Fourier transform infrared coupled to attenuated total reflectance (ATR–FTIR) spectra of curd cake samples were obtained using Nicolet iS5 FT-IR Spectrometer with the versatile iD5 ATR Accessory (Thermo Scientific, Waltham, MA, USA) as the result of the accumulation of 32 scans with a resolution of 4 cm^{-1} in the range of 4000–600 cm^{-1}. The processing of IR spectra and calculations were performed using the Fourier transforms (FFT filter) procedure followed by deconvolution of the second derivative using the PeakFit (Seasolve, San Jose, CA, USA), Origin2019 (OriginLab, Northampton, MA, USA) and KnowItAll Informatics System 2021 Academic Edition (Join Wiley Sons Co., Hoboken, NJ, USA) software. The quality of the results obtained was assessed using mean values, standard deviations, and coefficient of variation.

Water vapor adsorption/desorption isotherms were obtained and measured at 293 K in a range of relative humidity of 0.20–0.95 with an adsorption vacuum Mac-Ben apparatus with helical spring quartz scales. The moisture of the samples was determined using a Moisture Analyzer ADGS-50 (Axis, Gdank, Poland). Specific surface area and other monolayer parameters (capacity of monolayer, constant C of the adsorbent-adsorbate interaction) of the samples according to water vapor adsorption/desorption isotherms calculated with use equation BET [6].

3. Results and Discussions

3.1. Influence of Replacing Sugar with Sweetener Organoleptic Indicators

The organoleptic analysis of the obtained samples made it possible to assert that the products based on sugar and sucralose have the same level of sweetness (Table 1).

This confirms previous studies with fewer samples [7]. However, the introduction of increased amounts of cottage cheese and melange into the cottage cheese cake recipe does not completely solve the problem of compensating for the replacement of sugar. This technique made it possible to restore the total soluble solids of the semi-finished product for baking and increase the amount of protein in the finished product. However, the water content in these components and its significant loss during heat treatment leads to the formation of cavities in the texture. This was especially noticeable for samples S4 and S5. The sensory analysis allows us to conclude that the inexpediency of replacing more than 30–50% in the prescription composition of white crystalline sugar with sucralose.

Table 1. Organoleptic properties of curd cakes.

Indicator	S1	S2	S3	S4	S5
Form	Correct, corresponding to the form established in the recipe			Incorrect, lightly flattened	
Surface	The surface of the cupcakes with the presence of cracks and tears, which do not change the product				
Color	Dark brown	Dark brown	Brown	Light brown	Light brown
Appearance (section)	A well-baked curd cake with good porous structure			The presence of large voids	
Taste and smell	Intrinsic to this variety of cake without the foreign taste and smell				

3.2. Influence of Replacing Sugar with Sweetener on Physico-Chemical Properties

Moisture content and its state are important indicators of product quality when introducing new ingredients into product formulations. Figure 2a shows a fragment of the ATR-FTIR spectrum in the range of 3050–3600 cm^{-1} of the samples of curd paste, in the region of vibration of the OH group of water.

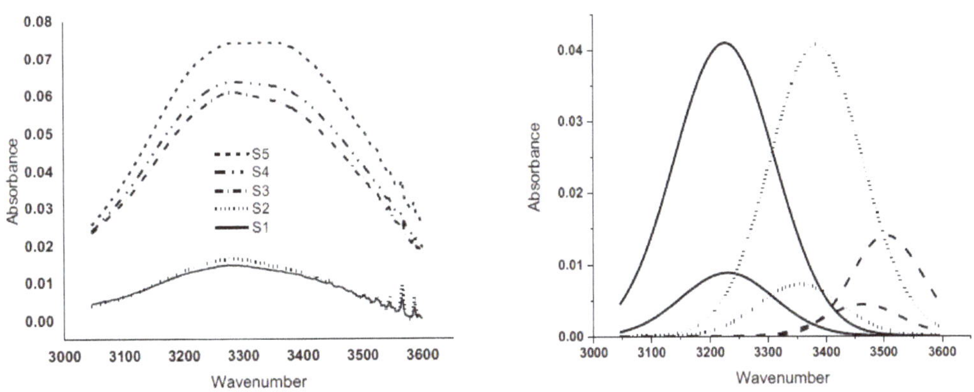

(a) (b)

Figure 2. (**a**) Fragment of ATR-FTIR spectrum in range 3000–3600 cm^{-1} of curd cares samples; (**b**) calculated water bands G1 (solid), G2 (point), and G3 (dashed) for samples S1 and S5.

The spectrum for sample S1 has one peak at 3275 cm^{-1}. This peak is also present in the spectra of samples with sweeteners. Samples S2–S5 are characterized by the existence of a second peak at 3355 cm^{-1}. The presence of a band with peaks in the range 3275–3355 cm^{-1} indicates that the water in the product is mainly in the form of network water [8]. A shift of this band to the region of large values is observed with an increase in the concentration of the sweetener. With an increase in the concentration of the sweetener, the band becomes wider, which indicates a wider energy variability of water molecules. The shape of the band resembles a plateau with two peaks, around 3270 and 3380 cm^{-1}. With an increase in the content of the sweetener, the ratio of the intensity of these two peaks undergoes changes in the direction of increasing the intensity of the band with a higher oscillation frequency. A similar positive correlation between the position of the band of stretching vibrations of the OH group and the concentration of maltodextrin was obtained [9].

A more detailed study of this issue is possible using the approach proposed by the authors [8,10]. The essence of the approach is the possibility of studying the state of moisture in systems that simultaneously contain water and carbohydrates by analyzing a wide band of stretching vibrations of the OH group of water in IR spectra. In such systems, this band is a combination of the contributions of ordered tetrahedral H-bonded structures (G1), deformed H-bonded structures (G2), and water molecules that have free hydroxyls or small-sized associates (G3). All the above facts indicate a change in the state of hydrogen bonds of water in the food matrix of curd cake when replacing sugar with a sweetener. For a more detailed analysis of the state of moisture in the product under study, the contributions of the above water structures G1, G2, and G3 were calculated for all samples. The calculation results for samples S1 and S5 are shown in Figure 2b. Three bands were obtained for each of the states of water, which in total characterizes the band of stretching vibrations of the OH group in the range of 3050–3600 cm^{-1}. For a structure bound by strong H-bonds G1, a band was studied near 3230 cm^{-1}, for water molecules with weak or deformed H-bonds in the range 3355–3385 cm^{-1}, and for free water at 3050–3600 cm^{-1}. The position of the bands shifts towards higher vibration frequencies with increasing sweeteners content. This fact corresponds to the weakening of the hydrogen bond of water, and from the point of view of texture regarding its loose packing. The latter is confirmed by the organoleptic control of the samples (Figure 1). This change in texture density explains the greater moisture content and water activity for samples S2–S5 compared to sample S1 (Table 2).

Table 2. Physico-chemical properties of samples of curd cakes.

Property	S1	S2	S3	S4	S5
Moisture content, %	19.48	19.90	20.71	22.14	24.54
Water activity	0.903	0.930	0.949	0.969	0.989
BET specific surface area, m^2/g	165	175	153	174	126
BET monolayer capacity	2.19	2.32	2.03	2.31	1.67
Constant C	2.96	2.79	3.11	3.40	5.86

To understand the state of water in the sample, sorption isotherms were studied, which give an idea of both the amount of water and its energy state, which characterize the activity of water. According to classical concepts, this is manifested by the adsorption of mesoporous solids with capillary-condensation hysteresis with three zones of water activity. The samples have almost a similar adsorption structure, since their adsorption curves coincide in shape (Figure 3). The presence of hysteresis on the curves indicates the presence in the system, in addition to the process of physical adsorption and capillary phenomena, the process of swelling of the corresponding components and chemisorption processes. The latter is indicated by the residual amount of water at the level of 1.5–2.0% of the amount adsorbed. A detailed analysis of adsorption isotherms suggests that the largest amount of water from the total is in a hygroscopic state. This number increases in the series of samples from S1 to S5.

This difference is due to polymolecular adsorption of moisture and capillary water, while monomolecular physical adsorption predominates for the control sample. The calculated values of the constant C of the BET equation for the samples have a magnitude of 2–5, which characterizes the weak adsorbent-adsorbate interaction. As the sweetener content increases, this interaction weakens. The value of the specific surface area has a similar behavior (Table 2).

It should be noted that the data of the applied physicochemical methods are in good agreement. The change in the width and intensity of the shape of the OH-group stretching vibration band with an increase in the sugar substitute content in the samples obtained in

ATR-FTIR studies positively correlates with the experimental data on sorption and water content in the samples.

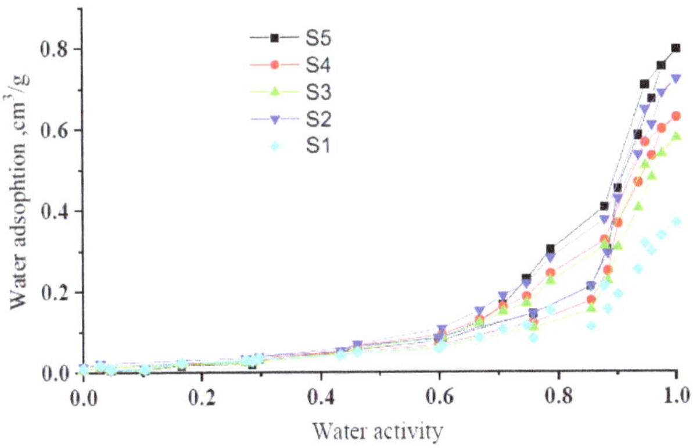

Figure 3. Isotherm water vapor adsorption/desorption of samples of curd cakes.

3.3. Influence of Replacing Sugar with Sweetener on Nutritional Value

The resulting product contains more protein, has fewer carbohydrates and fewer calories. Calculation of the energy value of samples S1, S2, and S3 based on the content of proteins, carbohydrates, and fats gives the following results 410, 386, and 369 kKal/100 g, respectively. Thus, the proposed technology makes it possible to reduce the calorie content of the product by 10–20%, depending on the replacement of sugar with a sweetener without deteriorating the nutritional value of the product.

4. Conclusions

The results of this investigation show that an encouraging novel formulation of curd cake production with a partial replacing sugar with sucralose was developed. Curd cake formulated with partial replacement of sucrose with up to 50% sweetener had organoleptic characteristics comparable with the same product prepared with 100% sucrose. When more than 50% sweetener replaced sugar in formulation, some negative sensory ratings such as poor appearance, incorrect and lightly flattened form, presence of large voids were noted. The last characteristic is confirmed by the data of the ATR-FTIR studies, which indicate the weakening of the hydrogen bond of water and the loose packing of the texture for the indicated samples. This fact is quantitatively confirmed by studies of the energy state of water by sorption/desorption isotherms. The calculation of the energy value of the obtained curd cake indicates a reduced calorie content of the product in comparison with the control sample on sugar. However, further research is needed to conduct an in vitro digestion assay to assess the possible reduction in predicted glycemic response from the developed curd cake.

Author Contributions: Conceptualization, E.A., V.E. and S.G.; Methodology, E.A. and S.G.; Software, S.G.; Validation, E.A.; Formal analysis, E.A. and V.E.; Writing—original draft preparation, E.A. and S.G.; Writing—review and editing, E.A., V.E. and S.G. All authors have read and agreed to the published version of the manuscript.

Funding: This research received no external funding.

Institutional Review Board Statement: Not applicable.

Informed Consent Statement: Not applicable.

Data Availability Statement: Data presented in this study are available on request from the corresponding author.

Conflicts of Interest: The authors declare no conflict of interest.

References

1. O'Brien Nabors, L. *Alternative Sweeteners*, 4th ed.; CRC Press: Boca Raton, FL, USA, 2011.
2. Clemens, R.A.; Jones, J.M.; Kern, M.; Lee, S.-Y.; Mayhew, E.J.; Slavin, J.L.; Zivanovic, S. Functionality of Sugars in Foods and Health. *Compr. Rev. Food Sci. Food Saf.* **2016**, *15*, 433–470. [CrossRef] [PubMed]
3. Aksonova, O.; Gubsky, S.; Torianik, D.; Murlykina, N. Tecnology of Curd Cake with Sucralose as Highly Effective Sweetener. *Prog. Tech. Technol. Food Prod. Enterp. Cater. Bus. Trade* **2020**, *1*, 7–20. [CrossRef]
4. van der Sman, R.G.M.; Renzetti, S. Understanding Functionality of Sucrose in Cake for Reformulation Purposes. *Crit. Rev. Food Sci. Nutr.* **2021**, *61*, 2756–2772. [CrossRef] [PubMed]
5. Pavlov, A.V. *Collection of Recipes for Flour Confectionery and Bakery Products for Public Catering Enterprises*; Gidrometeoizdat: Sankt-Petersburg, Russia, 1998.
6. International Organization for Standardization. *Determination of the Specific Surface Area of Solids by Gas Adsorption—BET Method (ISO 9277:2010(E))*; ISO: Geneva, Switzerland, 2010; 30p.
7. Aksonova, O.; Gubsky, S.; Torianik, D.; Evlash, V. The Technology of Curd Cake with Sucralose: A Infrared Spectroscopy Analysis. *BIO Web Conf.* **2021**, *30*, 01001. [CrossRef]
8. Gallina, M.E.; Sassi, P.; Paolantoni, M.; Morresi, A.; Cataliotti, R.S. Vibrational Analysis of Molecular Interactions in Aqueous Glucose Solutions. Temperature and Concentration Effects. *J. Phys. Chem. B* **2006**, *110*, 8856–8864. [CrossRef] [PubMed]
9. Sritham, E.; Gunasekaran, S. FTIR Spectroscopic Evaluation of Sucrose-Maltodextrin-Sodium Citrate Bioglass. *Food Hydrocoll.* **2017**, *70*, 371–382. [CrossRef]
10. Freda, M.; Piluso, A.; Santucci, A.; Sassi, P. Transmittance Fourier Transform Infrared Spectra of Liquid Water in the Whole Mid-Infrared Region: Temperature Dependence and Structural Analysis. *Appl. Spectrosc.* **2005**, *59*, 1155–1159. [CrossRef] [PubMed]

engineering proceedings

MDPI

Proceeding Paper

The Phytochemistry and Anticarcinogenic Activity of Noni Juice †

Janice S. Mani *, Joel B. Johnson and Mani Naiker

School of Health, Medical & Applied Sciences, Central Queensland University Australia, Rockhampton, QLD 4701, Australia; joel.johnson@cqumail.com (J.B.J.); m.naiker@cqu.edu.au (M.N.)

* Correspondence: janice.mani@cqumail.com

† Presented at the 2nd International Electronic Conference on Applied Sciences, 15–31 October 2021; Available online: https://asec2021.sciforum.net/.

Abstract: Noni juice, obtained from the fruit of the noni tree (*Morinda citrifolia* L.), is a popular commodity in the market, particularly in the South Pacific. It is widely used by consumers for the prevention of several lifestyle diseases. Although there is increasing interest in the potential therapeutic use of noni plants, there are no comparative studies on the various commercialized noni fruit juices available to decipher their phytochemical composition and properties against carcinomas. The present study, therefore, aims to fill this research gap and investigate the juice's anecdotal use as complementary alternative medicine to manage cancer. Five commercial brands of noni juice were included in this study, namely, Tahitian Organic Noni (TON), Cook Island Noni (CIN), Dynamic Health Noni (DHN), Fijian Noni (FN), and Life Health Noni (LHN). The juice samples were vacuum-filtered and freeze-dried to obtain crystal products for methanolic extraction. Total phenolic content (TPC) and antioxidant capacity (FRAP—ferric reducing antioxidant power) were determined on the methanolic extracts. The cytotoxicity of the noni juices was also tested on human cervical adenocarcinoma (HeLa cell lines) by dissolving 2 mg of the crystal product in sterile deionized water and diluting to 1000 μg/mL in the media culture. The final concentration of the extracts in the well plate was 500 μg/mL. The MTS cell viability assay was performed after the cells were incubated with the extracts for 48 h at 37 °C with 5% CO_2. The DHN and FN extracts were found to have the highest TPC of 5393 ± 298 and 5060 ± 23 mg gallic acid equivalent /100 g dry weight (DW), respectively, whereas the highest antioxidant capacity was seen in the CIN extract (6389 ± 49 mg Trolox equivalent/100 g DW). The CIN extract also showed the most promising effect with only 63 ± 1% cell viability, whilst the other extracts showed lower cytotoxic effects (76–90% cell viability) on the HeLa cell line. It is possible that greater cytotoxicity could be observed over long exposure times. The noni juice samples contain high levels of TP and antioxidant capacity and appear to show some level of cytotoxic activity, which were statistically different from the negative control. Further work involving more extensive in vitro and in vivo studies are necessary to elucidate its anticarcinogenic activities.

Keywords: noni juice; anti-cancer; total phenolic; antioxidant capacity

fv **Citation:** Mani, J.S.; Johnson, J.B.; Naiker, M. The Phytochemistry and Anticarcinogenic Activity of Noni Juice. *Eng. Proc.* **2021**, *11*, 16. https://doi.org/10.3390/ASEC2021-11154

Academic Editor: Bettina Wolf

Published: 15 October 2021

1. Introduction

Noni juice, obtained from the fruit of the noni tree (*Morinda citrifolia* L.), is a popular commodity in the herbal supplement market, particularly in the South Pacific. It is widely used by consumers for the prevention of several lifestyle diseases such as diabetes, high blood pressure, cardiopathy, and cerebral apoplexy caused by arteriosclerosis [1]. The noni plant is a small evergreen tree belonging to the Rubiaceae family. The genus *Morinda* consists of 80 species, which are known to grow primarily in coastal tropical regions up to 1330 feet above sea level, and is believed to have originated in Southeast Asia and spread across to Australia, the Pacific Basin, and the Caribbean [2]. As such, apart from "noni", which is the Hawaiian name, it is also known as "nono" in Tahiti, "kura" in the Fiji Islands, and cheese fruit in Australia [2–4]. The noni fruit has a lumpy surface, appears green when unripe, and turns yellow upon ripening. It is produced all year round and has an unpleasant taste and odor; thus, it is not commonly eaten.

At present, there is an increasing interest in potential anticancer activity of the noni fruit. Noni fruit juice, under the commercial name Tahitian fruit juice, has shown promising activity against breast cancer at the initiation stage of 7,12-dimethylbenz(a)anthracene-induced mammary tumorigenesis in female Sprague–Dawley rats [5]. Several other claims of breast cancer prevention at initiation stage in vitro and in vivo have been reported, and as such, noni fruit has been considered an alternative source of chemo preventive agent for breast cancer [6].

The chemoprotective activity may be attributed to several phenolic compounds, such as ursolic acid, kaempferol, quercetin, and rutin, contained in the noni fruit. Rutin, which is a glucoside of the flavonoid quercetin, exhibits substantial oxygen radical scavenging properties in both in vitro and in vivo [6]. In an in vivo study, the inhibitory effects on 12-O-tetradecanoylphorbol-13-acetate (TPA)-induced inflammation in mice and on the Epstein–Barr virus early antigen (EBV–EA) activation induced by TPA of ten known compounds isolated from methanol noni fruit extracts have been reported. The authors claim it to be only the third such report of anthraquinones in the noni fruits [1].

The role of antioxidant vitamins and some phytochemicals as adjuvants in cancer therapy due to its ability to selectively induce apoptosis in cancer cells and not in normal cells has been shown in experimental studies [7]. However, there are limited comparative studies on the various commercialized noni fruit juices available, which decipher their phytochemical compositions and properties against carcinomas. The present study therefore aims to fill this research gap and investigate its anecdotal use as complementary alternative medicine to manage cancer.

2. Methods

2.1. Noni Juice Samples

Five different commercial brands of noni juice of various origins were procured. The descriptions from the manufacturers' labels are provided in Table 1.

Table 1. Description of the noni juice samples included in this study.

Product Name	Origin (Where Fruits Were Obtained from)	Description
Tahitian Organic Noni (TON)	Tahiti	Never reconstituted—always fresh. No preservatives, coloring agents, or sugars.
Cook Island Noni Juice (CIN)	Cook Island	100% *M. citrifolia* fruit extract. Nonpasturized.
Dynamic Health Noni (DHN)	Tahiti	Contains no added sugar, artificial color, or preservatives. Due to the pure nature ingredients in this product, taste, color, and consistency may vary.
Fijian Noni (FN)	Fiji	Made from pure fruits (wild collection). No additives added. Pasteurized for optimum quality.
Life Health Noni (LHN)	New Zealand	100% Noni fruit juice.

2.2. Sample Preparation

The juice samples were vacuum filtered using 0.45 µm Advantec filter paper, and approximately 30 mL of each sample filtrate was frozen at −80 °C overnight. The samples were then freeze-dried using a FTS System-Flexi dry MP freeze drier operating at −55 °C, 400–500 mTorr for 72 h. Finally, the crystals were weighed and stored in the fridge covered in aluminum foil until required for further analysis.

2.3. Extraction Protocol and Measurement of TPC and FRAP

Methanolic extraction of polar phenolics from each crystal product was performed in duplicates as described previously [8]. The Folin–Ciocalteu assay was used to measure the total phenolic content (TPC), and the antioxidant capacity was determined using the ferric reducing antioxidant power (FRAP) method [8]. The results for the TPC and FRAP assays were reported as gallic acid equivalents (GAE) per 100 g (dry weight basis) and Trolox equivalents (TE) per 100 g (DWB), respectively.

2.4. Anticarcinogenic Activity

The anticancer potential of the extracts will be assessed using the MTS assay against the HeLa cell lines (human cervical carcinoma), using a modified method [9] developed in our laboratory. Briefly, at 80% cell confluency, cell suspension was diluted to obtain a final concentration of 5×10^4 cells/mL, which was inoculated into 96-well plate (100 µL cells/well). The plate was incubated for a period of 24 h at 37 °C with 5% CO_2, prior to the addition of the plant extracts. Subsequently, it was incubated again for a period of 48 h at 37 °C with 5% CO_2. The MTS assay protocol was then followed to assess cell viability [9]. Cisplatin (an anticancer drug) at a concentration of 10 µg/mL was used as the positive control, and the cell culture media DMEM (Dulbecco's Modified Eagle Medium) was used as negative control.

2.5. Data Analysis

Results were expressed as mean ± standard deviation of duplicate experiments. Statistical analyses were carried out using RStudio software.

3. Results and Discussion

3.1. Total Phenolic Content and Antioxidant Activity

The commercial noni juice extracts showed similar TPC (p-value > 0.05) ranging from 4700–5400 mg GAE/100 g DW, as shown in Figure 1. The DHN and FN extracts were found to have the highest TPC of 5393 ± 298 and 5060 ± 23 mg GAE/100 g dry weight (DW), respectively.

The observed results were significantly higher than previously reported value of 284.8 ± 25.9 mg GAE/100 g DW in mature noni fruit sourced from wild noni trees in Guam [9]. Fresh noni juice extracted from fruits sourced from Sri Lanka also showed lower TPC (0.28%) [10] compared to 4.4–5.6 % found in the current study. However, TPC were comparable to values reported in the ethanolic noni plant root extracts (4189 mg GAE/100 g) [11] and somewhat comparable to ethyl acetate fruit extracts (2370 mg GAE/100 g) [12].

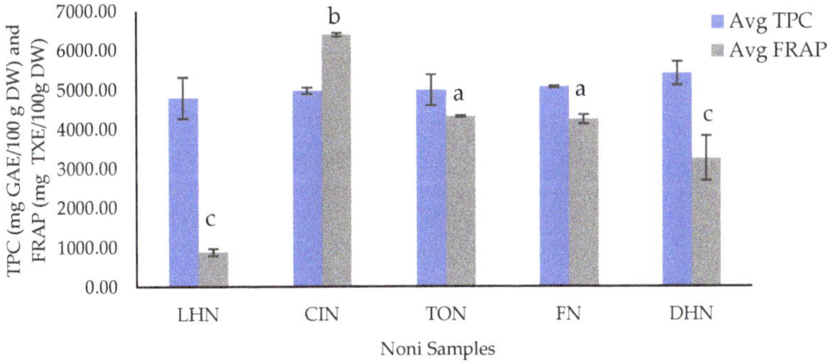

Figure 1. Total phenolic content and FRAP antioxidant capacity of the freeze-dried noni crude extracts. Different letters on the bar indicates significant difference between the tested sources of Noni crude extracts ($p < 0.05$).

The results obtained gave significantly different (p-value < 0.05) ferric reducing abilities for the different brands of noni juice (Figure 1). The antioxidant capacity of CIN was the highest (6389 ± 49 mg TXE/100 g DW), whereas the LHN was the lowest (873 ± 88 mg TXE/100 g DW). The observed results may be due to differences in interaction of polyphenols or the fact that the active compounds that contribute to the antioxidative activity of noni juice were probably non-polar in nature [13]. Moreover, the higher TPC did not always show the higher antioxidant activities, which were similar to trends observed in previous studies [14–17]. It may, however, be inferred that a higher antioxidant capacity would potentially result in high anticarcinogenic activity [18–20].

3.2. Anticarcinogenic Activity

A significantly different (p-value < 0.05) percentage cell viability of HeLa cells were observed upon treatment with the various Noni juice extracts (Figure 2). The CIN extracts showed the highest potency (cell viability of 63 ± 1%), significantly different from the negative control, whereas the FN, LHN, and DHN showed some degree of potency ranging from 76–90%. The effect of TON on the HeLa cells were not significantly different from that of the negative control, hence deeming it non- or the least cytotoxic. It is possible that longer cell exposure times (>48 h) to the Noni juice extracts would have resulted in greater reduction in cell viability.

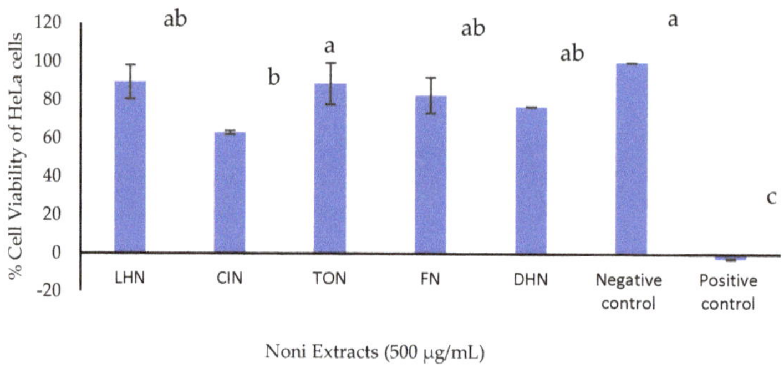

Figure 2. The percentage cell viability of HeLa cells treated with different brands noni juice extracts at 500 μg/mL. Different letters on the bar indicates significant difference between the tested sources of Noni crude extracts ($p < 0.05$).

Previously, Gupta et al. [21] reported that noni juice and cisplatin, either alone or in combination, were able to induce apoptosis through mitochondrial pathway on HeLa cells. Their findings also reported that cisplatin showed higher cell potency compared to Noni juice, comparative to the results of this study. However, the combination of noni juice and cisplatin showed additive effects, suggesting that it can be used as a chemo adjuvant in the treatment of cervical cancer [21,22]. Another study suggested that, in vitro, a "concentrated component" in noni juice and not the pure noni juice may firstly stimulate the immune system to "possibly" assist the body fight the cancer and then kill a small percentage (0–36%) of cancer depending on the type [2]. This may explain the low potency of the noni juice extracts (0–30%) (Figure 2) obtained in this study. Further fractionation and isolation of noni juice is required to identify the "component" responsible for the anti-carcinogenic activity.

4. Conclusions

From this study, it was found that commercial noni juice samples contain high levels of TP and antioxidant capacity and appear to show some level of cytotoxic activity, which were statistically different from the negative control. Whilst the DHN and FN juice extracts showed the highest TPC values, CIN extracts had the highest antioxidant capacity and, as such, also showed the greatest cytotoxicity against the HeLa cells. Further work utilizing more extensive in vitro and in vivo studies are necessary to verify its anticarcinogenic activities against various other cancer cell lines. Additionally, isolation and identification of bioactive compounds using high end analytical techniques such as column chromatography, fractionation, and mass spectroscopy is warranted.

Author Contributions: Conceptualization, J.S.M. and M.N.; methodology, J.S.M.; software, J.S.M.; validation, J.S.M.; formal analysis, J.S.M.; investigation, J.S.M. and J.B.J.; resources, J.S.M. and M.N.; data curation, J.S.M.; writing—original draft preparation, J.S.M.; writing—review and editing, J.S.M., J.B.J., and M.N.; visualization, J.S.M.; supervision, M.N.; project administration, M.N. All authors have read and agreed to the published version of the manuscript.

Funding: This research was funded by a New Staff Grant from CQ University, grant number RSH/5343. One of the authors (J.S.M.) also acknowledges support from the Australian Government in the form of a Research Training Program.

Institutional Review Board Statement: Not applicable.

Informed Consent Statement: Not applicable.

Data Availability Statement: The data presented in this study are available on request from the corresponding author.

Conflicts of Interest: The authors declare no conflict of interest.

References

1. Akihisa, T.; Matsumoto, K.; Tokuda, H.; Yasukawa, K.; Seino, K.I.; Nakamoto, K.; Kuninaga, H.; Suzuki, T.; Kimura, Y. Anti-inflammatory and potential cancer chemopreventive constituents of the fruits of Morinda citrifolia (Noni). *J. Nat. Prod.* **2007**, *70*, 754–757. [CrossRef] [PubMed]
2. Brown, A.C. Anticancer activity of Morinda citrifolia (noni) fruit: A review. *Phyther. Res.* **2012**, *26*, 1427–1440. [CrossRef] [PubMed]
3. Ahmad, M.S. Cancer preventive Effect of Morinda citrifolia (Noni) fruit juice against the AflatoxinB1-induced genotoxicity in human peripheral lymphocytes in vitro. *IOSR J. Pharm.* **2012**, *2*, 228–234. [CrossRef]
4. Pande, M.; Mills, G.; Singh, N.; Voro, T.; Dqg, S. The Kura Files: Qualitative Social Survey. *Pac. Health Surveill. Response* **2005**, *12*, 85–93. Available online: https://www.researchgate.net/profile/Mani-Naiker-4/publication/5667921_The_Kura_files_Qualitative_social_survey/links/53fdcf070cf22f21c2f84af5/The-Kura-files-Qualitative-social-survey.pdf (accessed on 14 February 2021).
5. Wang, M.Y.; Peng, L.; Anderson, G.; Nowicki, D. Breast cancer prevention with Morinda citrifolia (noni) at the initiation stage. *Funct. Foods Health Dis.* **2013**, *3*, 203–222. [CrossRef]
6. Boontha, S.; Kaewjaiboon, N.; Phitchayut, R.; Wanwalee, N.; Salisa Taolam, B.B.; Tasana, P. Cytotoxicity and Cell Migration Suppression by Noni Fruit Extract on Michigan Cancer Foundation-7 Human Breast Cancer Cells and Development of Topical Microemulsions. *Pharm. Mag.* **2019**, *15*, S38–S46. [CrossRef]
7. Gupta, R.K.; Singh, N. Morinda citrifolia (Noni) alters oxidative stress marker and antioxidant activity in cervical cancer cell lines. *Asian Pac. J. Cancer Prev.* **2013**, *14*, 4603–4606. [CrossRef]
8. Johnson, J.; Collins, T.; Walsh, K.; Naiker, M. Solvent extractions and spectrophotometric protocols for measuring the total anthocyanin, phenols and antioxidant content in plums. *Chem. Pap.* **2020**, *74*, 4481–4492. [CrossRef]
9. Erenler, R.; Meral, B.; Sen, O.; Elmastas, M.; Aydin, A.; Eminagaoglu, O.; Topcu, G. Bioassay-guided isolation, identification of compounds from Origanum rotundifolium and investigation of their antiproliferative and antioxidant activities. *Pharm. Biol.* **2017**, *55*, 1646–1653. [CrossRef]
10. Yang, J.; Gadi, R.; Thomson, T. Antioxidant capacity, total phenols, and ascorbic acid content of noni (*Morinda citrifolia*) fruits and leaves at various stages of maturity. *Micronesica* **2011**, *41*, 167–176.
11. Samarasiri, M.H.; Chandrasiri, T.A.; Wijesinghe, D.B.; Gunawardena, S.P. Antioxidant Capacity and Total Phenolic Content Variations against Morinda citrifolia L. Fruit Juice Production Methods. *ETP Int. J. Food Eng.* **2019**, *5*, 293–299. [CrossRef]
12. Pal, R.; Girhepunje, K.; Shrivastava, N.; Hussain, M.M.; Thirumoorthy, N. Antioxidant and free radical scavenging activity of ethanolic extract of Morinda citrifolia. *Res. J. Pharm. Technol.* **2011**, *4*, 1224–1226.
13. Haslaniza, H.; Wan Yaacob, W.A.; Osman, H.; Maskat, M.Y. Interaction of antioxidants and organic acid from noni (Morinda citrifolia L.) juice with ion exchange resins during deodorization via deacidification. *Der. Pharma Chem.* **2015**, *7*, 9–21.
14. Fidrianny, I.; Octaviani, G.D.; Kusmardiyani, S. Study of antioxidant profile and phytochemical content of different organs extracts of *Morinda citrifolia* L. *J. Pharm. Sci. Res.* **2018**, *10*, 2102–2105.
15. Bhardwaj, R.; Yadav, A.; Sharma, R. Phytochemicals and antioxidant activity in Boerhavia diffusa. *Int. J. Pharm. Pharm. Sci.* **2014**, *6*, 344–348.

16. Johnson, J.; Collins, T.; Power, A.; Chandra, S.; Skylas, D.; Portman, D.; Panozzo, J.; Blanchard, C.; Naiker, M. Antioxidative properties and macrochemical composition of five commercial mungbean varieties in Australia. *Legum Sci.* **2020**, *2*, e27. [CrossRef]
17. Johnson, J.B.; Collins, T.; Mani, J.S.; Naiker, M. Nutritional Quality and Bioactive Constituents of Six Australian Plum Varieties. *Int. J. Fruit Sci.* **2021**, *21*, 115–132. [CrossRef]
18. Sakulnarmrat, K.; Konczak, I. Composition of native Australian herbs polyphenolic-rich fractions and in vitro inhibitory activities against key enzymes relevant to metabolic syndrome. *Food Chem.* **2012**, *134*, 1011–1019. [CrossRef]
19. Sakulnarmrat, K.; Fenech, M.; Thomas, P.; Konczak, I. Cytoprotective and pro-apoptotic activities of native Australian herbs polyphenolic-rich extracts. *Food Chem.* **2013**, *136*, 9–17. [CrossRef]
20. Vuong, Q.V.; Hirun, S.; Phillips, P.A.; Chuen, T.L.K.; Bowyer, M.C.; Goldsmith, C.D.; Scarlett, C.J. Fruit-derived phenolic compounds and pancreatic cancer: Perspectives from Australian native fruits. *J. Ethnopharmacol.* **2014**, *152*, 227–242. [CrossRef]
21. Gupta, R.K.; Banerjee, A.; Pathak, S.; Sharma, C.; Singh, N. Induction of mitochondrial-mediated apoptosis by Morinda citrifolia (Noni) in human cervical cancer cells. *Asian Pac. J. Cancer Prev.* **2013**, *14*, 237–242. [CrossRef] [PubMed]
22. Ali, M.; Kenganora, M.; Manjula, S.N. Health benefits of morinda citrifolia (Noni): A review. *Pharm. J.* **2016**, *8*, 321–334. [CrossRef]

engineering proceedings

MDPI

Proceeding Paper

Analysis of Air Pollutant Emission Inventory from Farm Tractor Operations in Korea †

Gyu-Gang Han [1], Jun-Hyuk Jeon [2], Myoung-Ho Kim [2,3,4] and Seong-Min Kim [1,2,3,4,*]

1 Department of Agricultural Convergence Technology, Graduate School, Jeonbuk National University, Jeonju 54896, Korea; dt200v@jbnu.ac.kr
2 Department of Agricultural Machinery Engineering, Graduate School, Jeonbuk National University, Jeonju 54896, Korea; splinter9608@jbnu.ac.kr (J.-H.J.); myoung59@jbnu.ac.kr (M.-H.K.)
3 Department of Bioindustrial Machinery Engineering, College of Agriculture and Life Sciences, JeonBuk National University, Jeonju 54896, Korea
4 Institute for Agricultural Machinery & ICT Convergence, Jeonbuk National University, Jeonju 54896, Korea
* Correspondence: smkim@jbnu.ac.kr

† Presented at the 2nd International Electronic Conference on Applied Sciences, 15–31 October 2021; Available online: https://asec2021.sciforum.net/.

Abstract: Due to the decline in the agricultural labor force and rapid aging of farmers, agricultural machinery is becoming larger, higher-performance, and diversified. In this study, an air pollutant emission inventory for agricultural tractors was analyzed and compared with the inventory developed by a national agency. Agricultural tractors include walking and riding tractors and, further, riding tractors were divided into three subcategories based on their engine size. In addition, tractor emissions were classified according to the usage time of each operation. Seven air pollutants, such as CO, NOx, SOx, TSP, VOCs (PM_{10}), $PM_{2.5}$, and NH_3, were included in the inventory. The results showed that the total yearly emissions in 2017 were 3300 Mg, 9110 Mg, 4 Mg, 567 Mg, 522 Mg, 759 Mg, and 33 Mg for CO, NOx, SOx, TSP, VOCs, PM_{10}, $PM_{2.5}$, and NH_3, respectively. The most emitted air pollutant in the transporting operation using walking tractors is NOx, and the amount of emission is 1023 Mg/y. Riding tractors mainly emit a large amount of NOx, in the order of medium, large, and small tractors. The NOx emissions from medium, large, and small riding tractors are 1103 Mg/y, 676 Mg/y, and 322 Mg/y, respectively, from harrowing operations and are 445 Mg/y, 273 Mg/y, and 130 Mg/y, respectively, from tilling operations. The results also showed that the total pollutant emissions from tractors were increased 10% compared to the emission inventory developed by a national agency due to categorizing riding tractors into three subcategories. A geographic information system (GIS) was used to spatially assign air pollutant variables to 17 provinces and metropolitan cities in Korea.

Keywords: agricultural tractor; diesel emission inventory; air pollutant; geographic information system

fv **Citation:** Han, G.-G.; Jeon, J.-H.; Kim, M.-H.; Kim, S.-M. Analysis of Air Pollutant Emission Inventory from Farm Tractor Operations in Korea. *Eng. Proc.* **2021**, *11*, 17. https://doi.org/10.3390/ASEC2021-11187

Academic Editor: Nunzio Cennamo

Published: 15 October 2021

1. Introduction

Tractors are the main power machines used in agricultural work. Farm tractors are used in a variety of agricultural operations, with most work equipment attached and used in agriculture. Tasks such as farmland cultivation, leveling, sowing, fertilizing, and composting will begin at the beginning of the year. During crop cultivation and harvesting, tractors perform tasks such as loading, bailing, and transportation. At the end of 2020, the total number of small, medium, and large tractors exceeded 300,000, and the diesel consumption of agricultural, free-tax oil in 2019 reached 824,935 kL [1,2].

The production process of agricultural products has a great impact on the environment. Most of the impact is related to mechanization, especially tractor emissions [3]. Agricultural machinery is an important non-road vehicle source that can emit multiple pollutants and make a primary and secondary contribution to air pollution [4].Non-road vehicles need a large quantity of fossil fuels and their emissions cause significant air pollution problems.

These types of vehicles mostly use diesel fuel, which has proven to be a major source of nitrogen compounds (NOx) and particulate matter (PM) [5]. The engine of a tractor operating in agriculture burns a large amount of fuel and emits combustion gas [6].

Air pollutants, such as PM, NOx, CO, volatile organic compounds (VOCs), etc., emitted by agricultural machinery and diesel internal combustion engines have a great impact on the surrounding environment and human health [7,8]. Since the exhaust gas of the internal combustion engine is not applied when evaluating industrial indicators and economic indicators, it is not possible to immediately know the numerical values and influence. However, when the air that people breathe is polluted and agricultural products are cultivated on contaminated agricultural land or when polluted water is used, human health is adversely affected [9,10].

Algirdas Janulevičius [6] collected data on engine load, fuel consumption, and operating modes to study emissions characteristics during tractor operations and presented the average fuel consumption and CO, CO_2, and NOx emissions of the engine. Daniela Lovarelli analyzed [11] air pollutants emitted from ploughing, spike harrowing, sowing, and rolling operations with an engine exhaust gases emissions analyzer (CO_2, CO, and NOx).

To calculate the emissions of air pollutants in agricultural machinery, the National Institute of Environmental Research of the Republic of Korea uses the Tier 3 methodology of the EMEP/EEA (European Monitoring and Evaluation Programme/European Environment Agency) Guidebook, which is technology stratified by equipment. The type and number of agricultural machinery holdings, average annual activity, load factor, average rated power, etc., determine the amount of agricultural machinery air pollutants emitted. The tractor holdings used to calculate the air pollution emissions of tractors is not classified according to small, medium, and large size; the total number of tractors is used, so the accuracy of emission drops. In addition, the average rated output is fixed at 33.1 kW, which was studied in 1999, and does not reflect the average rated output due to the automation and upsizing of tractors. Reflecting these matters, this research intends to make advancement in the air pollution emissions of internal combustion engines of agricultural machinery.

In this study, the inventory of air pollutants generated by farm tractor operations, including walking and riding tractors, were calculated and analyzed. Riding tractors were further divided into three subcategories according to their engine power outputs for more precise investigation. Seven types of air pollutants, including CO, NOx, SOx, total solid particle (TSP), $PM_{2.5}$, VOCs, and NH_3, were calculated using the agricultural tractor inventory data and operating hours in 2017, and the spatial distribution of the pollutants was visualized by a geographic information system (GIS).

2. Materials and Methods

2.1. Estimation Method of Air Pollutant Emissions and Emission Factor

In this study, the emission of air pollutants, including CO, NOx, SOx, TSP, $PM_{2.5}$, VOCs, and NH_3, from farm tractors in the Korea was estimated for the year 2017 by using the EMEP/EEA's Tier 3 methodology. For air pollutants, including CO, NOx, TSP, $PM_{2.5}$, VOCs, and NH_3, the amount of emissions was calculated using the equation given below [12]:

$$E_{i,j,k} = \sum\{N_{i,k} \times HP_i \times LF \times HRS_i \times EF_{i,j}\} \quad (1)$$

where, $E_{i,j,k}$ is the total amount of air pollutant emitted from a specific region (kg/y); $N_{i,k}$ is the number of tractors in a specific region (each); HP_i is the average rated power of tractor (kw); LF is the load factor of agricultural practice (=0.48); HRS_i is the average annual activity of tractors (hr/y); $EF_{i,j}$ is the emission factor (kg/(kWh-each)); i is agricultural tractor type (i = 1, ... , 4); j is the type of air pollutant (j = 1, ... , 7); and k is region (k = 1, ... , 17).

For SOx emissions, the fuel consumption coefficient, according to the rated output of the farm tractor, is applied to the sulfur content, and the emission factor is calculated by the following Equation (2):

$$EF_i = FF_i\ (g/kWh\text{-}each)/1000 \times m \times \text{Fuel sulfur weight percent } (\%)/100 \quad (2)$$

where EF_i is emission factor (kg/(kWh-each)); FF_i is fuel factor (g/(kWh-each)); m is constant (=2.0) (grams of SOx formed from one gram of sulfur); and i is farm tractor type (i = 1, . . . , 4).

Table 1 shows the air pollutant emission factors of agricultural tractors.

Table 1. Emission factors of farm tractors.

Machinery	Emission Factor (kg/kWh-Unit)						
	CO	NOx	TSP	$PM_{2.5}$	VOCs	NH_3	SOx
Walking Tractor (Power Tiller)	6.80	13.60	1.36	1.251	0.48	0.00004	5.42
Riding Tractor	2.48	7.84	0.39	0.359	0.48	0.00003	5.38 5.30 5.30

2.2. *Average Annual Activity Hours of Agricultural Tractors*

Activity hours of agricultural tractors were obtained from the survey on the utilization of agricultural machinery and farm-work mechanization rates published by the Rural Development Administration (RDA) in Korea. Table 2 shows the types of agricultural tractors and the annual activity hours associated with various agricultural operations [13].

Table 2. Average annual activity hours for various operations of agricultural tractors.

Operation Type [1]	Average Activity Hours (hr/y)	
	Walking Tractor (Power Tiller)	Riding Tractor
TL	1.8	20.4
LL	-	16.1
HW	4.9	50.6
FS	-	8.1
PP	18.3	-
CS	-	6.1
SY	22.9	-
LD	-	30.6
BL	-	3.4
TP	41.3	15.9
Others	2.2	13.7
Total	91.4	164.8

[1] TL: Tilling, LL: Leveling, HW: Harrowing, FS: Fertilizer spreading, PP: Pumping, CS: Compost spreading, SY: Spraying, LD: Loading, BL: Baling, TP: Transporting.

2.3. *Number of Holdings and Average Rated Power of Agricultural Tractor*

The number of agricultural tractors was directly obtained from the Agricultural Machinery Holdings Survey yearbook [14]. As of 2017, the total number of agricultural machines, such as tractors, rice transplanters, combine harvesters, etc., registered in Korea was 1,918,745 units. Among them, agricultural tractors were 857,216 units, accounting for 45% of total agricultural machinery in Korea. The holding status of agricultural tractors is shown in Table 3.

Agricultural tractors include a two-wheeled walking tractor (power tiller) and a four-wheeled riding tractor. In this study, air pollutants emitted from walking and riding tractors were calculated and riding tractors were further divided into 3 subcategories according to their engine power outputs for more precise investigation. The riding tractor can be divided into small, medium, and large, according to the diesel engine power. The range of engine power for small, medium, and large riding tractors is less than 29.4 kw, equal and more than 29.4 kw and less than 44.1 kw, and equal and more than 44.1 kw, respectively. Average rated power is defined as a weighted average value with a normal distribution and is calculated by the number of tractors and the rated power.

Table 3. Average rated power and the number of agricultural tractors in Korea, as of 2017.

Machinery	Size [1] (ARP Range)	ARP [2] (kW)	Unit (ea)
Walking Tractor (Power Tiller)	-	6.7	567,070
Riding Tractor	S < 29.4 kW	23.0	73,403
	29.4 kW ≤ M < 44.1 kW	39.0	148,538
	44.1 kW ≤ L	52.1	68,205
	Subtotal		290,146

[1] S: Small, M: Medium, L: Large. [2] ARP: Average rated power. ARP is defined as a weighted average value and calculated from the number of tractors and their rated power.

2.4. Geographic Information System (GIS)

To visualize the spatial distribution of total air pollutant emissions from agricultural tractors in Korea, a piece of open-source geographic information system (GIS) software (QGIS, Windows 10) was used for 9 provinces and 8 metropolitan cities.

3. Results and Discussion

The air pollutant emission inventory for agricultural tractor usage in Korea was refined by categorizing the rated power of tractors and the types of operation tractors routinely perform. Table 4 and Figure 1 show the calculated inventory. In 2017, yearly amounts of CO, NOx, SOx, TSP (PM_{10}), $PM_{2.5}$, VOCs, and NH_3 emitted from agricultural tractors were calculated as 3300 Mg, 9110 Mg, 4 Mg, 567 Mg, 522 Mg, 759 Mg, and 33 Mg, respectively. The yearly amounts of total air pollutants emitted from one unit of walking tractors and small, medium, and large riding tractors were estimated to be 7.0 kg, 20.5 kg, 34.6 kg, and 46.3 kg, respectively.

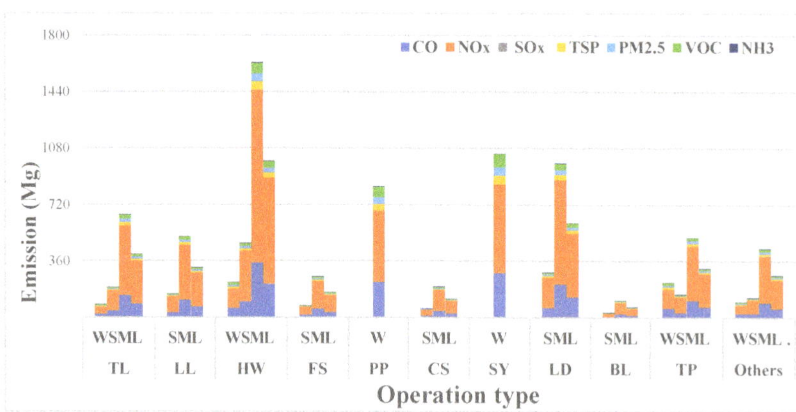

Figure 1. Calculated amount of air pollutant substance various farming practices by agricultural tractors. TL: Tilling, LL: Leveling, HW: Harrowing, FS: Fertilizer spreading, PP: Pumping, CS: Compost spreading, SY: Spraying, LD: Loading, BL: Baling, TP: Transporting, W: Walking tractor, S: Small riding tractor, M: Medium riding tractor, L: Large riding tractor.

Looking at the average activity hours by type of agricultural tractor, walking tractors are mainly used for transporting, pumping, and spraying operations, and riding tractors are mainly used for tilling, harrowing, and loading operations. As the area of cultivated land increases and the size of the tractor increases, the riding tractor replaces the tilling and harrowing operations that the walking tractor used to do in the past. The most emitted air pollutant in the transporting operations, which is the main work of the walking tractor, is NOx, and the amount of emission is 1023 Mg/y, and the amount of emission of $PM_{2.5}$, which is the main concern of air pollution, is 153.5 Mg/y. Riding tractors mainly emit a large amount of CO and NOx, in the order of medium, large, and small tractors. The NOx emissions from medium, large, and small riding tractors are 1103 Mg/y, 676 Mg/y, and 322 Mg/y, respectively, from harrowing operations and are 445 Mg/y, 273 Mg/y, and 130 Mg/y, respectively, from tilling operations. Utilization of riding tractors needs to be done efficiently to reduce air pollutant emissions, especially from harrowing operations.

Table 4. Calculated amounts of air pollutant substances emitted from agricultural tractor operations in 2017 (Mg/y).

Machinery		CO	NOx	SOx ($\times 10^{-2}$)	TSP	$PM_{2.5}$	VOCs	NH_3 ($\times 10^{-1}$)
Walking Tractor (Power Tiller)		1132	2260	60.5	226	208	340	66.6
Riding Tractor	Small	332	1049	48.2	52.2	48.0	64.2	40.1
	Medium	1137	3590	162.7	178.7	164.5	220	137.5
	Large	697	2200	99.8	109.6	100.9	134.9	84.3
	Subtotal	2170	6850	311	341	313	419	262
Total		3300	9110	371	567	522	759	329

Our air pollutant emission inventory results for agricultural tractors were 10% more than those established by the NIER. The discrepancy could be due to the way values were assigned for the average rated power of riding tractors. The NIER method used a single value of 33.1 kW for all 209,149 tractors, while values of 23, 39, and 52.1 kW were used to represent 73,403 small-size, 148,538 middle-size, and 68,205 large-size tractors, respectively, in this study. The spatial distribution of the total amount of air pollutant emissions from agricultural tractors in Korea was generated at the province/metropolitan city level using a GIS technique, as shown in Figure 2.

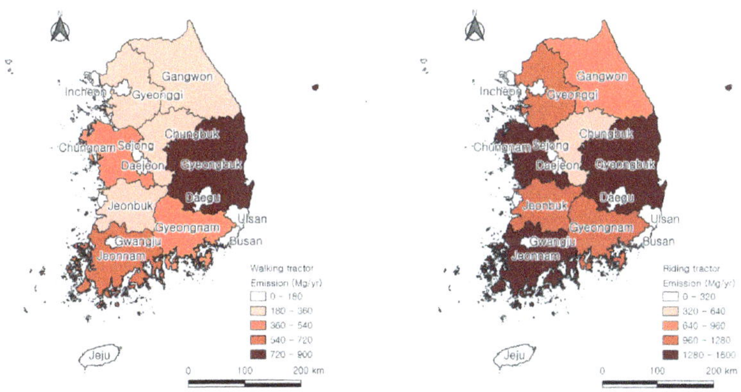

Figure 2. Spatial distribution of the total calculated air pollutant substance emitted from walking tractors (**left**) and riding tractors (**right**) over 9 provinces and 8 metropolitan cities in Korea.

4. Conclusions

In this study, the current air pollutant emission inventory of Korean agricultural tractors was refined by using the EEA Tier 3 methodology. The air pollutant emission inventory for agricultural tractor usage in Korea was calculated by categorizing the rated power of tractors into four subcategories. Yearly amounts of CO, NOx, SOx, TSP (including PM_{10}), $PM_{2.5}$, VOCs, and NH_3 emitted from agricultural tractors were calculated as 3298 Mg, 9110 Mg, 3.7 Mg, 567 Mg, 522 Mg, 756 Mg, and 33 Mg, respectively. Among the non-road vehicle pollutants calculated by the National Institute of Environmental Research (NIER) Air Policy Support System (CAPSS, Clean Air Policy Support System) in 2017, the emissions of agricultural machinery were 3018 Mg/y of CO, 8223 Mg/y of NOx, 3 Mg/y of SOx, 523 Mg/y of TSP (including PM_{10}), 481 Mg/y of $PM_{2.5}$, 705 Mg/y of VOCs, and 29 Mg/y of NH_3 [15]. Our air pollutant emission inventory results for agricultural tractors were 10% more than those established by the NIER. The discrepancy could be due to the way values were assigned for the average rated power of riding tractors.

Walking tractors emitted the most diesel emissions during transportation operation, and riding tractors emitted the most air pollutant emissions during the tilling and harrowing operations. In order to reduce the air pollutants emitted by inefficient agricultural operations, it is necessary to analyze agricultural practices utilizing agricultural tractors in detail. It is necessary to predict future air pollutant emissions through past agricultural tractor air pollutant inventory analysis.

Author Contributions: Conceptualization, G.-G.H. and S.-M.K.; software, G.-G.H.; validation, G.-G.H., M.-H.K. and S.-M.K.; investigation, J.-H.J.; resources, G.-G.H.; writing—original draft preparation, G.-G.H.; writing—review and editing, M.-H.K. and S.-M.K.; visualization, J.-H.J.; supervision, S.-M.K.; project administration, S.-M.K.; funding acquisition, S.-M.K. All authors have read and agreed to the published version of the manuscript.

Funding: This work was carried out with the support for "Study on Particulate Matter Outbreak Source Characteristics during Agricultural Practice and Inventory Integration (Project No. PJ01428301)" by Rural Development Administration, Korea.

Conflicts of Interest: The authors declare no conflict of interest.

References

1. MAFRA. *Agricultural Machinery Holdings Survey 2020*; Ministry of Agriculture, Food and Rural Affairs: Sejong, Korea, 2020; p. 10.
2. NHCF. *Agricultural Cooperative Yearbook 2020*; National Agricultural Cooperative Federation: Seoul, Korea, 2020; p. 209.
3. Bacenetti, J.; Lovarelli, D.; Facchinetti, D.; Pessina, D. An environmental comparison of techniques to reduce pollutants emissions related to agricultural tractors. *Biosyst. Eng.* **2018**, *171*, 30–40. [CrossRef]
4. Lang, J.; Tian, J.; Zhou, Y.; Li, K.; Chen, D.; Huang, Q.; Xing, X.; Zhang, Y.; Cheng, S. A high temporal-spatial resolution air pollutant emission inventory for agricultural machinery in China. *J. Clean. Prod.* **2018**, *183*, 1110–1120. [CrossRef]
5. Zhao, Y.; Qiu, L.P.; Xu, R.Y.; Xie, F.J.; Zhang, Q.; Yu, Y.Y.; Nielsen, C.P.; Qin, H.X.; Wang, H.K.; Wu, X.C.; et al. Advantages of a city-scale emission inventory for urban air quality research and policy: The case of Nanjing, a typical industrial city in the Yangtze River Delta, China. *Atmos. Chem. Phys.* **2015**, *15*, 12623–12644. [CrossRef]
6. Janulevicius, A.; Juostas, A.; Pupinis, G. Engine performance during tractor operational period. *Energy Convers. Manag.* **2018**, *68*, 11–19. [CrossRef]
7. Zhou, W.; Chen, C.; Lei, L.; Fu, R.; Sun, Y. Temporal variations and spatial distributions of gaseous and particulate air pollutants and their health risks during 2015–2019 in China. *Environ. Pollut.* **2021**, *272*, 116031. [CrossRef] [PubMed]
8. Beak, K.M.; Kim, M.J.; Kim, J.Y.; Seo, Y.K.; Beak, S.O. Characterization and health impact assessment of hazardous air pollutants in residential areas near a large iron-steel industrial complex in Korea. *Atmos. Pollut. Res.* **2020**, *11*, 1754–1766. [CrossRef]
9. Kim, L.S.; Jeon, J.W.; Son, J.Y.; Kim, C.S.; Ye, J.; Kim, H.J.; Lee, C.H.; Hwang, S.M.; Choi, S.D. Nationwide levels and distribution of endosulfan in air, soil, water, and sediment in South Korea. *Environ. Pollut.* **2020**, *265*, 115035. [CrossRef] [PubMed]
10. Enyoh, C.E.; Verla, A.W.; Qingyue, W.; Ohiagu, F.O.; Ohiagu, F.O.; Chowdhury, A.H.; Enyoh, E.C.; Chowdhury, T.; Verla, E.N.; Chinwendu, U.P. An overview of emerging pollutants in air Method of analysis and potential public health concern from human environmental exposure. *Trends Environ. Anal. Chem.* **2020**, *28*, e00107. [CrossRef]
11. Lovarelli, D.; Fiala, M.; Larsson, G. Fuel consumption and exhaust emissions during on-field tractor activity: A possible improving strategy for the environmental load of agricultural mechanization. *Comput. Electron. Agric.* **2018**, *151*, 238–248. [CrossRef]
12. NIER. *A Handbook of Method for National Air Pollutant Emission Estimation*; National Institute of Environmental Research: Sejong, Korea, 2013; pp. 199–203.

13. RDA. *Survey on the Utilization of Agricultural Machinery and Farmwork Mechanization Rate*; Rural Development Administration: Jeonju, Korea, 2018; pp. 23–27.
14. MAFRA. *Agricultural Machinery Holdings Survey 2017*; Ministry of Agriculture, Food and Rural Affairs: Sejong, Korea, 2017; p. 12.
15. National Air Pollutants Emission Service. Available online: https://airemiss.nier.go.kr/user/boardList.do?handle=160&siteId=airemiss&id=airemiss_030500000000 (accessed on 5 May 2021).

engineering proceedings

MDPI

Proceeding Paper

Development of a Predictive Model for Mild Cognitive Impairment in Parkinson's Disease with Normal Cognition Using Kernel-Based C5.0 Machine Learning Blending: Preliminary Research †

Haewon Byeon

Department of Medical Big Data, College of AI Convergence, Inje University, Gimhae 50834, Korea; bhwpuma@naver.com; Tel.: +82-10-7404-6969

† Presented at the 2nd International Electronic Conference on Applied Sciences, 15–31 October 2021; Available online: https://asec2021.sciforum.net/.

Abstract: This preliminary study mainly compared the performance for predicting mild cognitive impairment in Parkinson's disease (PDMCI) between single machine learning and hybrid machine learning. This study analyzed 185 patients with Parkinson's disease (75 Parkinson's disease) patients with normal cognition, and 110 patients with PDMCI. PDMCI, an outcome variable, was divided into "with PDMCI" and "with normal cognition" according to the diagnosis of the neurologist. This study used 48 variables (diagnostic data), including motor symptoms of Parkinson's disease, non-motor symptoms of Parkinson's disease, and sleep disorders, as explanatory variables. This study developed seven machine learning models using blending (three hybrid models (polydot + C5.0, vanilladot + C5.0, and RBFdot + C5.0) and four single machine learning models (polydot, vanilladot, RBFdot, and C5.0)). The results of this study showed that the RBFdot + C5.0 was the model with the best performance to predict PDMCI in Parkinson's disease patients with normal cognition (AUC = 0.88) among the seven machine learning models. We will develop interpretable machine learning using C5.0 in a follow-up study based on the results of this study.

Keywords: hybrid machine learning; blending approach; mild cognitive impairment in Parkinson's disease; SVM; C5.0

Citation: Byeon, H. Development of a Predictive Model for Mild Cognitive Impairment in Parkinson's Disease with Normal Cognition Using Kernel-Based C5.0 Machine Learning Blending: Preliminary Research. *Eng. Proc.* **2021**, *11*, 18. https://doi.org/10.3390/ASEC2021-11147

Academic Editor: Nunzio Cennamo

Published: 15 October 2021

1. Introduction

It has been reported that mild cognitive impairment (MCI), known as the preclinical phase of dementia, may last up to seven years and appropriate therapeutic interventions in the MCI stage can delay the progression to dementia approximately five years [1]. As a result, many studies [2,3] have focused on detecting MCI, known as an intermediate stage between normal aging and Alzheimer's disease, as soon as possible. As longitudinal studies [4,5] on Parkinson's disease have reported that patients with Parkinson's disease frequently suffer from cognitive impairment, recent studies [6,7] have paid more attention to mild cognitive impairment in Parkinson's disease (PDMCI) as well as Alzheimer's MCI. Although PDMCI occurs frequently in patients with Parkinson's disease, the characteristics of PDMCI are known much less than those of Alzheimer's MCI and those of vascular MCI.

Although a number of previous studies [8,9] have reported that the most critical characteristic of PDMCI is executive function impairment due to frontal lobe dysfunction found at an early stage, it is hard to detect it only with the degree of executive function because early-stage MCI due to Alzheimer disease or vascular dementia shows executive function impairment [10]. In particular, since Parkinson's disease progresses slowly and symptoms appear little by little, patients and caregivers can perceive the cognitive problems caused by PDMCI as the cognitive frailty in the normal aging process. Therefore, it is hard to diagnose it early.

MCI is diagnosed based on an interview, evaluation of cognitive function through standardized neuropsychological tests, and brain imaging. However, brain imaging has limitations in its use for early diagnosis purposes because although it can detect the presence of cerebrovascular disease and brain atrophy, it can find them only when these symptoms are very advanced. Therefore, neuropsychological tests also evaluating cognitive function are known to be effective screening tests for detecting MCI early [11].

On the other hand, studies in the medical field have steadily predicted the risk probability or high-risk groups of a disease using data mining in recent years [12,13]. However, it is challenging to accurately predict diseases with single machine learning (learner). For example, the artificial neural network technique has the limitation of not being able to explain the derived results but it offers high prediction accuracy. On the other hand, the decision tree technique allows clinicians to easily interpret the results derived from it, but it is exposed to a higher overfitting risk than other machine learning algorithms such as SVM, the results of it can be altered by the type and order of input variables, and the accuracy of it can be lowered depending on them.

To overcome these limitations, a hybrid model combining Support Vector Machine (SVM) and decision tree model has been used recently to develop a model that has higher predictive power and explanatory power compared to single machine learning [14]. This study developed a PDMCI predictive model considering health behaviors, environmental factors, medical history, physical function, depression, and cognitive level using a hybrid model combining C-SVM and C5.0 and provided baseline data for the prevention and early management of Parkinson's disease. This preliminary study mainly compared the performance for predicting PDMCI between single machine learning and hybrid machine learning. We will develop interpretable machine learning using C5.0 in a follow-up study based on the results of this study.

2. Method

2.1. Data Source

It is a secondary data analysis study that analyzed Parkinson's Disease Epidemiologic (Parde) Data after receiving an approval (No. KBN-2019-005) from the Distribution Committee and an approval (No. KBN-2019-1327) from the Research Ethics Review Committee of the Korea Centers for Disease Control and Prevention and National Biobank of Korea. The design and administration of Parde data are described in detail elsewhere [12]. This study analyzed 185 patients with Parkinson's disease (75 Parkinson's disease) patients with normal cognition, and 110 patients with PDMCI.

2.2. Measurement

PDMCI, an outcome variable, was divided into "with PDMCI" and "with normal cognition" according to the diagnosis of the neurologist. This study used 48 variables (diagnostic data), including motor symptoms of Parkinson's disease, non-motor symptoms of Parkinson's disease, and sleep disorders, as explanatory variables.

2.3. Model Blending Based on Machine Learning

In this study, a PDMCI prediction model was developed using the blending approach (base model = SVM; meta model = C5.0). This study chose "C5.0" implemented by Kuhn et al. (2013) for the decision tree algorithm and "kernel-based machine learning (kernlab)" implemented by Karatzoglou et al. (2016) for the SVM to develop a PDMCI predictive model. The kernlab algorithm includes a polynomial kernel function (polydot), a linear kernel function (vanilladot), and a radial basis kernel function (RBFdot) that enable nonlinear SVM analysis. This study developed seven machine learning models using blending (three hybrid models (polydot + C5.0, vanilladot + C5.0, and RBFdot + C5.0) and four single machine learning models (polydot, vanilladot, RBFdot, and C5.0)). The structure of the blending model in this study is presented in Figure 1.

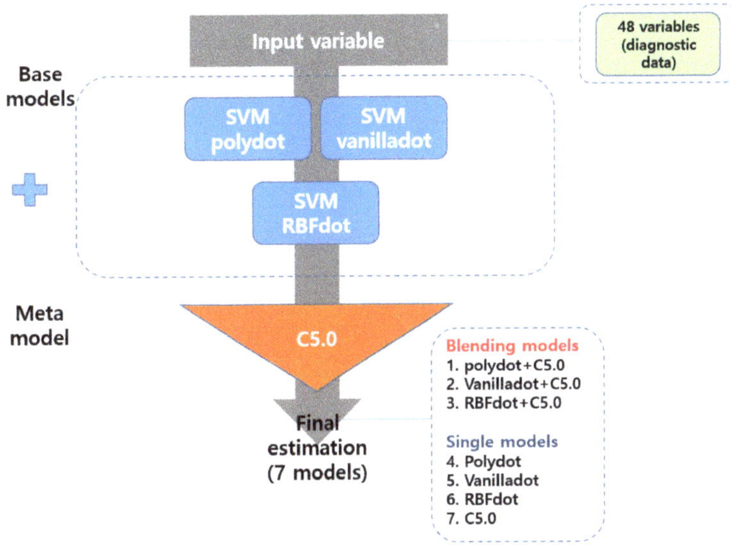

Figure 1. The structure of the prediction for PDMCI.

This study compared the predictive performance (general accuracy, F1-score, area under the curve (AUC), recall, precision) of these developed models using the 10-fold cross-validation method.

The performance evaluation of the machine learning model is different from the regression model and the classification model. In the case of the regression model, the average error value between the actual value and the predicted value is calculated, which is evaluated as MAE, MSE, and R^2. However, in the case of the classification model, the majority is calculated, which is calculated through general accuracy, F1-score, area under the curve (AUC), recall, and precision.

In this study, general accuracy, F1-score, AUC, recall, and precision are presented as performance evaluation indicators because the model is trained using binary data. In particular, recall and precision may have extreme values when data are unbalanced. Here, it is effective to use AUC or to use F1-score, which is a combination of recall and precision, as a performance evaluation. The formula of F1-SCORE is presented in Equation (1).

$$\mathrm{F1} = 2\frac{\mathrm{Precision} \times \mathrm{Recall}}{\mathrm{Precision} + \mathrm{Recall}} \tag{1}$$

This study assumed that a model with the highest AUC was the best predictive performance. If the AUC was the same, a model with the highest F1-score was assumed as the optimal model.

3. Results

3.1. General Characteristics of Subjects

Among 185 patients with Parkinson's disease, 59.5% (108 subjects) had PDMCI. The results of chi-square test showed that PDMCI and Parkinson's disease patients with normal cognition had significantly different REM and RBD, Motor score of UPDRS, Total score of UPDRS, Global CDR, K-MoCA, K-MMSE, Sum of boxes in CDR, H&Y staging, K-IADL, and Schwab and England ADL ($p < 0.05$).

3.2. Comparing the Predictive Performance of Single Model and That of Blending Model

The results of this study showed that the RBFdot + C5.0 was the model with the best performance to predict PDMCI in Parkinson's disease patients with normal cognition

(AUC = 0.88) among the seven machine learning models. The AUC and F1-scores of the seven machine learning models analyzed in this study are presented in Figures 2 and 3, respectively.

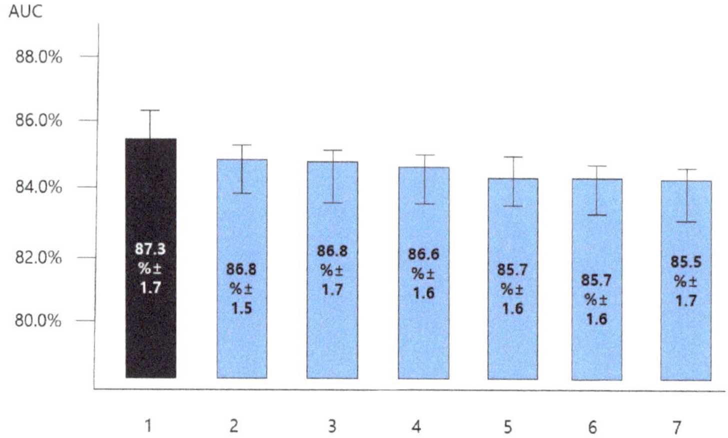

Figure 2. The comparison of AUC for seven machine learning models. 1 = RBFdot + C5.0; 2 = polydot + C5.0; 3 = vanilladot + C5.0; 4 = RBFdot + C5.0; 5 = C5.0; 6 = vanilladot; 7 = polydot.

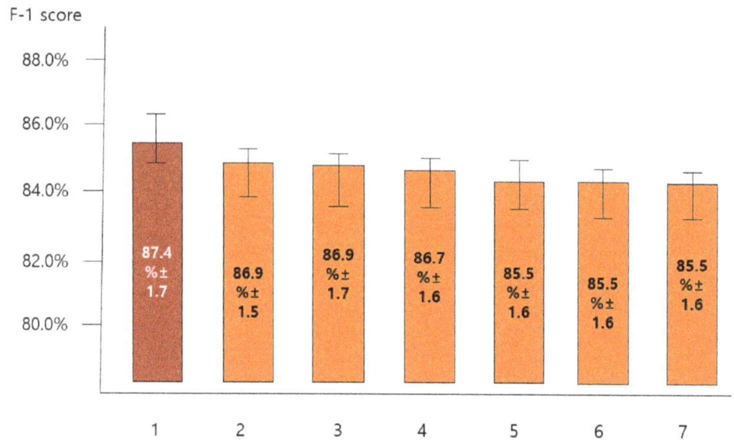

Figure 3. The comparison of F-1 score for seven machine learning models. 1 = RBFdot + C5.0; 2 = polydot + C5.0; 3 = vanilladot + C5.0; 4 = RBFdot + C5.0; 5 = C5.0; 6 = vanilladot; 7 = polydot.

4. Conclusions

The results of this study showed that the RBFdot + C5.0 was the model with the best performance to predict PDMCI in Parkinson's disease patients with normal cognition (AUC = 0.88) among the seven machine learning models. It is necessary to develop a customized screening program for detecting PDMCI in Parkinson's disease patients with normal cognition early based on the results of this study.

When developing a system to predict the morbidity of PDMCI from Parkinson's Disease with Normal Cognition in the future, it will be possible to predict more accurately with the RBFdot + C5.0 model proposed in this study than single machine learning such as SVM. We will develop a machine learning model that can explain the characteristics of high PDMCI risk groups based on the results of this study.

Supplementary Materials: The following are available online at https://www.mdpi.com/article/10.3390/ASEC2021-11147/s1, S1: Development of a Predictive Model for Mild Cognitive Impairment in Parkinson's Disease with Normal Cognition Using Kernel-Based C5.0 Machine Learning Blending: Preliminary Research.

Funding: This research was funded by Basic Science Research Program through the National Research Foundation of Korea (NRF) funded by the Ministry of Education, grant number 2018R1D1A1B07041091, 2021S1A5A8062526.

Institutional Review Board Statement: The study was conducted according to the guidelines of the Declaration of Helsinki, and approved by the Institutional Review Board (or Ethics Committee) of National Biobank of Korea under Korea Centers for Disease Control and Prevention (protocol code KBN-2019-1327).

Informed Consent Statement: Informed consent was obtained from all subjects involved in the study.

Conflicts of Interest: The author declares no conflict of interest.

References

1. Brookmeyer, R.; Gray, S.; Kawas, C. Projections of Alzheimer's disease in the United States and the public health impact of delaying disease onset. *Am. J. Public Health* **1998**, *88*, 1337–1342. [CrossRef] [PubMed]
2. Gustavson, D.E.; Elman, J.A.; Sanderson-Cimino, M.; Franz, C.E.; Panizzon, M.S.; Jak, A.J.; Reynolds, C.A.; Neale, M.C.; Lyons, M.J.; Kremen, W. Extensive memory testing improves prediction of progression to MCI in late middle age. *Alzheimer's Dement. Diagn. Assess. Dis. Monit.* **2020**, *12*, e12004. [CrossRef] [PubMed]
3. Lim, S.J.; Lee, Z.; Kwon, L.N.; Chun, H.W. Medical health records-based Mild Cognitive Impairment (MCI) prediction for effective dementia care. *Int. J. Environ. Res. Public Health* **2021**, *18*, 9223. [CrossRef] [PubMed]
4. Monastero, R.; Cicero, C.E.; Baschi, R.; Davì, M.; Luca, A.; Restivo, V.; Zangara, C.; Fierro, B.; Zappia, M.; Nicoletti, A. Mild cognitive impairment in Parkinson's disease: The Parkinson's disease cognitive study (PACOS). *J. Neurol.* **2018**, *265*, 1050–1058. [CrossRef] [PubMed]
5. Domellöf, M.E.; Ekman, U.; Forsgren, L.; Elgh, E. Cognitive function in the early phase of Parkinson's disease, a five-year follow-up. *Acta Neurol. Scand.* **2015**, *132*, 79–88. [CrossRef] [PubMed]
6. Baiano, C.; Barone, P.; Trojano, L.; Santangelo, G. Prevalence and clinical aspects of mild cognitive impairment in Parkinson's disease: A meta-analysis. *Mov. Disord.* **2020**, *35*, 45–54. [CrossRef] [PubMed]
7. Wen, M.C.; Chan, L.L.; Tan, L.C.; Tan, E.K. Mild cognitive impairment in Parkinson's disease: A distinct clinical entity? *Transl. Neurodegener.* **2017**, *6*, 1–7. [CrossRef] [PubMed]
8. Martinez-Horta, S.; Kulisevsky, J. Mild cognitive impairment in Parkinson's disease. *J. Neural Transm.* **2019**, *126*, 897–904. [CrossRef] [PubMed]
9. Roheger, M.; Kalbe, E.; Liepelt-Scarfone, I. Progression of cognitive decline in Parkinson's disease. *J. Parkinsons Dis.* **2018**, *8*, 183–193. [CrossRef] [PubMed]
10. Zokaei, N.; Husain, M. Working Memory in Alzheimer's Disease and Parkinson's Disease. In *Processes of Visuospatial Attention and Working Memory*; Springer: Cham, Switzerland, 2019.
11. Biundo, R.; Weis, L.; Pilleri, M.; Facchini, S.; Formento-Dojot, P.; Vallelunga, A.; Antonini, A. Diagnostic and screening power of neuropsychological testing in detecting mild cognitive impairment in Parkinson's disease. *J. Neural Transm.* **2013**, *120*, 627–633. [CrossRef] [PubMed]
12. Byeon, H. Application of machine learning technique to distinguish Parkinson's disease dementia and Alzheimer's dementia: Predictive power of Parkinson's disease-related non-motor symptoms and neuropsychological profile. *J. Pers. Med.* **2020**, *10*, 31. [CrossRef] [PubMed]
13. Przybyszewski, A.W.; Chudzik, A.; Szlufik, S.; Habela, P.; Koziorowski, D.M. Comparison of different data mining methods to determine disease progression in dissimilar groups of Parkinson's patients. *Fundam. Inform.* **2020**, *176*, 167–181. [CrossRef]
14. Sun, L.; Fu, S.; Wang, F. Decision tree SVM model with fisher feature selection for speech emotion recognition. *Eurasip J. Audio Speech Music Process.* **2019**, *2019*, 1–14. [CrossRef]

engineering proceedings

MDPI

Proceeding Paper

Aptamer-Based Plasmonic Plastic Optical Fiber Biosensors: A Focus on Relevant Applications †

Laura Pasquardini [1,*], Nunzio Cennamo [2] and Luigi Zeni [2]

1 Indivenire srl, Via Alla Cascata 56/C, 38123 Trento, Italy
2 Department of Engineering, University of Campania "L. Vanvitelli", Via Roma 29, 81031 Aversa, Italy; nunzio.cennamo@unicampania.it (N.C.); luigi.zeni@unicampania.it (L.Z.)
* Correspondence: l.pasquardini@indiveni.re
† Presented at the 2nd International Electronic Conference on Applied Sciences, 15–31 October 2021.

Abstract: Optical detection is one of the most used transduction methods in biosensors and apart from the commercially available instruments based on surface plasmon resonance (SPR), an emerging class of devices, based on both silica and plastic optical fibers (POFs), is finding its route. On the other hand, aptamers represent the next-generation biorecognition elements for biosensor implementation, thanks to several characteristics making them more appealing with respect to the conventional antibodies. The joint exploitation of plasmon resonance in plastic optical fibers and aptamers is here reported, focusing the attention on various relevant biological applications (e.g., thrombin, vascular endothelial growth factor (VEGF), and SARS-CoV-2 spike protein).

Keywords: surface plasmon resonance (SPR); aptamers; plastic optical fiber; biosensors

Citation: Pasquardini, L.; Cennamo, N.; Zeni, L. Aptamer-Based Plasmonic Plastic Optical Fiber Biosensors: A Focus on Relevant Applications. *Eng. Proc.* **2021**, *11*, 19. https://doi.org/10.3390/ASEC2021-11158

Academic Editor: Takayoshi Kobayashi

Published: 15 October 2021

1. Introduction

Surface plasmon resonance, or localized surface plasmon resonance (LSPR), represents a gold standard in the optical characterization of biomolecular interaction, due to its high sensitivity, and it is also widely used as a transduction method in biosensor implementation. In particular, SPR/LSPR systems based on plastic optical fibers (POF) represent an emerging field which is paving the way for the development of a new class of sensors [1–3], not only for the POFs' intrinsic characteristics such as excellent flexibility, large diameter and great numerical aperture, but also for the possibility for easily implementing different geometries, such as U-bent, D-shaped, side-polished, and tapered configurations. Furthermore, these kinds of optical fiber sensors can be used to realize small-size and low-cost optical biosensors and can be coupled to a variety of molecular recognition elements (MRE), such as antibodies, molecular imprinted polymers or aptamers (aptasensors).

Among MREs, the aptamers are an emerging class of molecules characterized by several advantages. Aptamers are short single-stranded DNA or RNA fragments selected to bind a wide range of analytes, ranging from very small molecules (pesticides, toxins) up to entire microorganisms; they exhibit an affinity constant in the nanomolar range minimizing the probability of false-positive results [4]. Moreover, they can be easily modified and are characterized by a high batch to batch reproducibility, a high resistance in acidic environments and at high temperatures. Due to the above characteristics, the aptamers find applications in different fields [5], ranging from the detection of small molecules [6–8] to point-of-care diagnostic systems [9,10], being suitable for the detection of bacteria [11] or of circulating tumor cells [12].

Here we focus our attention on POF-SPR biosensors, using aptamers as molecular recognition elements, for biomedical applications and in particular for the detection of: THR, a clinical marker of the blood coagulation cascade, VEGF, a circulating protein potentially associated with cancer, and the SARS-CoV-2 Spike protein.

2. Methods

The materials and methods were extensively reported in our previously published papers [13–15]. Here we briefly summarize the main steps for each application. A scheme of the optical platform with the aptamer layer is reported in Figure 1a and the different developed interfaces are reported in Figure 1b–d.

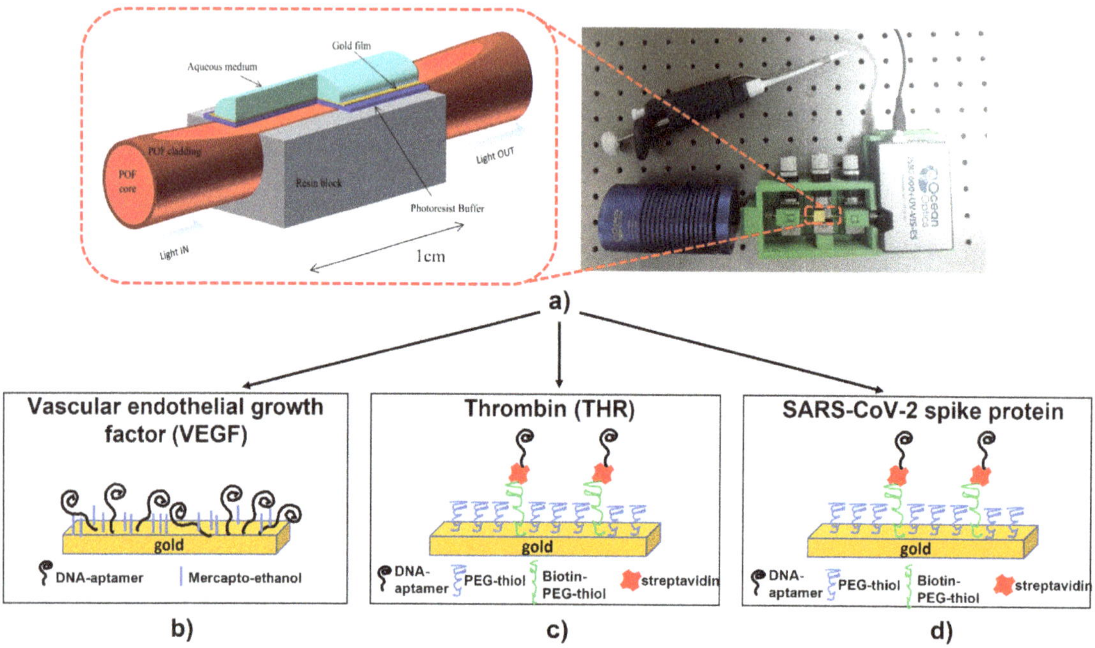

Figure 1. (**a**) Image of the sensor system with a zoom of the plasmonic optical sensor and different aptamer-based interfaces: (**b**) vascular endothelial growth factor (VEGF) [13], (**c**) thrombin [14], and (**d**) SARS-CoV-2 spike protein [15].

2.1. VEGF-Aptasensor Preparation

The aptasensor for VEGF detection was prepared according to the procedure reported in [13]. Briefly, after cleaning the gold surface through an argon plasma (6.8 W of power to the RF coil for one minute), a 1 μM aptamer (5′-HO-$(CH_2)_3$-S-S-$(CH_2)_3$-CC CGT CTT CCA GAC AAG AGT GCA GGG-3′) solution in 1 M potassium phosphate buffer pH 7 for one hour was applied followed by passivation in 1 mM mercaptoethanol solution in the same buffer for 30 min (Figure 1b).

2.2. THR-Aptasensor Preparation

The THR-aptasensor was prepared according to [14]. Briefly, after cleaning the gold surface (argon plasma, 6.8 W of power to the RF coil for one minute), a water solution of 0.2 mM of PEGthiol:BiotinPEGlipo in an 8:2 molar ratio was incubated overnight. After washing, 5 μg/mL streptavidin solution in phosphate buffer (10 mM phosphate buffer, 138 mM NaCl, 2.7 mM KCl, pH 7.4) was applied for one hour. Finally, 10 μM of biotin-TBA29 aptamer (5′-/5BiotinTEG/AG TCC GTG GTA GGG CAG GTT GGG GTG ACT-3′) was incubated for three hours in the same phosphate buffer (Figure 1c).

2.3. SARS-CoV-2 Spike Protein Aptasensor Preparation

A protocol similar to the one developed for THR was applied for the SARS-CoV-2 spike protein aptasensor implementation [15]. The only difference was related to the aptamer sequence which, in this case, was the following: 5′-/5BiotinTEG/CAG CAC CGA CCT TGT GCT TTG GGA GTG CTG GTC CAA GGG CGT TAA TGG ACA-3′ (Figure 1d).

2.4. Optical Measurements

The optical platform used for all the measurements is the one shown in Figure 1a. The polymer cladding over the PMMA core (980 µm) of the POF was removed and a Microposit S1813 photoresist was spun on it. A 60 nm thick gold layer was finally sputtered onto the photoresist layer. A halogen lamp (360 nm to 1700 nm) was used as light source and an Ocean Optics USB2000+VIS–NIR spectrometer (330 nm to 1100 nm) was employed to analyze the transmitted spectrum. The sensing experiments on the SPR-POF aptasensor were performed by dropping about 70 µL of solution over the sensing region. After 10 min, the solution was removed and washing in buffer was performed. After the dropping of fresh buffer, the transmission spectrum was recorded and normalized to air.

3. Results and Discussion

In the last decade, different MREs, such as antibodies, molecularly imprinted polymers (MIP), and aptamers, have been exploited on POF-based devices so proving the high versatility of this kind of sensing platform [2]. With respect to conventional MREs, aptamers represent a new emerging class increasingly employed to realize biosensors. Up to now, very few examples of coupling between aptamers and POF-based biosensors have been reported in the literature [13–18], and most of them are from our research group.

In 2015, we developed an aptamer-based POF-SPR sensor for the detection of vascular endothelial growth factor (VEGF), selected as a circulating protein potentially associated with cancer [13]. A thiolated aptamer was directly immobilized on the gold film deposited in the POF's sensing region. Typical dose-response transmission spectra, achieved by incubating different amounts of VEGF in 20 mM Tris–HCl buffer pH 7.4, are reported in Figure 2a. The obtained results suggested that the direct immobilization of the aptamers on the gold surface can negatively affect their recognition capability and that passivation is an important step of the interface layer build-up. In fact, proper passivation assures a better aptamer conformation and allows a limit of detection of 0.8 nM [13] to be reached. Even if our aptasensor exhibits similar performances with respect to other detection systems, the dissociation constant measured on our sensor was two orders of magnitude lower. This is probably caused by a loss of affinity in the immobilization procedure.

For the above reason, in the subsequent works, we changed the approach and developed an interface based on short polyethylene glycol (PEG). The idea was to keep away the aptamer from the surface in order to ensure a better sequence conformation. So, the detection of thrombin (THR), a clinical marker of the blood coagulation cascade, was performed modifying the gold-coated POF with a mixed interface (a short-PEG and a biotinylated-PEG) and immobilizing, through avidin-biotin chemistry, a THR binding aptamer [14] (see Figure 1c). Figure 2b reports an example of the dose-response curve recorded by POF-SPR measurement for different THR concentrations in buffer (Tris 50 mM, EDTA 1 mM, $MgCl_2$ 1 mM, KCl 150 mM pH 7.4) [14]. The good performance of the obtained interface were confirmed, resulting in a detection limit of 1.6 nM and in the increase of the dissociation constant of one order of magnitude.

Recently, we modified the PEG-based interface, previously developed for thrombin detection, in order to detect the receptor-binding domain (RBD) of the SARS-CoV-2 spike glycoprotein [15]. To this purpose, we changed the aptamer sequence, as reported in Section 2.3. Figure 2c reports an example of the dose-response curve obtained by incubating different amounts of protein in buffer (136.8 mM NaCl, 10.1 mM Na_2HPO_4, 2.7 mM KCl, 1.8 mM KH_2PO_4, 0.55 mM $MgCl_2$, pH 7.4). The aptasensor was tested, not only on the specific target, but also on aspecific targets (BSA, AH1H1 hemagglutinin protein and MERS spike protein) and in diluted human serum (50%). A limit of detection in the nanomolar range was achieved, confirming the good performance of this aptamer-based optical sensor.

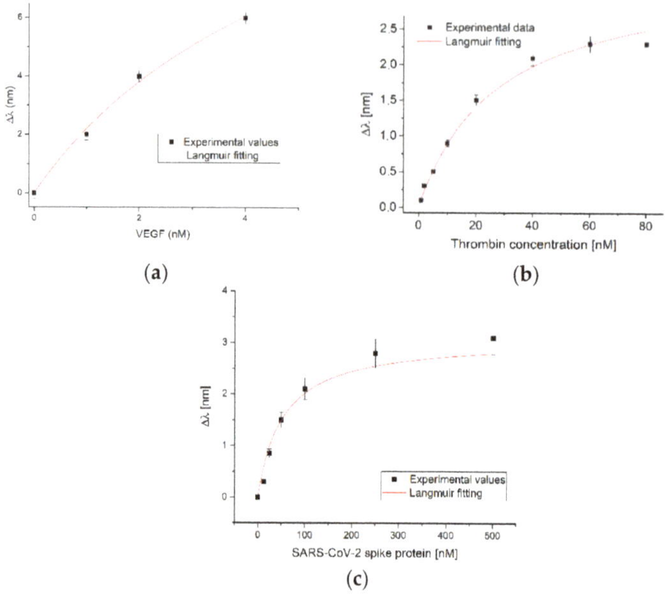

Figure 2. Dose-response curves obtained incubating different concentrations of VEGF (**a**), of thrombin, reprinted from [14] (**b**), and of SARS-CoV-2 spike protein (**c**). Langmuir fits are reported as well.

4. Conclusions

Aptamers represent an emerging class of biorecognition elements, more and more exploited in the development of optical biosensors. Their performance makes them the ideal elements to be immobilized on plasmonic optical fiber-based devices. On the other hand, the POF-based platform exhibits excellent flexibility, making it extremely interesting for coupling with aptamers in order to develop sensitive biosensors, which can be easily integrable in portable, small-size, simple-to-manufacture devices for clinical applications.

Author Contributions: Conceptualization, writing—original draft, writing—review and editing, L.P. and N.C.; writing—review and editing L.Z. All authors have read and agreed to the published version of the manuscript.

Funding: This research received no external funding.

Institutional Review Board Statement: Not applicable.

Informed Consent Statement: Not applicable.

Data Availability Statement: The data are available on reasonable request from the corresponding author.

Acknowledgments: This work was supported by the VALERE program of the University of Campania "Luigi Vanvitelli" (Italy).

Conflicts of Interest: The authors declare no conflict of interest.

References

1. Jin, Y.; Granville, A. Polymer Fiber Optic Sensors—A Mini Review of their Synthesis and Applications. *J. Biosens. Bioelectron.* **2016**, *7*, 194. [CrossRef]
2. Cennamo, N.; Pesavento, M.; Zeni, L. A review on simple and highly sensitive plastic optical fiber probes for bio-chemical sensing. *Sens. Actuators B Chem.* **2021**, *331*, 129393. [CrossRef]
3. Kadhim, R.A.; Abdul, A.K.K.; Yuan, L. Advances in Surface Plasmon Resonance-Based Plastic Optical Fiber Sensors. *IETE Tech. Rev.* **2020**, 1–18. [CrossRef]

4. Tombelli, S.; Minunni, M.; Mascini, M. Analytical applications of aptamers. *Biosens. Bioelectron.* **2005**, *20*, 2424–2434. [CrossRef] [PubMed]
5. Yan, S.R.; Foroughi, M.M.; Safaei, M.; Jahani, S.; Ebrahimpour, N.; Borhani, F.; Rezaei Zade Baravati, N.; Aramesh-Boroujeni, Z.; Foong, L.K. A review: Recent advances in ultrasensitive and highly specific recognition aptasensors with various detection strategies. *Int. J. Biol. Macromol.* **2020**, *155*, 184–207. [CrossRef] [PubMed]
6. Feng, C.; Dai, S.; Wang, L. Optical aptasensors for quantitative detection of small biomolecules: A review. *Biosens. Bioelectron.* **2014**, *59*, 64–74. [CrossRef] [PubMed]
7. Chen, W.; Lei, W.; Xue, M.; Xue, F.; Meng, Z.H.; Zhang, W.B.; Qu, F.; Shea, K.J. Protein recognition by a surface imprinted colloidal array. *J. Mater. Chem. A* **2014**, *2*, 7165–7169. [CrossRef]
8. Cunha, I.; Biltes, R.; Sales, M.G.F.; Vasconcelos, V. Aptamer-based biosensors to detect aquatic phycotoxins and cyanotoxins. *Sensors* **2018**, *18*, 2367. [CrossRef] [PubMed]
9. Hassan, E.M.; DeRosa, M.C. Recent advances in cancer early detection and diagnosis: Role of nucleic acid based aptasensors. *Trends Anal. Chem.* **2020**, *124*, 115806. [CrossRef]
10. Citartan, M.; Tang, T.H. Recent developments of aptasensors expedient for point-of-care (POC) diagnostics. *Talanta* **2019**, *199*, 556–566. [CrossRef] [PubMed]
11. Majdinasab, M.; Hayat, A.; Marty, J.L. Aptamer-based assays and aptasensors for detection of pathogenic bacteria in food samples. *Trends Anal. Chem.* **2018**, *107*, 60–77. [CrossRef]
12. Safarpour, H.; Dehghani, S.; Nosrati, R.; Zebardast, N.; Alibolandi, M.; Mokhtarzadeh, A.; Ramezani, M. Optical and electrochemical-based nano-aptasensing approaches for the detection of circulating tumor cells (CTCs). *Biosens. Bioelectron.* **2020**, *148*, 111833. [CrossRef] [PubMed]
13. Cennamo, N.; Pesavento, M.; Lunelli, L.; Vanzetti, L.; Pederzolli, C.; Zeni, L.; Pasquardini, L. An easy way to realize SPR aptasensor: A multimode plastic optical fiber platform for cancer biomarkers detection. *Talanta* **2015**, *140*, 88–95. [CrossRef] [PubMed]
14. Cennamo, N.; Pasquardini, L.; Arcadio, F.; Vanzetti, L.E.; Bossi, A.M.; Zeni, L. D-shaped plastic optical fibre aptasensor for fast thrombin detection in nanomolar range. *Sci. Rep.* **2019**, *9*, 18740. [CrossRef] [PubMed]
15. Cennamo, N.; Pasquardini, L.; Arcadio, F.; Lunelli, L.; Vanzetti, L.; Carafa, V.; Altucci, L.; Zeni, L. SARS-CoV-2 spike protein detection through a plasmonic D-shaped plastic optical fiber aptasensor. *Talanta* **2021**, *233*, 122532. [CrossRef] [PubMed]
16. Galatus, R.M.; Cristea, C.; Feier, B.; Cennamo, N.; Zeni, L. SPR based hybrid electro-optic biosensor for β-lactam antibiotics determination in water. In *Remote Sensing and Modeling of Ecosystems for Sustainability XIV*; International Society for Optics and Photonics: Bellingham, WA, USA, 2017; Volume 10405. [CrossRef]
17. Sun, J.; Jiang, S.; Xu, J.; Li, Z.; Li, C.; Jing, Y.; Zhao, X.; Pan, J.; Zhang, C.; Man, B. Sensitive and selective surface plasmon resonance sensor employing a gold-supported graphene composite film/D-shaped fiber for dopamine detection. *J. Phys. D Appl. Phys.* **2019**, *51*, 195402. [CrossRef]
18. Sanjay, M.; Singh, N.K.; Ngashangva, L.; Goswami, P. A smartphone-based fiber-optic aptasensor for label-free detection of: Plasmodium falciparum glutamate dehydrogenase. *Anal. Methods* **2020**, *12*, 1333–1341. [CrossRef]

engineering proceedings

MDPI

Proceeding Paper

A Review on the Effect of Fly Ash, RHA and Slag on the Synthesizing of Coal Bottom Ash (CBA) Based Geopolymer †

Nor Farhana Binti Ab Gulam [1,*], A. B. M. Amrul Kaish [1], Abir Mahmood [1], Sudharshan N. Raman [2], Maslina Jamil [3] and Roszilah Hamid [1]

1 Department of Civil Engineering, Faculty of Engineering and Built Environment, Universiti Kebangsaan Malaysia, Bangi 43600, Selangor, Malaysia; amrul.kaish@ukm.edu.my (A.B.M.A.K.); P108539@siswa.ukm.edu.my (A.M.); roszilah@ukm.edu.my (R.H.)

2 Civil Engineering Discipline, School of Engineering, Monash University Malaysia, Bandar Sunway 47500, Selangor, Malaysia; sudharshan.raman@monash.edu

3 Department of Architecture and Built Environment, Faculty of Engineering and Built Environment, Universiti Kebangsaan Malaysia, Bangi 43600, Selangor, Malaysia; maslinajamil@ukm.edu.my

* Correspondence: P112072@siswa.ukm.edu.my; Tel.: +60-1-8965-3282

† Presented at the 2nd International Electronic Conference on Applied Sciences, 15–31 October 2021.

Abstract: Geopolymerization is widely used in the construction sector for its characteristics of strong compressive strengths, quick hardening, long-term durability, fire resistance, and erosion resistance. This paper has gone through the geopolymer performances utilizing coal bottom ash (CBA), CBA blended with fly ash (FA), CBA mixed together with slag, and CBA with rice husk ash (RHA). CBA shows a better performance than FA in the compressive strength. This paper has discovered several elements that influence geopolymerization, the curing time, the curing temperature, the silicate and hydroxide ratio, and grinding CBA surfaces. The combination of CBA and RHA is suitable for lightweight concrete, as the range of the volumetric weight is within 1192 kg/m^3 to 1655 kg/m^3. The slump result decreases, as the ratio of CBA and slag increases. Slag particles are uneven in shape, which increases water consumption and leads to a honeycombed structure, whereas CBA particles are spherical in shape, which enhances workability.

Keywords: geopolymer; CBA-based geopolymer; slag; rice husk ash; fly ash; water glass solution (WGS)

fv **Citation:** Gulam, N.F.B.A.; Kaish, A.B.M.A.; Mahmood, A.; Raman, S.N.; Jamil, M.; Hamid, R. A Review on the Effect of Fly Ash, RHA and Slag on the Synthesizing of Coal Bottom Ash (CBA) Based Geopolymer. *Eng. Proc.* **2021**, *11*, 20. https://doi.org/10.3390/ASEC2021-11164

Academic Editor: Samuel B. Adeloju

Published: 15 October 2021

Publisher's Note: MDPI stays neutral with regard to jurisdictional claims in published maps and institutional affiliations.

1. Introduction

Geopolymer concrete is a type of concrete that is made by reacting materials containing aluminates and silicates with a caustic alkali activator. Waste materials such as coal ash or slag from iron and metal production are often used to help achieve a cleaner environment. This is because the waste is actually encapsulated in concrete and does not have to be disposed of during use. Geopolymer concrete does not require heating to be manufactured and does not produce carbon dioxide. The standard Portland cement-based concrete or Ordinary Portland Cement (OPC) requires heat and carbon dioxide. There are nine different types of geopolymer, but the largest potential application category for transportation infrastructure is made up of aluminosilicates materials and can be used to completely replace Portland cement in concrete buildings [1]. These geopolymers are based on thermally activated natural materials or industrial by-products (coal bottom ash (CBA), fly ash (FA), or slag) to provide with sources of silicon (Si) and aluminum (Al), which are dissolved in an alkaline activation solution, and the polymer chains and networks are then polymerized to form a hardened binder. This system is commonly referred to as alkali-activated cement or inorganic polymer cement [1].

In recent years, people's awareness of the quantity and diversity of hazardous solid waste and its impact on human health has continued to increase. Increasing attention to the environmental consequences of waste treatment has led to investigations into new ways of using it. The biggest problem facing the industry is the safe and efficient disposal of by-products such as emissions, sludge, and a large amount of coal ash generated during the combustion of coal for power generation. It is estimated that the amount of FA produced will be about 780 million tonnes annually. While the US Environmental Protection Agency and EPA have reported that, it estimated that 140 million tons of coal ash are produced annually. This makes coal ash the second largest industrial waste stream in the United States, after mining waste [2].

2. Chemical Composition of CBA

In keeping similarity with [3], among the analyzed chemical composition of the CBA, the highest percentage is 29.15% silica (SiO_2), followed by 26.685% alumina (Al_2O_3). Therefore, there may be a slightly lower sulphur trioxide (SO_3) content in the bottom ash. This may be due to the low porosity of the bottom ash particles [4], which makes the CBA classified in the F-class based on the ASTM C 61803. Due to the different sources of coal used, there are slightly differences in the chemical composition of the CBA. This classification is strengthened by [5], where the total percentage of SiO_2, Al_2O_3, and Fe_2O_3 in the CBA is more than 70%. The following Table 1 is the chemical compositions of CBA in different power plants.

Table 1. Chemical compositions of the coal bottom ash (CBA) in different power plants.

Power Plant Station/Chemical Composition (%)	Spanish Power Plant	TNB Electric Power Plant, Perak, Malaysia	Tanjung Bin Power Station, Johor, Malaysia	Guru Hargobind Power Plant Bathinda, India	Seocheon Coal-Fired Power Plant, South Korea
SiO_2	52.30	54.80	29.15	56.44	44.2
Al_2O_3	25.14	28.50	26.68	29.24	31.5
Fe_2O_3	9.23	8.49	7.28	8.44	8.9
CaO	2.37	4.20	16.36	0.75	2.0
MgO	1.84	0.35	1.51	0.40	2.6
Na_2O	0.66	0.08	1.15	0.09	-
K_2O	3.72	0.45	0.53	1.29	-
TiO_2	1.45	2.71	-	3.36	2.4

3. Performance Comparison of CBA with Other Pozzolans

3.1. Performance of the CBA Geopolymer

According to [6], it has been stated that the average compressive strengths for pure CBA geopolymer paste was 13.58 MPa, 18.34 MPa, 24.06 MPa, and 22.77 MPa after the pastes were cured at 70 °C and tested after 3, 7, 14, and 28 days. The geopolymer paste reached its maximum compressive strength after 14 days, according to general observations; these findings also indicated that as the curing period lengthens, mortar strength increases [7]. At an elevated temperature, the microstructure of CBA appears to be weakened after 14 days [8]. However, the gap in the compressive strengths between the 14th and 28th days is not statistically significant. In order to check the quality of the geopolymer in a different temperature, 100% CBA specimens were mixed, prepared separately and were cured until 5th, 10th, 15th, and 25th days. The compressive strengths after these days were recorded as shown in Figure 1 and the highest strength was found to be 6.95 MPa. This means that the curing temperature has a significant impact on the strength of the geopolymer concrete. Another important idea claims that increasing the concentration of alkali contained in Na and K metallic ions or decreasing the silicate SiO_2 concentration increases the compressive strength. Singh and Bhardwaj [9] determines that increasing the ratio of sodium silicate to sodium hydroxide has an impact on the geopolymer compressive strength performance. The compressive strength increases, as the Si/Al ratio increases with the increasing percentage of NaOH. Furthermore, the use of finer CBA (4.3 mm) increased the compressive strength due to the inherent pore refinement action of finer particles filling the pores in the paste, increasing hydration products formed during pozzolanic reactions [10]. Physically and chemically, ground CBA resembles FA. Almost all investigations have revealed that adding grinded CBA reduces the compressive strength at early ages. It was discovered that replacing cement with bottom ash had a poor initial curing performance. The performance of the combinations continued to improve even at the age of seven days. When the curing time was increased to 28 days, the performance of the combinations improved dramatically. All of the cement mixtures' compressive strength values are higher than the control sample. Due to the natural pore refinement activity, finer particles filled the pores in the paste, increasing hydration products generated during pozzolanic reactions, and the incorporation of finer CBA with a size of 4.3 mm showed an improvement in the compressive strength. Previous researcher [10,11] has found that the by reducing the particle size of CBA in concrete, the qualities of the concrete are enhanced. The compressive strength of CBA is influenced by the grinding time of CBA with a highball mill. Most research recommended that, the grinded CBA has a potential to be a good pozzolanic material by the increase in fineness. Figure 2 below shows the SEM picture of the 24 h ground bottom ash paste. The particles are closely connected together in the left image of Figure 2. The geopolymer matrix generated by the dissolving of bottom ash and a mixing of sodium hydroxide and sodium silicate solution interlinks the bottom ash particles and the sand particles. The failure zone is depicted in the right image of Figure 2. The bottom ash particles, which are not dissolved in the geopolymer matrix, seem to be split from the gel itself in the failure zone.

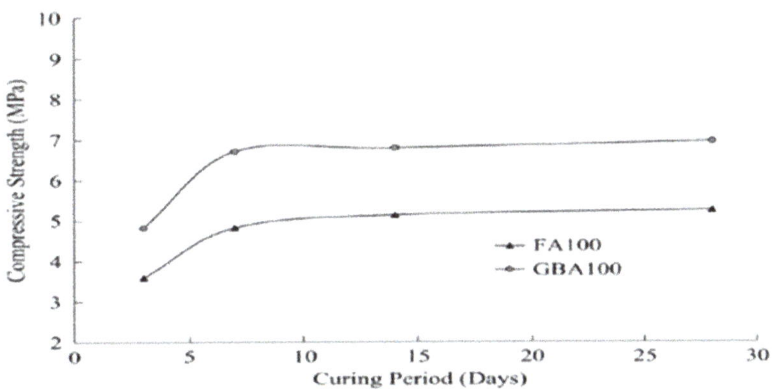

Figure 1. Compressive strengths of geopolymer paste samples made "separately" using 100 percent fly ash (FA) and 100 percent of ground bottom ash to study the effects of the room-temperature curing.

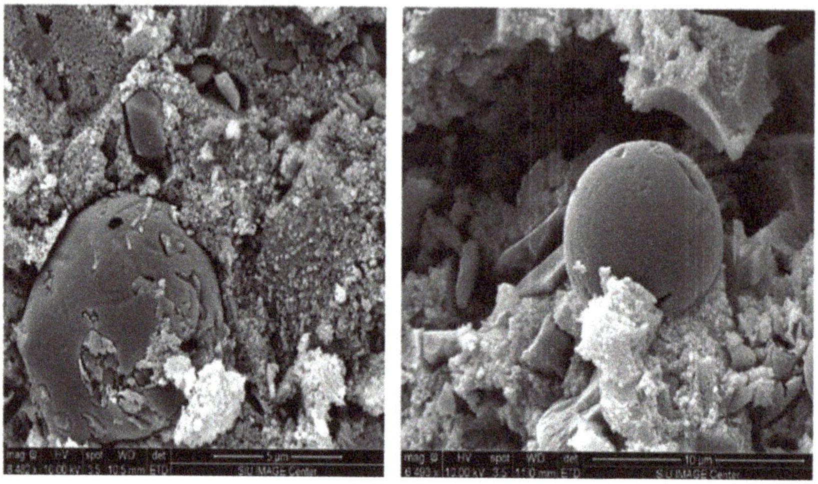

Figure 2. Figure left side: The particles are closely connected together. While, Figure right side: The bottom ash particles, which are not dissolved in the geopolymer matrix, seem to be split from the gel itself in the failure zone.

3.2. Performance of the CBA and Rice Husk Ash (RHA) Geopolymer

As mentioned before, geopolymer is an inorganic polymer material made up of alumino-silicate networks, which are developed when alumino-silicate materials react in a high alkaline environment., which is the result of reactions between alumino-silicate materials in a high alkaline condition. Coal bottom ash (CBA) and rice husk ash (RHA) are combined, with the CBA serving as the principal source of reactive alumina and silicate and the RHA serving as the key source of reactive silica.

According to Van Phuc & Thang [12], geopolymers with an average compressive strength of 17.4 MPa after 28 days, a water absorption of 259.9 kg/m^3, and a volumetric weight of 1655 kg/m^3 were produced using a solid powder mix of 50% CBA and 50% RHA and alkaline activated with 28% (by weight of solids) of water glass (silica modulus of 2.5). After a period of time, with the same portion of solid powder mix as in Van Phuc & Thang [12] with alkaline activated by using the concentration of sodium silicate to 30%, after an average of 28 days, the compressive strength of CBA with RHA was determined to be 37.41 MPa, with a water absorption of 129.94 kg/m^3 and a volumetric weight of 1192 kg/m^3 [13]. Following the same procedure, with 35% CBA, 35% of RHA, and 30% of a water glass solution, it achieved the best performance where the compressive strength was 17.41 MPa, the volumetric weight was 1485.30 kg/m^3, and the water absorption reached 189.94 kg/m^3 [14]. These results from three different papers were in good compliance with the ASTM C55 and C90 requirements for the development of lightweight concrete. The SEM images of coal bottom ash and the blended rice husk ash can be seen in the Figure 3. The summary of the specimens performances can be shown in the Table 2.

Table 2. Engineering properties of the geopolymer (CBA + rice husk ash (RHA)).

Mixture Proportions	Volumetric Weight (kg/m^3)	Water Absorption (kg/m^3)	Compressive Strength (MPa)
50% CBA + 50% RHA 28% sodium silicate	1655	259.9	17.4
50% CBA + 50% RHA 30% sodium silicate	1192	129.94	37.41
35% CBA + 35% RHA 30% sodium silicate	1485.30	189.94	17.41

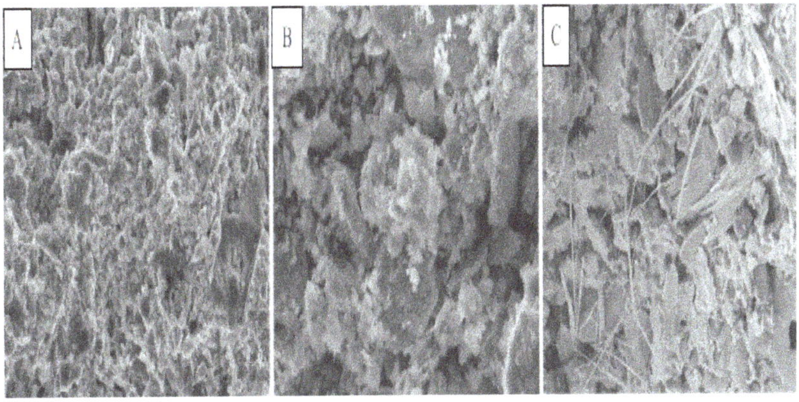

Figure 3. SEM image of RHA, (**A**) coal bottom ash, (**B**) and geopolymer with mixture proportions 35% of coal bottom ash, (**C**) 35% of rice husk ash with 30% of sodium silicate, as can be seen that appearance of a new phase with many long and thin rods [as shown in Figure 3C] as this morphology did not appear in any of the raw materials in image A and B.

3.3. Performance of the CBA and FA Geopolymer

Apart from that, the combination of CBA and coal FA as the geopolymer paste shows very satisfied performances in terms of the compressive strength. According to [15], the 90% of CBA and 10% of coal FA (CBA90FA10) combination generated the highest compressive strength among all the coal fly ash and ground bottom ash combinations, i.e., 22.44 MPa after 14 days, while the compressive strength of the combination of 50% of CBA and 50% of coal FA (CBA50FA50) was found to be 20.82 MPa after 14 days of the curing period. On the other hand, the geopolymer with 70% of CBA and 30% of coal FA (CBA70FA30) recorded, after 14 days of the curing period, the compressive strength achieved to be around 22.13 MPa. The results revealed that as the curing period is lengthened, the strength increases. In addition, the curing temperature has a significant impact on the strength of geopolymer concrete. The compressive strength appears to decrease after 14 days in many situations. Nevertheless, the difference of the compressive strengths between 14 and 28 days may not be statistically significant. Besides that, as the ratio of sodium silicate to sodium hydroxide increases, the compressive strength increases. This increasing pattern of compressive strength is due to the excess sodium silicate hinders water evaporation and the structure formation. From [16], 10M of NaOH concentration is suitable for both raw materials. It can also be seen that during the curing times of 14 and 28 days, even if the proportion is increased, no significant increase in the compressive strength of the paste is recorded. This theory supported by Paija et.al [6], which has found that the CFA/CBA ratio and the concentration of the activating solution may have a considerable impact on the mechanical characteristics of geopolymer made from FA and bottom ashes. The following Table 3 is the engineering properties of the geopolymer (CBA + FA) specimens.

Table 3. Engineering properties of the geopolymer (CBA + FA) specimens.

Mixture Proportions	Compressive Strength at the Room-Temperature Curing	Compressive Strength (after 14th Days at the Elevated-Temperature Curing)
50% CBA + 50% FA	-	20.82 MPa
70% CBA + 30% FA	-	22.13 MPa
100% CBA	6.95 MPa	24.06 MPa
100% FA	5.25 MPa	20.46 MPa

3.4. Performance of the CBA and Slag Geopolymer

The engineering characteristics of concrete including CBA and granulated blast furnace slag have been studied. It was discovered that when the mix proportion of granulated blast furnace slag + coal bottom ash increases, the workability of new concrete decreases. This might be related to the particle form of the substance, according to [17]. CBA particles are spherical in shape, which improves workability, whereas slag particles are irregular in shape, which increases water consumption and leads to a honeycombed structure, which may impact workability performance, and aggregate porosity may also influence workability. Although the water content in the mixture was increased, the workability of the mortar decreased when 75% of bottom ash was combined with it. This was related to the angular form and irregular texture of bottom ash impacting high-interparticle friction [18]. Concrete's water absorption capacity is influenced by its permeability and porosity. Because the replacement components have a higher water absorption capacity than sand, concrete permeability is important. Porosity is vital for the concrete's surface [19]. The GBFS and CBA particles have a distinct surface texture than sand. The creation of a stronger connection between the aggregates and the cement paste is aided by a rougher texture. As a result, the potential replacement of granulated blast furnace slag combined with coal bottom ash for the concrete geopolymer development should be in a less ratio, or new precautions to lessen water absorption capacity should be considered. Second, replacing CBA and slag as fine particles in concrete reduces the compressive strength, due to the fact that the blend of CBA+ slag +FA is higher in compressive strength than the combination of CBA + slag [20]. It is mentioned that the existence of FA in the mixture exerts a balancing effect to some extent as can be seen in Table 4 below, the last proportions contained fly ash (FA) has significantly affecting the strength of the specimens. It is previously been reported that FA contributes to the compressive strength and improves the durability of the concrete. The compressive strength of the concrete is determined by the curing time and temperature, since as the curing time and temperature increase, the compressive strength increases.

Table 4. Engineering properties of the geopolymer CBA with granulated blast furnace slag (GBFS) specimens.

Mixture Proportions	Measured Slump (cm)	Water Absorption (%)	Compressive Strength after the 7th Day (MPa)	Compressive Strength after the 28th Day (MPa)
100% Slag cement 100% aggregates	14	4.14	24.25	37.77
100% slag cement 30% GFBS + 30% CBA	6	6.87	14.14	21.91
100% slag cement 15% GFBS + 15% CBA	10	6.11	17.34	27.54
100% slag cement 25% GFBS + 25% CBA	6	6.66	15.63	25.67
95% slag cement + 5% FA 5% GFBS 5% + 5% CBA	13	4.4	21.30	33.22

4. Conclusions

The conclusions that can be drawn throughout this study are there are several factors affecting the development of geopolymers. The CBA geopolymer and the CBA + FA geopolymer investigation has found that the curing period lengthens, the mortar strength increases and the increase in strength for curing periods beyond the 14th day is not very significant. Because prolonged curing at elevated temperatures breaks the granular structure of the geopolymer mixture, the compressive strength decreases at higher temperatures for longer periods of time. Secondly, the curing temperature is found to be a vital factor in geopolymerization as explained in Section 3.1. Thirdly, increasing the concentration of alkali contained in Na and K metallic ions or decreasing silicate SiO_2 increases the compressive strength. This is because excess sodium silicate hinders water evaporation and structure formation. The matrix activated with potassium silicate KOH obtains the greatest compressive strength, while sodium silicate/NaOH-activated matrixes are generally weaker followed by potassium silicate.

Author Contributions: Conceptualization, N.F.B.A.G. and A.B.M.A.K.; methodology, N.F.B.A.G.; validation, M.J. and A.B.M.A.K.; formal analysis, N.F.B.A.G. and A.B.M.A.K.; investigation, N.F.B.A.G., R.H. and A.B.M.A.K.; resources, S.N.R. and M.J.; data curation, N.F.B.A.G. and A.B.M.A.K.; writing—original draft preparation, N.F.B.A.G.; writing—review and editing, N.F.B.A.G., A.M. and A.B.M.A.K.; visualization, A.B.M.A.K.; supervision, R.H., M.J., S.N.R. and A.B.M.A.K.; project administration, M.J. and A.B.M.A.K.; funding acquisition, S.N.R., M.J., R.H. and A.B.M.A.K. All authors have read and agreed to the published version of the manuscript.

Funding: This research is funded by the Ministry of higher Education, Malaysia through the Fundamental Research Grant Scheme (FRGS/1/2019/TK01/UKM/02/2).

Institutional Review Board Statement: Not applicable.

Informed Consent Statement: Not applicable.

Data Availability Statement: Not applicable.

Acknowledgments: The authors acknowledge Universiti Kebangsaan Malaysia and Ministry of higher Education, Malaysia for providing the necessary opportunities and funding through "Research Graduate Assistance" scheme under the project number FRGS/1/2019/TK01/UKM/02/2.

Conflicts of Interest: The authors declare that there is no conflict of interest.

References

1. Khale, D.; Chaudhary, R. Mechanism of geopolymerization and factors influencing its development: A review. *J. Mater. Sci.* **2007**, *42*, 729–746. [CrossRef]
2. An Article Entitled "Coal Ash: Hazardous to Human Health". Available online: https://www.psr.org/wp-content/uploads/2018/05/coal-ash-hazardous-to-human-health.pdf (accessed on 25 November 2021).
3. Deraman, L.M.; Abdullah, M.M.A.B.; Ming, L.Y.; Hussin, K.; Yahya, Z.; Kadir, A.A. Utilization of bottom ash for alkali-activated (Si-Al) materials: A review. *J. Eng. Appl. Sci.* **2006**, *10*, 8351–8357.
4. Rafieizonooz, M.; Mirza, J.; Salim, M.R.; Hussin, M.W.; Khankhaje, E. Investigation of coal bottom ash and fly ash in concrete as replacement for sand and cement. *Constr. Build. Mater.* **2016**, *116*, 15–24. [CrossRef]
5. Hannan, N.I.R.R.; Shahidan, S.; Ali, N.; Maarof, M.Z. A comprehensive review on the properties of coal bottom ash in concrete as sound absorption material. *MATEC Web Conf.* **2017**, *103*, 01005. [CrossRef]
6. Paija, N.; Kolay, P.K.; Mohanty, M.; Kumar, S. Ground bottom ash application for conventional mortar and geopolymer paste. *J. Hazard. Toxic Radioact. Waste* **2020**, *24*, 04019025. [CrossRef]
7. ul Haq, E.; Padmanabhan, S.K.; Licciulli, A. Synthesis and characteristics of fly ash and bottom ash based geopolymers—A comparative study. *Ceram. Int.* **2014**, *40*, 2965–2971. [CrossRef]
8. Van Jaarsveld, J.G.S.; Van Deventer, J.S.; Lukey, G.C. The effect of composition and temperature on the properties of fly ash-and kaolinite-based geopolymers. *Chem. Eng. J.* **2002**, *89*, 63–73. [CrossRef]
9. Singh, N.; Bhardwaj, A. Reviewing the role of coal bottom ash as an alternative of cement. *Constr. Building. Mater.* **2020**, *233*, 117276. [CrossRef]
10. Basirun, N.F.; Wan Ibrahim, M.H.; Jamaludin, N.; Putra Jaya, R. A Review: The Effect of Grinded Coal Bottom Ash on Concrete. *MATEC Web Conf.* **2017**, *103*, 01007. [CrossRef]
11. Ryu, G.S.; Koh, K.T.; An, G.H.; Lee, J.H. Evaluation Of Shrinkage And Durability Of Geopolymer Concrete Using F-Class Coal Ashes. In *Sustainable Solutions in Structural Engineering and Construction*; ISEC Press: Fargo, ND, USA, 2014.
12. Van Phuc, N.; Thang, N.H. Evaluation on Engineering Properties of Geopolymers from Bottom Ash and Rice Husk Ash. 2017. Available online: https://jstf.hufi.edu.vn/uploads/files/so-tap-chi/nam-2017/so-12/so-12-81-88.pdf (accessed on 25 November 2021).
13. Thang, N.H.; Hoa, N.N.; Quyen, P.V.T.H.; Tuyen, N.N.K.; Anh, T.V.T.; Kien, P.T. Engineering properties of lightweight geopolymer synthesized from coal bottom ash and rice husk ash. *AIP Conf. Proc.* **2018**, *1954*, 040009. [CrossRef]
14. Nguyen, H.T.; Pham, T.K.; Promentilla, M.A. Development of geopolymer-based materials from coal bottom ash and rice husk ash with sodium silicate solutions. In *Congrès International de Géotechnique–Ouvrages–Structures*; Springer: Singapore, 2017; pp. 402–410. [CrossRef]
15. Li, Q.; Xu, H.; Li, F.; Li, P.; Shen, L.; Zhai, J. Synthesis of geopolymer composites from blends of CFBC fly and bottom ashes. *Fuel* **2012**, *97*, 366–372. [CrossRef]

16. Chindaprasirt, P.; Jaturapitakkul, C.; Chalee, W.; Rattanasak, U. Comparative study on the characteristics of fly ash and bottom ash geopolymers. *Waste Manag.* **2009**, *29*, 539–543. [CrossRef] [PubMed]
17. Pyo, S.; Kim, H.K. Fresh and hardened properties of ultra-high performance concrete incorporating coal bottom ash and slag powder. *Constr. Build. Mater.* **2017**, *131*, 459–466. [CrossRef]
18. Deraman, L.M.; Al Bakri Abdullah, M.M.; Liew, Y.M.; Hussin, K.; Yahya, Z. A Review on Processing and Properties of Bottom Ash Based Geopolymer Materials. *Key Eng. Mater.* **2015**, *660*, 3–8. [CrossRef]
19. Ling, Y.; Wang, K.; Wang, X.; Hua, S. Effects of mix design parameters on heat of geopolymerization, set time, and compressive strength of high calcium fly ash geopolymer. *Constr. Build. Mater.* **2019**, *228*, 116763. [CrossRef]
20. Özkan, Ö.; Yüksel, I.; Muratoğlu, Ö. Strength properties of concrete incorporating coal bottom ash and granulated blast furnace slag. *Waste Manag.* **2007**, *27*, 161–167. [CrossRef] [PubMed]

engineering proceedings

Proceeding Paper

Mini-Review of the Importance of Hydrazides and Their Derivatives—Synthesis and Biological Activity †

Suraj N. Mali [1], Bapu R. Thorat [2,*], Deepa Rani Gupta [2] and Anima Pandey [1]

1 Department of Pharmaceutical Sciences and Technology, Birla Institute of Technology, Mesra, Jharkhand 835215, India; suraj1695@gmail.com (S.N.M.); apandey@bitmesra.ac.in (A.P.)
2 Department of Chemistry, Government College of Arts and Science, Aurangabad 431001, India; deepagupta@gmai.com
* Correspondence: bthorat78@gmail.com

† Presented at the 2nd International Electronic Conference on Applied Sciences, 15–31 October 2021; Available online: https://asec2021.sciforum.net/.

Abstract: Organic acid hydrazides include a vast group of organic derivatives of hydrazines containing the active functional group (-C(=O)NHNH$_2$). Acid hydrazones were important bidentate ligands and show keto-enol (amido-iminol) tautomerism. They usually exist in keto form in the solid-state while in equilibrium between keto and enol forms in solution state. Such hydrazones were synthesized in the laboratory by heating substituted hydrazides or hydrazines with corresponding aldehydes or ketones in different organic solvents such as ethanol, methanol, butanol, tetrahydrofuran, etc., and some cases with the ethanol-glacial acetic acid or acetic acid alone. Hydrozones are very important intermediates for the synthesis of heterocyclic compounds and also have different biological activities. The organic chemist would have more interest in the synthesis of acid hydrazides and their derivatives because of their properties. These derivatives having wide applications as chemical preservers for plants, drugs, for manufacturing polymers, glues, etc., in industry and many other purposes. These acid hydrazides and their derivatives were found to be useful synthons for various heterocyclic five, six or seven-membered rings with one or more heteroatoms that were exhibited great biological, pharmacological and industrial applications. This paper will present a review of the chemistry and pharmacological potentials of hydrazide-hydrazones. The various synthetic routes for hydrazone, as well as antibacterial, antifungal and antiviral potentials, have been elaborated in brief.

Keywords: hydrazones; medicinal chemistry; synthesis; pharmacology; organic chemistry

Citation: Mali, S.N.; Thorat, B.R.; Gupta, D.R.; Pandey, A. Mini-Review of the Importance of Hydrazides and Their Derivatives—Synthesis and Biological Activity. *Eng. Proc.* **2021**, *11*, 21. https://doi.org/10.3390/ASEC2021-11157

Academic Editor: Samuel Adeloju

Published: 15 October 2021

1. Introduction

Organic acid hydrazides include the vast group of organic derivatives of hydrazine containing the active functional group (-C(=O)-NH-NH$_2$). Acid hydrazones are important bidentate ligands, which show keto-enol (amido-iminol) tautomerism. They usually exist in the keto form in the solid-state while it retains an equilibrium between keto and enol, when in a solution state (Figure 1).

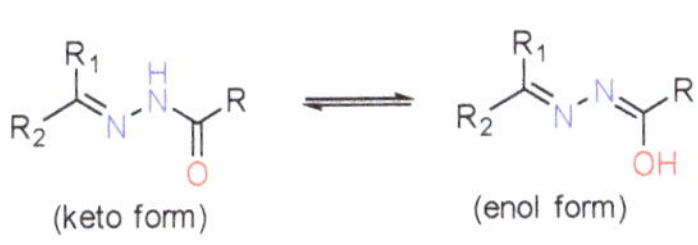

Figure 1. Schematic representation of keto and enol form of acid hydrazones.

Acid hydrazones can react with both electrophiles and nucleophiles simultaneously. They are widely used for the synthesis of heterocyclic compounds. Those compounds contain nucleophilic nitrogen (imine and amino type), both electrophilic and nucleophilic imine carbon atoms and acidic N-H proton (Figure 2).

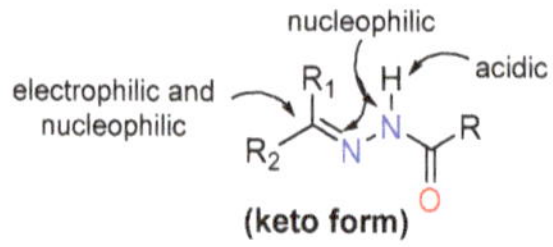

Figure 2. Schematic representation of possible attacking positions in keto form of acid hydrazones.

Such hydrazones were previously synthesized in the laboratory by heating substituted hydrazides or hydrazine with aldehydes or ketones in different organic solvents such as ethanol, methanol, butanol, tetrahydrofuran and, in some cases, with ethanol-glacial acetic acid or acetic acid alone. Hydrazones are very important intermediates for the synthesis of various heterocyclic compounds and usually have wider biological activities. These derivatives have broad applications such as chemical preservers for plants, drugs, for manufacturing polymers, glues, in industry and for many other purposes [1]. These acid hydrazides and their derivatives were found to be useful synthons for the synthesis of various heterocyclic five, six or seven-membered rings with one or more heteroatoms. These compounds were previously exhibited excellent biological, pharmacological and industrial applications such as antibacterial agents, pharmaceuticals, herbicides, antimalarial, antimycobacterial, anticonvulsant, anti-inflammatory, antidepressant, anticancer, antimicrobial activities and dyes [2–12]. Hydrazides and their derivatives could be transformed into various heterocyclic compounds either by cyclisation or cyclo-addition with numerous reagents.

2. Medicinal Chemistry

2.1. Antibacterial Activity

In the past few decades, bacterial and fungal strains have developed resistance towards conventional drugs and, therefore, multidrug-resistant bacterial and fungi infections are becoming serious threats to healthcare settings all over the world. Therefore, for medicinal chemists, the search for new antimicrobial agents is a never-ending and important task. Chemists are constantly looking for different pharmacophores, among them, acid hydrazones/hydrazides are one of the challenging synthons. Numbers of acid hydrazones and their derivatives were synthesized, characterized and evaluated for their antimicrobial activity. There were various reports having hydrazone motif bearing imidazoles (**1**), different thiazolidinone derivatives (**2**, **3**), 1,3,4-thiadiazole based hydrazone derivative (**4**), benzimidazole bearing hydrazone derivative (**5**); benzofuran based hydrazones (**6**); and quinoline-pyridine nucleus containing hydrazones (**7**). These derivatives were also screened for their antibacterial activity against different bacterial strains [13–18] (Figure 3).

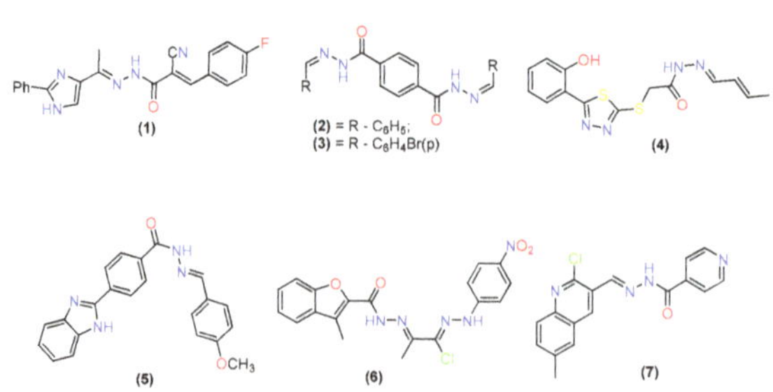

Figure 3. Anti-bacterial compounds (**1–7**) based on hydrazide scaffold.

Some amide containing hydrazones (**8**) and (**9**); some piperidine/pyridine based hydrazones (**10**) and (**11**); heterocyclic ring containing hydrazones (**12**); such as Nifuroxazide; thiophene based hydrazones (**17**); imidazo[2,1-b]thiazole based hydrazones (**13**); imidazo[1,2-a]pyridine based hydrazones (**14**); nitrofurans based hydrazones (**15**); biphenyl-hydrazones (**16**); chloropyrrole based aroylhydrazone (**18**) and (**19**); aryloxyacetic acid hydrazide (**20**); cholic acid-based hydrazones (**21**); benzyledine-hydrazides (**22**); and imidazole bearing hydrazones (**23, 24**) were demonstrated good antibacterial activities [11,19–33] (Figure 4).

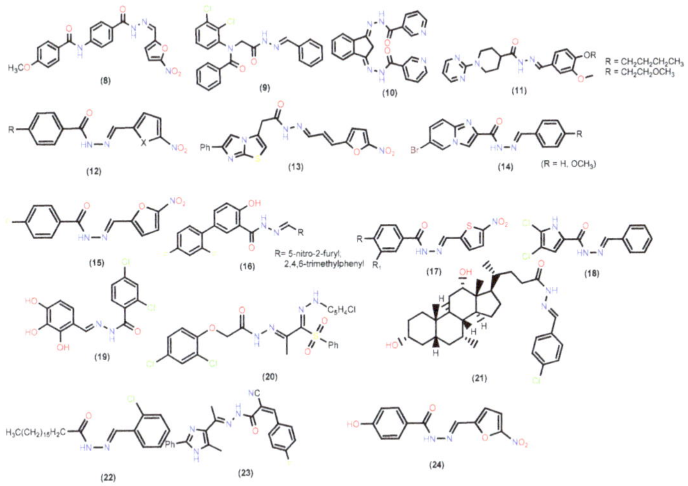

Figure 4. Antimycobacterial compounds (**8–24**) based on hydrazide scaffold.

2.2. Anti-Fungal Activities

Many fungal species may cause many superficial or systemic infections in plants, animals, human beings and also in livestock. Today, synthetic chemists are involved in identifying newer antifungal agents with unique mechanisms. There are many synthesized hydrazone derivatives available in the literature which were also studied for their antifungal activity. Some of them were found to be potent antifungal agents or showed promising antifungal activities against different fungi strains, which included scaffolds such as imidazo[1,2-a]pyridine derivative (**25**), tetrazole based acid hydrazide (**26**), benzofuran based hydrazone (**27**) and 5-bromothiophene-2-yl based hydrazones (**28**) [34–37] (Figure 5).

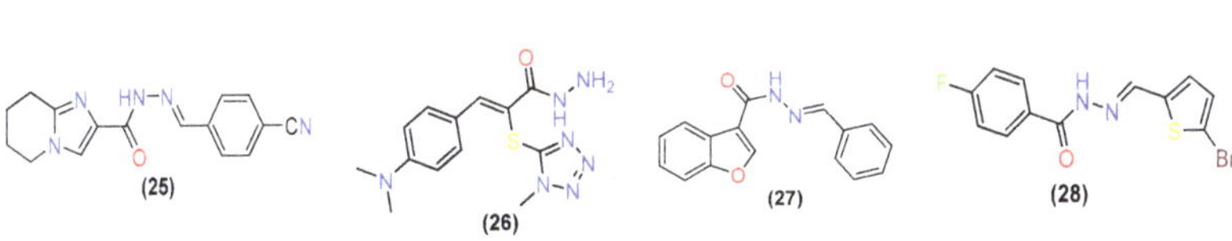

Figure 5. Antifungal compounds (**25–28**) based on hydrazide scaffold.

2.3. Antiviral Activity

The virus is a small infectious agent, which can replicate only inside the living cell of an organism. They can cause immense harmful effects to the host body. They mostly infect all types of organisms, including humans, animals and plants. Several reported hydrazones were showed potent antiviral activities against different viral strains or had lower MIC values, such as imidazole-amide containing acid hydrazones (**29–31**) and sulfonamide containing acid hydrazones (**32, 33**) [38–40] (Figure 6).

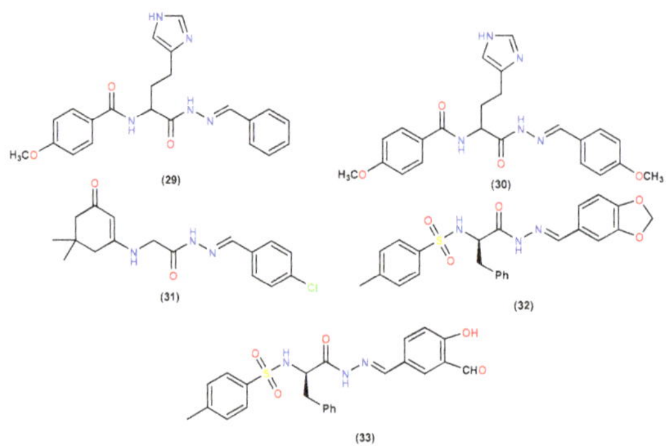

Figure 6. Antiviral compounds (**29–33**) based on hydrazide scaffold.

2.4. Antitubercular Activity

Tuberculosis is a highly spreading infectious, chronic and most prevalent disease. It causes more than three million deaths every year [41]. Different strains of *Mycobacterium tuberculosis* can cause infections in the different parts of the body especially in the lungs, liver and bones. Tuberculosis becomes a serious health problem because of the developed resistance to front-line TB drugs such as isoniazid and rifampin. This indicates the need for more effective drugs for the efficient management of tuberculosis. Some hydrazones (**33–57**) were synthesized and studied for their anti-TB activity against various strains of *mycobacterium tuberculosis*. These included derivatives such as isoniazid derived hydrazones; pyridylmethyleneamino derivatives of isonicotinoylhydrazones; imidazo[4,5-b]pyridine based hydrazones; 5-nitro-2-furyl based hydrazones; 2-substituted 5-(Pyridine-2-yl)-1,3,4-thiadiazole based hydrazones; 1,2,4-triazole-3-mercaptoacetic acid hydrazones; 5-nitro-thiophene containing arylhydrazone; and diclofenac acid hydrazones [42–57]. Some other hydrazones such as aryloxyhydrazone derivatives, benzofuran-3-carbohydrazide derivatives and 2-substituted quinoline based hydrazones (**52**) were also synthesized, characterized and studied for their anti-TB activities. Several acid hydrazones were also demonstrated potent antimycobacterial activity. Compound (**53**) was the most active (MIC = 1.56 μg/mL, IC_{50} = 5.06 μg/mL and SI = 401) among differently synthesized hydrazone derivatives. Compound (**53**) was found to be better than those of "first-line" or "second line" drugs commonly used to treat TB. Compounds (**54,55**), (**56**) and (**57**) were displayed significant and promising antitubercular activity (Figure 7) [42–64].

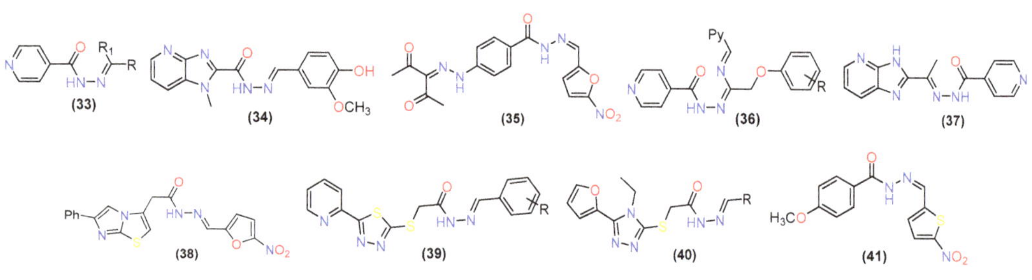

Figure 7. *Cont.*

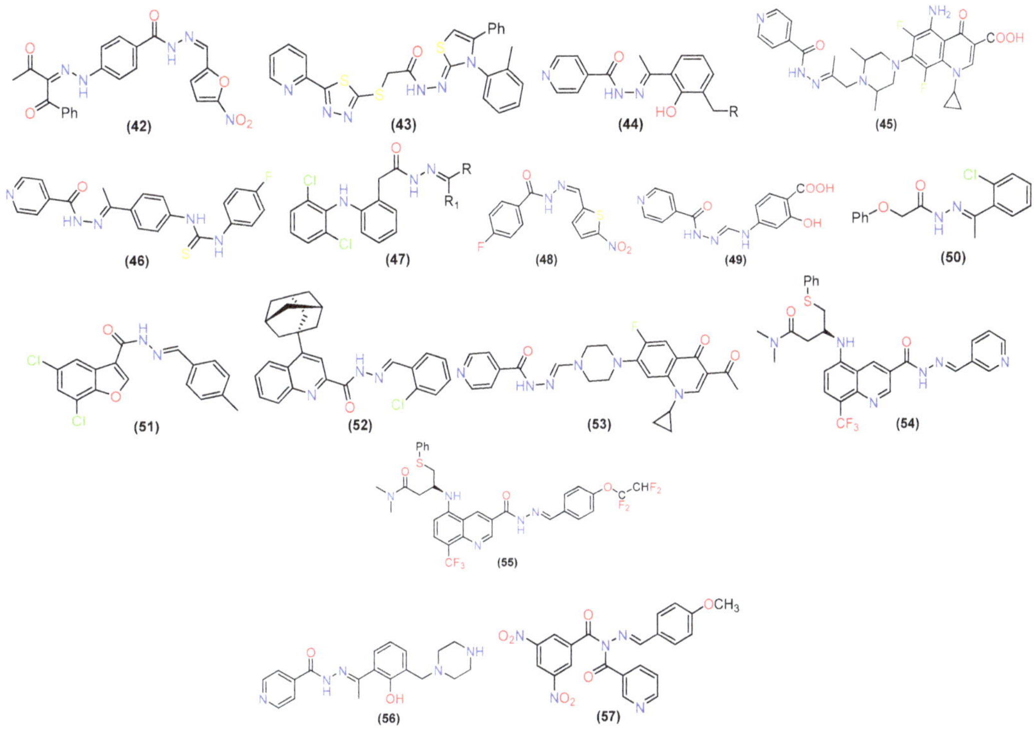

Figure 7. Antimycobacterial compounds (**34–57**) based on hydrazide scaffold.

3. Different Synthetic Routes of Hydrazine, Hydrazones and Their Derivatives

Fenamic hydrazides (**60**) were synthesized from corresponding fenamic acids (**58**) through esters intermediates (**59**). Fenamic acids were previously esterified in methanol by using sulfuric acid and under reflux conditions for 12–18 h. These esters were then treated with hydrazine hydrate (under reflux for 1.5–12 h) [65,66] (Scheme 1).

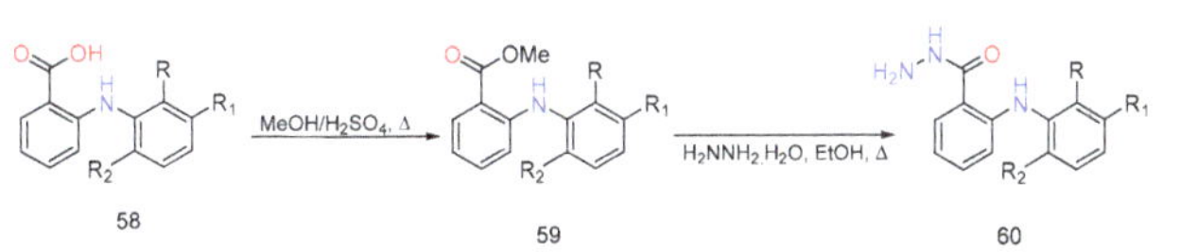

Scheme 1. Synthesis of fenamic acid hydrazides from the corresponding fenamic acids (**58**).

Acyl hydrazones act as mono-nucleating hydrazone ligands, which are easily obtained by condensing acyl hydrazides **62** and substituted aldehydes **61** in ethanol under reflux conditions. These ligands act as a powerful bidentate ligand **63**, **64** and forms complexes with different transition metals. One of the acyl hydrazones, 2-hydroxy-5-chloro-4-methylacetophenone-4-nitrobenzoylhydrazone **63** was synthesized from 4-nitrobenzoyl hydrazide **62** and 2-hydroxy-5-chloro-4-methyl acetophenone **61** in ethanol and got easily coordinated as a tridentate ligand **64** to Cr(III), Mn(III), Ti(III), VO(IV), Fe(III) and Zr(IV) under reflux conditions [67] (Scheme 2).

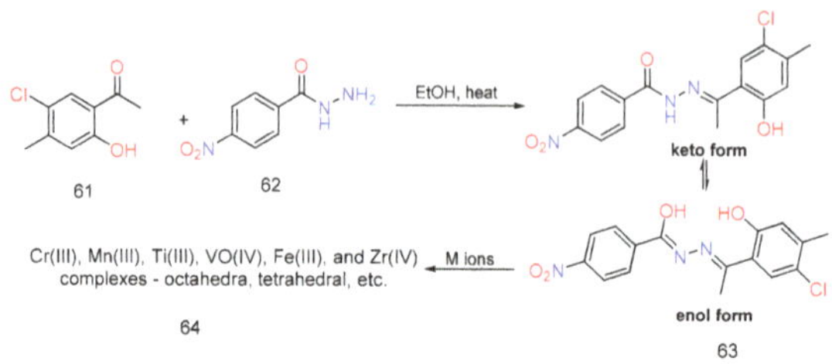

Scheme 2. Synthesis of coordinated tridentate ligands **64**.

A series of substituted ibuprofen-based acyl hydrazones **67** was synthesized under microwave irradiations and by conventional methods using a small quantity of methanol from ibuprofen hydrazide **65** and aryl aldehydes **66** [68] (Scheme 3).

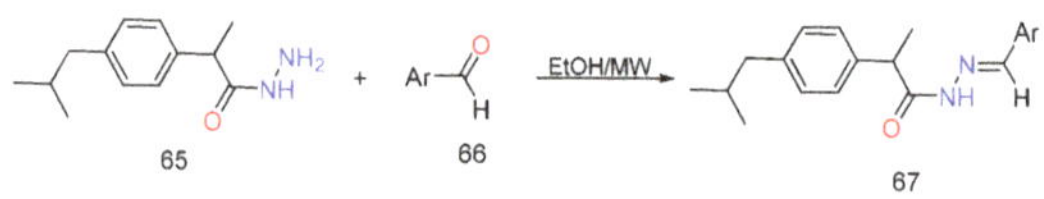

Scheme 3. Synthesis of ibuprofen-based acyl hydrazones.

Several biphenyl-4-carboxylic acid hydrazide-hydrazones **70** were prepared from biphenyl-4-carboxylic acid hydrazide **69** and substituted benzaldehyde using methanol and glacial acetic acid combinations. All synthesized compounds were noted promising antimicrobial activity. Varieties of phenylacetohydrazones were synthesized by using HCl as a catalyst. N-Arylhydrazone derivatives of N-phenyl anthranilic acid were synthesized by condensing 2-(phenylamino) benzohydrazide with various aromatic ketones and aldehydes [10,69,70] (Scheme 4).

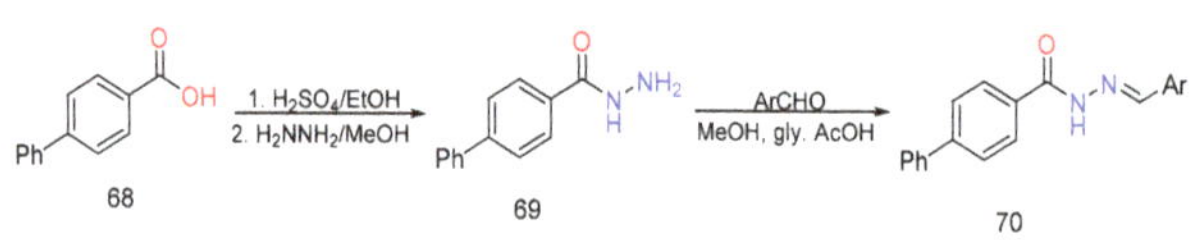

Scheme 4. Synthesis of N-Arylhydrazone derivatives.

Acylhydrazide Schiff base derivatives were prepared by acetic acid-catalyzed condensation of acylhyrazide with different aromatic aldehydes and acetophenones in ethanol under reflux conditions [71] (Scheme 5).

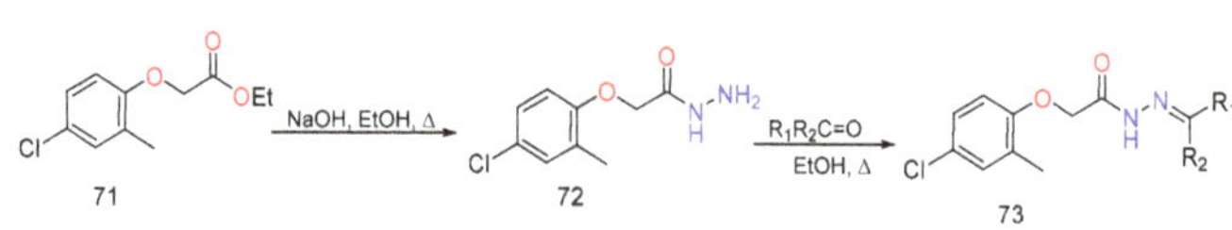

Scheme 5. Synthesis of Acyl hydrazide Schiff base derivatives.

4. Conclusions

To summarize, hydrazone coupled motifs are having an immense pharmacological potential and can be used for synthesizing newer novel motifs with higher potencies. We have also summarized various synthetic routes to synthesis these derivatives.

Author Contributions: Conceptualization, S.N.M., A.P., D.R.G. and B.R.T.; methodology, B.R.T.; software, S.N.M.; writing—review and editing, B.R.T. and D.R.G.; visualization, S.N.M. and A.P.; supervision, A.P. All authors have read and agreed to the published version of the manuscript.

Funding: We wish to thank the Department of Pharmaceutical Sciences, Birla Institute of Technology, Mesra, India for financial assistance. S.N.M. is also thankful for the provision of IRF (PhD/PH/10006/20) (Ref. No. GO/Estb/Ph.D./IRF/2020-21/) provided by BIT, Mesra, India.

Institutional Review Board Statement: Not applicable.

Informed Consent Statement: Not applicable.

Data Availability Statement: Not applicable.

Acknowledgments: The authors would like to thank the Head, Department of Pharmaceutical Sciences and Technology, BIT, Mesra and Department of Chemistry, Government of Maharashtra's Ismail Yusuf College of Arts, Science and Commerce, Mumbai–60 for providing the research facilities for performing the current study.

Conflicts of Interest: The authors declare no conflict of interest.

References

1. Thiyagarajan, S.; Gunanathan, C. Direct Catalytic Symmetrical, Unsymmetrical N,N-Dialkylation and Cyclization of Acylhydrazides Using Alcohols. *Organ. Lett.* **2020**, *22*, 6617–6622. [CrossRef]
2. DeMarinis, R.M.; Hoover, J.R.E.; Dunn, G.L.; Actor, P.; Uri, J.V.; Weisbach, J. A new parenteral cephalosporin, SK & F 59962: Chemistry and structure activity relationships. *J. Antibiot.* **1975**, *28*, 463–470.
3. Elnagdi, M.H.; Erian, A.W. New routes to polyfunctionally substituted pyridine, pyridopyridine, quinoline, and pyridazine derivatives. *Arch. Pharm.* **1991**, *324*, 853–858. [CrossRef]
4. Costales, M.J.; Kleschick, W.A.; Ehr, R.J.; Weimer, M.R.U.S. N-(1-Ethyl-4-pyrazolyl)triazoloazinesulfonamide Herbicides. Patent 5,763,359, 9 June 1998.
5. Gemma, S.; Kukreja, G.; Fattorusso, C.; Persico, M.; Romano, M.P.; Altarelli, M.; Savini, L.; Campiani, G.; Fattorusso, E.; Basilico, N.; et al. Synthesis of N1-arylidene-N2-quinolyl-and N2-acrydinylhydrazones as potent antimalarial agents active against CQ-resistant *P. falciparum* strains. *Bioorgan. Med. Chem. Lett.* **2006**, *16*, 5384–5388. [CrossRef] [PubMed]
6. Bijev, A. New heterocyclic hydrazones in the search for antitubercular agents: Synthesis and in vitro evaluations. *Lett. Drug Des. Discov.* **2006**, *3*, 506–512. [CrossRef]
7. Ragavendran, J.V.; Sriram, D.; Patel, S.K.; Reddy, I.V.; Bharathwajan, N.; Stables, J.; Yogeeswari, P. Design and synthesis of anticonvulsants from a combined phthalimide–GABA–anilide and hydrazone pharmacophore. *Eur. J. Med. Chem.* **2007**, *42*, 146–151. [CrossRef]
8. Todeschini, A.R.; de Miranda, A.L.P.; da Silva, K.C.M.; Parrini, S.C.; Barreiro, E.J. Synthesis and evaluation of analgesic, antiinflammatory and antiplatelet properties of new 2-pyridylarylhydrazone derivatives. *Eur. J. Med. Chem.* **1998**, *33*, 189–199. [CrossRef]
9. Ergenç, N.; Günay, N.S.; Demirdamar, R. Synthesis and antidepressant evaluation of new 3-phenyl-5-sulfonamidoindole derivatives. *Eur. J. Med. Chem.* **1998**, *33*, 143–148. [CrossRef]
10. Deep, A.; Jain, S.; Sharma, P.C.; Verma, P.; Kumar, M.; Dora, C.P. Design and biological evaluation of biphenyl-4-carboxylic acid hydrazide-hydrazone for antimicrobial activity. *Synthesis* **2010**, *182*, 183OC.
11. Masunari, A.; Tavares, L.C. A new class of nifuroxazide analogues: Synthesis of 5-nitrothiophene derivatives with antimicrobial activity against multidrug-resistant Staphylococcus aureus. *Bioorgan. Med. Chem.* **2007**, *15*, 4229–4236. [CrossRef]
12. Fahmy, S.M.; Badran, A.H.; Elnagdi, M.H. Synthesis of some new azopyrazole dyes. *J. Chem. Technol. Biotechnol.* **1980**, *30*, 390–395. [CrossRef]
13. Abdel-Wahab, B.F.; Awad, G.E.; Badria, F.A. Synthesis, antimicrobial, antioxidant, anti-hemolytic and cytotoxic evaluation of new imidazole-based heterocycles. *Eur. J. Med. Chem.* **2011**, *46*, 1505–1511. [CrossRef] [PubMed]
14. Palekar, V.S.; Damle, A.J.; Shukla, S.R. Synthesis and antibacterial activity of some novel bis-1,2,4-triazolo[3,4-b]-1,3,4-thiadiazoles and bis-4-thiazolidinone derivatives from terephthalic dihydrazide. *Eur. J. Med. Chem.* **2009**, *44*, 5112–5116. [CrossRef]
15. Zhong, N.J.; Wang, Y.Z.; Cheng, L.; Wang, D.; Liu, L. Recent advances in the annulation of Morita–Baylis–Hillman adducts. *Organ. Biomol. Chem.* **2018**, *16*, 5214–5227. [CrossRef] [PubMed]
16. Özkay, Y.; Tunalı, Y.; Karaca, H.; Işıkdağ, İ. Antimicrobial activity and a SAR study of some novel benzimidazole derivatives bearing hydrazone moiety. *Eur. J. Med. Chem.* **2010**, *45*, 3293–3298. [CrossRef]
17. Abdel-Aziz, H.A.; Mekawey, A.A. Stereoselective synthesis and antimicrobial activity of benzofuran-based (1E)-1-(piperidin-1-yl)-N2-arylamidrazones. *Eur. J. Med. Chem.* **2009**, *44*, 4985–4997. [CrossRef] [PubMed]
18. Verma, G.; Marella, A.; Shaquiquzzaman, M.; Akhtar, M.; Ali, M.R.; Alam, M.M. A review exploring biological activities of hydrazones. *J. Pharm. Bioallied Sci.* **2014**, *6*, 69–80.
19. Küçükgüzel, Ş.G.; Oruç, E.E.; Rollas, S.; Şahin, F.; Özbek, A. Synthesis, characterisation and biological activity of novel 4-thiazolidinones, 1, 3, 4-oxadiazoles and some related compounds. *Eur. J. Med. Chem.* **2002**, *37*, 197–206. [CrossRef]
20. Asif, M. Pharmacologically potentials of hydrazonone containing compounds: A promising scaffold. *Int. J. Adv. Chem.* **2014**, *2*, 85–103. [CrossRef]

21. Rollas, S.; Gulerman, N.; Erdeniz, H. Synthesis and antimicrobial activity of some new hydrazones of 4-fluorobenzoic acid hydrazide and 3-acetyl-2,5-disubstituted-1,3,4-oxadiazolines. *Il Farmaco* **2002**, *57*, 171–174. [CrossRef]
22. Jubie, S.; Meena, S.; Ramaseshu, K.V.; Jawahar, N.; Vijayakumar, S. Synthesis and biological evaluation of some hydrazones and carbazones of indane-1,3-dione. *Indian J. Chem.* **2010**, *49*, 1261–1263.
23. Govindasami, T.; Pandey, A.; Palanivelu, N.; Pandey, A. Synthesis, characterization and antibacterial activity of biologically important vanillin related hydrazone derivatives. *In. J. Organ. Chem.* **2011**, *1*, 71. [CrossRef]
24. Tavares, L.C.; Chiste, J.J.; Santos, M.G.; Penna, T.C. Synthesis and biological activity of nifuroxazide and analogs. *Il Bollettino Chimico Farmaceutico* **1999**, *138*, 432–436.
25. Ulusoy, N.; Çapan, G.; Otük, G.; Kiraz, M. Synthesis and antimicrobial activity of new 6-phenylimidazo[2,1-b]thiazole derivatives. *Bollettino Chimico Farmaceutico* **2000**, *139*, 167–172. [PubMed]
26. Turan-Zitouni, G.; Blache, Y.; Güven, K. Synthesis and antimicrobial activity of some imidazo-[1,2-a]pyridine-2-carboxylic acid arylidenehydrazide derivatives. *Bollettino Chimico Farmaceutico* **2001**, *140*, 397–400.
27. Küçükgüzel, S.G.; Mazi, A.; Sahin, F.; Öztürk, S.; Stables, J. Synthesis and biological activities of diflunisal hydrazide–hydrazones. *Eur. J. Med. Chem.* **2003**, *38*, 1005–1013. [CrossRef]
28. Rane, R.A.; Telvekar, V.N. Synthesis and evaluation of novel chloropyrrole molecules designed by molecular hybridization of common pharmacophores as potential antimicrobial agents. *Bioorgan. Med. Chem. Lett.* **2010**, *20*, 5681–5685. [CrossRef]
29. Lee, J.Y.; Jeong, K.W.; Shin, S.; Lee, J.U.; Kim, Y. Discovery of novel selective inhibitors of Staphylococcus aureus β-ketoacyl acyl carrier protein synthase III. *Eur. J. Med. Chem.* **2012**, *47*, 261–269. [CrossRef]
30. Abdel-Wahab, B.F.; Khidre, R.E.; Awad, G.E. Regioselective synthesis and antimicrobial activities of some novel aryloxyacetic acid derivatives. *Eur. J. Med. Chem.* **2012**, *50*, 55–62. [CrossRef]
31. Ali, M.R.; Marella, A.; Alam, M.T.; Naz, R.; Akhter, M.; Shaquiquzzaman, M.; Saha, R.; Tanwar, O.; Alam, M.M.; Hooda, J. Review of biological activities of hydrazones. *Indones. J. Pharm.* **2012**, *23*, 193–202.
32. Kumar, D.; Judge, V.; Narang, R.; Sangwan, S.; De Clercq, E.; Balzarini, J.; Narasimhan, B. Benzylidene/2-chlorobenzylidene hydrazides: Synthesis, antimicrobial activity, QSAR studies and antiviral evaluation. *Eur. J. Med. Chem.* **2010**, *45*, 2806–2816. [CrossRef]
33. Singh, S.; Sharma, B.; Siddiqi, N.J. HYDRAZONES AS PROSPECTIVE ANTI-HIV AGENT. *Eur. J. Biomed. Pharm. Sci.* **2016**, *3*, 454–458.
34. Devi, T.S.; Rajitha, G. Microwave Assisted Synthesis and Evaluation of N-cinnamoyl aryl hydrazones for Cytotoxic and Antioxidant Activities. *Orient. J. Chem.* **2016**, *32*, 1703–1709. [CrossRef]
35. Altıntop, M.D.; Özdemir, A.; Turan-Zitouni, G.; Ilgın, S.; Atlı, Ö.; İşcan, G.; Kaplancıklı, Z.A. Synthesis and biological evaluation of some hydrazone derivatives as new anticandidal and anticancer agents. *Eur. J. Med. Chem.* **2012**, *58*, 299–307. [CrossRef] [PubMed]
36. Telvekar, V.N.; Belubbi, A.; Bairwa, V.K.; Satardekar, K. Novel N′-benzylidene benzofuran-3-carbohydrazide derivatives as antitubercular and antifungal agents. *Bioorgan. Med. Chem. Lett.* **2012**, *22*, 2343–2346. [CrossRef]
37. Koçyiğit-Kaymakçıoğlu, B.; Oruç-Emre, E.E.; Ünsalan, S.; Tabanca, N.; Khan, S.I.; Wedge, D.E.; İşcan, G.; Demirci, F.; Rollas, S. Synthesis and biological activity of hydrazide–hydrazones and their corresponding 3-acetyl-2,5-disubstituted-2,3-dihydro-1,3,4-oxadiazoles. *Med. Chem. Res.* **2012**, *21*, 3499–3508. [CrossRef]
38. Hiremathad, A.; Patil, M.R.; Chethana, K.R.; Chand, K.; Santos, M.A.; Keri, R.S. Benzofuran: An emerging scaffold for antimicrobial agents. *RSC Adv.* **2015**, *5*, 96809–96828. [CrossRef]
39. El-Sabbagh, O.I.; Rady, H.M. Synthesis of new acridines and hydrazones derived from cyclic β-diketone for cytotoxic and antiviral evaluation. *Eur. J. Med. Chem.* **2009**, *44*, 3680–3686. [CrossRef]
40. Tian, B.; He, M.; Tang, S.; Hewlett, I.; Tan, Z.; Li, J.; Jin, Y.; Yang, M. Synthesis and antiviral activities of novel acylhydrazone derivatives targeting HIV-1 capsid protein. *Bioorgan. Med. Chem. Lett.* **2009**, *19*, 2162–2167. [CrossRef] [PubMed]
41. African American Cultural Collaborative of Mercer County. Available online: https://www.taacf.com/ (accessed on 6 September 2021).
42. Maccari, R.; Ottanà, R.; Vigorita, M.G. In vitro advanced antimycobacterial screening of isoniazid-related hydrazones, hydrazides and cyanoboranes: Part 14. *Bioorgan. Med. Chem. Lett.* **2005**, *15*, 2509–2513. [CrossRef]
43. Shindikar, A.V.; Viswanathan, C.L. Novel fluoroquinolones: Design, synthesis, and in vivo activity in mice against Mycobacterium tuberculosis H37Rv. *Bioorgan. Med. Chem. Lett.* **2005**, *15*, 1803–1806. [CrossRef]
44. Sriram, D.; Yogeeswari, P.; Madhu, K. Synthesis and in vitro antitubercular activity of some 1-[(4-sub) phenyl]-3-(4-{1-[(pyridine-4-carbonyl) hydrazono] ethyl}phenyl) thiourea. *Bioorgan. Med. Chem. Lett.* **2006**, *16*, 876–878. [CrossRef]
45. Imramovský, A.; Polanc, S.; Vinšová, J.; Kočevar, M.; Jampílek, J.; Rečková, Z.; Kaustová, J. A new modification of anti-tubercular active molecules. *Bioorgan. Med. Chem.* **2007**, *15*, 2551–2559. [CrossRef]
46. Cocco, M.T.; Congiu, C.; Onnis, V.; Pusceddu, M.C.; Schivo, M.L.; De Logu, A. Synthesis and antimycobacterial activity of some isonicotinoylhydrazones. *Eur. J. Med. Chem.* **1999**, *34*, 1071–1076. [CrossRef]
47. Bukowski, L.; Janowiec, M. 1-Methyl-1H-2-imidazo [4, 5-b] pyridinecarboxylic acid and some of its derivatives with suspected antituberculotic activity. *Die Pharmazie* **1996**, *51*, 27–30. [CrossRef]
48. Bukowski, L.; Janowiec, M.; Zwolska-Kwiek, Z.; Andrzejczyk, Z. Synthesis and some reactions of 2-acetylimidazo[4,5-b]pyridine. Antituberculotic activity of the obtained compounds. *Die Pharmazie* **1999**, *54*, 651–654. [CrossRef] [PubMed]

49. Küçükgüzel, İ.; Tatar, E.; Küçükgüzel, Ş.G.; Rollas, S.; De Clercq, E. Synthesis of some novel thiourea derivatives obtained from 5-[(4-aminophenoxy)methyl]-4-alkyl/aryl-2,4-dihydro-3H-1,2,4-triazole-3-thiones and evaluation as antiviral/anti-HIV and anti-tuberculosis agents. *Eur. J. Med. Chem.* **2008**, *43*, 381–392. [CrossRef]
50. Küçükgüzel, Ş.G.; Rollas, S. Synthesis, characterization of novel coupling products and 4-arylhydrazono-2-pyrazoline-5-ones as potential antimycobacterial agents. *Il Farmaco* **2002**, *57*, 583–587. [CrossRef]
51. Mamolo, M.G.; Falagiani, V.; Zampieri, D.; Vio, L.; Banfi, E. Synthesis and antimycobacterial activity of [5-(pyridin-2-yl)-1,3,4-thiadiazol-2-ylthio] acetic acid arylidene-hydrazide derivatives. *Il Farmaco* **2001**, *56*, 587–592. [CrossRef]
52. Mamolo, M.G.; Falagiani, V.; Zampieri, D.; Banfi, L.V.E.; Scialino, G. Synthesis and antimycobacterial activity of (3,4-diaryl-3H-thiazol-2-ylidene)-hydrazide derivatives. *Il Farmaco* **2003**, *58*, 631–637. [CrossRef]
53. Ulusoy, N.; Gürsoy, A.; Ötük, G. Synthesis and antimicrobial activity of some 1,2,4-triazole-3-mercaptoacetic acid derivatives. *Il Farmaco* **2001**, *56*, 947–952. [CrossRef]
54. Rando, D.G.; Sato, D.N.; Siqueira, L.; Malvezzi, A.; Leite, C.Q.; Ferreira, E.I.; Tavares, L.C. Potential tuberculostatic agents. Topliss application on benzoic acid [(5-nitro-thiophen-2-yl)-methylene]-hydrazide series. *Bioorgan. Med. Chem.* **2002**, *10*, 557–560. [CrossRef]
55. Bedia, K.K.; Elçin, O.; Seda, U.; Fatma, K.; Nathaly, S.; Sevim, R.; Dimoglo, A. Synthesis and characterization of novel hydrazide–hydrazones and the study of their structure–antituberculosis activity. *Eur. J. Med. Chem.* **2006**, *41*, 1253–1261. [CrossRef]
56. Saeed, A.; Bolte, M. 1-(4-Chlorophenyl)-4,4,6-trimethyl-3,4-dihydropyrimidine-2 (1H)-thione. *Acta Crystallogr. Sect. E: Struct. Rep. Online* **2010**, *66*, o440. [CrossRef] [PubMed]
57. Raja, A.S.; Agarwal, A.K.; Mahajan, N.; Pandeya, S.N.; Ananthan, S. Antibacterial and antitubercular activities of some diphenyl hydrazones and semicarbazones. *Indian J. Chem. Sect. B.* **2010**, *49*, 1384–1388.
58. Rahaman, M.; Hossain, M.M. Convenient Preparation of 3-Ethoxycarbonyl Benzofurans from Salicylaldehydes and Ethyl Diazoacetate. *Org. Synth.* **2019**, *96*, 98–109. [CrossRef]
59. Nayyar, A.; Monga, V.; Malde, A.; Coutinho, E.; Jain, R. Synthesis, anti-tuberculosis activity, and 3D-QSAR study of 4-(adamantan-1-yl)-2-substituted quinolines. *Bioorgan. Med. Chem.* **2007**, *15*, 626–640. [CrossRef]
60. Poggi, M.; Barroso, R.; Costa-Filho, A.J.; Barbosa de Barros, H.; Pavan, F.; Queico Leite, C.; Gambino, D.; Helvecia Torre, M. New isoniazid complexes, promising agents against Mycobacterium tuberculosis. *J. Mex. Chem. Soc.* **2013**, *57*, 198–204. [CrossRef]
61. Eswaran, S.; Adhikari, A.V.; Pal, N.K.; Chowdhury, I.H. Design and synthesis of some new quinoline-3-carbohydrazone derivatives as potential antimycobacterial agents. *Bioorgan. Med. Chem. Lett.* **2010**, *20*, 1040–1044. [CrossRef] [PubMed]
62. Sriram, D.; Yogeeswari, P.; Madhu, K. Synthesis and in vitro and in vivo antimycobacterial activity of isonicotinoyl hydrazones. *Bioorgan. Med. Chem. Lett.* **2005**, *15*, 4502–4505. [CrossRef]
63. Kumar, P.; Narasimhan, B.; Yogeeswari, P.; Sriram, D. Synthesis and antitubercular activities of substituted benzoic acid N′-(substituted benzylidene/furan-2-ylmethylene)-N-(pyridine-3-carbonyl)-hydrazides. *Eur. J. Med. Chem.* **2010**, *45*, 6085–6089. [CrossRef] [PubMed]
64. Baran, A.U. Comparative study of microwave-assisted and conventional synthesis of ibuprofen-based acyl hydrazone derivatives. *Turk. J. Chem.* **2013**, *37*, 927–935. [CrossRef]
65. Narsinghani, T.; Chaturvedi, S.C. QSAR analysis of meclofenamic acid analogues as selective COX-2 inhibitors. *Bioorgan. Med. Chem. Lett.* **2006**, *16*, 461–468. [CrossRef]
66. Lehmann, J.; Kraft, G. Amphiphile Verbindungen, 1. Mitt. Zur Synthese von 1-Aryl-, 1-Aroyl-und 1-Benzyl-2,3,4-5-tetrahydro-1H-1,4-benzodiazepinen. *Arch. Pharmazie* **1984**, *317*, 595–606. [CrossRef]
67. Reddy, L.V.; Suman, A.; Beevi, S.S.; Mangamoori, L.N.; Mukkanti, K.; Pal, S. Design and synthesis of 1-aroyl-2-ylidene hydrazines under conventional and microwave irradiation conditions and their cytotoxic activities. *J. Brazilian Chem. Soc.* **2010**, *21*, 98–104. [CrossRef]
68. Yaul, A.R.; Dhande, V.V.; Aswar, A.S. Synthesis and Characterization of Dioxotungsten (VI) and Dioxomolybdenum (VI) Complexes of Chelating Hydrazone via Their Oxoperoxo Complexes. *World J. Chem.* **2013**, *8*, 38–41.
69. Gunjan, J.; Awadh, D. Synthesis and Biological Evaluation of Some Phenyl Acetic Acid Hydrazone Derivatives. *Int. Res. J. Pharm. (IRJP)* **2011**, *2*, 110–112.
70. Parashar, B.; Punjabi, P.B.; Gupta, G.D.; Sharma, V.K. Synthesis of some novel N-arylhydrazone derivatives of N-phenyl anthranilic acid. *Int. J. ChemTech Res.* **2009**, *1*, 1022–1025.
71. Jamil, W.; Perveen, S.; Shah, S.A.A.; Taha, M.; Ismail, N.H.; Perveen, S.; Ambreen, N.; Khan, K.M.; Choudhary, M.I. Phenoxyacetohydrazide Schiff bases: β-Glucuronidase inhibitors. *Molecules* **2014**, *19*, 8788–8802. [CrossRef]

engineering proceedings

MDPI

Proceeding Paper

Proposal of a Computational Algorithm for Calculating Material Ratio of Surface Texture †

Hirokazu Machida [1], Ichiro Yoshida [2,*] and Yuki Kondo [2]

[1] Mechanical Engineering, Graduate School of Science and Engineering, HOSEI University, Tokyo 184-8584, Japan; hirokazu.machida.7t@stu.hosei.ac.jp
[2] Department of Mechanical Engineering, Faculty of Science and Engineering, HOSEI University, Tokyo 184-8584, Japan; yuki.kondoh.57@hosei.ac.jp
* Correspondence: yoshida.ichiro@hosei.ac.jp; Tel.: +81-042-387-6033
† Presented at the 2nd International Electronic Conference on Applied Sciences, 15–31 October 2021; Available online: https://asec2021.sciforum.net/.

Abstract: The material ratio curve (hereafter referred to as MRC) of ISO 13565-2 and ISO 4287 is widely used in industrial fields. The computational algorithm of MRC proposed in ISO has a problem of long calculation time, because of a method of slicing the roughness profile. Therefore, in this study, a sort method was proposed as a computational algorithm for time reduction. However, depending on the form of the surface profile, the algorithm of the proposed sort method has a problem in that calculation errors occur. Therefore, in this paper, we report a new improved algorithm that solves this problem. In this paper, a new and improved algorithm for calculating MRC has been researched and developed. The proposed algorithm in this paper succeeded in reducing the computing time to derive MRC compared with the calculating algorithm of MRC proposed in the ISO standard. This algorithm is expected the efficiency improvement of quality control.

Keywords: material ratio curve; Abbott–Firestone curve; roughness; surface texture; computational algorithm

Citation: Machida, H.; Yoshida, I.; Kondo, Y. Proposal of a Computational Algorithm for Calculating Material Ratio of Surface Texture. *Eng. Proc.* **2021**, *11*, 22. https://doi.org/10.3390/ASEC2021-11168

Academic Editor: Filippo Berto

Published: 15 October 2021

1. Introduction

Surface textures are geometric features that collectively refer to surface roughness and scratches. This is described in the ISO standard [1]. Since the surface textures are greatly related to the geometrical specifications of the product from an industrial point of view, their quality is required to be controlled quantitatively.

An example of where surface textures play an important role is in bearings. Bearings are used in automobiles, aircraft, and household appliances such as refrigerators and air conditioners because they have the role of preventing energy loss due to frictional resistance by rotating the shaft smoothly.

Because the frictional resistance of a bearing depends on the surface texture, the bearing is closely related to the surface texture.

ISO 4287 [1] contains various surface texture parameters. Typical parameters of surface texture include arithmetic mean roughness Ra, maximum height Rz, etc. Among them, the material ratio curve (MRC) and MRC parameters can be mentioned as effective methods for evaluating surfaces with excellent lubrication and friction characteristics [2,3]. MRC is a curve that expresses the ratio of the material and void parts of the surface profile with respect to the height direction [4]. MRC and MRC parameters are used in the industrial world as an effective evaluation method for the quality control of automobile parts and bearings.

A method for calculating MRC from the roughness profile has been proposed in ISO 13565-2 [5]. This calculation method has a problem that it takes a long time to calculate as the number of times of slicing the roughness profile increases. Therefore, in a previous paper [6], the sort method, which is a new MRC calculation algorithm that can shorten the calculation time compared to the ISO standard method, has been reported. However, it was clarified that this calculation method causes problems in the derivation of MRC depending on the profile shape of the measured data [6]. Therefore, in this paper, we report a new improved algorithm that does not cause problems in deriving the MRC. If the improved algorithm by this research is completed, it can be expected to contribute to the efficiency of quality control in actual industrial sites.

2. The Calculation Method of MRC by ISO 13565-2

The calculation method of MRC proposed by the ISO standard is as follows: First, a slicing level *ZL* is set parallel to the reference line for the evaluation of the roughness curve. The slicing level is the height of slice (hereafter referred to as the slice height) [5,7]. Next, the ratio of the material and void parts when the surface profile is sliced at a certain height is calculated, and this calculation is performed sequentially over the whole height direction (henceforth referred to as the slicing method). In this research, the algorithm for calculating MRC with the image shown in Figure 1 [7] is coded using MATLAB.

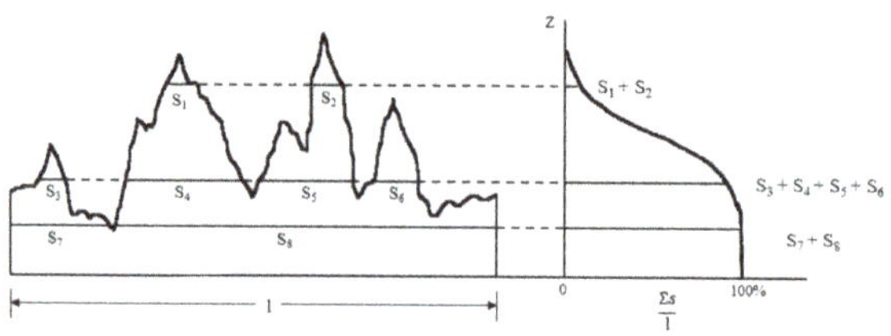

Figure 1. Slice method: derivation method of MRC [5,7].

The procedure of the slicing method is as follows.

(1) The slice height is set.
(2) The intersection point between the roughness profile and the slice height line is calculated.
(3) The length of the material part between the intersections is obtained by linear interpolation.
(4) The lengths of material parts *li* are summed over the evaluation length *l*. Next, the material ratio is calculated by dividing the sum of the lengths *li* by the evaluation length *l*.
(5) Steps (1) to (4) are performed sequentially, changing the slice height at arbitrary ΔZ intervals.

Figure 2 shows an adaptation of the slice method procedure (1) to the roughness profile. The blue line shows the roughness curve (extract) and the red line shows the slice height in Figure 2. The slicing method has a problem in that it is time consuming when the number of slicing levels increases.

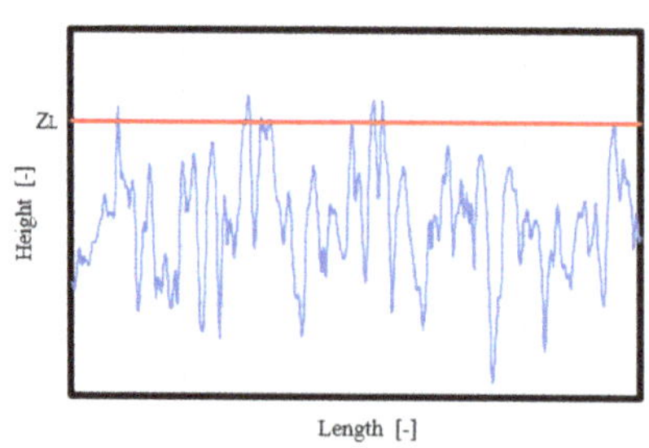

Figure 2. Roughness profile (blue line) and slice height (red line).

3. New Proposed Algorithm for Calculating MRC

3.1. Sort Method for MRC

In this research, the sort method is proposed as a new method of calculating MRC which can reduce the calculation time [6,8]. The sort method is a method for sorting the height data of the roughness profile in ascending order. Figure 3 shows the MRC of the data points sorted by the sort method. The procedure of the sort method is as following steps.

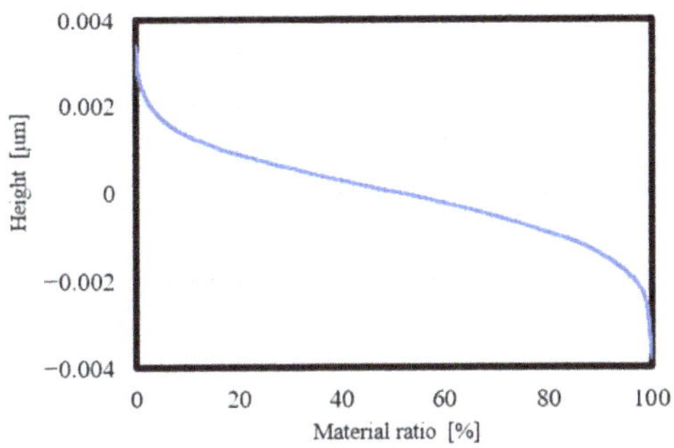

Figure 3. MRC using the sort method.

(1) Draw the intersection of a horizontal line and a roughness profile at a certain height. The positions of the intersections are then patterned.
(2) Calculate the total number of data points that are higher than a certain height value.
(3) Repeat step (2), changing the height in descending order.
(4) Sort the calculated data points in ascending order.

The measured data of a roughness profile may have the same value as height data. In the sort method algorithm, it was clarified that if the same height value exists in the measured data, counting is performed multiple times, which causes a problem in the calculating of MRC (hereinafter referred to as the overlap counting). The overlap counting is one of the factors that obstructs the correct calculation of MRC in the sort method.

Figure 4 shows MRC where the overlap counting occurs. Since the height values from 0.22 to 0.24% are the same, the sort method results in the same number of data points. Therefore, the values from 0.23 to 0.24% are not output on MRC in Figure 4. Therefore, this research develops a new algorithm to solve the problem of overlap counting.

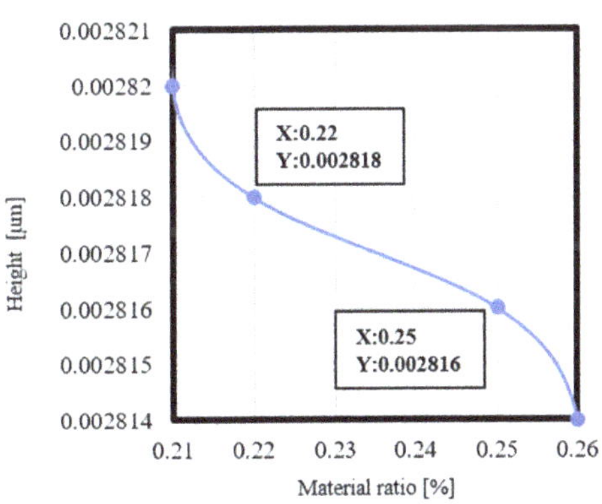

Figure 4. Enlarged view of MRC with overlap counting.

3.2. Improved Sort Method for MRC

The time to calculate MRC was successfully reduced by using the sort method [6]. However, some conditions need to be satisfied so as to use the sort method. If the values of the surface profile are randomly arranged as shown in Figure 5a, the sort method can be applied. If the sort method is applied to the case in which values of the same height occur consecutively, as shown by the red dots in Figure 5b, a problem occurs in the derivation of MRC.

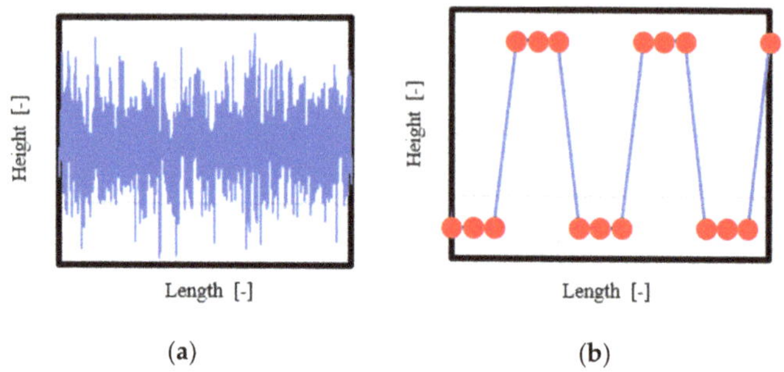

Figure 5. Random and rectangular waves: (**a**) random wave; (**b**) rectangular wave (Red dots show that values of the same height occur consecutively).

Therefore, in this research, in order to solve these problems, we develop an algorithm that does not cause problems in the derivation of MRC even if data with the same height value exists. The green circles in Figure 6 show the measured data where the same three values appeared consecutively. In this study, when the same values appear consecutively as shown in Figure 6, we have developed an algorithm that does not cause problems in the derivation of MRC even if up to three points appear consecutively.

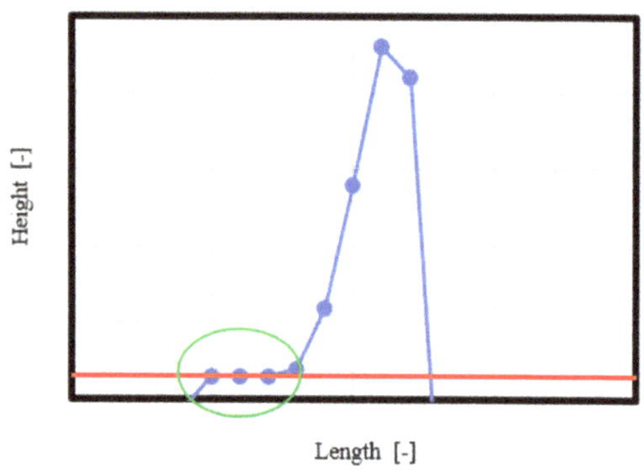

Figure 6. Example of three consecutive points with equal values (green circle) at a certain slice height (red line).

4. Experiment

4.1. Experimental Results by the Sort Method

Figure 7a shows MRC calculated by the slice method and MRC calculated by the sort method before the improvement of the overlap counting. Figure 7b is an enlarged view of the part where the overlap counting occurs in Figure 7a. The blue line in Figure 7b is MRC by the slice method, and the red line is MRC by the sort method. The slice method (blue line) shows that MRC is parallel to the *X*-axis from 0.22 to 0.24%; therefore, measured data with the same height values exist. On the other hand, the sort method (red line) shows that MRC is a diagonal straight line from 0.22 to 0.25%.

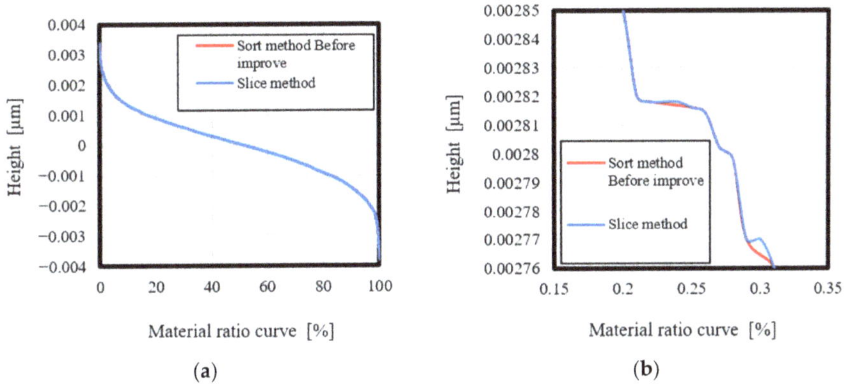

Figure 7. Comparison of calculated MRCs: (**a**) MRC by sort method before improvement and slice method; (**b**) enlarged view of (**a**).

The above results show MRC by the sort method is different from that by the slicing method when the measurement data of the same height value exist.

4.2. Experimental Results Obtained Using the Improved Sort Method

In the sort method before improvement, the locations with the same value of height and the locations where the same value occurs consecutively are recorded in a variable by specifying the condition. In addition, the sort method calculates the number of data points between data points. Therefore, the sort method is improved such that the derivation of MRC does not have problems by modifying the algorithm not to recognize the data points where the same value appears consecutively.

Figure 8a shows the deviations between MRC by the sort method before improvement and the theoretical values based on the total number of data points. The red circle shown in Figure 8a is the error when three same values appear consecutively. Figure 8a shows the deviations between MRC by the improved sort method and the theoretical values. The results in Figure 8b show that the error in Figure 8a has disappeared.

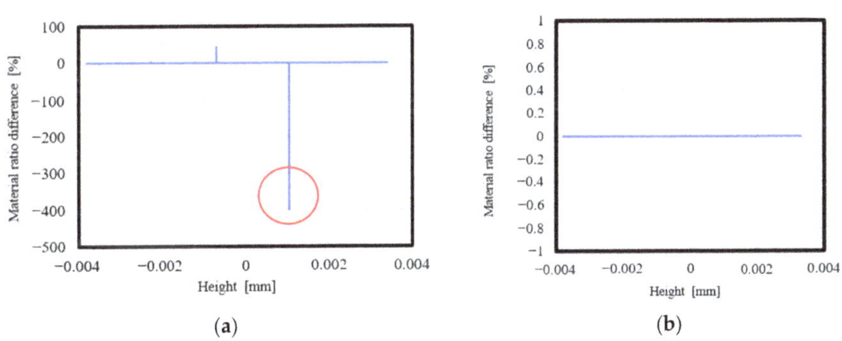

Figure 8. Deviations from theoretical value: (**a**) before improvement (Red circle shown the error when three same values appear consecutively); (**b**) after improvement.

From the above results, the development of an algorithm that does not cause problems in the derivation of MRC is succeeded when the same value that appears continuously is up to three. However, if four or more of the same values appear consecutively, a problem still occurs in the derivation of MRC; therefore, further improvement of the algorithm is necessary.

5. Conclusions

The results and new knowledge of this study can be summarized as follows:

(1) An improved sort algorithm was developed which solves the problem of the sort method. The improved sort algorithm also succeeded in reducing the computing time to derive MRC compared with the calculating algorithm of MRC proposed in the ISO standard.
(2) The developed improved sort algorithm succeeded in the derivation of MRC without causing problems when the same value appearing continuously is up to three.
(3) The improved sort algorithm caused a problem in the derivation of MRC when four or more of the same values appear consecutively; therefore, we will develop a new sort algorithm to solve these problems in the future.

Author Contributions: Conceptualization, I.Y.; methodology, H.M. and I.Y.; software, H.M. and I.Y.; validation, H.M. and I.Y.; formal analysis, H.M., I.Y. and Y.K.; investigation, I.Y. and H.M.; resources, I.Y.; data curation, H.M.; writing—original draft preparation, H.M. and I.Y.; writing—review and editing, H.M., I.Y. and Y.K.; visualization, H.M.; supervision, I.Y.; project administration, I.Y.; funding acquisition, I.Y. All authors have read and agreed to the published version of the manuscript.

Funding: This research was funded by the Precise Measurement Technology Promotion Foundation (PMTP-F) grant number 1-38.

Institutional Review Board Statement: Not applicable.

Informed Consent Statement: Not applicable.

Data Availability Statement: Not applicable.

Conflicts of Interest: The authors declare no conflict of interest.

References

1. ISO 4287. *Geometrical Product Specifications (GPS)-Surface Texture: Profile Method-Terms, Definitions and Surface Texture Parameters*; ISO: Geneva, Switzerland, 1997.
2. Pawlus, P.; Reizer, R.; Wieczorowski, M.; Krolczyk, G. Material ratio curve as information on the state of surface topography—A review. *Precis. Eng.* **2020**, *65*, 240–258. [CrossRef]
3. Yoshida, I.; Tsukada, T.; Arai, Y. Characterization of three-stratum surface textures. *Jpn. J. Tribol.* **2008**, *53*, 99–111. Available online: https://www.researchgate.net/publication/297936036_Characterization_of_three-stratum_surface_textures (accessed on 1 September 2021).
4. Yoshida, I. Surface Roughness-Part 2, How to Use and Clues of the Surface Texture Parameters. *J. Japan Soc. Precis. Eng.* **2013**, *79*, 405–409. [CrossRef]
5. ISO 13565-2. *Geometrical Product Specifications (GPS)-Surface Texture: Profile Method; Surfaces Having Stratified Functional Properties-Part 2: Height Characterization Using the Linear Material Ratio Curve*; ISO: Geneva, Switzerland, 1996.
6. Machida, H.; Yoshida, I. Proposal of Calculating Method for Material Ratio Curve of Surface Texture. In Proceedings of the 5th STI-Gigaku 2020, Nagaoka, Japan, 30–31 October 2020; STI-9-31.
7. Okamoto, J.; Nakayama, K.; Sato, M. *Introduction to Tribology*, 1st ed.; Saiwai Shobo: Tokyo, Japan, 1990; pp. 10–11. (In Japanese)
8. Machida, H.; Nagai, S.; Yoshida, I. Study on a Computational Algorithm for Material Ratio of Surface Texture. In Proceedings of the 6th STI-Gigaku 2021, Nagaoka, Japan, 20–22 October 2021; STI-9-28.

engineering proceedings

Proceeding Paper

Surface Roughness Evaluation of the Inner Surface of Automobile Engine Bores by RANSAC and the Least Squares Method †

Sho Nagai [1], Ichiro Yoshida [2,*] and Ryo Sakakibara [3]

1 Mechanical Engineering, Graduate School of Science and Engineering, HOSEI University, Tokyo 184-8584, Japan; sho.nagai.8v@stu.hosei.ac.jp
2 Department of Mechanical Engineering, Faculty of Science and Engineering, HOSEI University, Tokyo 184-8584, Japan
3 Mechanical Engineering, Graduate School of Science and Engineering, Formerly of HOSEI University, Tokyo 184-8584, Japan; ryo.sakakibara.1011@gmail.com
* Correspondence: yoshida.ichiro@hosei.ac.jp; Tel.: +81-042-387-6033
† Presented at the 2nd International Electronic Conference on Applied Sciences, 15–31 October 2021; Available online: https://asec2021.sciforum.net/.

Abstract: Analysis methods for plateau surfaces have been described in the ISO standards, JIS, and previous studies. The authors of a previous study proposed a method based on the concept of random sample consensus (RANSAC). This method achieved high analysis accuracy for plateau surfaces by setting detailed conditions. However, the process of setting optimal conditions is performed manually, which reduces productivity due to the manpower and man-hours required. In this study, we propose a new method for automating the setting of conditions. This method, which does not require human intervention, is expected to contribute to the improvement of productivity at production sites.

Keywords: material ratio curve; material probability curve; plateau surface; least squares method; RANSAC

Citation: Nagai, S.; Yoshida, I.; Sakakibara, R. Surface Roughness Evaluation of the Inner Surface of Automobile Engine Bores by RANSAC and the Least Squares Method. *Eng. Proc.* **2021**, *11*, 23. https://doi.org/10.3390/ASEC2021-11169

Academic Editors: Filippo Berto and Nunzio Cennamo

Published: 15 October 2021

1. Introduction

The inner surface of an automobile engine bore requires high sliding properties because the piston slides inside the cylinder. For this reason, the bore surface is finished by plateau honing, as shown in Figure 1. The plateau-honed surface has a plateau region and a valley region. Each region has a different requirement specification and is therefore evaluated using a material probability curve. The material probability curve shows the ratio of the real part of surface roughness to the void part, relative to the height direction. The material probability curve is represented on a normal probability plot. The slope of the straight line that is fitted to the slope of the part of the material probability curve corresponding to the plateau and valley regions is the parameter value [1]. The degree of fit of the straight line to the material probability curve affects the roughness parameter value of the plateau surface. Therefore, several analytical methods were proposed, because it is important to detect the slope of a straight line with high validity [2–8].

The method for evaluating plateau surfaces, described by the ISO standard [2,3], requires an arcane computational algorithm. Therefore, a simple calculation algorithm was proposed. Reference [4] proposed a method based on image processing techniques. Reference [5] proposed a method that complies with the ISO standard. Reference [6] proposed a method that allows the evaluation of plateau surfaces, which the ISO standard regards as unanalyzable. This study proposes a new linear fitting method based on the concept of random sample consensus (RANSAC) and the least squares method. In addition, because the industrial world demands improved productivity, this study aims to develop a method that does not require human intervention.

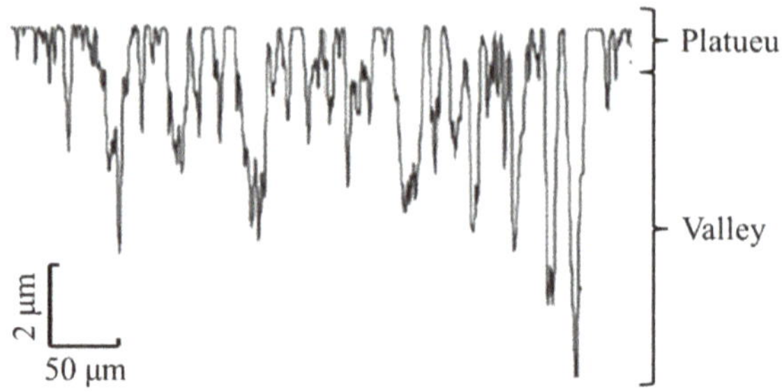

Figure 1. Plateau-honed surface.

2. Proposed Method Applying RANSAC and the Least Squares Method

The algorithm for the proposed method, which applies the RANSAC concept, is as follows:

1. Two points are randomly extracted from the material probability curve, and the model line is calculated from the two randomly extracted points.
2. The number of data points within the tolerance (inliers) from the model line is counted.
3. The best model is the one in which the number of inliers is greater than the specified value and the total error between the acceptable data and the model line is the smallest.

This step is performed at each slope of the plateau and valley regions. The three conditions, to be set in advance, are the boundary position between the plateau and valley regions on the material probability curve, the tolerance, and the number of inliers. In Sections 2.1 and 2.2, we describe the determination of these three conditions.

2.1. Determination of Boundary Position Using the Least Squares Method

The calculation procedure of the method developed by this study for locating the boundary is as follows:

1. The material probability curve is scanned one point at a time from the edge, and the straight line that is fitted to each range is calculated using the least squares method.
2. The change in the slope of the line produces an extreme value.
3. The local maximum point with the largest difference from the neighboring local minimum point, among several detected extremes, is set as a feature point.
4. The point on the material probability curve corresponding to the feature point is determined to be the boundary position.

The boundary positions on the material probability curve corresponding to the feature points are shown in Figure 2 [7].

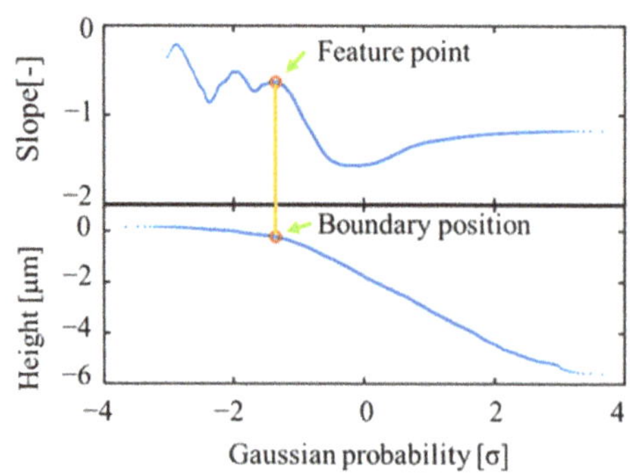

Figure 2. Detection result of feature point.

2.2. Automization of Setting of Tolerances and Number of Inliers

The proposed method based on the RANSAC concept seeks manually the best combination to achieve the property that the accuracy of straight-line detection varies depending on the combination of tolerances and the number of inliers. Therefore, the straight line is detected for all combinations of each value. Automization of the setting of tolerances and the number of inliers is achieved by selecting a model line, with an even smaller error, from the model lines calculated for each combination.

3. Experiment

In this study, the validity of the newly developed algorithm was verified by applying it to the surface profile data of samples with sliding times from 0 to 10,000. As the roughness profile of the plateau surface changes with the sliding time, the material probability curve also changes. It is necessary to apply the proposed method at each sliding time to verify the validity of the algorithm. Figure 3 shows the results of applying the proposed method and fitting straight lines to the material probability curve for the following number of times: 0, 5, 10, 50, 100, and 500. Figure 3 shows that the straight lines coincide with the plateau and valley regions.

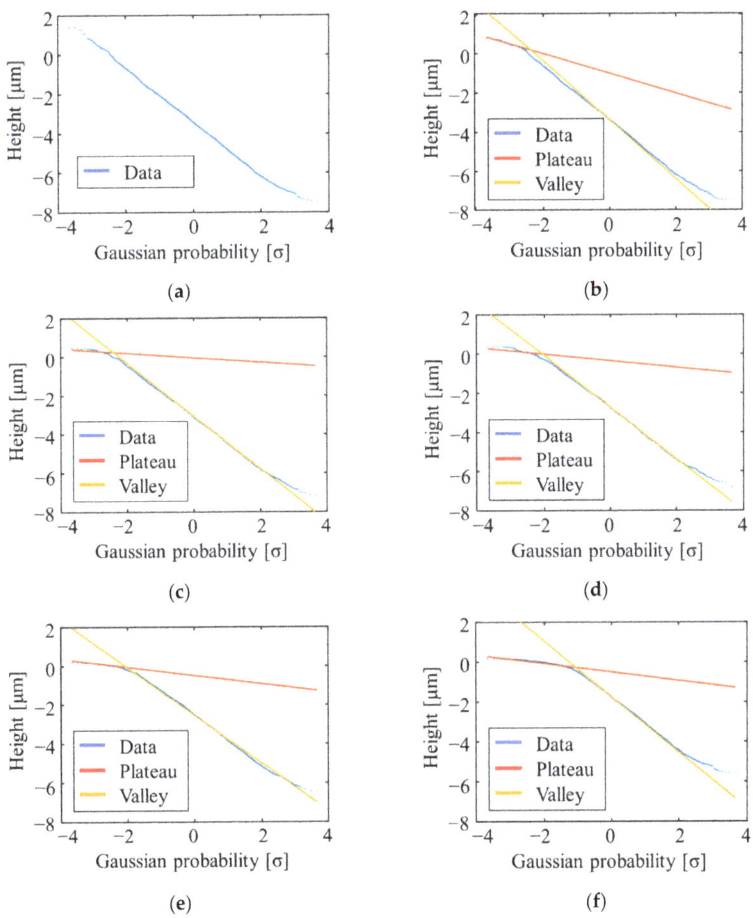

Figure 3. Detection result of straight lines for the following sliding times: (**a**) 0; (**b**) 5; (**c**) 10; (**d**) 50; (**e**) 100; and (**f**) 500.

4. Conclusions

This study developed a method that applied the concept of the least squares method and RANSAC to contribute to the improvement of productivity at production sites. Because this method requires the setting of three conditions, we developed an algorithm to identify these conditions. The experimental results in this study showed that RANSAC can be applied to evaluate plateau surfaces. However, the developed algorithm for identifying the boundary position has a problem, in that it is not fully automated, as the final decision to identify the boundary position is made by visual evaluation. In the future, we will contribute to the improvement of productivity in manufacturing plants by developing a new method to solve this problem.

Author Contributions: Conceptualization, I.Y.; methodology, I.Y. and S.N.; software, S.N. and R.S.; validation, S.N., R.S. and I.Y.; formal analysis, I.Y.; investigation, I.Y., S.N. and R.S.; resources, I.Y.; data curation, S.N.; writing—original draft preparation, S.N. and I.Y.; writing—review and editing, S.N., I.Y. and R.S.; visualization, S.N. and I.Y.; supervision, I.Y.; project administration, I.Y.; funding acquisition, I.Y. All authors have read and agreed to the published version of the manuscript.

Funding: This research received no external funding. This work was partly supported by the Precise Measurement Technology Promo-tion Foundation (PMTP-F) grant number 1-38.

Institutional Review Board Statement: Not applicable.

Informed Consent Statement: Not applicable.

Data Availability Statement: Not applicable.

Conflicts of Interest: The authors declare no conflict of interest.

References

1. Yoshida, I.; Tsukada, T.; Arai, Y. Characterization of three-stratum surface textures. *Jpn. J. Tribol.* **2008**, *53*, 99–111. Available online: https://www.researchgate.net/publication/297936036_Characterization_of_three-stratum_surface_textures (accessed on 1 September 2021).
2. ISO 13565-3:1998 Geometrical Product Specifications (GPS)—Surface Texture: Profile Method; Surfaces Having Stratified Functional Properties-Part 3: Height Characterization Using the Material Probability Curve. Available online: https://www.iso.org/obp/ui/#iso:std:iso:13565:-3:ed-1:v1:en (accessed on 25 November 2021).
3. ISO 21920-2:2021 Geometrical product specifications (GPS)—Surface Texture: Profile-Part 2: Terms, Definitions, and Surface Texture Parameters. Available online: https://www.iso.org/standard/72226.html (accessed on 25 November 2021).
4. Yoshida, I.; Kondo, Y.; Wakatsuki, G.; Numada, M. Study of a Plateau Surface Evaluation Method Using a Fast M-Estimation Type Hough Transform, International Journal of Automation Technology. *Int. J. Autom. Technol.* **2019**, *13*, 118–123. [CrossRef]
5. Sakakibara, R.; Yoshida, I.; Kondo, Y.; Numada, M.; Yamashita, K. A Proposal of Hyperbolic Fitting Method by Applying the Properties of Functions for Plateau Surface Analysis in ISO13565-3. *Nanomanufacturing Metrol.* **2020**, *3*, 1–11.
6. Sakakibara, R.; Yoshida, I.; Nagai, S.; Kondo, Y.; Yamashita, K. Surface roughness evaluation method based on roughness parameters in ISO 13565-3 using the least-squares method for running-in wear process analysis of plateau surface. *Tribol. Int.* **2021**, *163*, 1–9. [CrossRef]
7. Nagai, S.; Yoshida, I.; Machida, H.; Kondo, Y.; Sakakibara, R.; Yamashita, K. Study on Surface Roughness Evaluation Method of Inner Surface of Engine Bore Using RANSAC Method and Least Squares Method. In Proceedings of the Abrasive Technology Conference 2021, Hokkaido, Japan, 30 November–3 December 2021; ABTEC-U-C44, p. 215.
8. Nagai, S.; Yoshida, I.; Sakakibara, R. Study on Surface Roughness Evaluation Method for Engine Bore Surface Using RANSAC and Least Square Method. In Proceedings of the 6th STI-Gigaku 2021, Nagaoka, Japan, 20–22 October 2021, STI-9-19.

engineering proceedings

MDPI

Proceeding Paper

Computational Intelligence Model of Orally Disintegrating Tablets: An Attempt to Explain Disintegration Process †

Jakub Szlęk *, Adam Pacławski, Natalia Czub and Aleksander Mendyk

Department of Pharmaceutical Technology and Biopharmaceutics, Jagiellonian University Medical College, Medyczna 9, 30-688 Kraków, Poland; adam.paclawski@uj.edu.pl (A.P.); natalia.czub@doctoral.uj.edu.pl (N.C.); aleksander.mendyk@uj.edu.pl (A.M.)

* Correspondence: j.szlek@uj.edu.pl; Tel.: +48-12-620-56-04

† Presented at the 2nd International Electronic Conference on Applied Sciences, 15–31 October 2021; Available online: https://asec2021.sciforum.net/.

Abstract: We obtained a curated database based on the database published elsewhere. Chemical descriptors were introduced as characteristics of active pharmaceutical ingredients (APIs). We used H2O AutoML platform in order to develop a Deep Learning model and SHAP method to explain its predictions. Obtained results were satisfactory with *NRMSE* of 8.1% and R2 of 0.84. Finally, we identified critical parameters affecting the process of disintegration of directly compressed ODTs.

Keywords: machine learning model; computational intelligence; AutoML; orally disintegrating tablets; ODTs; disintegration time

Citation: Szlęk, J.; Pacławski, A.; Czub, N.; Mendyk, A. Computational Intelligence Model of Orally Disintegrating Tablets: An Attempt to Explain Disintegration Process. *Eng. Proc.* **2021**, *11*, 24. https://doi.org/10.3390/ASEC2021-11163

Academic Editors: Nunzio Cennamo and Takayoshi Kobayashi

Published: 15 October 2021

1. Introduction

Traditional tablets are not an ideal drug dosage form. Many groups of patients, e.g., pediatric or geriatric patients, have problems with swallowing or simply are not willing to take tablets. As a consequence, all these factors may reduce a patient's compliance. In order to overcome the inconvenience of conventional tablet use, orally disintegrating tablets (ODTs) were introduced into the drug market. One of the methods of preparing ODTs is direct compression, which is cost efficient and simple. It involves comparatively fewer stages than compression preceded by wet or dry granulation. In brief, powders are grinded, if necessary, and blended. Then, the mixture is compressed into the tablets. Although the process is quite simple, there are many factors that influence the characteristics of the ODTs. Among the crucial factors is the disintegration time.

One of the methods used to solve problems with many factors, where the hypothesis governing the phenomenon is unknown or the whole process is complex, is machine learning (ML). Automated machine learning (AutoML) is currently in focus as a branch of ML automating the time-consuming, iterative tasks of model development. AutoML enables machine-driven building of large-scale, high-performance, and superb predictability models with minimal human intervention.

The Motivation of this study was a limited knowledge of relationships between excipients, APIs, and process parameters of direct compression and their influence on disintegration of ODTs. Knowing such behavior would enhance the design and development of novel drug dosage forms. In this work we applied a concept of AutoML-based heuristic model development for prediction of the disintegration time based on the quantitative and qualitative composition of powder mixtures.

2. Materials and Methods

Our database was built based on the database presented by Han et al. [1]. First, we curated the existing database [1], neglecting any unclear or uncertain data records. We put emphasis on the occurrence of the ODTs' characteristic and process parameters, such as tablet hardness, thickness, and dimension of tablet press die. Moreover, we performed a literature survey in order to enhance the database. Scopus® database was searched for publications fulfilling the following criteria: the direct compression method of ODTs should be used in processing, the amount of all excipients should be present, tablet characteristics (hardness, thickness, and die dimension) should be present, and the compendial disintegration test should be applied (Ph. Eur. or USP).

After data scrapping, we calculated APIs' two-dimensional (2D) molecular descriptors using mordred-descriptor v.1.2.1a1 Python package [2] and included them in the curated database. Excipients' types and amounts were encoded in a topological manner. The only output was the time needed for the disintegration of tablets.

A Computational experiment was performed according to the scheme presented in Figure 1. In brief, a preprocessed database was passed to the Python script [3] performing at the first stage a feature selection. Then, the final model was built according to a 10-fold cross-validation scheme. All available algorithms in H2O implementation of AutoML were used [4]: Distributed Random Forest (DRF), Extremely Randomized Trees (XRT), Generalized Linear Model (GLM), Extreme Gradient Boosting Machine, (XGBoost), Gradient Boosting Machine (GBM), Deep Learning (fully connected multi-layer artificial neural network), and Stacked Ensemble models.

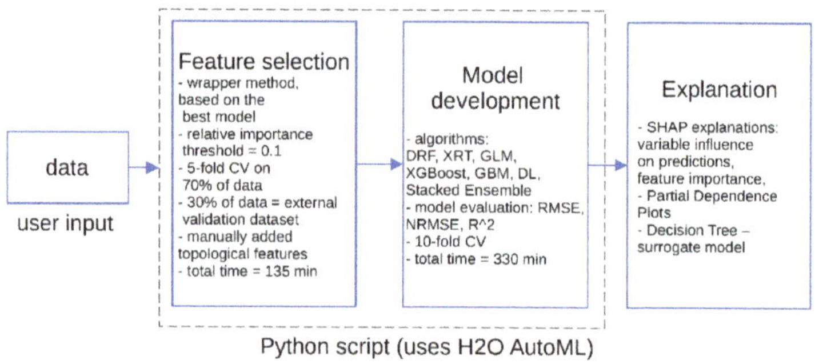

Figure 1. Scheme of computational experiment design.

Model performance was assessed according to the 10-fold cross-validation (10-CV) and expressed by three goodness of fit metrics: root-mean-square error ($RMSE$), normalized root-mean-square error ($NRMSE$), and coefficient of determination (R^2). For reference, please see Equations (1)–(3).

$$RMSE = \sqrt{\frac{\sum_{i=1}^{n}(pred_i - obs_i)^2}{n}} \tag{1}$$

where obs_i and $pred_i$ are observed and predicted values, i is the data record number, and n is the total number of records.

$$NRMSE = \frac{RMSE}{obs_{max} - obs_{min}} \times 100\% \tag{2}$$

where $RMSE$ is the root-mean-square error and obs_{max} and obs_{min} are observed minimal and maximal values.

$$R^2 = 1 - \frac{SS_{res}}{SS_{tot}} = 1 - \frac{\sum_{i=1}^{n}(pred_i - obs)^2}{\sum_{i=1}^{n}(obs_i - obs)^2} \tag{3}$$

where R^2 is the coefficient of determination, SS_{res} is the sum of squares of the residual errors, SS_{tot} is the total sum of the errors, obs_i and $pred_i$ are observed and predicted values, and obs is the arithmetical mean of observed values.

Predictions of the best model were explained with the use of another Python wrapper [5] implementing among others, including the SHapley Additive exPlanations (SHAP) method by Lundberg et al. [6].

3. Results

Each record of the curated database represented one formulation of ODTs. It consisted of 633 chemical descriptors encoding API, 28 inputs encoding amounts of excipients, and 9 inputs characterizing drug dosage form. A single, independent variable was the disintegration time. The database consisted of 243 records (formulations), of which only 52 records (~21%) overlapped the Han et al. database [1].

In the feature selection stage, inputs' number was reduced to 39, among which there were 28 inputs (amount of 2-hydroxypropyl-beta-cyclodextrin [%], Aerosil [%], Amberlite IRP 64-69 [%], API [%], beta-cyclodextrin [%], calcium silicate [%], camphor [%], colloidal silicon dioxide [%], croscarmellose sodium [%], crospovidone [%], cyclodextrin methacrylate [%], Eudragit EPO [%], hydroxy propyl methyl cellulose [%], lactose [%], low-substituted hydroxy propyl cellulose [%], magnesium stearate [%], mannitol [%], microcrystalline cellulose [%], Poloxamer 188 [%], polyvinyl alcohol [%], polyvinylpyrrolidone [%], pregelatinized starch [%], sodium bicarbonate [%], sodium carboxymethyl starch [%], sodium lauryl sulphate [%], sodium starch glycolate [%], sodium stearyl fumarate [%], and talc [%]) responsible for encoding the quantity of excipients and API, 8 molecular descriptors characterizing API (API Geary autocorrelation of lag 7 weighted by ionization potential, API topological charge index of order 7, API Geary autocorrelation—lag 7/weighted by polarizabilities, API modified information content index, API Moran autocorrelation of lag 4 weighted by polarizability, API negative logarithm of the partition (oil/water) coefficient, API number of 12-membered rings (includes counts from fused rings), API number of 8-membered fused rings containing heteroatoms (N, O, P, S, or halogens)), and 3 inputs characterizing the drug dosage form (diameter of die or tablet [mm], hardness of ODT [N], thickness of ODT [mm]). A list of selected features along with their type and relative importance is presented in Table 1.

Table 1. The First 15 selected features and their relative importance.

Feature	Type	Relative Importance
CC_Na_perc	Amount of croscarmellose sodium [%]	1.0000
Crospovidone_perc	Amount of crospovidone [%]	0.8013
SSG_perc	Amount of sodium starch glycolate [%]	0.7341
Hardness_N	Hardness of ODT [N]	0.6564
Eudragit_EPO_perc	Amount of Eudragit EPO [%]	0.5620
Mg_stearate_perc	Amount of magnesium stearate [%]	0.5008
Aerosil_perc	Amount of Aerosil [%]	0.3991
GATS7i	API Geary autocorrelation of lag 7 weighted by ionization potential	0.3441
MCC_perc	Amount of microcrystalline cellulose [%]	0.3394
Colloidal_silicon_dioxide_perc	Amount of colloidal silicon dioxide [%]	0.2336
Mannitol_perc	Amount of mannitol [%]	0.2335
Pregelatinized_starch_perc	Amount of pregelatinized starch [%]	0.2009
PVA_perc	Amount of polyvinyl alcohol [%]	0.1618
Thickness_mm	Thickness of ODT [mm]	0.1482
CD_methacrylate_perc	Amount of cyclodextrin methacrylate [%]	0.1253
(...)	(...)	(...)
Disintegration_time_sec	Disintegration time [s]	output

The best results were obtained by a Deep Learning (DL) model, which had *RMSE* = 10.9, *NRMSE* = 8.1%, and R^2 = 0.84. The model had two hidden layers with 100 neurons in each layer and a rectifier with dropout as an activation function. A plot of predicted versus observed disintegration values is presented in Figure 2.

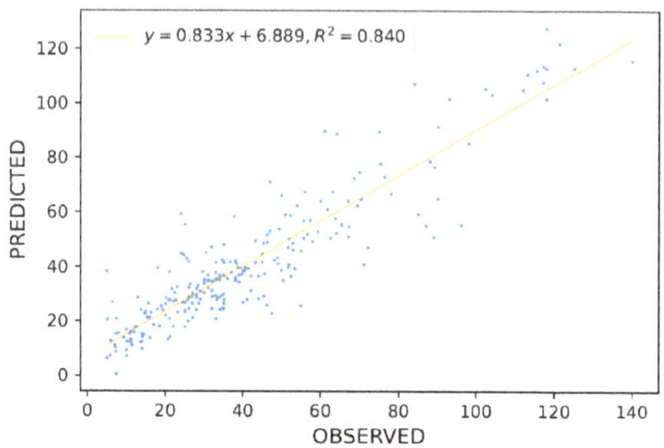

Figure 2. Predicted vs. Observed values for disintegration time for Deep Learning model.

Following model development, a procedure of the SHAP method was applied. Then, selected plots were analyzed, and conclusions were drawn (Figure 3).

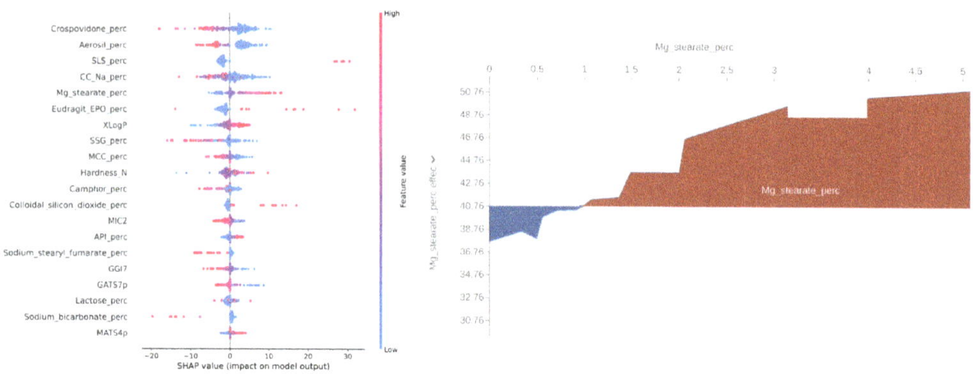

(**a**) (**b**)

Figure 3. Results of the model's explanation: (**a**) Summary plot of impact on model output and feature value; (**b**) Effect of magnesium stearate amount [%] on the average model's prediction.

4. Discussion

Based on the obtained prediction metrics (*RMSE*, *NRMSE*, R^2), it was concluded that the model was satisfactory in terms of generalization. The 10-fold cross-validation technique was used as a golden standard. The mean error of the model was 10.9 (*NRMSE* = 8.1%); therefore it was possible to optimize a formulation with its use. Moreover, in Figure 3a, critical parameters and their impact on the disintegration time were identified. It seems that a high amount of sodium lauryl sulphate, magnesium stearate, Eudragit EPO, and colloidal silicon dioxide could increase the disintegration time of ODTs. On the other hand, a high amount of crospovidone, Aerosil, croscarmellose, sodium starch glycolate, or sodium stearyl fumarate could lead to a decreased disintegration time. Looking more closely at the variable effects, a percolation threshold could be found. For example, in Figure 3b, at a magnesium stearate value of about 1% a reverse in effects could be observed. This observation was consistent with the findings of previous studies [7]. It is believed that magnesium stearate in higher amounts than 1%, besides the usual action as a lubricant, could form a hydrophobic film around API particles and could prevent water from penetrating into the core of the tablet. Using similar reasoning, the XLogP limit was determined for the API, the value of which will increase the disintegration time of ODTs (Figure 4). The general conclusion is that a more hydrophobic API with XLogP higher than 3.5 would negatively affect the disintegration time by increasing it.

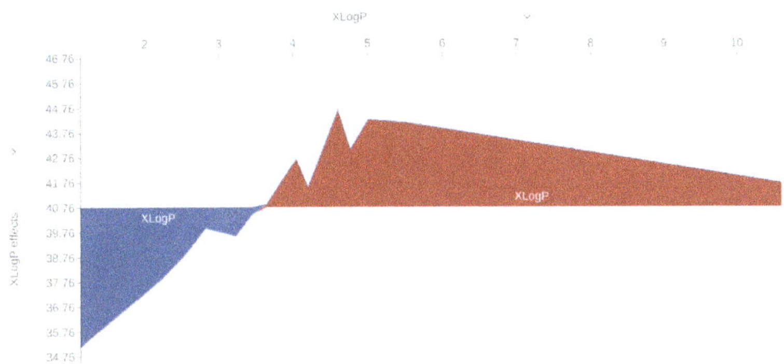

Figure 4. Effect of XLogP (calculated partition coefficient of API) on the average model's prediction.

Author Contributions: Conceptualization, J.S. and A.M.; methodology, J.S., A.P., N.C. and A.M.; software, J.S.; validation, J.S., A.P., N.C. and A.M.; formal analysis, J.S.; investigation, J.S.; resources, A.M.; data curation, J.S.; writing—original draft preparation, J.S., A.P., N.C. and A.M.; writing—review and editing, J.S., A.P., N.C. and A.M.; visualization, J.S.; supervision, J.S.; project administration, J.S. All authors have read and agreed to the published version of the manuscript.

Funding: This research was funded by Uniwersytet Jagielloński Collegium Medicum, grant number N42/DBS/000205.

Institutional Review Board Statement: Not applicable.

Informed Consent Statement: Not applicable.

Data Availability Statement: Data available upon written request.

Acknowledgments: Not applicable.

Conflicts of Interest: The funders had no role in the design of the study; in the collection, analyses, or interpretation of data; in the writing of the manuscript; or in the decision to publish the results. The authors declare no conflict of interest.

References

1. Han, R.; Yang, Y.; Li, X.; Ouyang, D. Predicting oral disintegrating tablet formulations by neural network techniques. *Asian J. Pharm. Sci.* **2018**, *134*, 336–342. [CrossRef] [PubMed]
2. Moriwaki, H.; Tian, Y.-S.; Kawashita, N.; Takagi, T. Mordred: A molecular descriptor calculator. *J. Cheminform.* **2018**, *10*, 1–14. [CrossRef] [PubMed]
3. Szlęk, J. h2o_AutoML_Python, Python Script for AutoML in H2O. Available online: https://github.com/jszlek/h2o_AutoML_Python (accessed on 10 August 2021).
4. LeDell, E.; Poirier, S. H2O AutoML: Scalable automatic machine learning. In Proceedings of the 7th ICML Workshop on Automated Machine Learning, Vienna, Austria, 17–18 July 2020.
5. Szlęk, J. Model Interpretation. 2021. Available online: https://github.com/jszlek/MODEL_INTERPRETATION (accessed on 10 August 2021).
6. Lundberg, S.M.; Erion, G.G.; Lee, S.I. Consistent Individualized Feature Attribution for Tree Ensembles. *arXiv* **2018**, arXiv:1802.03888.
7. Wang, J.; Wen, H.; Desai, D. Lubrication in tablet formulations. *Eur. J. Pharm. Biopharm.* **2010**, *75*, 1–15. [CrossRef] [PubMed]

engineering proceedings

MDPI

Proceeding Paper

Concise Review of Classical Guitar Modelling Technologies †

Alexandre M. Löw [1,*], Herbert M. Gomes [1] and César M. A. Vasques [2]

1 Departamento de Engenharia Mecânica, Universidade Federal de Rio Grande do Sul, Porto Alegre 90040-060, Brazil; herbert@mecanica.ufrgs.br
2 proMetheus, Escola Superior de Tecnologia e Gestão, Instituto Politécnico de Viana do Castelo, Rua Escola Industrial e Comercial de Nun'Álvares, 4900-347 Viana do Castelo, Portugal; cmavasques@gmail.com
* Correspondence: alexandre.low@ufrgs.br
† Presented at the 2nd International Electronic Conference on Applied Sciences, 15–31 October 2021; Available online: https://asec2021.sciforum.net/.

Abstract: Analytical modeling and numerical simulation of multiphysics coupled systems is an exciting research area, even when it comes to intrinsically linear or linearized formulations, as is usually the case with coupled vibroacoustic problems. The combined effect of many localized geometrical miss-modeling with significant uncertainty in mechanical characterization of some organic materials yields large discrepancies in the natural frequencies and mode shapes obtained. The main goal of this work is to compare two basic approaches for the modeling of stringed musical instruments in the frequency domain: simplified lumped-parameter analytic modeling, considering only the most influential degrees of freedom, and discretized finite element modal analysis. Thus, the emphasis is on a review of some key references in this field, including previous work by the authors, which may shed light on some of the most relevant issues surrounding this problem.

Keywords: classical guitar; lumped-parameter modeling; finite element method; vibroacoustics; frequency domain; musical instruments; luthier

Citation: Löw, A.M.; Gomes, H.M.; Vasques, C.M.A. Concise Review of Classical Guitar Modelling Technologies. *Eng. Proc.* **2021**, *11*, 25. https://doi.org/10.3390/ASEC2021-11179

Academic Editor: Filippo Berto

Published: 15 October 2021

1. Introduction

Due to the diversity of phenomena involved in the modeling of musical instruments and the inherent difficulties in formulating vibroacoustics problems, including the various underlying possible dissipating effects, and despite the availability of some classical and computational approaches, this task remains a significant challenge after many years. It is amazing that despite the huge global popularity of plucked stringed instruments like the guitar, scientific research into the piano and violin began much earlier, paving the way for the application of similar principles to the modeling and study of classic guitars.

One may wonder, however, to what extent are analytical models truly exact? Though convergence towards the analytic solution is mathematically granted for methods such as the finite or boundary element methods [1], to what extent can we call this result correct? This article is intended to discuss the above questions contrasting two major approaches when it comes to acoustic guitar modelling. On one hand, a quantitative analysis can follow a familiar approach among engineers and more practically oriented researchers: the quest for an equivalent model capable of yielding a sufficiently accurate approximate result, given that model parameters lie inside a broad-enough range of permitted variations—this paradigm is eventually called gray box modelling, because it does not rely entirely neither on experimental identification nor on analytical modelling. Usefulness has proven to be the key feature here, with the loss of accuracy perceived as a minor drawback in comparison with some gains expected for a computationally inexpensive model. On the other hand, lies the demand for high-fidelity descriptions of physical reality, where a number of numerical methods could be employed to discretize the system into a set of algebraic linear equations—with finite element (FE) models standing by far as the preferred option. Thus, bigger descriptions can be obtained, even though they should unavoidably require greater sets of input data, certainly contaminated by some amount of uncertainty. Thus, model updating has become a standard practice both in academic research and industry (commercial software) applications [2]. It is important to note from its definition, however, that the whole model uncertainty remains distributed over the parameters in an unpredictable, possibly overlapped manner, consequently raising serious concerns about the uniqueness of obtained solutions [3]. These issues become evidently more severe for huge complex models with possibly hundreds of free parameters. Thus, it is clear that a delicate trade-off emerges at this point for those situations where a smaller, less detailed model, could still bring useful results while providing valuable gains from narrower bounds on uncertainty propagation.

2. Simplified Guitar Models

Even as a complex and truly intriguing system, the construction of musical instruments still relies almost solely on traditional knowledge in the great majority of famous lutheries, with an increasing number of academic researchers trying to establish this bridge out of a personal passion [4]. However, things were slightly different for a guitar's close relative: the violin. For a long time, being a learned instrument, it has naturally attracted researchers' attention earlier than its plucked-string cousin. Thus, Schelleng [5] proposed, already in the 1960's, an analogue 2nd order electro-magnetic resistance–inductance–capacitance circuit that could emulate the elasto-acoustic interactions of top plate mechanical properties—mass distribution, stiffness and damping—with the air confined inside and surrounding the resonant chamber. Firth [6] adapted the analogue circuit to the dynamics of acoustic guitar, relating electrical quantities to the most clearly important structural and acoustical parameters. While valuable for those familiar with electrical–mechanical analogy (a standard theoretical apparatus in classical control theory) this approach eventually gave place to a rather direct description of the coupled phenomena. Thus, Caldersmith [4] devised, constructed and measured a simple resonator composed by parts with similar functions as those pointed in [6], and Christensen [7], relying upon these concepts and related equations [4–6], presented a simplified 2 degrees of freedom (DOF) model derived from purely mechanical considerations, focusing on the relations between 1st and 2nd coupled modes and their intermediate Helmholtz anti-resonance, where almost all model parameters (with exception of those related to damping) can be obtained from simple measurement of these 3 frequencies. This work greatly simplified the resulting model, thereby bringing timely applicability and physical insight to subsequent developments. In this way, Caldersmith [8] introduced a straightforward yet important modelling refinement by taking into account the contribution of back plate flexibility in the coupled dynamics, resulting in a 3 DOF system for the resonant chamber, and French [9,10] tested a procedure for identification of model parameters and prediction of structural modification based on the sensitivities of eigenvalues for both 2 and 3 DOF models. Finally, the last extension to this simplified modelling approach was presented in [11], where additional mass and motion DOF were considered for the ribs, neglecting, however, any associated stiffness in this region. Following [7], here as well, model parameters can be obtained, with the exception of dissipative coefficients, from simple measurements. At the top and back plate, bending is represented by that of a thin axisymmetric disk corresponding approximately to the region circumscribed by the lower bout, in an average condition between hinged and clamped. In this way, theoretical approximate expressions could be obtained for the enclosed air volume variation as the product of equivalent plate area and displacement.

2.1. Formulation of Simplified Models and Definition of Parameters

In what follows, the equations of these simplified models discussed above will be presented, trying to keep clarity and conciseness as possible. The meaning of each index adopted in the development of the equations below is shown in Table 1 (for an explanation of guitar components, refer to Figure 1 below). Table 2 presents the parameters appearing in the equations of the 4 DOF model [11]; it should be noted that those appearing in the equations for 2 and 3 DOF models [7,8,10] are also covered in this table, but some table entries do not apply for the simpler models. Damping coefficients are presented separately on the last line because these parameters are to be defined through experimental modal analysis.

Table 1. Meaning of the indexes used in the various vibroacoustic models of the classic guitar.

Index	Meaning
t	Top plate
a	Air column through the sound hole
b	Back plate
r	Ribs (i.e., the remaining parts moving as a whole)

Table 2. Equivalent parameters of all 3 models and damping coefficients.

Symbol	Description	Units
V	Air volume inside the resonant chamber	m^3
m_a	Equivalent mass of the air column	kg
m_t	Equivalent mass of the top plate	kg
m_b	Equivalent mass of the back plate	kg
m_r	Equivalent mass of the ribs	kg
A_a	Equivalent area of the air column	m^2
A_t	Equivalent area of the top plate	m^2
A_b	Equivalent area of the back plate	m^2
k_t	Equivalent stiffness of the top plate	N/m
k_b	Equivalent stiffness of the back plate	N/m
ζ_i, η_i	i-th mode damping coefficient ($i = 1, \ldots, 4$)	–

About the damping parameters in Table 2, it is important to emphasize at this point that these coefficients alone would imply, for instance, a purely viscous, or viscoelastic/hysteretic dissipation behaviour, which is clearly a non-physical assumption considering the predicted forms of energy loss in the modelling of vibro-acoustic coupled systems; i.e., the energy balance in this cases should include, at least, some sort of viscous, viscoelastic and acoustic radiation forms of damping [12,13]. Indeed, for those parameters related to energy dissipation, that equivalent modelling approach mentioned earlier is almost always adopted—with damping ratios, loss factors or another dissipation quantifier accounting alone for all forms of energy loss.

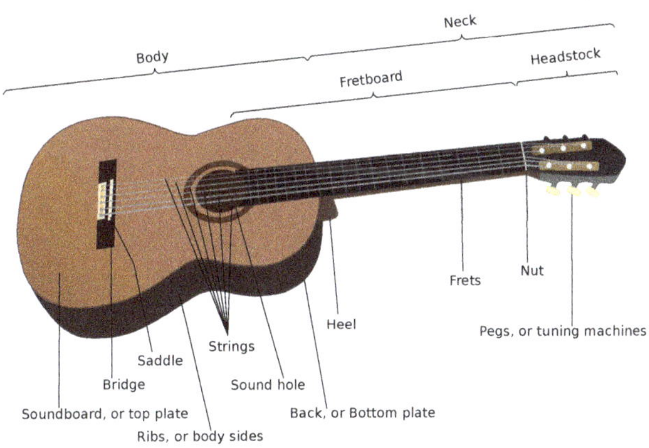

Figure 1. Acoustic guitar parts. (Source: adapted from original image due to William Crochot, distributed under CC BY-SA 3.0 license; Wikimedia Commons).

2.1.1. Simplified 2 Degrees-of-Freedom Model

Now, suppose the tensions in guitar strings transmit a force of magnitude f through the bridge (see Figure 1), that can be represented as acting pointwise, up and down, in the centroid of top plate. Thus, the equations of motion representing the dynamic coupling with the air column DOF displayed in Figure 2 can be written as

$$\begin{aligned} m_t\ddot{x}_t + k_t x_t - \Delta p A_t &= f, \\ m_a\ddot{x}_a - \Delta p A_a &= 0. \end{aligned} \tag{1}$$

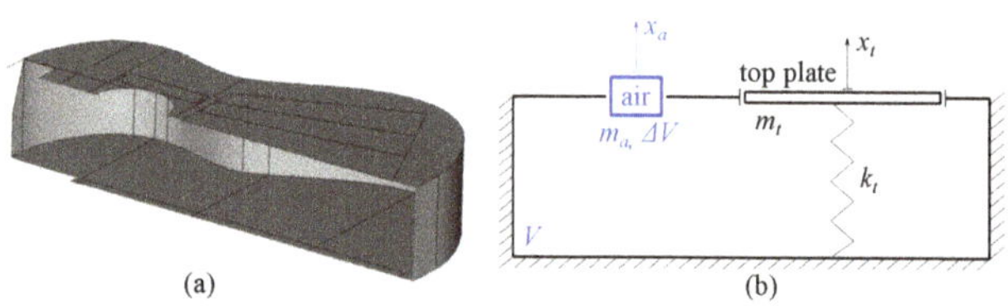

Figure 2. Schematic of the simple 2 DOF model: (**a**) Cut plane view of a guitar resonant chamber 3D CAD model, constructed with shells, planes and lines for subsequent FE analysis. (**b**) Representation as a simple 2 DOF model, considering only the motion of top plate and air column.

Considering the adiabatic, linearized form, of the equations of acoustics, the pressure variation Δp may be expressed in terms of the total volume V inside the resonant chamber, the sound velocity c and mass density ρ of the surrounding air, and of the (idealized) chamber volume increment $\Delta V = A_t x_t + A_a x_a$ [14], as

$$\Delta p = -\rho c^2 \frac{A_t x_t + A_a x_a}{V}. \tag{2}$$

Thus, substituting Equation (2) into (1) and writing it in matrix form, i.e., $M\ddot{x} + Kx = f$, yields

$$\begin{bmatrix} m_t & 0 \\ 0 & m_a \end{bmatrix}\begin{bmatrix} \ddot{x}_t \\ \ddot{x}_a \end{bmatrix} + \begin{bmatrix} k_t + \dfrac{\rho c^2 A_t^2}{V} & \dfrac{\rho c^2 A_t A_a}{V} \\ \dfrac{\rho c^2 A_t A_a}{V} & \dfrac{\rho c^2 A_a^2}{V} \end{bmatrix}\begin{bmatrix} x_t \\ x_a \end{bmatrix} = \begin{bmatrix} f \\ 0 \end{bmatrix}. \tag{3}$$

Quantities appearing on the left-hand side of Equation (3) could be viewed separately into two classes: those described in Table 2 depend on geometric and/or material properties of guitar components, and thus shall be treated as adjustable parameters of the equivalent system, while the remaining ones (speed of sound and mass density) are physical properties of the surrounding air, and so it is more reasonable to consider them as system-independent constants. On the other hand, the driving force f would be truly impossible to obtain, both experimentally or analytically, because this excitation motion is rather imposed by an effectively distributed conjugate on the interfacing area between the top plate and the bridge; i.e., its much higher stiffness impels it to rotate when forced to and fro by the tension of guitar strings, acting some distance above the plate. Theoretically, once the plate is adequately approximated (to some level of accuracy) as a clamped or hinged plate [11] an equivalent pivot distance could be considered, via static equilibrium of moments, in order to compare the driving force over the top plate with the tension in guitar strings. This approach has not been explored up to this moment, however, to the authors' knowledge. Nevertheless, analytical values for modal parameters can be readily extracted from this matrix equation, and in this way it has been extensively used for model comparison [4,7,9], structural modifications prediction based on eigenvalues sensitivity [9,10] or frequency domain model updating [15].

Unlike the mentioned references, the previous equations were written in undamped form. While this choice was made primarily to simplify the presentation, it has a more profound justification stemming from the authors' previous works [16] that is: given the levels of uncertainty in these simplified models, the use of more complex, non-proportional damping formulations would not cause any perceivable alteration in measurable outputs. Thus, since a proportionally damped analytical system produces the same eigenproperties as its undamped version, and taking into consideration the prominence of modal parameters over the forced response for the envisaged applications, damping factors can be confidently considered here as fine tuning adjustments. Among the myriad of available methods to extract damping coefficients (along with mode shapes and natural frequencies) from measured responses, Subspace-Stochastic Identification (SSI) has proven to be a very convenient, output-only, alternative [16].

2.1.2. Simplified 3 and 4 Degrees-of-Freedom Models

From the definitions presented above for 2 DOF, modelling extensions can be readily devised. Figure 3 shows schematic representations for two of these possible expansions, resulting in 3 and 4 DOF models, respectively. The most obvious approach then is to consider the flexibility of back plate in a symmetrical manner as was made for the top plate. This in turn adds up three more free parameters: the mass, the stiffness and one more modal damping coefficient, and the system can now vibrate with three linearly independent modes. Thus, following [10], the resultant displacements vector and the corresponding mass and stiffness matrices for the 3 DOF model are given as

$$\mathbf{x} = \begin{bmatrix} x_t \\ x_a \\ x_b \end{bmatrix}, \ \mathbf{M} = \begin{bmatrix} m_t & 0 & 0 \\ 0 & m_a & 0 \\ 0 & 0 & m_b \end{bmatrix},$$

$$\mathbf{K} = \begin{bmatrix} k_t + \dfrac{\rho c^2 A_t^2}{V} & \dfrac{\rho c^2 A_t A_a}{V} & \dfrac{\rho c^2 A_t A_b}{V} \\ \dfrac{\rho c^2 A_t A_a}{V} & \dfrac{\rho c^2 A_a^2}{V} & \dfrac{\rho c^2 A_a A_b}{V} \\ \dfrac{\rho c^2 A_t A_b}{V} & \dfrac{\rho c^2 A_a A_b}{V} & k_b + \dfrac{\rho c^2 A_b^2}{V} \end{bmatrix}. \quad (4)$$

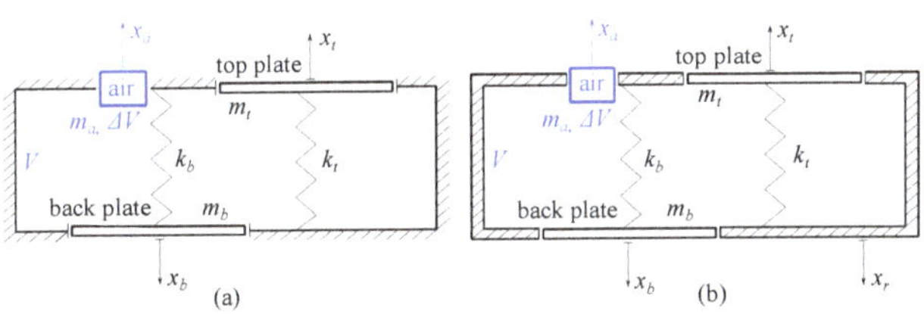

Figure 3. Schematic of 3 and 4 DOF simplified models; (**a**) 3 DOF: allowing for motions on top and back plates, and air column through the hole. (**b**) 4 DOF: basically allowing for the same motions of the 3 DOF model, plus one for the entire structure (free condition).

The next modeling extension, considering mass and motion for the ribs, is a little bit trickier and thus needs deeper attention. First of all, the motion allowed for the 'ribs' in fact means that the whole guitar is free to move. Thus, this is not exactly about freeing a motion in the same sense as it was done for the back plate, but rather allowing a fully free boundary condition. While the governing equations of the 4 DOF model in [11] are not explicitly presented in matrix form, the matrix form was derived and the model is here written in matrix form as

$$\mathbf{x} = \begin{bmatrix} x_t \\ x_a \\ x_b \\ x_r \end{bmatrix}, \ \mathbf{M} = \begin{bmatrix} m_t & 0 & 0 & 0 \\ 0 & m_a & 0 & 0 \\ 0 & 0 & m_b & 0 \\ 0 & 0 & 0 & m_r \end{bmatrix},$$

$$\mathbf{K} = \begin{bmatrix} k_t + \frac{\rho c^2 A_t^2}{V} & \frac{\rho c^2 A_t A_a}{V} & \frac{\rho c^2 A_t A_b}{V} & -k_t - \frac{\rho c^2 A_t^2}{V} + \frac{\rho c^2 A_t A_b}{V} - \frac{\rho c^2 A_t A_a}{V} \\ \frac{\rho c^2 A_t A_b}{V} & \frac{\rho c^2 A_a A_b}{V} & k_b + \frac{\rho c^2 A_b^2}{V} & k_b + \frac{\rho c^2 A_b^2}{V} - \frac{\rho c^2 A_a A_b}{V} - \frac{\rho c^2 A_t A_b}{V} \\ -k_t & 0 & k_b & k_t + k_b \\ \frac{\rho c^2 A_t A_b}{V} & \frac{\rho c^2 A_a A_b}{V} & \frac{\rho c^2 A_b^2}{V} & \frac{\rho c^2 A_b^2}{V} - \frac{\rho c^2 A_a A_b}{V} - \frac{\rho c^2 A_t A_b}{V} \end{bmatrix}. \quad (5)$$

3. Finite Element Modelling of Acoustic Guitar Vibroacoustics

Many factors have contributed to the popularity of FE modelling in the last two or three decades, but most of all, possibly, the greater availability of computational resources. Thus, it is not surprising that nowadays, the bulk of academic research in vibro-acoustics of musical instruments makes use of it to a greater or lesser extent. Brooke [17] was a pioneering work on the application of FE to guitar coupled vibro-acoustics. The simplified models were employed alongside the FE modelling of the top plate and boundary element (BE) model of surrounding air, thus achieving an organic coupling between lumped parameters and domain discretization approaches. Elejabarrieta, Ezcurra and Santamaría [18,19] performed fluid-structure interaction FE analyses, both for fluid and structural domains. The fluid domain was considered only in a small air column, in a similar fashion as was done for the lumped models discussed earlier. Neck and headstock participations were disregarded. 8474 brick elements were used to model the structure, while the mesh on the fluid domain was adjusted to give sufficiently accurate results up to 1 kHz. Comparison of coupled and uncoupled mode shapes confirmed the physical insights brought by lumped models and the importance of Helmholtz frequency to the lower range coupled dynamics. Chaigne and collaborators [20–22] successfully employed FE and finite differences in a full non-linear transient simulation, where considerable effort is put into precise definition of purely mathematical aspects related to the problem's well-posedness and stability of the proposed solution procedure. The fictitious domain method was used to avoid what they called an 'ectoplasm', referring to the rather artificial definition of boundaries in the fluid domain. Structural flexibility of the resonant chamber is considered only within the top plate, disregarding fan bracing. Both these simplifications may leave aside important aspects of top plate dynamic behaviour. It is interesting to note that, despite the clear relevance of such a complete description, these important features are especially focused on by the lumped-parameter models presented above. Other instruments, closely related to the acoustic guitar, were successfully modelled via FE analysis as well, and we pinpoint here references [23], on the Brazilian viola caipira, and [24], on the Colombian bandola. Both present detailed equations and solution procedures for the full coupled FE analysis and solve equations using ANSYS commercial software, but in [24] a convenient symmetrized form for the coupling matrices is introduced and an effective length is considered for the air column [18,19].

4. Results and Discussion

This section is intended to show some results obtained by the authors with both kinds of modelling techniques discussed up to this point. Updating/identification of the 3 DOF model parameters was performed by Löw [15], following [10], but using some heuristic optimization algorithms instead to minimize the sum of squared differences between analytical and experimental FRFs. Masses were measured with 0.1 g accuracy, and stiffness parameters were previously estimated by static force/displacement testing, as suggested by [6,7,9]. Figure 4 below compares the updated theoretical model with experimental data; a modified particle swarm algorithm was used in error minimization. Experimental data was acquired on a low quality, large-scale manufactured instrument.

For the sake of comparison, some more recent results [16] obtained for a Giannini GWNE15, a Brazilian mid quality, partially hand-made guitar, are presented below, this time parametrically modelled in ANSYS commercial software. The mesh used 17,001 shell elements, considering wood plates orthotropic properties. Fan bracing, struts and neck were modelled with shell elements as well. Figure 5 below shows the FE mesh, and Figure 6 shows structural (in vacuo) modes of top and back plates. Table 3 compares experimental and analytical natural frequencies for both models showed earlier. Bigger discrepancies in FE model are primarily due to not considering the air-structure coupling, so it is difficult to establish mode matching. Even so, it is clear that the greater number of available vibration modes can be used to better represent effects associated with higher frequency range dynamics, such as the timber of instruments, and radiation efficiency.

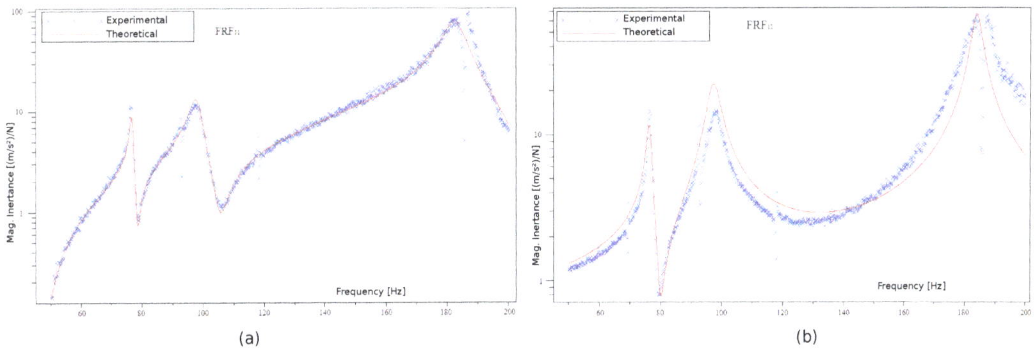

Figure 4. 3 DOF model analytical vs. experimental FRFs: (**a**) top plate; (**b**) forced on top plate and measured on the back plate.

Table 3. Natural frequencies of the classic guitar with the updated 3DOF and FE models: analytical/experimental results (relative deviation).

3 DOF [Hz]	FE [Hz]
76.7/76.6 (0.1%)	75.9/71.88 (5.5%)
97.7/98.0 (−0.3%)	110.2/108.4 (1.6%)
182.6/187.2 (−2.4%)	169.3/209.4 (−19.1%)

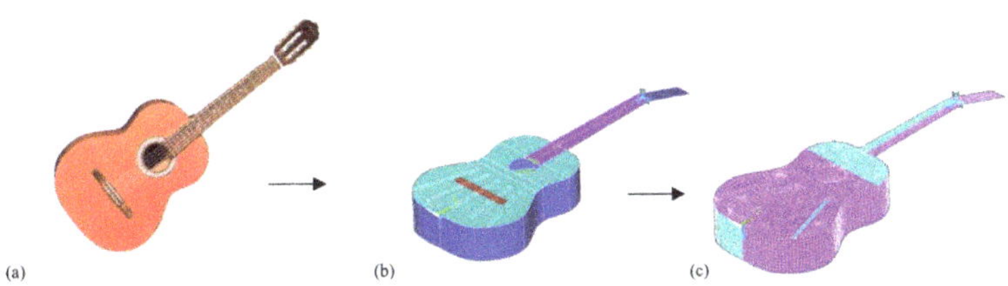

Figure 5. (**a**) Giannini catalog photo for model GWNE15; (**b**) 3D CAD model; and (**c**) resultant mesh for numerical analysis of Giannini guitar.

The simplicity and low computational burden of the lumped parameter 3D model permitted an easy application of an optimization routine to curve-fit structural parameters to reproduce low-frequency range dynamic behaviour with good agreement. The phase angles were not taken into account in the error function and some acceptable agreement was verified as well, thus bringing reliability to the obtained parameters values. On the other hand, a numerical approximate solution of partial differential equations makes it easy to visualize mode shapes, obtain dozens of natural frequencies, maybe more than one hundred if one needs and has sufficient computational resources, accounting for global and local contributions as well. It should be mentioned, however, that FE results are largely affected by a significant number of purely numerical issues, such as element distortion (close to singular, or even negative jacobian determinant), and spurious modes due to artificial stiffness in FE formulation.

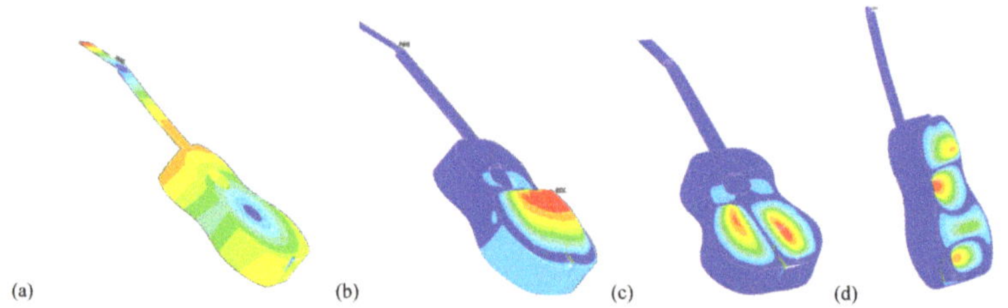

Figure 6. Undamped structural modes: (**a**) in phase top plate mode (75.9 Hz); (**b**) anti phase top plate mode (110.2 Hz); (**c**) first dipole top plate mode (169.3 Hz) and (**d**) quadrupole back plate mode (174.1 Hz).

5. Conclusions

The use of FE modeling is unquestionably advantageous for systems with a high degree of complexity, coupling diverse physical phenomena, where interesting applications for numerical analysis of musical instruments have been presented. While many of the FE codes are very user-friendly and non-commercial versions may be found, they are not accessible to the majority of the luthiers and the process is far too complex for them. This article describes the classical lumped mass approach and the widely used FE analysis for modelling and simulation of the acoustic guitar, highlighting the advantages and disadvantages of each. There is no undisputed standard for coupled vibroacoustic analysis, especially among users and manufacturers of acoustic guitars.

Author Contributions: Conceptualization, all; methodology, all; software, A.M.L. and H.M.G.; validation, all; formal analysis, all; investigation, all; writing—original draft preparation, A.M.L.; writing—review and editing, A.M.L. and C.M.A.V.; funding acquisition, H.M.G. and C.M.A.V. All authors have read and agreed to the published version of the manuscript.

Funding: The first two authors gratefully acknowledge the support by Coordenação de Aperfeiçoamento de Pessoal de Nível Superior, CAPES, Brazil. The third author gratefully acknowledges the support granted by the Foundation for Science and Technology (FCT) of Portugal, in the scope of the project of the Research Unit on Materials, Energy and Environment for Sustainability (proMetheus), Ref. UID/05975/2020, financed by national funds through the FCT/MCTES.

Institutional Review Board Statement: Not applicable.

Informed Consent Statement: Not applicable.

Data Availability Statement: Not applicable.

Conflicts of Interest: The authors declare no conflict of interest.

References

1. Rektorys, K. *Variational Methods in Mathematics, Science and Engineering*, 2nd ed.; D. Reidel Publ. Co.: Dordrecht, The Nederlands, 1980.
2. Friswell, M.; Mottershead, J.E. *Finite Element Model Updating in Structural Dynamics*; Springer: Dordrecht, The Nederlands, 1995.
3. Maia, N.M.M.; Silva, J.M.M. *Theoretical and Experimental Modal Analysis*; Research Studies Press: Baldock, UK, 1997.

4. Caldersmith, G. Guitar as a reflex enclosure. *J. Acoust. Soc. Am.* **1978**, *63*, 1566–1575. [CrossRef]
5. Schelleng, J.C. The violin as a circuit. *J. Acoust. Soc. Am.* **1963**, *35*, 326–338. [CrossRef]
6. Firth, I.M. Physics of the guitar at the Helmholtz and first top-plate resonances. *J. Acoust. Soc. Am.* **1977**, *61*, 588–593. [CrossRef]
7. Christensen, O.; Vistisen, B.B. Simple model for low-frequency guitar function. *J. Acoust. Soc. Am.* **1980**, *68*, 758–766. [CrossRef]
8. Caldersmith, G. Fundamental coupling and its musical importance. *Catgut. Acoust. Soc. Newsl.* **1981**, *36*, 21–27.
9. French, R.M. Structural modification of stringed instruments. *Mech. Syst. Sign. Proc.* **2007**, *21*, 98–107. [CrossRef]
10. French, R.M. *Engineering the Guitar: Theory and Practice*; Springer: New York, NY, USA, 2008.
11. Popp, J.E. Four mass coupled oscillator guitar model. *J. Acoust. Soc. Am.* **2012**, *131*, 829–836. [CrossRef] [PubMed]
12. Gaul, L. Tutorial Guideline VDI 3830: Damping of Materials and Members. In *Modal Analysis Topics, Volume 3, Proceedings of the 29th IMAC, Jacksonville, FL, USA, 31 January–3 February 2011*; Springer: New York, NY, USA, 2011; pp. 17–25.
13. Cremer, L.; Heckl, M. *Structure-Borne Sound: Structural vibrations and Sound Radiation at Audio Frequencies*; Springer: Berlin, Germany, 1973.
14. Pierce, A.D. *Acoustics—An Introduction to Its Physical Principles and Applications*, 3rd ed.; Acoustical Society of America: Melville, NY, USA, 1989.
15. Löw, A.M. Identificação Modal Experimental da Caixa Acústica de um Violão Clássico. Master Thesis, Universidade Federal do Rio Grande do Sul, Porto Alegre, Brasil, 2012. (In Portuguese)
16. Löw, A.M.; Gomes, H.M.; Prado, G.S.; Vasques, C.M.A. A Generalized Proportional Damping Model Applied to a Guitar Model. In Proceedings of the 25th International Congress of Mechanical Engineering, Uberlândia, Brasil, 20–25 October 2019.
17. Brooke, M. Numerical Simulation of Guitar Radiation Fields Using the Boundary Element Method. Ph.D. Thesis, University of Wales, Cardiff, Wales, 1992.
18. Elejabarrieta, M.J.; Ezcurra, A.; Santamaría, C. Coupled modes of the resonance box of the guitar. *J. Acoust. Soc. Am.* **2002**, *131*, 829–836. [CrossRef] [PubMed]
19. Ezcurra, A.; Elejabarrieta, M.J.; Santamaría, C. Fluid-structure coupling in the guitar box: Numerical and experimental comparative study. *Appl. Acoust.* **2005**, *66*, 411–425. [CrossRef]
20. Derveaux, G.; Chaigne, A.; Joly, P.; Becache, E. Time-domain simulation of a guitar: Model and method. *J. Acoust. Soc. Am.* **2003**, *114*, 107–126. [CrossRef] [PubMed]
21. Becache, E.; Chaigne, A.; Derveaux, G.; Joly, P. Numerical simulation of a guitar. *Comput. Struct.* **2005**, *83*, 3368–3383. [CrossRef]
22. Chaigne, A.; Kergomard, J. *Acoustics of Musical Instruments*; Springer: New York, NY, USA, 2016.
23. Paiva, G.O. Análise Modal Vibroacústica da Caixa de Ressonância de uma Viola Caipira. Master Thesis, Universidade Estadual de Campinas, Campinas, Brasil, 2013. (In Portuguese)
24. Rodríguez, S.E. *Numerical Analysis of the Modal Coupling at Low Resonances in a Colombian Andean Bandola in C Using the Finite Element Method*; Undergraduate Report; Universidad EAFIT: Medellín, Colombia, 2012.

engineering proceedings

MDPI

Proceeding Paper

First Results for the Selection of Repeating Earthquakes in the Eastern Tien Shan (China) †

Kseniia Nepeina

Research Station RAS in Bishkek, Bishkek 720049, Kyrgyzstan; nepeina.k@mail.ru; Tel.: +996-312-613-140

† Presented at the 2nd International Electronic Conference on Applied Sciences, 15–31 October 2021; Available online: https://asec2021.sciforum.net/.

Abstract: This research is the first stage of seeking repeating earthquakes sequences (RES) in the modern orogeny active zones. The main idea is to find the possible influence of space weather parameters on the seismic process. This is the reason why I am interested in the satellite CSES-01 data. It is a tool that has monitored Earth's seismo-electromagnetic activity since 2018. Presuming the "ionosphere-atmosphere-lithosphere" relation exists, it is necessary to involve both satellite and ground-based observational data. The seeking of the triggering mechanism still requires additional analysis of consistent geophysical ground-based networks (geomagnetic and seismic). The stations' coordinates and instruments are presented. In this work, an earthquake catalog (NEIC) of 400 earthquakes with 2.5+ magnitude from 2015 to 2020 was used. The earthquakes epicenters are illustrated on Google Earth basemap (Landsat image) with geologic linear faults. It could help to find any correlation with relief surface or shear zones, which could be areas of nucleation. Some earthquake clusters were found in the Eastern Tien Shan (region of China), on the border with Kazakhstan and Kyrgyzstan, due to the K-means algorithm. Clustering helps group earthquakes into small families for further cross-correlation of seismic waveforms and the best match selection between the neighbors.

Keywords: repeating earthquakes; earthquake cycle; geomagnetic station; seismic station

Citation: Nepeina, K. First Results for the Selection of Repeating Earthquakes in the Eastern Tien Shan (China). *Eng. Proc.* **2021**, *11*, 26. https://doi.org/10.3390/ASEC2021-11122

Academic Editor: Nicholas Vassiliou Sarlis

Published: 15 October 2021

1. Introduction

Recent studies indicate that repeated earthquakes occur all over the world [1–5]. Authors report repeating earthquakes at Parkfield in [5] and say, comparing to (e.g., [6–8]): "A characteristic repeating earthquake sequence (RES) is defined as a group of events with nearly identical waveforms, locations, and magnitudes that represent repeated ruptures of effectively the same patch of fault". The result shows that the sequence of earthquakes could be extended in time, even those of low magnitude. The observations of RES prove that there are aseismic slips at depth loads of repeating ruptures. In this case, the view of China Mainland is very attractive as an example, especially the north-western mountainous part. The hazard map for PGA corresponding to a 10% probability of exceedance in 50 years is presented in Figure 1.

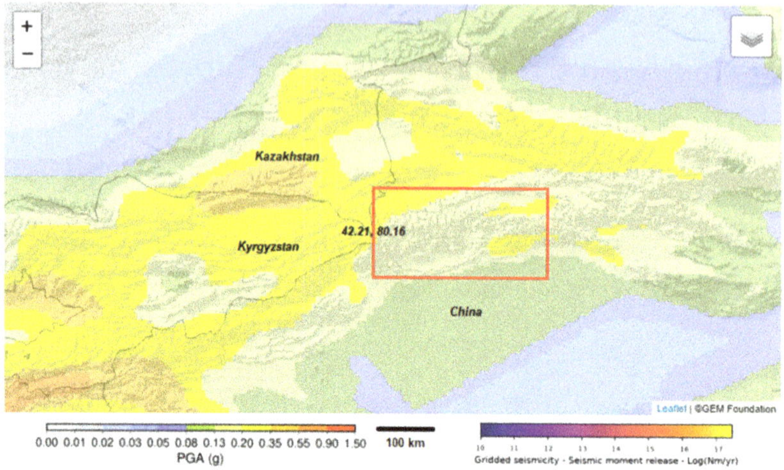

Figure 1. The Global Hazard Mosaic coverage of China, based on the 2015 seismic hazard model of China (GB18306-2015) for peak ground acceleration (PGA) and spectral acceleration (SA) at 0.2, 0.5, 1.0, and 2 s [9].

Some research about repeating earthquake sequences (RES) is able to monitor volcanic activity [10]. The RES comparison results from different authors we could find in [1]. Sometimes repeating earthquakes are called duplet events and families or clusters if they are located inside the same area. Upon research from Uchida and Bürgmann [7], well-characterized RES are mostly small (M < 4) but can be larger than M6 and show long-term slip rates increasing for several years to a decade. I chose the area (marked as a red rectangle in Figure 1) for the analysis for several reasons. Firstly, it is a part of the Central Asian Orogenic Belt (CAOB). CAOB is characterized by complicated intracontinental processes and active tectonics [11–14]. This part of CAOB is crossed by three neighboring countries: Kyrgyzstan, China and Kazakhstan. The detailed geological description and tectonic features are given in [12,14]. This seems a potential advantage because more monitoring tools from different countries could be involved in future earthquakes analysis. RES is a tool for studying the earthquake preparation process with cycles in modern geodynamics and helps in tracking fault zones at the surface [15]. They "illuminated earthquake triggering mechanism and revealed systematic changes in rupture characteristics as a function of loading rate" [7]. Some RES studies for the same area are under consideration of different research groups at the same time in 2021, e.g., [16].

For the seeking process of RES events, the accurate hypocenter determinations are of high importance. Hypocenter location and waveform similarity are two main methods to identify repeating earthquakes to obtain the highest cross correlation coefficient (CC) between each waveform pair [1]. The most popular method is cross correlation of regional seismic waveforms. Their waveform characteristics provide important insights into frictional fault mechanics, earthquake source heterogeneity, and Earth structure changes. That is the reason why the China catalogs from 1985 to 2005 near 5623 events were relocated by Schaff et al. [17]. In such a way, Schaff and Richards [18] found that 10% of seismic events in and around China are repeating earthquakes (with no more than 1 km from each other), whereas 64% of postshocks have smaller magnitudes than the preceding RES events [5]. Recent novelty in epicenter location is satellite data usage to reduce uncertainty. Geodetic observations and imaging geodesy observations complete the NEIC catalogs [19,20]. Statistical correlation of seismicity and geodetic strain rate in the Chinese Mainland is given in [21].

The aim of this research is to provide a number of samples of repeating earthquakes to study the "ionosphere–atmosphere–lithosphere" system in seismic-prone regions and the relationship between the earthquake source, effect on deformation processes and space weather parameters based on data of space monitoring by the Chinese seismo-electromagnetic satellite CSES-01, and the possibility of earthquake triggering by strong bursts of geomagnetically induced currents in conducting seismogenic faults of the Earth crust. There are some proofs of the space weather influence triggering seismic activity [22,23].

2. Materials and Methods

For the RES catalog forming, firstly, we need to use open worldwide catalogs, e.g., NEIC [20], because they accumulate regional bulletins. We need to choose the interesting area and form a text format catalog and save it for further analysis. Afterward, we need to compare hypocenter locations and find any clusters or repeating sources. Usually, the RES distance is no more than 1 km from each other. The next step is to find any close seismic stations, where the seismograms for these cluster events are recorded. As I said before, the only reason to collect waveforms is to correlate them with each other. Thus, I need to download earthquake waveforms for each event recorded by the long-term seismic segment. Usually, they should have a 100 Hz sampling rate. Eventually, we group repeating pairs into clusters using median CC value ≥0.9 with at least two stations [1,24]. The additional information (e.g., the number of stations, station distribution, and timing accuracy of station clocks) is outside of our attention. The CC time windows are very sensitive [25]. The CC time length is compared to ~5000 samplings, starting 0.5 s before the P/S wave onset [24,26]. The process and results should be similar to the research by Deng et al. [27,28].

Actually, for all those steps, firstly, we need to find and accumulate information about seismic and observatories, monitoring the Earth's magnetic field. The previous data for the China Digital Network was reviewed in [29]. Permanent stations on China's National Digital Seismograph Network (CNDSN) are: Beijing (BJI), Enshi (ENH), Hailar (HIA), Kunming (KMI), Lhasa (LSA), Lanzhou (LZH), Mudanjiang (MDJ), Qiongzhong (QIZ), Shanghai (SSE), Urumqi (WMQ), Xi'an (XAN) [29]. There is also the China National Seismic Network at the Institute of Geophysics [30]. There were only two permanent stations in China near the chosen region that are useful for the study: Urumqi (ENH), Kashi (KSH). The other seismic stations are located in Kyrgyzstan and Kazakhstan for studying seismicity [31–36]. Nine seismic stations with their coordinates were selected for further analysis and are given in Table 1.

Table 1. The nearest seismic stations to the selected area.

Station Code	Station Name	Country	Latitude, °N	Longitude, °E	Network Name/Data Center(s)
ASAI	Aksay	Kyrgyzstan	40.9178	76.521	CAIAG/GEOFON
ENEL	Enylcheck	Kyrgyzstan	42.1529	79.455	CAIAG/GEOFON
MRZ1	Lake Merzbacher	Kyrgyzstan	42.2246	79.8597	CAIAG/GEOFON
TARG	Taragay	Kyrgyzstan	41.7291	77.8048	CAIAG/GEOFON/IRISDMC
PRZ	Prjevalsk	Kyrgyzstan	42.5	78.4	KRNET/IRISDMC
PRZ1	Karakol	Kyrgyzstan	42.5	78.400002	KRNET/IRISDMC
PDGK	Podgonoye	Kazakhstan	43.3276	79.4849	KNDC/IRISDMC
WMQ	Urumqi	China	43.821098	87.695	CDSN/IRISDMC
KSH	Kashi	China	39.516998	75.922997	CNDSN/IRISDMC

After RES searching in the catalog and waveform CC, as mentioned before, we are interested in solar storms, and consequently, strong bursts of geomagnetic indices. Therefore, I list below geomagnetic stations (Tables 2–4), which could be potentially useful to find any relationship between RES and indices variations. Upon new numerous studies there are certain triggering mechanisms that we observe in geomagnetic indices [37,38].

Table 2. Local stationary geomagnetic stations in Kyrgyzstan [33].

Station	Latitude, °N	Longitude, °E	Type
Ak-Suu	42.603	74.008	transient electro- magnetic sounding method (TEM) geomagnetic-variation modular system "MB-07" developed by RS RAS
Shavay	42.617	74.222	
Chonkurchak	42.626	74.608	
Tash-Bashat	42.667	74.770	
Issyk-Ata	42.638	74.960	
Kegety	42.613	75.157	

Table 3. INTERMAGNET observatories (the global network of observatories, monitoring the Earth's magnetic field) near the target area [39].

Code	Name	Country	Latitude, °N	Longitude, °E	Institute	Elevation, m	Instruments
AAA	Alma Ata	Kazakhstan	46.8	76.9	IIRK	1300	Variations: Fluxgate magnetometer LEMI-008, Overhauser proton magnetometer POS-1, Absolutes: DI-fluxgate 3T2KP LEMI-203, Overhauser Proton Magnetometer POS-1
WMQ	Urumqi	China	46.19	87.71	CEA	908	Variations: Continuously Recording • Vector Magnetometer DMI FGE • Scalar Magnetometer GSM-90F Absolutes: DI Fluxgate Theodolite, Minregion DIM, Hungary and Proton Magnetometer G856AX

Table 4. Meridian project station locations in China, along with the types of observations and instruments deployed at each [40].

Station	Latitude, °N	Longitude, °E	Instruments
Mohe	53.5	122.4	magnetometer, digisonde, TEC [a] monitor/ionospheric scintillation monitor
Manzhouli	49.6	117.4	magnetometer, ionosonde
Changchun	44.0	125.2	magnetometer, ionosonde
Beijing	40.3	116.2	magnetometer, digisonde, lidar, [b] all-sky imager, Fabry–Perot interferometer, mesosphere-stratosphere-thermosphere radar, interplanetary scintillation monitor, cosmic ray monitor, TEC monitor/ionospheric scintillation monitor, high-frequency Doppler frequency shift monitor
Xinxiang	34.6	113.6	magnetometer, ionosonde, TEC monitor/ionospheric scintillation monitor
Hefei	33.4	116.5	lidar
Wuhan	30.5	114.6	magnetometer, digisonde, lidar, mesosphere-stratosphere-thermosphere radar, meteor radar, TEC monitor/ionospheric scintillation monitor, high-frequency Doppler frequency shift monitor
Guangzhou	23.1	113.3	magnetometer, ionosonde, cosmic ray monitor, TEC monitor/ionospheric scintillation monitor
Hainan	19.0	109.8	magnetometer, digisonde, TEC monitor/ionospheric scintillation monitor, lidar, all-sky imager, very high frequency radar, sounding rockets, meteor radar

Table 4. *Cont.*

Station	Latitude, °N	Longitude, °E	Instruments
Zhongshan	69.4	76.4	magnetometer, digisonde, high-frequency coherent scatter radar, aurora spectrometer
Shanghai	31.1	121.2	Magnetometer, TEC monitor
Chongqing	29.5	106.5	magnetometer, ionosonde
Chengdu	31.0	103.7	magnetometer, ionosonde
Qujing	25.6	103.8	incoherent scatter radar
Lhasa	29.6	91.0	magnetometer, ionosonde

[a] Total electron content. [b] Light detection and ranging.

3. Results

In this work, an earthquake catalog (NEIC) 2015–2020 from [20] of 400 earthquakes with magnitudes m_b 2.5–6.3 was used. The magnitude consistency is shown in Figure 2.

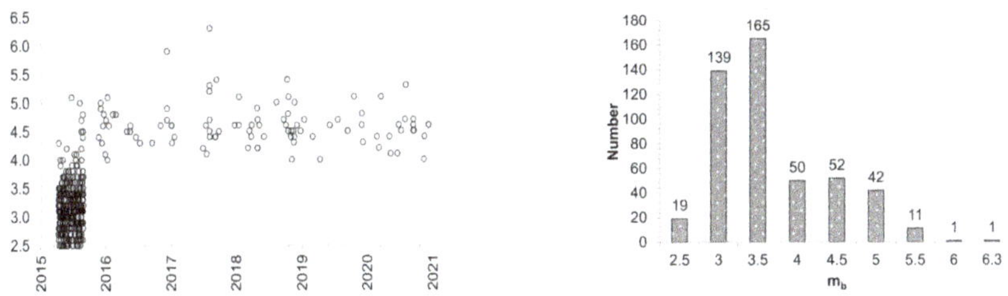

(**a**) (**b**)

Figure 2. Final locations for 5623 events well distributed throughout China—3689 for all of China from 2015 to 2020.

The overview of the events' coordinates gives an impression of seismic areas by distribution density. By K-means clustering algorithm in Origin 9 [41], I obtained six main clusters after ten iterations (Figure 3a). For the detailed analysis, I decided to look in clusters 3, 4 and 6. I applied K-means clustering again to separate small families of adjacent events (Figure 3b). I zoomed in to the area from 41.5° N, 81° E to 43° N, 85° E (118 events). This procedure helps separate each location and group them into families. Inside these 28 families, I could start to download seismic waveforms for cross-correlation for repeated earthquakes.

On the Landsat image, the epicenter distribution is sparse (Figure 4). The hypocenter depth is shown in color. It is obvious that some epicenters group near trench zones and orogenesis. Google Earth is an open source for geospatial data visualization. One of its advantages is a relief basemap. It helps a lot with understanding Earth's surface features where the earthquakes occur. Future steps could touch-up the geospatial imaginary datasets with Google Earth Engine Tools (open-source platform for satellite data processing). This area (Urumqi) is mountainous with intense surface elevations.

Finally, by checking each of the 28 families, we could find potential RES events. For example, for the first cluster, K-means creates a cluster of five events. Only two of them (4 December 2019, T_0 = 12:48:30 λ = 41.7243° N, ϕ = 81.7066° E, and 17 January 2019, T_0 = 13:32:37, λ = 41.7386 ° N, ϕ = 81.6901° E) are close to each other (distance ~2.5 km) (Supplement Table S1). Both have occurred at the same depth 10 km, which intermediately indicates the similarity of sources.

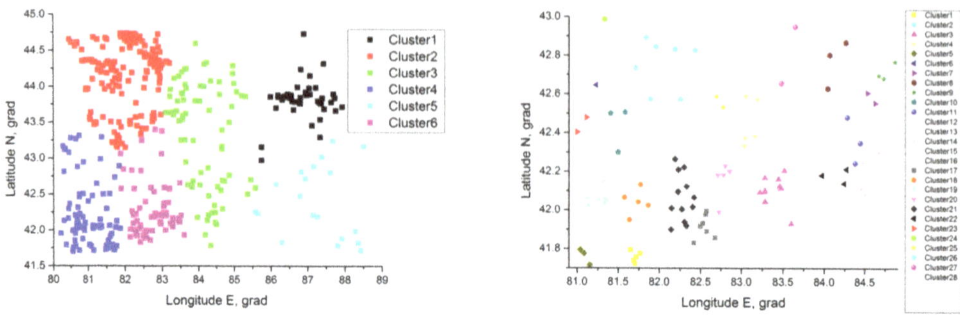

(**a**) (**b**)

Figure 3. Clustered epicenter locations for events from 2015 to 2020 distributed throughout China: (**a**) all events for the target area; (**b**) only events for 3, 4, 6 clusters from (**a**).

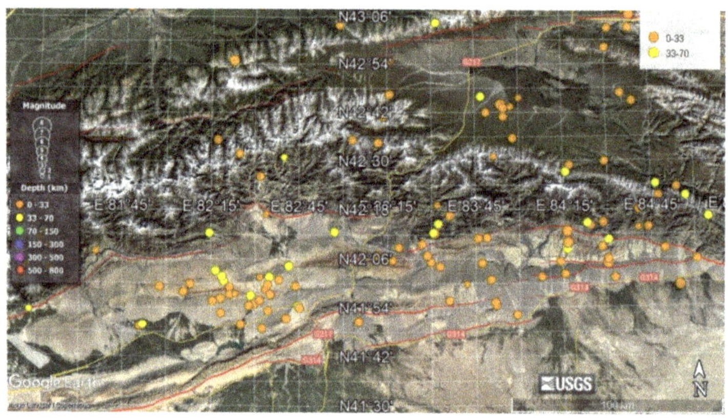

Figure 4. Earthquakes epicenters on ©Google Earth Landsat image [42]. Red lines—fault from database [43].

4. Conclusions

I present here principle moments in the repeating earthquake sequences search using K-means algorithm and epicenter coordinates from the NEIC seismic catalog. The first results give us an idea of the possibility of finding similar events over a long period of time. The idea of repeating earthquakes statistics could be useful for seismic hazard maps and earthquake repeating maps, which are obviously used for construction and building engineering. The hypothesis about the "ionosphere–atmosphere–lithosphere" relation as the possibility of earthquake triggering by strong bursts of geomagnetically induced currents requires additional analysis, using data from geomagnetic stations indicated in the study and Chinese seismo-electromagnetic satellite CSES-01 data. The geomagnetic index amplitude variations, the choice of specific indices or their relation and the accuracy of the location of the phenomena also claim special attention and would play a critical role.

Supplementary Materials: The following are available online at https://www.mdpi.com/article/10.3390/ASEC2021-11122/s1, Table S1: K-means clustering results for the area from 41.5° N, 81° E to 43° N, 85° E (118 events).

Funding: This research was funded by the Russian Foundation of Basic Research (RFBR) and the National Scientific Fund of China (NSFC) grant number 21-55-53053.

Institutional Review Board Statement: Not applicable.

Informed Consent Statement: Not applicable.

Data Availability Statement: Data available on request due to restrictions, e.g., privacy or ethics.

Acknowledgments: The research was carried out with the financial support of the Russian Foundation for Basic Research RFBR and NSFC in the framework of scientific project No. 21-55-53053. I thank anonymous reviewers for their valuable opinions and comments.

Conflicts of Interest: The authors declare no conflict of interest. The funders had no role in the design of the study; in the collection, analyses, or interpretation of data; in the writing of the manuscript, or in the decision to publish the results.

References

1. Uchida, N. Detection of Repeating Earthquakes and Their Application in Characterizing Slow Fault Slip. *Prog. Earth Planet. Sci.* **2019**, *6*, 1–21. [CrossRef]
2. Satake, K. Geological and historical evidence of irregular recurrent earthquakes in Japan. *Philos. Trans. R. Soc. A Math. Phys. Eng. Sci.* **2015**, *373*, 20140375. [CrossRef]
3. Wu, C.; Gomberg, J.; Ben-Naim, E.; Johnson, P. Triggering of repeating earthquakes in central California. *Geophys. Res. Lett.* **2014**, *41*, 1499–1505. [CrossRef]
4. Mesimeri, M.; Karakostas, V. Repeating Earthquakes in Western Corinth Gulf (Greece): Implications for Aseismic Slip near Locked Faults. *Geophys. J. Int.* **2018**, *215*, 659–676. [CrossRef]
5. Chen, K.H.; Bürgmann, R.; Nadeau, R.M. Do earthquakes talk to each other? Triggering and interaction of repeating sequences at Parkfield. *J. Geophys. Res. Solid Earth* **2013**, *118*, 165–182. [CrossRef]
6. Nadeau, R.M.; Johnson, L.R. Seismological studies at Parkfield VI: Moment release rates and estimates of source parameters for small repeating earthquakes. *Bull. Seismol. Soc. Am.* **1998**, *88*, 790–814.
7. Uchida, N.; Bürgmann, R. Repeating Earthquakes. *Annu. Rev. Earth Planet. Sci.* **2019**, *47*, 305–332. [CrossRef]
8. Chen, T.; Lapusta, N. Scaling of Small Repeating Earthquakes Explained by Interaction of Seismic and Aseismic Slip in a Rate and State Fault Model. *J. Geophys. Res. Solid Earth* **2009**, *114*, B01311. [CrossRef]
9. Gao, M.; Chen, G.; Xie, F.; Xu, X.; Li, X.; Yu, Y.; Li, S.; Zhou, B.; Pan, H.; Lu, Y.; et al. The Global Hazard Mosaic Coverage of China Is Based on the 2015 Seismic Hazard Model of China (GB18306-2015). Available online: https://hazard.openquake.org/gem/models/CHN/ (accessed on 29 August 2021).
10. Wellik, J.J.; Prejean, S.G.; Syahbana, D.K. Repeating Earthquakes During Multiple Phases of Unrest and Eruption at Mount Agung, Bali, Indonesia, 2017. *Front. Earth Sci.* **2021**, *9*, 653164. [CrossRef]
11. Xiao, W.; Windley, B.F.; Allen, M.B.; Han, C. Paleozoic multiple accretionary and collisional tectonics of the Chinese Tianshan orogenic collage. *Gondwana Res.* **2013**, *23*, 1316–1341. [CrossRef]
12. Zhong, L.; Wang, B.; de Jong, K.; Zhai, Y.; Liu, H. Deformed continental arc sequences in the South Tianshan: New constraints on the Early Paleozoic accretionary tectonics of the Central Asian Orogenic Belt. *Tectonophysics* **2019**, *768*, 228169.
13. Cai, K.; Long, X.; Chen, H.; Sun, M.; Xiao, W. Accretionary and collisional orogenesis in the south domain of the western Central Asian Orogenic Belt (CAOB). *J. Asian Earth Sci.* **2018**, *153*, 1–8. [CrossRef]
14. He, Z.; Wang, B.; Ni, X.; De Grave, J.; Scaillet, S.; Chen, Y.; Liu, J.; Zhu, X. Structural and kinematic evolution of strike-slip shear zones around and in the Central Tianshan: Insights for eastward tectonic wedging in the southwest Central Asian Orogenic Belt. *J. Struct. Geol.* **2021**, *144*, 104279. [CrossRef]
15. Chalumeau, C.; Agurto-Detzel, H.; De Barros, L.; Charvis, P.; Galve, A.; Rietbrock, A.; Alvarado, A.; Hernandez, S.; Beck, S.; Font, Y.; et al. Repeating Earthquakes at the Edge of the Afterslip of the 2016 Ecuadorian M W 7.8 Pedernales Earthquake. *J. Geophys. Res. Solid Earth* **2021**, *126*, e2021JB021746. [CrossRef]
16. Tang, L.L.; Zhang, Y.F.; Yao, D.P. Tempo-spatial characteristics of repeating seismic events in the middle of Tianshan orogenic belt. *Acta Seismol. Sin.* **2021**, *43*, 551–568. [CrossRef]
17. Schaff, D.P.; Richards, P.G.; Slinkard, M.; Heck, S.; Young, C. Lg-Wave Cross Correlation and Epicentral Double-Difference Location in and near China. *Bull. Seismol. Soc. Am.* **2018**, *108*, 1326–1345. [CrossRef]
18. Schaff, D.P.; Richards, P.G. Repeating Seismic Events in China. *Science* **2004**, *303*, 1176–1178. [CrossRef]
19. Barnhart, W.D.; Hayes, G.P.; Wald, D.J. Global Earthquake Response with Imaging Geodesy: Recent Examples from the USGS NEIC. *Remote Sens.* **2019**, *11*, 1357. [CrossRef]
20. NEIC. USGS Earthquake Catalog Search. Available online: https://earthquake.usgs.gov/earthquakes/search/ (accessed on 26 July 2021).
21. Wu, Y.; Jiang, Z.; Pang, Y.; Chen, C. Statistical Correlation of Seismicity and Geodetic Strain Rate in the Chinese Mainland. *Seismol. Res. Lett.* **2022**, *93*, 268–276. [CrossRef]
22. Liperovsky, V.A.; Pokhotelov, O.A.; Meister, C.-V.; Liperovskaya, E.V. Physical models of coupling in the lithosphere-atmosphere-ionosphere system before earthquakes. *Geomagn. Aeron.* **2008**, *48*, 795–806. [CrossRef]
23. Novikov, V.; Ruzhin, Y.; Sorokin, V.; Yaschenko, A. Space weather and earthquakes: Possible triggering of seismic activity by strong solar flares. *Ann. Gephi.* **2020**, *63*, PA554. [CrossRef]

24. Hatakeyama, N.; Uchida, N.; Matsuzawa, T.; Okada, T.; Nakajima, J.; Matsushima, T.; Kono, T.; Hirahara, S.; Nakayama, T. Variation in high-frequency wave radiation from small repeating earthquakes as revealed by cross-spectral analysis. *Geophys. J. Int.* **2016**, *207*, 1030–1048. [CrossRef]
25. Nepeina, K. The Alternative Approach for Seismic Monitoring Data Identification Excluding "Master" Events. In *Proceedings of 2ECCES*; 2014. Available online: http://www.eaee.org/Media/Default/2ECCES/2ecces_eaee/560.pdf (accessed on 8 September 2021).
26. Adushkin, V.V.; Kitov, I.O.; Konstantinovskaya, N.L.; Nepeina, K.S.; Nesterkina, M.A.; Sanina, I.A. Detection of Ultraweak Signals on the Mikhnevo Small-Aperture Seismic Array by Using Cross-Correlation of Waveforms. *Dokl. Earth Sci.* **2015**, *460*, 189–191. [CrossRef]
27. Deng, Y.; Peng, Z.; Liu-Zeng, J. Systematic Search for Repeating Earthquakes Along the Haiyuan Fault System in Northeastern Tibet. *J. Geophys. Res. Solid Earth* **2020**, *125*, e2020JB019583. [CrossRef]
28. Deng, Y.; Peng, Z.; Liu-Zeng, J. Dataset for Deng's 2020 JGR Paper. *Dataset* **2020**. [CrossRef]
29. Mangino, S.; Priestley, K.; Ebel, J. The receiver structure beneath the China Digital Seismograph Network stations. *Bull. Seismol. Soc. Am.* **1999**, *89*, 1053–1076.
30. FDSN Portal. China National Seismic Network. Available online: https://www.fdsn.org/networks/detail/CB/ (accessed on 10 September 2021).
31. Parolai, S.; Boxberger, T.; Pilz, M.; Fleming, K.; Haas, M.; Pittore, M.; Petrovic, B.; Moldobekov, B.; Zubovich, A.; Lauterjung, J. Assessing Earthquake Early Warning Using Sparse Networks in Developing Countries: Case Study of the Kyrgyz Republic. *Front. Earth Sci.* **2017**, *5*, 74. [CrossRef]
32. Medved, I.; Bataleva, E.; Buslov, M. Studying the Depth Structure of the Kyrgyz Tien Shan by Using the Seismic Tomography and Magnetotelluric Sounding Methods. *Geosciences* **2021**, *11*, 122. [CrossRef]
33. Research Station RAS in Bishkek (RS RAS). Available online: http://www.gdirc.ru/en/ (accessed on 26 August 2021).
34. The Institute of Geophysical Research Republic Kazakhstan (IGR RK). Available online: https://www.igr.kz/en (accessed on 8 September 2021).
35. Berezina, A.; Fyen, J.; Abdrakhmatov, K.; Schweitzer, J.; Mykkelveit, S. The new seismic network KRNET: Perspectives and capacity development. *Sci. Technol.* **2011**. T5-P11.
36. Wang, C. Research on Kyrgyzstan seismic activity and monitoring network. *Prog. Earthquake Sci.* **2021**, *51*, 65–71. (In Chinese) [CrossRef]
37. Zhima, Z.; Hu, Y.; Shen, X.; Chu, W.; Piersanti, M.; Parmentier, A.; Zhang, Z.; Wang, Q.; Huang, J.; Zhao, S.; et al. Storm-Time Features of the Ionospheric ELF/VLF Waves and Energetic Electron Fluxes Revealed by the China Seismo-Electromagnetic Satellite. *Appl. Sci.* **2021**, *11*, 2617. [CrossRef]
38. Gou, X.; Li, L.; Zhang, Y.; Zhou, B.; Feng, Y.; Cheng, B.; Raita, T.; Liu, J.; Zhima, Z.; Shen, X. Ionospheric Pc1 waves during a storm recovery phase observed by the China Seismo-Electromagnetic Satellite. *Ann. Geophys.* **2020**, *38*, 775–787. [CrossRef]
39. INTERMAGNET Project. Available online: https://www.intermagnet.org (accessed on 26 August 2021).
40. Wang, C. New Chains of Space Weather Monitoring Stations in China. *Sp. Weather* **2010**, *8*, S08001. [CrossRef]
41. Origin Lab 9. Available online: https://www.originlab.com (accessed on 8 September 2021).
42. Google Earth Service. Available online: https://earth.google.com/web (accessed on 29 August 2021).
43. Database of Active Faults of Eurasia. RU 2019621553. Available online: http://neotec.ginras.ru/index/english/database_eng.html (accessed on 26 August 2021).

engineering proceedings

MDPI

Proceeding Paper

Proposal of Edge-Preserving, Image Noise Reduction Filter for Using L2-Norm †

Yudai Yamaguchi [1], Ichiro Yoshida [2,*] and Yuki Kondo [2]

1 Mechanical Engineering, Graduate School of Science and Engineering, HOSEI University, Koganei-shi 184-8584, Japan; yudai.yamaguchi.2b@stu.hosei.ac.jp
2 Department of Mechanical Engineering, Faculty of Science and Engineering, HOSEI University, Koganei-shi 184-8584, Japan; yuki.kondoh.57@hosei.ac.jp
* Correspondence: yoshida.ichiro@hosei.ac.jp; Tel.: +81-042-387-6033

† Presented at the 2nd International Electronic Conference on Applied Sciences, 15–31 October 2021; Available online: https://asec2021.sciforum.net/.

Abstract: Images taken by digital cameras include noise. The image recognition rate decreases with increasing noise. Reducing noise is essential for improving the accuracy of image recognition. Low-pass filters, such as a Gaussian filter (GF), are often used to reduce noise from images. Low-pass filters can reduce noise. However, low-pass filters always blur the edges. As the edge blur becomes stronger, the accuracy of edge and feature detection of image recognition worsens. Therefore, we propose a noise reduction filter for images that can preserve edges by combining the GF and the L2-norm. The proposed method is expected to improve the image quality and, consequently, the accuracy of image recognition.

Keywords: digital image processing; noise reduction; edge preserving; gaussian filter; L2-norm

Citation: Yamaguchi, Y.; Yoshida, I.; Kondo, Y. Proposal of Edge-Preserving, Image Noise Reduction Filter for Using L2-Norm. *Eng. Proc.* **2021**, *11*, 27. https://doi.org/10.3390/ASEC2021-11170

Academic Editor: Nunzio Cennamo

Published: 15 October 2021

1. Introduction

Images taken by digital cameras include noise [1]. Image quality reduces with increasing noise. In addition, the recognition rate of image recognition decreases with increasing noise. Currently, image recognition is used in security technology through face recognition and in image inspection at production sites. Therefore, the accuracy of image recognition needs to be improved. Reducing noise is essential for improving the accuracy of image recognition. Low-pass filters, such as a Gaussian filter (GF) [2], are often used to reduce noise in images. Low-pass filters can reduce noise; however, low-pass filters always blur the edges [3]. As the edge blur becomes stronger, the accuracy of the edge and the feature detection of image recognition worsens [4]. Therefore, noise reduction filters need to be able to preserve edges.

A non-local mean filter (NLMF) [5] was proposed as a noise reduction filter for images that can preserve edges. The NLMF has high denoising performance for weak noise. However, the NLMF has low denoising performance for strong noise. In addition, the NLMF has a problem in that the amount of calculation is large and the processing time is long.

In order to solve this problem, in this research, we propose a noise reduction filter for images that can preserve edges by combining the GF and the L2-norm [6]. The proposed method is expected to simultaneously achieve high denoising performance and edge-preservation performance compared to NLMF by using the L2-norm. In addition, the proposed method can shorten the processing time compared with NLMF. Therefore, the proposed method is expected to improve the image quality and, consequently, the accuracy of image recognition. The remainder of this paper is organized as follows: we describe the proposed method in Section 2, verification experiments are presented in Section 3, and the paper is summarized in Section 4.

2. Proposed Method

The filter proposed in this research is a filter for images that combines the L2-norm and GF (hereinafter referred to as L2GFI). The algorithm for the proposed method is shown in the following bullet points. In addition, the software we used to develop the proposed method is MATLAB.

1. The reference points are set in the cells corresponding to each coordinate in the original image;
2. The Gaussian function is applied in the *xy*-direction;
3. In the *z*-coordinate, the range where the value of the weight does not become zero consecutively is recognized as a mountain of weight;
4. The mountain where the sum of weights is the maximum is selected;
5. The output value is calculated through a convolution of the normalized weights and the pixel value *z* in the range of the selected mountain.

The specific processing procedure for the L2GFI is explained. The L2GFI needs to assign an array to pixel value *z* in addition to the *xy*-coordinate. Therefore, processing the L2GFI requires a three-dimensional array in the case of two-dimensional grayscale images. The L2GFI plots reference points in the corresponding cells of the *x*-coordinate, *y*-coordinate, and pixel value *z* of each pixel in the image. The GF is applied in the *xy*-direction for the plotted reference points for each pixel. At that time, the weight array of GF is added to the corresponding cell. After applying the GF in the calculation of the L2GFI algorithm, the distribution of weights in the *z* direction is checked for each pixel. The range where the value of the weight does not become zero consecutively is recognized as a mountain of weight. At that time, if the distance between the neighboring mountains is small, the neighboring mountains are recognized as one mountain. The output value is calculated only from the range of the mountain where the sum of weights is the maximum. The output value is calculated as the sum of the product for the normalized weights and pixel value *z*. Let the pixel values in the calculation range be $z1, z2, \ldots, zn$, the weights of the Gaussian function be $w1, w2, \ldots, wn$, and the output value be o, and then the output of L2GFI can be shown as Equation (1):

$$o = \sum_{i=1}^{n} z_i w_i \quad (1)$$

A comparison of the output processing when the target pixel is a normal value and when the target pixel is noise is shown in Figure 1. Figure 1a shows the target pixel and surrounding pixels when the target pixel is a normal value. Figure 1b shows the target pixel and its surrounding pixels when the target pixel is noise. Figure 1c shows the weight distribution in the *z*-direction, the output process of the GF, and the output process of the L2GFI after applying GF to Figure 1a. Figure 1d shows the weight distribution in the *z*-direction, the output process of the GF, and the output process of the L2GFI after applying the GF to Figure 1b. Comparing the output values of the GF in Figure 1c,d, the output value of Figure 1d shows a larger difference from the normal value than the output value of Figure 1c. The output value of the GF is affected by noise because the GF calculates from all ranges in the *z*-direction. Comparing the output values of the L2GFI in Figure 1c,d, it can be seen that the output values of the L2GFI are not different. This is because the output values of the L2GFI are calculated only from the range of the mountain where the sum of weights is maximum. Therefore, L2GFI is expected to achieve high denoising performance. In addition, when L2GFI (with size 5 × 5) is applied to an area that includes edges, such as in Figure 1a, the weight distribution has two mountains, as shown in Figure 1c. The sum of the weights tends to be larger for the mountains that include the target pixel due to the property of the Gaussian function. As a result, the L2GFI is able to preserve the edges because the output value is calculated only from the pixel values of the edge that includes the target pixel. Furthermore, as shown in Figure 1b,d, the L2GFI can remove noise without blurring the edges, even when noise is included in the edges. The L2GFI can also be expected to achieve high edge-preserving performance by processing other pixels in the same way.

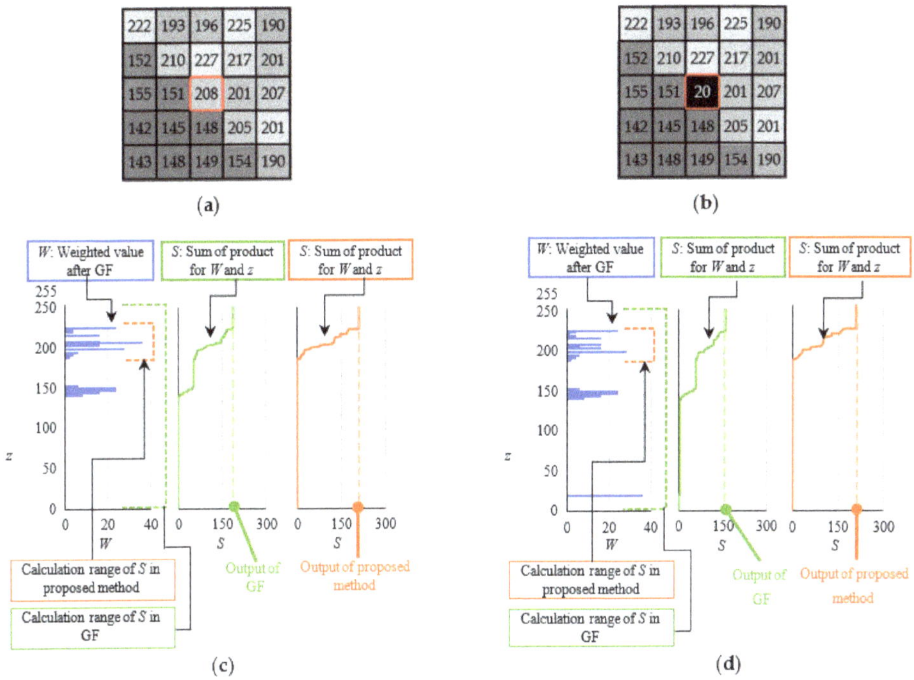

Figure 1. Comparison of output processing of normal value and noise when using GF and L2GFI; (**a**) sample image of only normal values; (**b**) sample image with noise; (**c**) output processing of GF and L2GFI for (**a**); (**d**) output processing of GF and L2GFI for (**b**).

3. Experiments

3.1. Noise Reduction Experiment

The denoising and edge-preserving performances of the L2GFI algorithm are verified here. The experiment compares the denoising and edge-preserving performances of the L2GFI and the NLMF on a noise-added sample image using visual inspection and root mean square error (RMSE). As the value of RMSE becomes lower, it can be said that the reproducibility of the original image becomes higher. A higher reproducibility of the original image indicates higher denoising and edge-preservation performances. Three types of standard images are used as sample images for the experiment. The size of the sample image is 256 × 256 pixels. We apply each filter with a filter size of 5 × 5 pixels. In this experiment, to avoid implementation errors, the NLMF is used as the function of OpenCV 4.5.3 [7]. The parameter values of the NLMF use the recommended values of the function.

Figures 2–4 show the original image, noise-added image, and result of applying each filter to the noise-added image for each sample image. The output images of each filter in Figures 2–4 are compared through a visual inspection. The output image of the NLMF showed no major changes in any of the samples compared to the noise-added image. It can be confirmed that the NLMF has a high edge-preserving performance, and a low denoising performance for strong noise. The output image for the L2GFI showed that noise is removed in all the samples. Therefore, it can be seen that the L2GFI has higher denoising and edge-preservation performances compared to the NLMF.

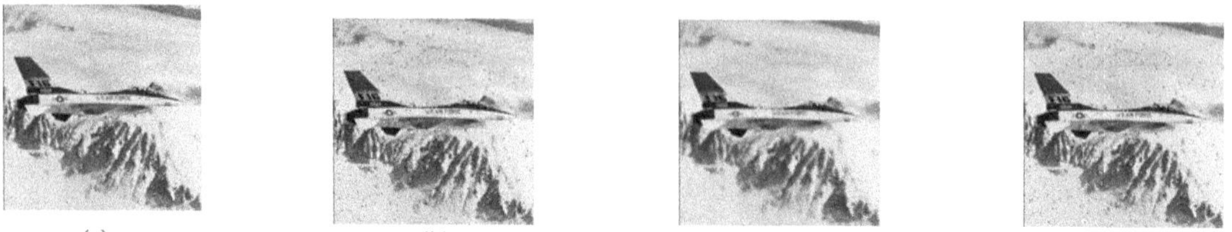

(a) (b) (c) (d)

Figure 2. Filtering results for sample image "Airplane"; (**a**) input image; (**b**) image with noise added to (**a**); (**c**) filtering result of (**b**) by L2GFI; (**d**) filtering result of (**b**) by NLMF.

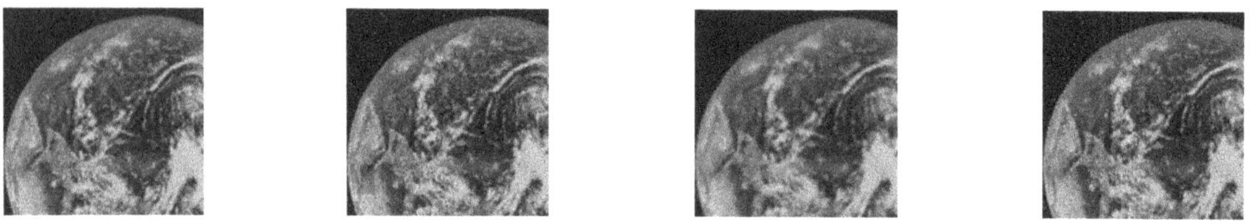

(a) (b) (c) (d)

Figure 3. Filtering results for sample image "Earth"; (**a**) input image; (**b**) image with noise added to (**a**); (**c**) filtering result of (**b**) by L2GFI; (**d**) filtering result of (**b**) by NLMF.

(a) (b) (c) (d)

Figure 4. Filtering results for sample image "Girl"; (**a**) input image; (**b**) image with noise added to (**a**); (**c**) filtering result of (**b**) by L2GFI; (**d**) filtering result of (**b**) by NLMF.

Table 1 lists the RMSE values calculated using each filter (first row) and deviation ratio (%) from the RMSE values calculated using the L2GFI (second row) for each sample image. The first row of Table 1 shows that the RMSE value of NLMF is larger than the RMSE value of L2GFI. The second row of Table 1 shows how large the RMSE value of NLMF is compared to the RMSE value of L2GFI. From Table 1, the RMSE values of the L2GFI are 34 (%) or lower than those of the NLMF for all sample images. Therefore, it can be said that the L2GFI has higher denoising and edge-preservation performances than the NLMF in the evaluation of RMSE values.

Table 1. RMSE values calculated using each filter (upper section) and deviation ratio (%) from RMSE values calculated using L2GFI (lower section).

Sample Image	L2GFI	NLMF
Airplane	9.47	14.55
	-	34.88
Earth	5.94	13.56
	-	56.23
Girl	5.20	15.04
	-	65.38

3.2. Processing Time

The high-speed processing performance of L2GFI is verified. The experiment compares the processing times of the L2GFI and NLMF. We used the functions provided by the software to measure the processing time. The processing time in the comparison experiment is the average time when the filter is applied 10 times. The sample images and parameter values of each filter are the same as those in the experiment in Section 3.1. As a result of the experiment, the average processing time of the L2GFI is 0.2 s, whereas the average processing time of the NLMF is 17.6 s. The processing time of L2GFI is 88 times faster than that of the processing time of the NLMF. In order to verify the uncertainty of the measurement, we calculated the standard deviation of the data for 10 measurement times. The standard deviation of the calculation time for the L2GFI is 0.007 (s); therefore, the calculation time for the L2GFI is 0.2 ± 0.014 s, with a coverage factor $k = 2$. The standard deviation of the calculation time for the NLMF is 0.01 (s); therefore, the calculation time for the NLMF is 17.6 ± 0.02 s ($k = 2$). The uncertainty of these measurements are, therefore, small. Furthermore, in terms of high-speed processing performance, the L2GFI is better than the NLMF.

4. Conclusions

In this research, we proposed the L2GFI, a noise reduction filter for images that can preserve edges that combines the GF and L2-norm. The results of this research are as follows.

1. A visual comparison of the experimental results showed that the L2GFI has better denoising and edge-preservation performances than the NLMF.
2. As a result of experimental verification, L2GFI had RMSE values that were 34 (%) or lower than those of the NLMF. Consequently, it can be said that the reproducibility of the original image of the L2GFI is at a higher level than that of the NLMF.
3. The verification results showed that the processing time of the L2GFI was faster than that of the NLMF. Therefore, the L2GFI is better than NLMF in high-speed processing performance.

From these results, it can be said the L2GFI combines high denoising, edge-preservation, and high-speed processing performances compared to the NLMF. We considered that the performance of the L2GFI needs to be further validated for practical applications. Therefore, in the future, we will verify the performance of the L2GFI for more complex sample images and other types of noise.

Author Contributions: Conceptualization, I.Y.; methodology, Y.Y., and I.Y.; software, Y.Y., and I.Y.; validation, Y.Y., and I.Y.; formal analysis, Y.Y., and I.Y., Y.K.; investigation, I.Y., and Y.Y.; resources, I.Y.; data curation, Y.Y.; writing—original draft preparation, Y.Y., and I.Y.; writing—review and editing, Y.Y., and I.Y., Y.K.; visualization, Y.Y.; supervision, I.Y.; project administration, I.Y.; funding acquisition, I.Y. All authors have read and agreed to the published version of the manuscript.

Funding: This research received no external funding.

Institutional Review Board Statement: Not applicable.

Informed Consent Statement: Not applicable.

Conflicts of Interest: The authors declare no conflict of interest.

References

1. Singh, H.; Kommuri, S.V.R.; Kumar, A.; Bajaj, V. A new technique for guided filter based image denoising using modified cuckoo search optimization. *Expert Syst. Appl.* **2021**, *176*, 114884. [CrossRef]
2. Gonzalez, R.C.; Woods, R.E. *Digital Image Processing*, 3rd ed.; Prentice-Hall: Hoboken, NJ, USA, 2006; p. 298.
3. Wang, M.; Zheng, S.; Li, X.; Qin, X. A new image denoising method based on Gaussian filter. In Proceedings of the 2014 International Conference on Information Science, Electronics and Electrical Engineering (ISEEE), Sapporo, Japan, 26–28 April 2014.

4. Kondo, Y.; Yoshida, I.; Numada, M.; Koshimizu, H. A study on edge preserving noise reduction filter using fast m-estimation method. *J. Jpn. Soc. Precis. Eng.* **2020**, *86*, 1034–1041. (In Japanese) [CrossRef]
5. Buades, A.; Coll, B.; Morel, J.M. A non-local algorithm for image denoising. *CVPR* **2005**, *2*, 60–65.
6. Omidikia, N.; Ghaffari, M.; Rajko, R. Sparse non-negative multivariate curve resolution: L0, L1, or L2 norms? *Chem. Intell. Lab. Syst.* **2020**, *199*, 103969. [CrossRef]
7. OpenCV Denoising. Available online: https://docs.opencv.org/2.4/modules/photo/doc/denoising.html (accessed on 3 August 2021).

engineering proceedings

Proceeding Paper

A Numerical Analysis on the Cyclic Behavior of 316 FR Stainless Steel and Fatigue Life Prediction †

Ikram Abarkan [1,*], Abdellatif Khamlichi [2] and Rabee Shamass [3]

1 Department of Physics, Faculty of Sciences, Abdelmalek Essaâdi University, Tetouan 93002, Morocco
2 Department of Industrial and Civil Sciences and Technologies, National School of Applied Sciences, Abdelmalek Essaâdi University, Tetouan 93000, Morocco; khamlichi7@yahoo.es
3 Division of Civil and Building Services Engineering, School of the Built Environment and Architecture, London South Bank University, London SE1 0AA, UK; shamassR@Lsbu.ac.uk
* Correspondence: aberkan.ikraam@gmail.com
† Presented at the 2nd International Electronic Conference on Applied Sciences, 15–31 October 2021; Available online: https://asec2021.sciforum.net/.

Abstract: The present work aims to predict the cyclic behavior and fatigue life of 316 FR stainless steel specimens at 650 °C. First, the samples were modeled using finite element analysis under different strain amplitudes, and the obtained numerical hysteresis loops were compared against experimental results available in the literature. Then, the fatigue life was estimated using different fatigue life prediction models, namely the Coffin–Manson model, Ostergren's damage function, and Smith–Watson–Topper model, and was compared to the experimental fatigue life. The obtained results revealed that the numerical cyclic stress–strain data are in good agreement with those obtained experimentally. In addition, the predicted fatigue lives using the previously mentioned fatigue life models and based on the provided equation parameters are within a factor of 2.5 of the experimental results. Accordingly, it is suggested that they can be used to predict the fatigue life of 316 FR stainless steel.

Keywords: cyclic stress–strain behavior; fatigue life prediction; finite element analysis; low cycle fatigue; 316 FR stainless steel

Citation: Abarkan, I.; Khamlichi, A.; Shamass, R. A Numerical Analysis on the Cyclic Behavior of 316 FR Stainless Steel and Fatigue Life Prediction. *Eng. Proc.* **2021**, *11*, 28. https://doi.org/10.3390/ASEC2021-11116

Academic Editor: Nunzio Cennamo

Published: 15 October 2021

Publisher's Note: MDPI stays neutral with regard to jurisdictional claims in published maps and institutional affiliations.

1. Introduction

Advanced Gas-cooled Reactors (AGR) in the nuclear power plant industry are designed to operate at severe temperatures [1], resulting in thermal stresses occurring simultaneously with mechanical loads. Basically, the frequent start-up and shut-down procedures, as well as the change in power level owing to the daily energy consumption, are the main reason for these components experiencing combined mechanical and thermal cyclic loadings. The resulting repetitive loads cause microscopic damage to the material, which leads to fatigue crack initiation, propagation, and eventually, failure.

Over time, several fatigue life prediction approaches for the Low Cycle Fatigue (LCF) regime have been published in the literature; the most popular are the plastic strain-based approaches such as the Coffin–Manson model [2,3], strain energy-based criteria such the Smith–Watson–Topper (SWT) damage model [4], which can be used for both low- and high-cycle fatigue conditions, and Ostergren's equation [5]. The two latest ones (i.e., SWT and Ostergren models) consider the effect of mean stress on fatigue life. Another well-known strain energy-based method worth mentioning is Golos and Ellyin's total strain energy density approach [6] as it is valid for low and high cycle fatigue regimes, as well as for both Masing and non-Masing material response. Moreover, when using these equations, the accuracy of the stress–strain data is also important for estimating accurate low-cycle fatigue life. Finite Element Analysis (FEA) is one of the most effective tools to use, since it has been shown to be precise and accurate [7,8].

Materials with good cyclic characteristics are generally required for use in AGRs to withstand severe low cycle fatigue loadings. 316FR Stainless Steel (SS) is identical to 316LN SS [7,8], a low-carbon increased nitrogen grade of austenitic stainless steel that is typically selected for this sort of application due to its extending mechanical, low-cycle fatigue and creep properties at higher temperatures [1].

Many research investigations have been undertaken in the last few years to examine the durability of 316 SS under low-cycle fatigue at room temperature, in particular [7–9], but few studies have been dedicated to low-cycle fatigue at higher temperatures. Hormozi [1], for example, performed thorough experimental and numerical investigations of isothermal and in-phase thermomechanical low cycle fatigue of 316 FR SS with and without hold time. As a result, he developed a substantial number of findings related to the analysis of stress–strain data, cyclic plasticity behavior, and creep-fatigue damage evolution for low-cycle fatigue and thermomechanical fatigue conditions.

Recently, a vast majority of investigations have been conducted to study the accuracy of the widely used low-cycle fatigue life equations, namely the Coffin–Manson [2,3], Smith–Watson–Topper [4], and Ostergren models [5] at room temperature, in particular [10,11]. However, only a few studies have focused on determining the parameters and evaluating the accuracy of these LCF models, for 316 stainless steel, at higher temperatures. In the present paper, the cyclic stress–strain curves have been generated based on finite element analysis and have been compared with the experimental ones found by Hormozi [1]. Then, an examination of the aforementioned low-cycle fatigue life prediction equations, i.e., the Coffin–Manson [2,3], Smith–Watson–Topper (SWT) [4], and Ostergren equations [5], has been made and the parameters of these equations have been proposed for dumbbell specimens made of 316 FR SS at 650 °C. The predicted fatigue lives have been compared with the test data provided by Hormozi [1], Hong et al. [12], and Tak et al. [13].

2. Experimental Conditions

Hormozi [1] conducted a fully reversed uniaxial low-cycle fatigue experiment on four polished dumbbell specimens made of 316 FR SS, the chemical composition of which is indicated in Table 1 (in weight percent). The specimens have a gauge diameter and length of 8 and 16 mm, respectively, as shown in Figure 1. The LCF experiments were performed under different mechanical strain amplitude levels, namely ±0.4, ±0.8, ±1.0, and ±1.2%, and at a constant temperature of 650 °C. All the tests were undertaken in the air environment with a frequency of 0.01 Hz.

Table 1. Chemical Composition of 316FR SS in weight (%) [1].

As	B	C	Co	Cr	Cu	Mn	Mo	N	Nb
0.02	0.003	0.05	0.08	18.08	0.1	1.88	2.22	0.048	0.01
Ni	**P**	**S**	**Si**	**Sn**	**Ti**	**V**	**W**	**Zr**	
11.8	0.023	0.006	0.38	0.02	0.01	0.1	0.04	0.01	

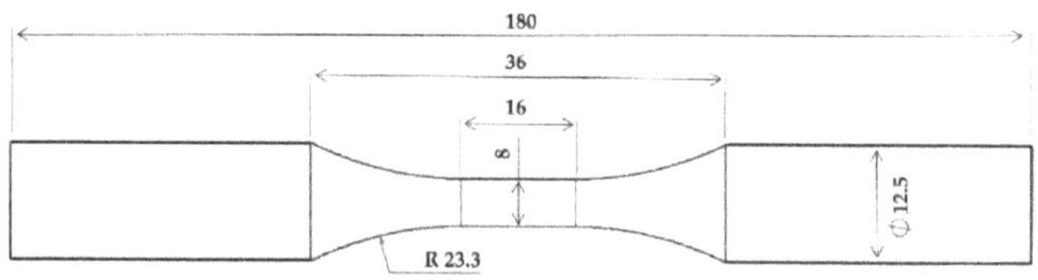

Figure 1. Specimens shape and dimensions (in mm).

3. Finite Element Analysis

Finite element analysis has been conducted on four cylindrical specimens using ABAQUS software [14]. The 2D-axisymmetric model, with a radius of 4 mm and a height of 6.25 mm, has been created to illustrate the gauge section of the samples under study.

As shown in Figure 2a, symmetry boundary conditions have been generated along the gauge length and gauge diameter, and prescribed cyclic displacement has been applied to the higher extremity of the 2D model in a symmetrical triangular waveform as illustrated in Figure 2b. Moreover, the temperature was fixed and set to 650 °C. The CAX4R elements have been considered in the mesh section. The kinematic and isotropic plasticity data from [1], as well as other material properties, represented in Figure 3 and Table 2, respectively, have been implemented in the FEA software's property section.

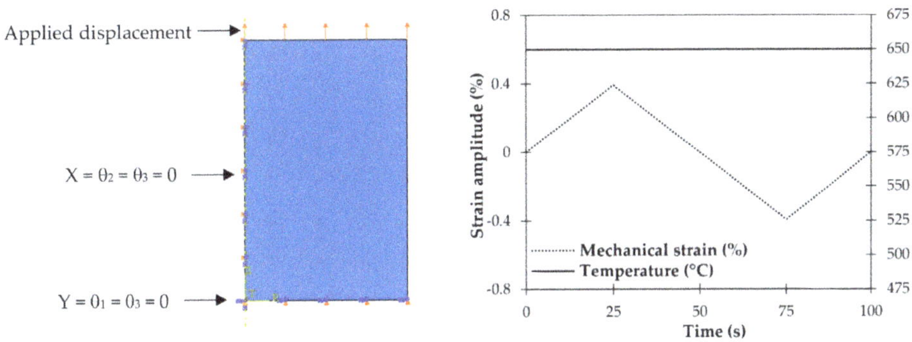

(a) (b)

Figure 2. Representation of the finite element model on Abaqus; (**a**) boundary conditions, and (**b**) applied loads waveform.

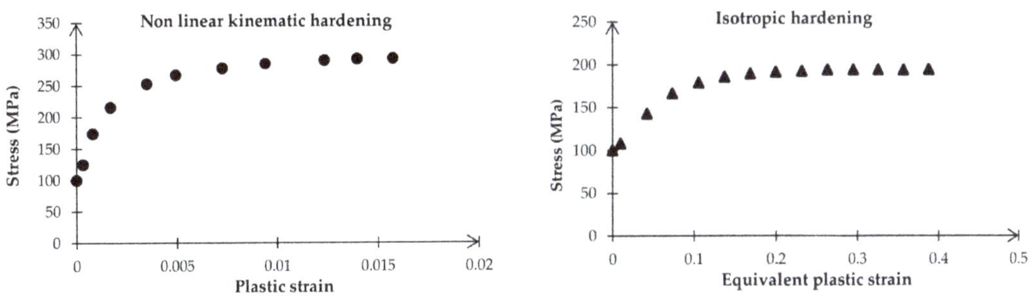

(a) (b)

Figure 3. Plasticity data of (**a**) non-linear kinematic hardening, and (**b**) isotropic hardening of 316 FR SS, at 650 °C, for a strain amplitude of ±1.0% [1].

Table 2. Material properties of 316 FR SS at 650 °C [1].

Young's Modulus (MPa)	Yield Strength (MPa)	Thermal Conductivity ($Wm^{-1}\,°C^{-1}$)	Coefficient of Thermal Expansion ($10^{-6}\,°C^{-1}$)
160,000	100	23	21

4. Results & Discussion

4.1. Cyclic Stress–Strain Response

The estimated cyclic stress–strain data at ±0.4% have been compared to Hormozi's experimental results [1]. As shown in Figure 4, the numerically estimated hysteresis loops are in good agreement with those found experimentally. As a result, the FE data are accurate and can be used to predict the low-cycle fatigue life of 316 FR SS.

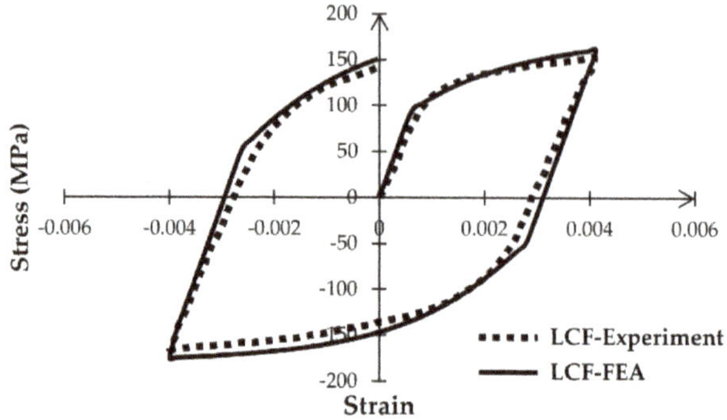

Figure 4. Comparison between the numerical hysteresis loop and the experimental one provided by Hormozi [1] under ±0.4% strain amplitude.

4.2. Fatigue Life Prediction

This section covers the evaluation of the well-known fatigue life equations for estimating the low-cycle fatigue life for dumbbell specimens made of 316 FR SS at 650 °C. These include the Coffin–Manson, Ostergren, and Smith–Watson–Topper models [2–5]. The predicted fatigue lives found from the present study will be compared with the test results reported in [1,12,13].

4.2.1. Coffin–Manson Model

In the low-cycle regime, Coffin and Manson [2,3] independently established a log–log linear equation to consider the effect of plastic strain range $\Delta\varepsilon_p$ on the low-cycle fatigue life N_f. The well-known Coffin–Manson equation is given as follows:

$$\Delta\varepsilon_p = 2\varepsilon'_f\left(2N_f\right)^c \quad (1)$$

where ε'_f, and c are the fatigue ductility coefficient and fatigue ductility exponent, respectively. The values of these two material parameters at 650 °C, obtained by the least square regression technique with a coefficient of determination R^2 of 0.998, are listed in Table 3. The numerically and experimentally obtained plastic strain amplitudes at the saturation stage, for each applied mechanical strain amplitude, are provided in Table 4. In this table (i.e., Table 4), the Relative Error (RE) between the experimental and numerical plastic strain amplitude values, for all applied strain amplitudes, shows that the finite element model accurately predicts the plastic strains under LCF conditions. The estimated fatigue lives obtained using Equation (1) were compared to those provided by Hormozi [1] and are found to be conservative with an average relative error of −7.87% and lie very close to a factor of 1, as illustrated in Figure 5. For further validation, the obtained predicted fatigue lives from the Coffin–Manson model were also compared to 10 test data points provided by Hong et al. [12], and Tak et al. [13] for a mechanical strain amplitude between ±0.4 and ±0.8% and found to fall within a factor of 2.5 of the test results, as shown in Figure 5. Hence, the suggested parameters in Table 3 are valid and can be used to predict the fatigue life of the present study's used material when the temperature is 650 °C.

Table 3. Coffin–Manson, Ostergren and SWT equations parameters for 316 FR SS at 650 °C.

Coffin-Manson		Ostergren		SWT	
ε'_f	c	L (MPa)	n	C (MPa)	β
0.9121	−0.767	874.9	−0.949	7839	−0.378

Table 4. Relative error between the predicted and experimental [1] maximum stress and plastic strain amplitude.

Strain Amplitude	$\sigma_{max,pre}$	$\sigma_{max,exp}$	RE	$\Delta\varepsilon_{p,pre}/2$	$\Delta\varepsilon_{p,exp}/2$	RE
(%)	(MPa)	(MPa)	(%)	(%)	(%)	(%)
0.4	227	223	1.79	0.25	0.23	8.70
0.8	274	281	−2.49	0.62	0.59	5.08
1	288	297	−3.03	0.81	0.78	3.85
1.2	292	-	-	1.02	-	-

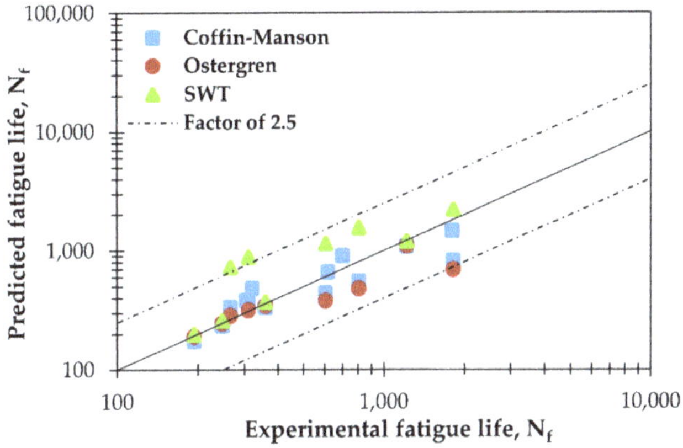

Figure 5. Comparison of the predicted fatigue life with the experimental results in [1,12,13].

4.2.2. Ostergren Damage Model

Ostergren [5] proposed a damage function that relates the plastic strain range $\Delta\varepsilon_p$ and the maximum stress σ_{max} to the fatigue life N_f as follows:

$$\sigma_{max}\,\Delta\varepsilon_p = LN_f^n \tag{2}$$

where L and n are material parameters. The obtained values for each by the least square regression technique with an R^2 of 0.997 are represented in Table 3.

The FE obtained maximum stress for each applied strain amplitude is listed in Table 4. The percentage relative error between the FE predicted peak stress and the experimental one further indicates that the FE results are in good agreement with those found experimentally. Moreover, the calculated fatigue life using Equation (2) are plotted against the experimental data in Figure 5. As can be seen, the estimated low cycle fatigue life using the Ostergren damage function [4] lies extremely close to the factor of 1, and the maximum relative error is only −9.41% at 0.4% strain amplitude. The comparison between the predicted fatigue lives using the Ostergren model and the experimental data of Hong et al. [12] shows that the fatigue life is conservative at 0.4, 0.5, and 0.6% strain amplitudes and with a more significant relative error than at 0.7 and 0.8% strain amplitude. By way of example, the relative error is −37.36 and 2.5% for strain amplitudes of 0.6 and 0.7%, respectively. Accordingly, the overall predicted results are within a factor of 2.5 of the test data. Thus, one can conclude that the Ostergren model along with the proposed parameters in Table 3 can well predict the low cycle fatigue life of 316 FR SS, at 650 °C.

4.2.3. Smith–Watson–Topper Damage Model

Smith et al. [4] presented the SWT parameter (i.e., $\sigma_{max}\,\Delta\varepsilon$) as a damage parameter that is related to cycle life in the following [15]:

$$\sqrt{E\sigma_{max}\Delta\varepsilon} = CN_f{}^{\beta} \tag{3}$$

where $\Delta\varepsilon$ is the mechanical strain range, and β and C are material constants.

The fatigue lives calculated using Equation (3) with the parameters listed in Table 3 (R^2 of 1), under different strain amplitudes, have been compared with those obtained experimentally [1]. As observed from Figure 5, the predicted LCF life by means of Equation (3) along with the SWT material parameters represented in Table 3, are in good agreement with the experimental ones (factor of 1), and the maximum relative error is only 4.71% at the strain amplitude of 1%. Further validation with test data from reference [12] was made, and the predicted fatigue lives are found to have a factor of 2.5 as shown in Figure 5. Therefore, it may be concluded that the Smith–Watson–Topper equation along with the present study's supplied parameters can correctly estimate the fatigue life for 316 FR SS at 650 °C.

5. Conclusions

In this work, the cyclic stress–strain response of 316 FR SS at 650 °C has been numerically obtained using FEA and compared to the experimental results in order to examine the accuracy of the finite element model. The fatigue life has been estimated for various applied strains at the same mentioned temperature and compared to the experimental data provided in the literature [1,12,13] to assess the accuracy of the commonly used fatigue life equations, namely the Coffin–Manson, Ostergren, and Smith–Watson–Topper models [2–5]. The following conclusions have been made; (1): the cyclic stress–strain data were found to be in good agreement with the experimental results. (2): The fatigue life equations parameters that were found using least square regression analysis have been supplied. (3): The fatigue life prediction models with the given parameters yielded results that were close to the experimental findings by a factor of 2.5. Hence, it is suggested that these fatigue life equations can be used to accurately estimate the fatigue life of 316 FR SS at 650 °C.

Supplementary Materials: The following supporting information can be downloaded at: https://www.mdpi.com/article/10.3390/ASEC2021-11116/s1.

Author Contributions: Conceptualization, methodology, formal analysis, investigation, writing—original draft preparation, I.A.; validation, writing—review and editing, A.K. and R.S. All authors have read and agreed to the published version of the manuscript.

Funding: This research received no external funding.

Institutional Review Board Statement: Not applicable.

Informed Consent Statement: Not applicable.

Data Availability Statement: Data can be made available on demand.

Conflicts of Interest: The authors declare no conflict of interest.

References

1. Hormozi, R. Experimental and Numerical Simulations of Type 316 Stainless Steel Failure under LCF/TMF Loading Conditions. Ph.D. Thesis, Imperial College, London, UK, 2014.
2. Coffin, L.F., Jr. *A Study of the Effects of Cyclic Thermal Stresses on a Ductile Metal*; Knolls Atomic Power Laboratory: Schenectady, NY, USA, 1953.
3. Manson, S.S. *Behavior of Materials under Conditions of Thermal Stress*; National Advisory Committee for Aeronautics: Washington, DC, USA, 1954.
4. Smith, K.N.; Watson, P.; Topper, T.H. A stress-strain function for the fatigue of metals. *J. Mater.* **1970**, *5*, 767–778.
5. Meltzer, R.; Fiorini, Y.; Horstman, R.; Moore, I.; Batik, A.; Ostergren, W. A Damage Function and Associated Failure Equations for Predicting Hold Time and Frequency Effects in Elevated Temperature, Low Cycle Fatigue. *J. Test. Eval.* **1976**, *4*, 327. [CrossRef]

6. Golos, K.; Ellyin, F. Total strain energy density as a fatigue damage parameter. In *Advances in Fatigue Science and Technology*; Springer: Dordrecht, The Netherlands, 1989; Volume 159, pp. 849–858. [CrossRef]
7. Abarkan, I.; Shamass, R.; Achegaf, Z.; Khamlichi, A. Numerical and analytical studies of low cycle fatigue behavior of 316 LN austenitic stainless steel. *J. Press. Vessel Technol.* **2020**. [CrossRef]
8. Abarkan, I.; Khamlichi, A.; Shamass, R. A Study on low cycle fatigue life assessment of notched specimens made of 316 LN austenitic stainless steel. *J. Press. Vessel. Technol.* **2022**, *144*, 021503. [CrossRef]
9. Roy, S.C.; Goyal, S.; Sandhya, R.; Ray, S. Low cycle fatigue life prediction of 316 L(N) stainless steel based on cyclic elasto-plastic response. *Nucl. Eng. Des.* **2012**, *253*, 219–225. [CrossRef]
10. Dutta, A.; Dhar, S.; Acharyya, S.K. Material characterization of SS 316 in low-cycle fatigue loading. *J. Mater. Sci.* **2010**, *45*, 1782–1789. [CrossRef]
11. Yuan, X.; Yu, W.; Fu, S.; Yu, D.; Chen, X. Effect of mean stress and ratcheting strain on the low cycle fatigue behavior of a wrought 316LN stainless steel. *Mater. Sci. Eng. A* **2016**, *677*, 193–202. [CrossRef]
12. Hong, S.G.; Yoon, S.; Lee, S.B. The effect of temperature on low-cycle fatigue behavior of prior cold worked 316L stainless steel. *J. Fatigue* **2003**, *25*, 1293–1300. [CrossRef]
13. Tak, N.H.; Kim, J.S.; Lim, J.Y. An Energy-Based Unified Approach to Predict the Low-Cycle Fatigue Life of Type 316L Stainless Steel under Various Temperatures and Strain-Rates. *Materials* **2019**, *12*, 1090. [CrossRef] [PubMed]
14. *ABAQUS/Standard*, version 6.14; Dassault Systèmes Simulia Corp.: Providence, RI, USA, 2014.
15. Bartošák, M. Phenomenological Models for Lifetime Prediction under Low-Cycle Fatigue and Thermo-Mechanical Fatigue Loading Conditions. Ph.D. Thesis, Czech Technical University, Prague, Czech Republic, 2019.

engineering proceedings

MDPI

Proceeding Paper

Normal Range of Motion of Lower Extremity Joints in Mongolian Subjects †

Batbayar Khuyagbaatar, Tserenchimed Purevsuren and Danaa Ganbat *

Biomechanical Research Laboratory, School of Mechanical Engineering and Transportation, Mongolian University of Science and Technology, Ulaanbaatar 14191, Mongolia; batbayarkh@must.edu.mn (B.K.); tserenchimed.p@must.edu.mn (T.P.)

* Correspondence: ganbatda@must.edu.mn; Tel.: +976-9914-7004

† Presented at the 2nd International Electronic Conference on Applied Sciences, 15–31 October 2021; Available online: https://asec2021.sciforum.net/.

Abstract: It is important to identify the normal range of motion (ROM) of the human joints for both biomechanical and clinical applications. For health care providers, including physicians and therapists, the restoration of normal ROM is a difficult task. The severity of impaired joint mobility or the postoperative rehabilitation process must be evaluated in comparison with a normal reference value. However, there are no studies that have reported the ROM of Mongolian subjects. In this study, we measured the hip, knee, and ankle joint angles using multiple wearable inertial sensors. Ten healthy young subjects participated. The three-dimensional (3D) motion data were collected while the subject were walking at normal speed. In our knowledge, this study is the first to analyze the normal ROM of Mongolian male subjects. The collected data can be used as reference values for evaluating the disability of the motion and performance in rehabilitation programs.

Keywords: normal range of motion; Mongolian; wearable sensors

Citation: Khuyagbaatar, B.; Purevsuren, T.; Ganbat, D. Normal Range of Motion of Lower Extremity Joints in Mongolian Subjects. *Eng. Proc.* **2021**, *11*, 29. https://doi.org/10.3390/ASEC2021-11140

Academic Editor: Roger Narayan

Published: 15 October 2021

1. Introduction

Identification of the natural gait characteristics of people is important from both biomechanical and clinical perspectives [1]. Many diseases and injuries can impair joint mobility, which results in a decline in ROM or changes the gait characteristics. Furthermore, abnormal gait characteristics are associated with aging and an abnormal lifestyle. Joint motion varies with age and is generally more restricted in the older age group [2].

For health care providers, including physicians and therapists, the restoration of normal ROM is a difficult task. The severity of impaired joint mobility or postoperative rehabilitation process must be evaluated in comparison with normal gait patterns. These normal gait parameters have been investigated extensively in a variety of countries including the United States [3], Sweden [2], Korea [1], Japan [4], and China [5]. However, there are no comparison studies on ROM in Mongolia with other countries. Furthermore, video-based motion analysis systems cannot provide the details of joint motions during movements. The 3D joint angle measurement is an important requirement, notably in the orthopedic and rehabilitation fields [6].

Recent developments in sensor technology allow us to precisely measure human movements. Several studies have used the inertial measurement unit (IMU) sensor for analyzing the kinematics of the lower extremity during normal walking and other motions in a variety of countries [6–9]. In this study, we investigated the joint angles of the hip, knee, and ankle during walking using wearable IMU sensors.

2. Materials and Methods

2.1. Participant Information

Ten male subjects (age, 26.1 ± 7.8 years; height, 177 ± 7 cm; weight, 76.4 ± 17.6 kg), who had not had any musculoskeletal injuries within the past year, were recruited in this study. All subjects were recruited with informed consent from the Mongolian University of Science and Technology (MUST).

2.2. Experiment Procedure

Each subject performed three bouts of walking under supervision. Subjects were asked to perform their walking with normal speed to minimize the speed differences. Prior to the experiment, each participant was asked to perform several bouts of walking as a warm-up. The experiment was conducted in the indoor laboratory at MUST. The subjects were wearing IMU sensors, a training suit, and sports shoes. In total, six IMU sensors (Wearnotch, Notch Interface Inc., New York, NY, USA) were used to record the right lower extremity motion during normal walking. The sensors were attached to the chest, stomach, right thigh, shank, and foot using the straps. The IMU sensor includes 3-axis acceleration and 3-axis gyroscope with ±16 g and ±2000°/s, with a sampling rate of 100 Hz. The sensors' locations are described in Figure 1.

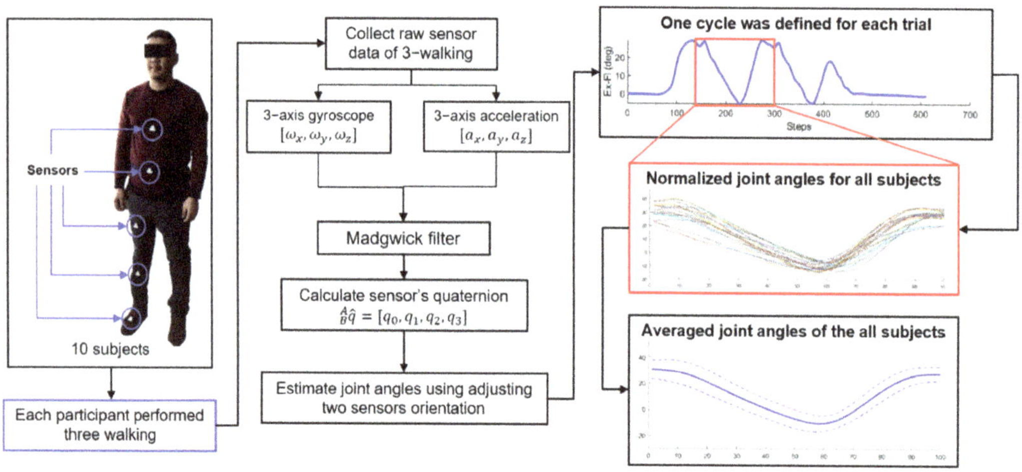

Figure 1. The experiment protocol and data processing.

2.3. Data Processing

At the beginning of the walking, subjects did a steady pose as a calibration. After sensor calibration, each participant performed normal walking. The sensor's raw data was sent to the server computer and processed using Matlab® R2015a (The Mathworks Inc., Natick, MA, USA) [10]. After data processing, the quaternions of the six sensors were estimated using the 3-axis acceleration and 3-axis gyroscope data based on the Madgwick filter algorithm [11]. Then, the joint angles were estimated based on the orientation difference between the adjusting two sensors. The joint angles were represented as the Euler angles of the distal segment reference frame relative to the proximal segment reference frame using the sensor's orientation [12]. The accuracy of the wearable sensor was compared against the conventional marker-based system in our previous study. The estimated error between the wearable sensors and marker-based system was approximately 5.8% [10]. The 3D rotations of the joints in sagittal, transverse, and coronal planes were expressed as extension-flexion (Ex-Fl), internal-external (Int-Ext) rotation, and adduction-abduction (Add-Abd), respectively (Figure 2). After calculating the 3D joint angles, the cycle of one bout of walking was defined and normalized. From the start to the end of the cycle was from a right heel strike to the next right heel strike.

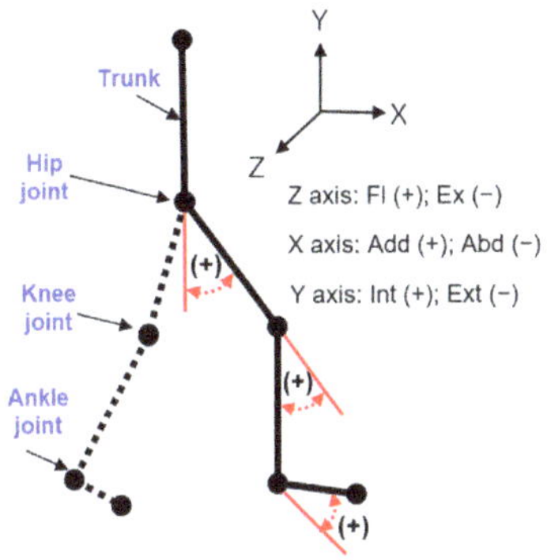

Figure 2. Joint angle definition for hip, knee, and ankle joints.

3. Results

3.1. Joint Angles of Hip, Knee, and Ankle Joint

The hip, knee, and ankle joint angles during walking are shown in Figure 3. The maximum extension-flexion angles of the hip and knee were 30.8 ± 3.5° and 55.2 ± 3.4°, respectively. In the ankle joint, the dorsiflexion-plantar flexion angle was 13.5 ± 6.0°. The maximum adduction-abduction angles of the hip, knee, and ankle joints were 5.3 ± 4.8°, 16.6 ± 10.2°, and 10.3 ± 5.8°, respectively. The maximum internal-external rotation angles of the hip, knee, and ankle joints were 5.5 ± 4.5°, 7.3 ± 10.2°, and 6.3 ± 3.8°, respectively (Figure 3).

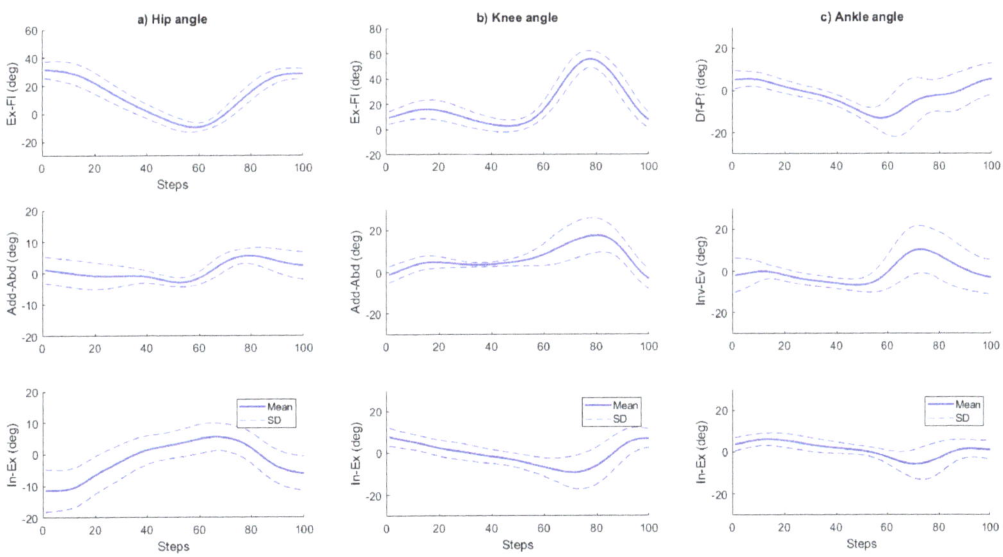

Figure 3. Joint angles of hip, knee, and ankle joint.

3.2. Comparison with Previous Studies

The maximum flexion angles of the hip, knee, and ankle joints were compared to the previous studies. The results are summarized in Table 1. The total joint angle data were similar to previous studies [1,5,13,14]. Of note, the hip and ankle flexion were similar to that data from the Chinese population, while the knee flexion was similar to all data except the Chinese. This is due to the anthropometric similarities between Asian countries.

Table 1. Maximum flexion angle comparison with previous studies.

	Hip	Knee	Ankle
Our study	$30.8 \pm 5.5°$	$55.2 \pm 3.4°$	$13.5 \pm 6.0°$
United States [13]	$38.2 \pm 3.1°$	$57.4 \pm 5.0°$	$10.6 \pm 3.9°$
Korea [1]	$24.2 \pm 6.6°$	$58.9 \pm 5.8°$	$10.8 \pm 4.1°$
China [5]	$34.7 \pm 3.0°$	$66.2 \pm 4.8°$	$13.7 \pm 4.3°$
Italy [14]	$43.4 \pm 8.0°$	$55.6 \pm 4.4°$	$19.7 \pm 8.1°$

Throughout flexion motions, the maximum difference of hip, knee, and ankle joints between the current and the previous studies was 12.6°, 11.0°, and 6.2°, respectively. The dissimilarity may be from the use of different motion capture systems. Our study utilized the wearable motion capture system, while other studies have used the conventional marker-based system. Of course, the difference occurred due to the gait characteristics of different nationalities.

4. Conclusions

In this study, we utilized a wearable IMU sensor for measuring the normal ROM of the lower extremity during walking for Mongolians. The wearable sensor technology can be applied to both indoor and outdoor environments without any restrictions. The raw sensor data was processed using the in-house developed algorithm based on the Mahony filter. To my knowledge, this study is the first to analyze the 3D normal ROM of the Mongolian subjects. But, the subjects were only male, and the number of participants was small. Therefore, future work will need to include a large number of subjects as well as different sexes and age groups. Future work will also focus on the gait characteristics of the nomadic peoples. This study provides fundamentals of the normal gait characteristics during walking, which can be reference values for evaluating the disability of the motion and performance in rehabilitation programs.

Author Contributions: Conceptualization, B.K. and T.P.; methodology, B.K. and T.P.; writing—original draft preparation, B.K.; writing—review and editing, D.G.; funding acquisition, D.G. All authors have read and agreed to the published version of the manuscript.

Funding: This work was supported by the "Mongolia-Japan Engineering Education Development" (MJEED) Project financed by the Japan International Cooperation Agency and executed by Ministry of Education and Science of Mongolia.

Informed Consent Statement: Informed consent was obtained from all subjects involved in the study.

Data Availability Statement: Not applicable.

Acknowledgments: The authors would like to acknowledge the financial support from the Mongolian University of Science and Technology.

Conflicts of Interest: The authors declare no conflict of interest.

References

1. Ryu, T.; Soon Choi, H.; Choi, H.; Chung, M.K. A comparison of gait characteristics between Korean and Western people for establishing Korean gait reference data. *Int. J. Ind. Ergon.* **2006**, *36*, 1023–1030. [CrossRef]
2. Roaas, A.; Andersson, G.B.J. Normal range of motion of the hip, knee and ankle joints in Male subjects, 30–40 years of age. *Acta Orthop.* **1982**, *53*, 205–208. [CrossRef] [PubMed]
3. Boone, D.C.; Azen, S.P. Normal range of motion of joints in male subjects. *J. Bone Jt. Surg.-Ser. A* **1979**, *61*, 756–759. [CrossRef]
4. Shimada, T.; Takemasa, S.; Kawakami, K.; Inoue, Y.; Susuki, K. Normal range of motion of joints in young Japanese people. *Bull. Allied Med. Sci.* **1988**, *61*, 756–759.
5. Han, S.; Cheng, G.; Xu, P. Three-dimensional lower extremity kinematics of Chinese during activities of daily living. *J. Back Musculoskelet. Rehabil.* **2015**, *28*, 327–334. [CrossRef] [PubMed]

6. Favre, J.; Aissaoui, R.; Jolles, B.M.; de Guise, J.A.; Aminian, K. Functional calibration procedure for 3D knee joint angle description using inertial sensors. *J. Biomech.* **2009**, *42*, 2330–2335. [CrossRef] [PubMed]
7. Kim, K.J.; Agrawal, V.; Bennett, C.; Gaunaurd, I.; Feigenbaum, L.; Gailey, R. Measurement of lower limb segmental excursion using inertial sensors during single limb stance. *J. Biomech.* **2018**, *71*, 151–158. [CrossRef] [PubMed]
8. Tadano, S.; Takeda, R.; Miyagawa, H. Three dimensional gait analysis using wearable acceleration and gyro sensors based on quaternion calculations. *Sensors* **2013**, *13*, 9321–9343. [CrossRef] [PubMed]
9. Vargas-Valencia, L.S.; Elias, A.; Rocon, E.; Bastos-Filho, T.; Frizera, A. An IMU-to-body alignment method applied to human gait analysis. *Sensors* **2016**, *16*, 2090. [CrossRef] [PubMed]
10. Choi, Y.C.; Khuyagbaatar, B.; Cheon, M.; Batbayar, T.; Lee, S.; Kim, Y.H. Kinematic Comparison of Double Poling Techniques Between National and College Level Cross-Country Skiers Using Wearable Inertial Measurement Unit Sensors. *Int. J. Precis. Eng. Manuf.* **2021**, *22*, 1105–1112. [CrossRef]
11. Madgwick, S. An efficient orientation filter for inertial and inertial/magnetic sensor arrays. *Rep. x-io Univ. Bristol (UK)* **2010**, *25*, 113–118.
12. Khuyagbaatar, B.; Purevsuren, T.; Park, W.M.; Kim, K.; Kim, Y.H. Interjoint coordination of the lower extremities in short-track speed skating. *Proc. Inst. Mech. Eng. Part H J. Eng. Med.* **2017**, *231*, 987–993. [CrossRef] [PubMed]
13. Kadaba, M.P.; Ramakrishnan, H.K.; Wootten, M.E. Measurement of lower extremity kinematics during level walking. *J. Orthop. Res.* **1990**, *8*, 383–392. [CrossRef] [PubMed]
14. Benedetti, M.G.; Catani, F.; Leardini, A.; Pignotti, E.; Giannini, S. Data management in gait analysis for clinical applications. *Clin. Biomech.* **1998**, *13*, 204–215. [CrossRef]

engineering proceedings

MDPI

Proceeding Paper

Analysis of NO_2 Pollution over Bangladesh between the Two COVID-19 Caused Lockdowns in 2020 and 2021 Using Sentinel-5P Products †

S.M.Sohel Rana *, Sheikh Mohammad Famim Ahmed and Hamida Akter

Department of Environmental Sciences, Jahangirnagar University, Savar, Dhaka 1342, Bangladesh; famim.stu2016@juniv.edu (S.M.F.A.); sahamida73@gmail.com (H.A.)

* Correspondence: sohel.stu20161@juniv.edu; Tel.: +880-1851986530

† Presented at the 2nd International Electronic Conference on Applied Sciences, 15–31 October 2021; Available online: https://asec2021.sciforum.net/.

Abstract: Due to the COVID-19 pandemic, all countries around the world have imposed nationwide lockdowns to control the spreading of the virus. During the lockdown period, many countries saw a drastic drop in air pollution. In Bangladesh, there were two nationwide lockdowns. The first lockdown was imposed on 26 March–30 May in 2020 and the second lockdown was imposed on 3 April until the study period of 31 May in 2021. This study aimed to analyze the NO_2 pollution over Bangladesh during the two periods of lockdown. Tropospheric NO_2 column spatial configuration was measured over Bangladesh using Sentinel-5P data. A map of the monthly average concentration of tropospheric NO_2 in 2020 and 2021 over Bangladesh was produced using the HARP toolkit and Python. Then, the map was compared with same period Sentinel-5P product's map for the same period in 2019. It was found that during the first lockdown in Bangladesh between 26 March and 30 May 2020, NO_2 concentration drastically decreased in April but increased in May. However, during the second lockdown from 3 April to 31 May in 2021, the NO_2 concentration was found to be much higher. Most of the pollution occurred in the Dhaka district. During the second lockdown, the restrictions were much lighter than those during the first one, which impacted the NO_2 concentration. This kind of study can be essential for the authorities to look closely at air quality and use sentinel data to improve air quality monitoring in the future.

Keywords: Sentinel-5P; air quality; COVID-19 lockdown; NO_2

Citation: Rana, S.M.S.; Ahmed, S.M.F.; Akter, H. Analysis of NO_2 Pollution over Bangladesh between the Two COVID-19 Caused Lockdowns in 2020 and 2021 Using Sentinel-5P Products. *Eng. Proc.* **2021**, *11*, 30. https://doi.org/10.3390/ASEC2021-11139

Academic Editor: Nunzio Cennamo

Published: 15 October 2021

Publisher's Note: MDPI stays neutral with regard to jurisdictional claims in published maps and institutional affiliations.

1. Introduction

The COVID-19 pandemic has caused significant changes in society, the economy, and the lifestyles of people worldwide. To contain the spread of the SARS-2 virus, countries all around the world have been imposing different types of measures such as wearing masks, maintaining physical distancing, and vaccines. Initially, lockdowns were the most used strategy for most of the countries around the world to stop the transmission of the COVID-19 virus. The temporary or total closure of international borders, educational institutions, and non-essential businesses and restrictions on citizen mobility, have been some of the measures taken during lockdowns [1]. These lockdowns have led to some positive impacts on the environment, especially in improving air quality. A number of studies have proved that lockdowns have temporarily improved the air quality around the globe [1–4]. Several papers have also suggested a significant reduction of air pollution in Bangladesh [5,6]. Most of these papers investigated multiple air pollutants (CO_2, O_3, NO_2, SO_2, $PM_{2.5}$, PM_{10}, etc.). Nitrogen dioxides (NO_2) are one of the important air constituents. The main sources of NO_2 gases are industry, power plants, residential heating, and vehicle exhausts in the form of nitric oxide (NO) [7]. This reddish-brown gas is produced from NO conversion through the oxidation process. NO_2 is a great indicator of human-made combustion activities and is a precursor of ozone (O_3) and aerosols. Respiratory diseases and asthma can be caused by NO_2. About 4 million cases of pediatric asthma annually result from exposure to NO_2 [8]. Moreover, it can lead to environmental deterioration by producing acid rain [9].

The objective of this study was to collect and analyze the air quality data of Bangladesh during the lockdown periods (March to May) in 2019, 2020, and 2021. The analysis of 2019 was the baseline to compare the trends of NO_2 found during the lockdown periods in 2020 and 2021.

2. Methods and Materials

2.1. Study Area

Our study was focused on Bangladesh. Bangladesh is located in South Asia at 20°34′ and 26°38′ N and 88°01′ and 92°42′ E [6]. Bangladesh is a densely populated country with a density of 1265 people per km^2 [10]. Geographically, Bangladesh, a low-lying riverine country, is most vulnerable to climate change [6]. Bangladesh was the most polluted country in 2019 according to the world air quality report by IQAir [10]. There is a lack of air pollution source inventory in Bangladesh [11]. The industrial sector of Bangladesh has been experiencing some expansion, and vehicular and industrial emissions are considered the main sources of air pollution in Bangladesh [12].

2.2. Data

The Sentinel-5 Precursor is an earth orbiting single satellite system dedicated to providing information and services on air quality, climate, and the ozone layer [13]. The satellite mission was launched to continue the data between previous missions such as SCIAMACHY, GOME-2, OMI, and the upcoming Sentinel-5 [14]. Sentinel-5P consists of the TROPOspheric Monitoring Instrument (TROPOMI) [9]. TROPOMI is a passive-sensing hyperspectral imager that allows acquisitions of 8-band imagery covering the domain of UV, visible to near-infrared, and shortwave infrared [9,14]. TROPOMI has a spatial resolution of 7×3.5 km^2, which is higher than that of all of its predecessors. The high spatial resolution of Sentinel-5P provides new potential for monitoring air pollution sources [9].

The data used for the study were downloaded from NASA's Earthdata website [4]. The data were downloaded using the subset data option available on the website. Data were downloaded only for the extent of Bangladesh. The data came in netCDF format, which stores multidimensional scientific information, including dimensions, variables/parameters, attributes, and coordinates [9]. A total of 276 offline L2 NO_2 products were used for the study. A total of 93, 90, and 93 products were used, respectively, for 2019, 2020, and 2021, covering 26–31 March and the full month of April and May for each year.

2.3. Methodology

The analysis was carried out following the methodology presented in the RUS Copernicus training in "Monitoring Pollution with Sentinel-5p", a case study in Italy during 2019–2020 [15], using the Python language with the HARP atmospheric toolbox provided by ESA. The NO_2 Level 2 product had been converted to Level 3 using the HARP tool in Python. The Sentinel-5P NO_2 Level 2 product was resampled to the spatial resolution of 0.01×0.01 degrees covering the extent of Bangladesh between 20 and 27 degrees of latitude. The data were filtered to a tropospheric NO_2 column density value quality over 75. This process was done to avoid errors due to cloud cover. During the conversion of the Level 2 product, tropospheric NO_2 column density was derived from the main product, and the unit was converted to Pmolec/cm^2 from the default unit of mol/m^2. Then, 6-day mosaics of average tropospheric NO_2 column density were produced for the month of March for each year. Monthly mosaics of average tropospheric NO_2 column density were produced in April and May for each year. The tropospheric NO_2 column density data for the time series analysis of specific cities (Dhaka and Chattogram, Bangladesh) were extracted from the Level 3 product using the Python programming language. The data were then plotted in a graph using WPS Office.

3. Results and Discussion

3.1. NO_2 Concentration Distribution over Bangladesh

This study focused on the NO_2 pollution over Bangladesh during the two lockdown periods in 2020 and 2021. It was found that the most NO_2 pollution occurs in Dhaka and its surrounding districts, such as Narayanganj and Gazipur. Most of the industries are situated in these districts, which could be the reason for the pollution. Additionally, Chattogram district also faced NO_2 pollution because of having an industrial zone. The comparative analysis of maps produced from the monthly average mosaic of the Sentinel-5P product for tropospheric NO_2 vertical column density showed various trends in each month of the lockdown. In March 2020, when the first lockdown was announced by the government of Bangladesh, the density of NO_2 declined sharply in other cities of Bangladesh compared to the pollution of the same period in 2019. However, in 2021, there was no lockdown in March, so it was expected and found that the concentration of NO_2 was high all over the country. The concentration of NO_2 was much higher than that of the same period in 2019. Figure 1 shows the distribution of NO_2 over the country between 26–31 March of each of the three studied years.

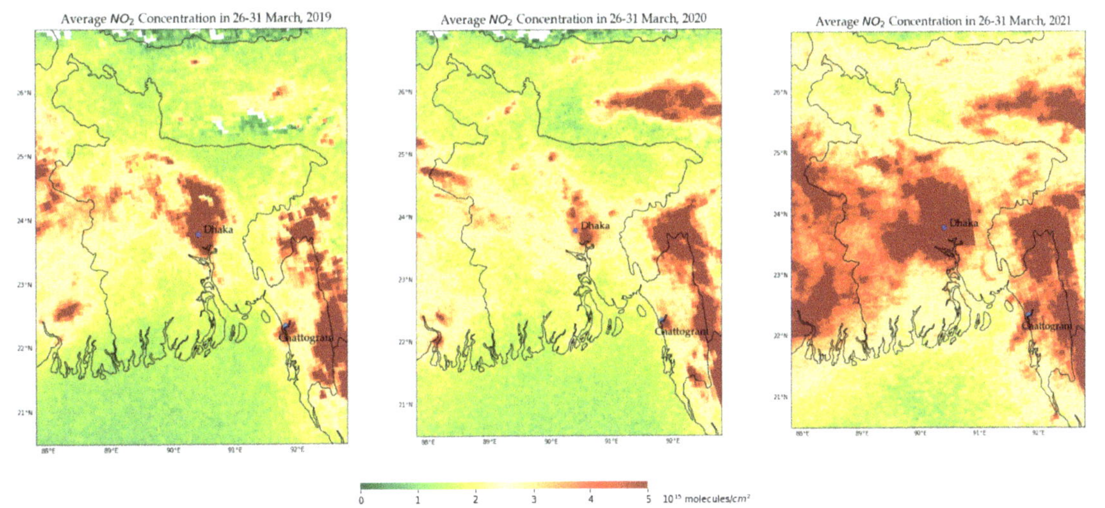

(a) (b) (c)

Figure 1. Average NO_2 pollution over Bangladesh in March (**a**) 2019, (**b**) 2020, and (**c**) 2021. March 2021 was more polluted than the same time period in 2019 and 2020 throughout Bangladesh.

In April 2020, the government strengthened the strictness of the lockdown and we can observe that there was a significant drop in NO_2 concentration in 2020. Compared to the concentration in 2019, the concentration of NO_2 dropped drastically all over the country, including in Dhaka and Chattogram city. In 2021, the lockdown was declared on 1 April, but the lockdown was not implemented strictly. This resulted in no such difference in NO_2 pollution in the major cities such as Dhaka, Chattogram, and Rajshahi. Rather, the concentration of NO_2 was higher compared to that in April 2019. Figure 2 shows the comparative cartography of NO_2 concentration over Bangladesh.

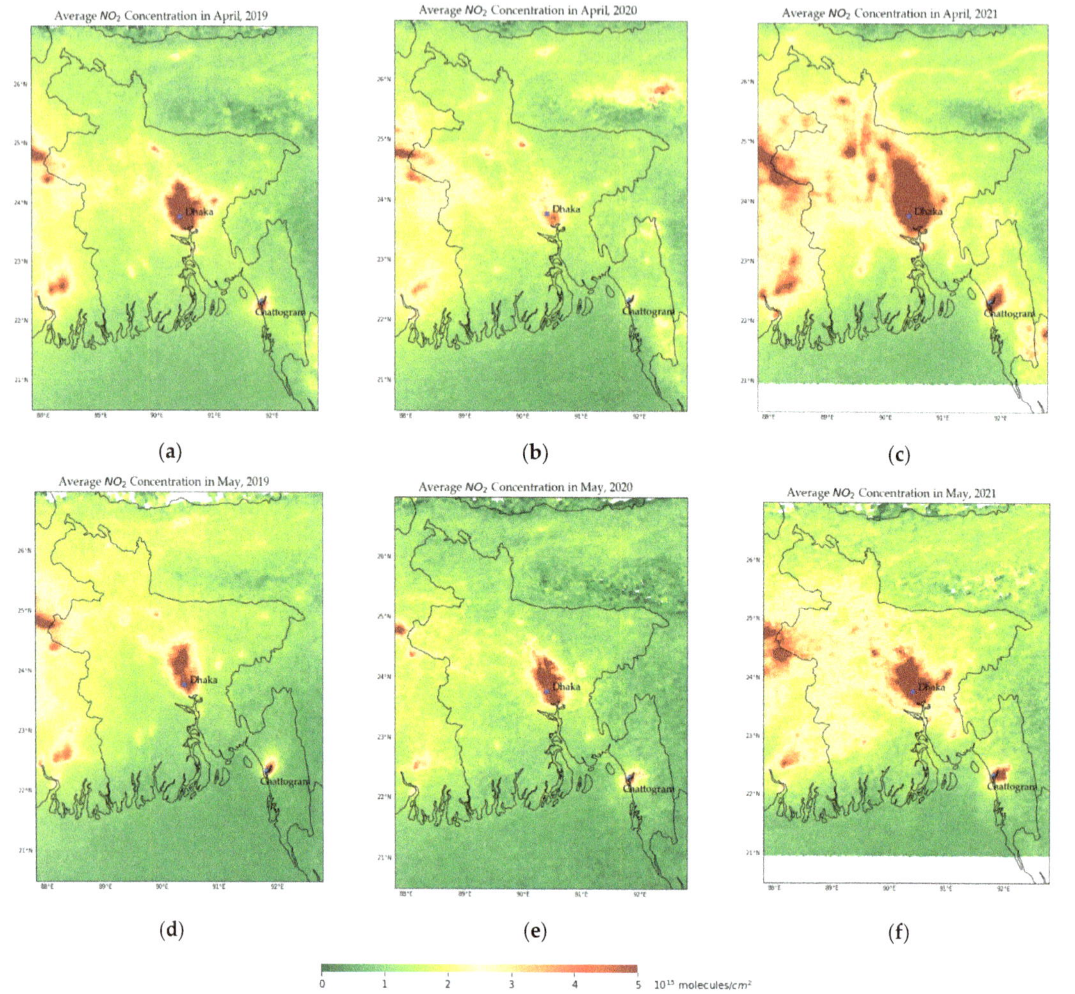

Figure 2. Average NO_2 pollution over Bangladesh in April (**a**) 2019, (**b**) 2020, and (**c**) 2021 and May (**d**) 2019, (**e**) 2020, and (**f**) 2021. Bangladesh saw a drastic drop of NO_2 pollution all over the country. Pollution decreased in April and May 2020 but increased in 2021.

In May 2020, the strictness of lockdown was compromised by the government. Transports such as private cars and buses began to return to the road and industries started their working process slowly. This was reflected in the NO_2 concentration across the country. The concentration of NO_2 increased compared to that in April 2020 but it was still less polluted compared to the same period of NO_2 pollution in 2019. In 2021, the lockdown was compromised and public transports started taking half of its capacity. This was reflected in the NO_2 pollution throughout the county.

3.2. NO_2 Concentration over Selected Cities: Dhaka and Chattogram

Dhaka and Chattogram districts were found to be hotspots of NO_2 pollution in Bangladesh as can be seen from Figures 2–4. We had analyzed the concentration of NO_2 in these two specific cities to understand the difference and impact of lockdown in detail. The time series of weekly average tropospheric NO_2 vertical column number density derived from the Sentinel-5P product shows that Dhaka city underwent a significant drop of NO_2 concentration from 26 March to 5 May 2020 compared to 2019. After 5 May, the concentration increased in the next week and it continued to increase until the end of the study period. In contrast, the concentration of NO_2 was higher than that in 2019 and 2020 from 26 March until 26 May. After 26 May, there was a drop of NO_2 concentration until the study period. Figure 3 shows the time series of NO_2 pollution over the Dhaka district.

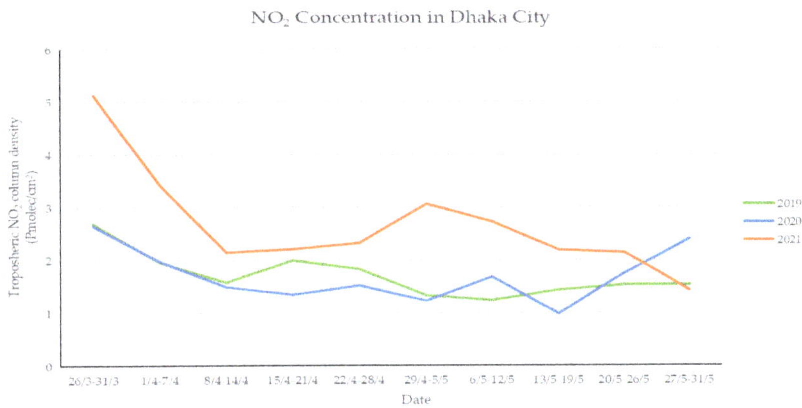

Figure 3. NO_2 Concentration in Dhaka city during the study period on weekly average.

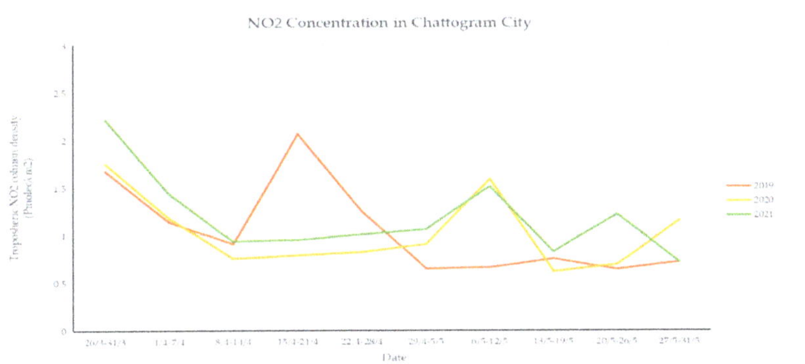

Figure 4. NO_2 concentration in Chattogram city during the study period on weekly average.

Chattogram district showed a different trend compared to Dhaka district. The concentration of NO_2 was observed to decline in 2020 compared to 2019 until the sixth week of the study period, which was 29 April–5 May. After that period, the concentration fluctuated compared with 2019.

In 2021, the concentration was very high in both 2019 and 2020 throughout the study period. There was not much of a decline observed in NO_2 concentration in 2021 compared to that in 2019 and 2020 in Chattogram. Figure 4 shows the time series of weekly average NO_2 concentration in Chattogram city.

4. Conclusions

This study was carried out to examine the variations in NO_2 concentrations during the lockdowns in the years 2020 and 2021. At the time of the strict lockdown in April 2020, there was a significant variation in the concentration of NO_2 compared with 2019. NO_2 levels had dropped dramatically across the country, including in Dhaka and Chattogram. However, when the lockdown was loosened by the government and people started their daily lives as before, the concentration of NO_2 again became higher than in the lockdown period.

Overall, it has been found that the strict lockdown imposed in 2020 showed a decrease in NO_2 pollution. However, during the second lockdown in 2021, the lockdown was only partially implemented and thus did not impact much on the NO_2 pollution.

Continuous data collection is quite difficult for developing countries such as Bangladesh as it requires a lot of effort and costs. The Sentinel-5P mission provides continuous and various types of data that can be useful for monitoring air quality and air pollution in Bangladesh. In the future, Sentinel-5P and remote sensing analysis can be implemented effectively to monitor and control air pollution in Bangladesh.

Author Contributions: Conceptualization, S.M.S.R. and S.M.F.A.; methodology, S.M.S.R.; software, S.M.S.R.; formal analysis, S.M.S.R.; investigation, S.M.S.R.; writing—original draft preparation, S.M.S.R., S.M.F.A. and H.A.; writing—review and editing, S.M.F.A. and H.A. All authors have read and agreed to the published version of the manuscript.

Funding: This research received no external funding.

Data Availability Statement: Data can be made available upon request.

Acknowledgments: We thank the ESA for providing Sentinel-5P products for free.

Conflicts of Interest: The authors declare no conflict of interest.

References

1. Venter, Z.; Aunan, K.; Chowdhury, S.; Lelieveld, J. COVID-19 lockdowns cause global air pollution declines. *Proc. Natl. Acad. Sci. USA* **2020**, *117*, 18984–18990. [CrossRef] [PubMed]
2. Fu, F.; Purvis-Roberts, K.; Williams, B. Impact of the COVID-19 Pandemic Lockdown on Air Pollution in 20 Major Cities around the World. *Atmosphere* **2020**, *11*, 1189. [CrossRef]
3. Kumari, P.; Toshniwal, D. Impact of lockdown on air quality over major cities across the globe during COVID-19 pandemic. *Urban Clim.* **2020**, *34*, 100719. [CrossRef] [PubMed]
4. Briz-Redón, Á.; Belenguer-Sapiña, C.; Serrano-Aroca, Á. Changes in air pollution during COVID-19 lockdown in Spain: A multi-city study. *J. Environ. Sci.* **2021**, *101*, 16–26. [CrossRef] [PubMed]
5. Islam, M.; Chowdhury, T. Effect of COVID-19 pandemic-induced lockdown (general holiday) on air quality of Dhaka City. *Environ. Monit. Assess.* **2021**, *193*, 343. [CrossRef] [PubMed]
6. Qiu, Z.; Ali, M.; Nichol, J.; Bilal, M.; Tiwari, P.; Habtemicheal, B.; Almazroui, M.; Mondal, S.; Mazhar, U.; Wang, Y.; et al. Spatiotemporal Investigations of Multi-Sensor Air Pollution Data over Bangladesh during COVID-19 Lockdown. *Remote Sens.* **2021**, *13*, 877. [CrossRef]
7. Dumka, U.C.; Tiwari, S.; Kaskaoutis, D.G.; Soni, V.K.; Safai, P.D.; Attri, S.D. Aerosol and Pollutant Characteristics in Delhi during a Winter Research Campaign. Environ. *Sci. Pollut. Res.* **2019**, *26*, 3771–3794. [CrossRef] [PubMed]
8. Venter, Z.; Aunan, K.; Chowdhury, S.; Lelieveld, J. Air pollution declines during COVID-19 lockdowns mitigate the global health burden. *Environ. Res.* **2021**, *192*, 110403. [CrossRef] [PubMed]
9. Vîrghileanu, M.; Săvulescu, I.; Mihai, B.; Nistor, C.; Dobre, R. Nitrogen Dioxide (NO_2) Pollution Monitoring with Sentinel-5P Satellite Imagery over Europe during the Coronavirus Pandemic Outbreak. *Remote Sens.* **2020**, *12*, 3575. [CrossRef]
10. Islam, M.S.; Tusher, T.R.; Roy, S.; Rahman, M. Impacts of nationwide lockdown due to COVID-19 outbreak on air quality in Bangladesh: A spatiotemporal analysis. *Air Qual. Atmos. Health* **2020**, *14*, 351–363. [CrossRef] [PubMed]
11. Mamun, M.I. The Seasonal Variability of Aerosol Optical Depth over Bangladesh Based on Satellite Data and HYSPLIT Model. *Am. J. Remote Sens.* **2014**, *2*, 20. [CrossRef]
12. Mahmood, S. Air pollution kills 15,000 Bangladeshis each year: The role of public administration and governments integrity. *J. Public Adm. Policy* **2011**, *3*, 129–140. [CrossRef]
13. TROPOMI. Available online: https://sentinel.esa.int/web/sentinel/user-guides/sentinel-5p-tropomi (accessed on 14 June 2021).
14. Ialongo, I.; Virta, H.; Eskes, H.; Hovila, J.; Douros, J. Comparison of TROPOMI/Sentinel-5 Precursor NO2 observations with ground-based measurements in Helsinki. *Atmos. Meas. Tech.* **2020**, *13*, 205–218. [CrossRef]
15. Serco Italia SPA. Monitoring Pollution with Sentinel-5p (Version 1.1). Retrieved from RUS Lectures. 2020. Available online: https://rus-copernicus.eu/portal/the-rus-library/learn-by-yourself (accessed on 10 September 2021).

engineering proceedings

MDPI

Proceeding Paper

Unveiling Naturally Occurring Green Tea Polyphenol Epigallocatechin-3-Gallate (EGCG) Targeting *Mycobacterium* DPRE1 for Anti-Tb Drug Discovery †

Suraj N. Mali * and Anima Pandey

Department of Pharmaceutical Sciences and Technology, Birla Institute of Technology, Mesra 835215, Jharkhand, India; apandey@bitmesra.ac.in

* Correspondence: mali.suraj1695@gmail.com; Tel.: +91-9657330138

† Presented at the 2nd International Electronic Conference on Applied Sciences, 15–31 October 2021; Available online: https://asec2021.sciforum.net/.

Abstract: The increasing rates of multidrug-resistant (MDR) and extremely drug-resistant (XDR) cases of tuberculosis (TB) strains are alarming, and eventually hampered the effective control of the pathogenic disease. Epigallocatechin gallate (EGCG) is a major polyphenolic constituent of green tea, which previously demonstrated in vitro potency against TB strains. However, efforts to elucidate the exact mechanism of the interaction are still ongoing. Aiming to elucidate the probable mechanism of its anti-TB action as decaprenylphosphoryl-beta-D-ribose 2′-epimerase (DprE) inhibition, we conducted a molecular modeling analysis. Our molecular docking analysis for a set of 65 bioactive compounds of tea realized that EGCG has the highest binding affinity (docking score: −142.98 kcal/mol) against DPRE (pdb id: 4p8c) from *Mycobacterium tuberculosis*. Further, a molecular dynamics analysis of 100 ns resulted in extreme stability of the ligand–protein complex. We further assessed the in silico pharmacokinetics and toxicities of the top three green tea polyphenols, based on the docking scores. Our results provide critical insights into the mechanism of action of EGCG and other green tea polyphenols, and their use as potential therapeutic agents (DprE1) against TB.

Keywords: tuberculosis; *Mycobacterium*; EGCG; green tea polyphenols; DprE1

Citation: Mali, S.N.; Pandey, A. Unveiling Naturally Occurring Green Tea Polyphenol Epigallocatechin-3-Gallate (EGCG) Targeting *Mycobacterium* DPRE1 for Anti-Tb Drug Discovery. *Eng. Proc.* **2021**, *11*, 31. https://doi.org/10.3390/ASEC2021-11185

Academic Editor: Bettina Wolf

Published: 15 December 2021

Publisher's Note: MDPI stays neutral with regard to jurisdictional claims in published maps and institutional affiliations.

1. Introduction

Tuberculosis (TB) is considered to be a public health crisis, which is hampering healthcare systems, especially in low-economic countries [1–5]. As per the latest WHO statistics, 1.4 million people died from TB in 2019 (10 million people fell ill) [2]. Furthermore, considering the increasing numbers of multidrug-resistant TB (MDR-TB), or multidrug- or rifampicin-resistant TB (MDR/RR-TB) cases, there is an urgent need to develop newer anti-TB agents with unique mechanisms of action. The mycobacterial cell wall is made up of mycolic acids, which are long fatty acids. The synthesis of mycolic acid is regulated via enzymes of the fatty acid synthase (FAS) complex. Enoyl reductase, pantothenate synthetase, and decaprenylphosphoryl-β-d-ribose 2′-epimerase (DprE1) are key attractive targets for the discovery of newer anti-TB agents. From our literature analysis, it is clear that polyphenols (Figure 1) from the leaves of green tea (GTPs) and black tea (*Camellia sinensis*) have significant pharmacological potentials against a variety of biological targets [3]. Recently, Anand et al. (2006) showed the potential of GTP, epigallocatechin-3-gallate, to inhibit *Mycobacterium tuberculosis* survival within human macrophages [5]. Their study suggested that epigallocatechin-3-gallate has the ability to down-regulate tryptophan aspartate-containing coat protein gene transcription [5]. Sun et al. (2015) proposed a probable mechanism for the effects of epigallocatechin gallate (EGCG) on the growth of *Mycobacterial smegmatis* mc(2)155 [4]. Moreover, their analysis revealed that EGCG had an impact on the integrity of the mycobacterial cell wall. The probable targets of EGCG for the inhibition of *Mycobacterium tuberculosis* have not been studied in detail, although a few reports have mentioned some antimycobacterial activity based on enzyme inhibition studies utilizing compounds such as triclosan. Considering these facts, we particularly aimed to explore the probable anti-TB drug target for epigallocatechin gallate (EGCG), a major bioactive compound from green tea extracts [6]. Furthermore, we have also calculated docking affinity scores for other GTPs. The top three GTPs were subjected to in silico absorption, distribution, metabolism, excretion (ADME) and toxicity (T) analyses. Finally, the best docked hit, epigallocatechin gallate (EGCG), the 4P8C complex, was simulated and analyzed using 100 ns molecular dynamics analysis.

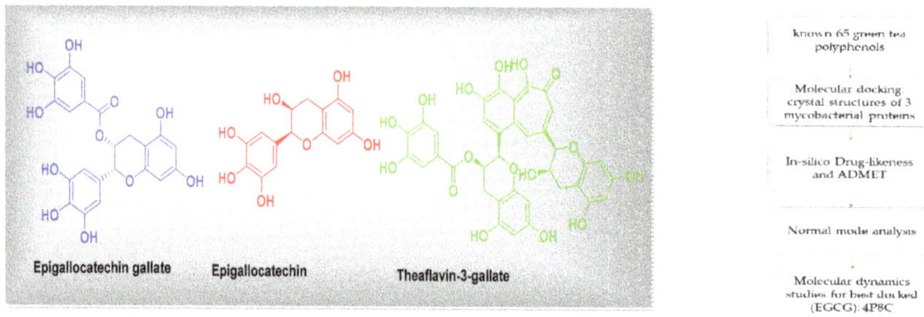

(**a**) (**b**)

Figure 1. (**a**) Chemical structures of green tea polyphenols (representatives) and (**b**) the graphical summary of workflow followed.

Lastly, we have signified a probable lead that could be developed as a drug candidate against mycobacterial targets.

2. Materials and Methods

2.1. Molecular Docking Simulations

For the current study, we have taken a set of 65 known green tea polyphenols or compounds [7]. Overall, molecular docking comprises the following 5 main steps: (1) protein preparation, (2) ligand preparation, (3) receptor grid generation, (4) ligand docking procedure and then (5) viewing the docking results. All necessary structures have been drawn using 'ChemSketch'. All the 3D crystal structures of 3 mycobacterial proteins (the enoyl reductase receptor protein (PDB IDs: 2 × 22), decaprenylphosphoryl-β-d-ribose 2′-epimerase (DprE1), 4p8c, and the pantothenate synthetase, 3IVX) were downloaded from the protein database bank (PDB database, www.rcsb.org, accessed on 26 February 2021). The data set including the inbound (co-crystallized) ligand was docked into the binding pocket of pantothenate synthetase and DprE1 (PDB-ID: 3IVX, 4p8c) target enzymes. For the reliability of the docking protocol, we evaluated it through the RMSD value. The grid was centered around active binding site residues Gly158, Met195, Pro38, etc. (for case of 3ivx). We have performed our docking analysis using 'iGemDock' as per standard procedures by making 15 Å radius from binding pocket, followed by visualization using Discovery Studio 2020 Visualizer [8,9].

2.2. In Silico Drug Likeness and ADMET (Absorption, Distribution, Metabolism, Excretion, Toxicity) Analysis

Top 3 higher-affinity GTPs were subjected to an in silico ADME analysis using SWISS tools (http://www.swissadme.ch, accessed on 21 April 2021). For the toxicity assessments, we used 'admetSAR' (http://lmmd.ecust.edu.cn:8000/, accessed on 21 April 2021).

2.3. Normal Mode Analysis

To gain more insights into the conformational flexibilities [8] of proteins with their best docked hits, we have performed normal mode analysis (NMA) with internal coordinates (IC) using a fast and easy server, iMODS (http://imods.chaconlab.org/, accessed on 21 April 2021).

2.4. Molecular Dynamics Studies

The docking studies do not take into account the flexible nature of the protein. For confirming the exact binding mode and stability, we have to study MD simulations with the Desmond program. The stability of epigallocatechin gallate (EGCG), 4P8C complex, was evaluated through 100 ns molecular dynamics simulations. The simulations provided the exact binding interaction of the docking complex with system embedded with water molecules, temperature and pressure. The standard NTP conditions were followed for MD simulation setup. The complex was originated in all proper binding poses with an acceptable RMSD value (<3 Å). Molecular dynamics (MD) simulation for a period of 100 ns was carried out for best docked hit with epigallocatechin gallate (EGCG), 4P8C complex, and this was achieved with the GROMACS simulation package, 2020, as per known literature protocols [9].

3. Results and Discussion

3.1. Molecular Docking Simulations

Our molecular docking analysis of 65 GTPs on 3 anti-TB targets suggested that epigallocatechin gallate (EGCG) had the highest affinity towards the DprE1 target, rather than 2X22 and 3IVX. The docking interactions depicted that this compound had interactions with GLN A:328, TYR A:60, GLY A:334, LYS A:418, VAL A:365, and LEU A:317 amino acids, with seven conventional hydrogen bonds present at the receptor site of the target 4P8C (Figure 2) [9,10]. VAL A:365 formed a pi–sigma interaction with the aromatic portion. Alkyl and pi–alkyl interactions were also observed for LEU A:363 and LEU A:317 amino acids. In Figure 2, it is also revealed that the binding mode of EGCG has favorable H-bond donors (purple colored) and acceptable (green colored) regions in the binding cavity. Tables 1 and 2 give better insights into the interaction profiles of the studied green tea/black tea molecules. From our docking analysis of 65 bioactive compounds of tea on 4p8c, we found the top three best docked hits to be theaflavin-3-gallate (docking score: −124.06 kcal/mol), epigallocatechin gallate (docking score: −142.98 kcal/mol), and epigallocatechin (docking score: −127.73 kcal/mol). The docking affinities for these three bioactive compounds were found to be greater than the standard drug ciprofloxacin* (docking score: −109.20 kcal/mol). The standard drug isoniazid interacted with LYS A:367, VAL A:365, CYS A:387, and ASN A:385 amino acid residues (docking score: −121.21 kcal/mol). The binding cavity residues for the inbound ligands were found to be VAL A: 365, LYS A: 418, ASN A: 385, and LYS A: 418 (salt bridge). Moreover, our re-docking validation protocol also resulted in an RMSD value below 2 Å, suggesting that the docking results are valid.

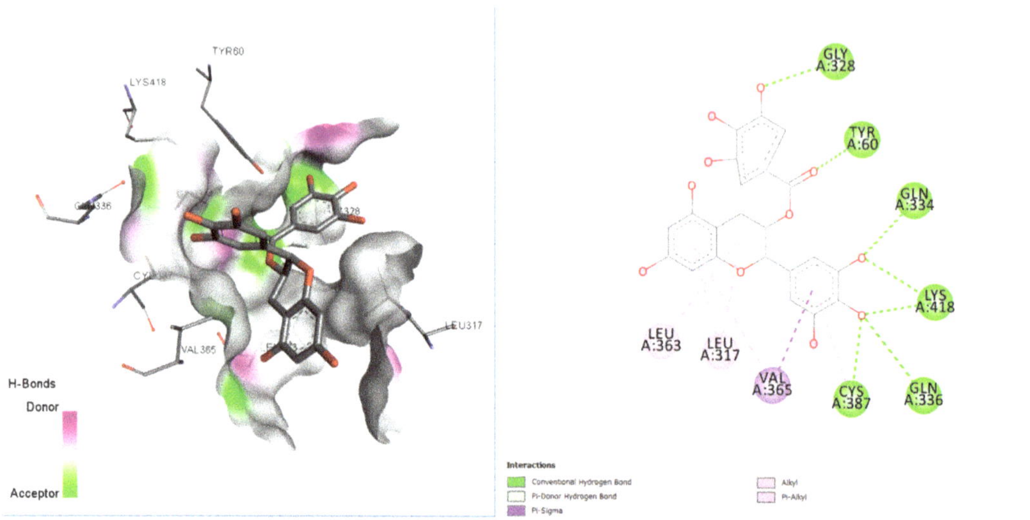

Figure 2. 2D and 3D interaction profiles for best docked epigallocatechin gallate with 4p8c.

Table 1. Docking interaction energies * of selected 65 bioactive molecules and 3 FDA-approved drugs for target protein 4P8C.

Molecules	-iGemDock Interaction Energy	Molecules	-iGemDock Interaction Energy
Oolonghomobisflavan A	−62.2219	Theaflavic Acid	−82.4934
Theasinensin D	−70.1619	Barrigenol R1	−86.4843
Theaflavin-3-gallate	−124.06	Barringtogenol	−52.0693
Isotheaflavin	−62.621	Camelliagenin	−95.1799
Epigallocatechin-3,5-Di-*O*-Gallate	−71.0176	Gallocatechin	−85.7374
Oolonghomobisflavan B	−75.4779	Catechin	−101.992
Cis-3-Hexenol	−63.5566	Epicatechin	−98.6033
Epigallocatechin-3,4-Di-*O*-Gallate	−92.6784	Epiafzelechin	−91.5357
Vicenin 2	−96.9806	Quercetin	−102.834
Epicatechin-3,5-Di-*O*-Gallate	−101.495	Cryptoxanthin	−92.1799
Rutin	−87.1416	Myricetin	−82.5936
Proanthocyanidin	−84.8129	Apigenin	−63.6163
Pheophytin	−90.2865	Nerolidol	−82.584
Benzaldehyde	−91.9877	Kaempferol	−89.1838
Epitheaflavic Acid 3′-Gallate	−65.361	Theanine	−73.9851
Epigallocatechin Gallate	−142.98	Ascorbic Acid	−80.1271
Theasinensin E	−62.6409	Quinic Acid	−75.3299
Myricitrin	−61.915	Succinic Acid	−85.5696
Theaflavin	−55.9704	Methyl Salicylate	−81.1848
Epicatechin Gallate	−72.5287	Theobromine	−72.7269
Kaempferitrin	−71.7401	Caffeine	−84.4502

Table 1. *Cont.*

Molecules	-iGemDock Interaction Energy	Molecules	-iGemDock Interaction Energy
Isoquercetin	−73.9058	Xanthine	−75.7595
Epiafzelechin 3-*O*-Gallate	−73.4119	Linalool Oxide	−83.9907
Pheophorbide	−71.1657	Phenylacetaldehyde	−87.8044
Epigallocatechin 3-*O*-P-Coumarate	−76.8643	Methylxanthine	−79.6185
Pheophorbide	−67.9266	Theophylline	−88.1319
Oxalic Acid	−82.9277	Geraniol	−95.2378
Cryptoxanthin	−81.2634	Hexanal	−95.8974
Isovitexin	−72.924	Diphenylamine	−93.4455
Vitexin	−55.6638	Trans−2−Hexenal	−94.076
Chlorogenic Acid	−49.7604	Linalool	−86.4307
Coumaroyl Quinic Acid	−94.7189	Phenylethanol	−101.468
Epigallocatechin	−127.73	Ciprofloxacin *	−109.20

* Docking scores have only been provided for the higher-affinity scored target protein.

Table 2. Energy contribution of the key residues computed by docking methodology.

Sr. No.	Molecules	Residues with Contribution Energy
1.	Isoniazide	LYS A:367; VAL A:365; CYS A:387; ASN A:385
2.	Pyrazinamide	VAL A: 365; LYS A: 418; ASN A: 385
3.	Ciprofloxacin	GLN A:117; VAL A:365; CYS A:387; LYS A:367; HIS A:132
4	Theaflavin-3-gallate (Best docked)	TRP A:16; LEU A:363; HIS A:315; THR A:118; LEU A:317
5	Epigallocatechin	GLN A:334; TYR A:60; CYS A:387; LYS A:418
6	Epigallocatechin Gallate (EGCG)	GLN A:328; TYR A:60; GLY A:334; LYS A:418; VAL A:365; LEU A:317
7	Inbound ligand (y22)	VAL A: 365; LYS A: 418; ASN A: 385; LYS A: 418 (salt bridge)

3.2. Molecular Dynamics Simulation and Normal Mode Analysis

The best docked hit, epigallocatechin gallate (EGCG), with the target 4P8C, was simulated for 100 ns molecular dynamics and normal mode analysis. The MD simulations showed that the obtained root mean square fluctuation (RMSF) values were within tolerable ranges (0.4 nm). The obtained root mean square deviation (RMSD) value was below 0.25 nm, suggesting that the complex was stable (Figure 3). From our NMA results, we observed that epigallocatechin gallate, with the protein 4P8C complex, was retained with good deformability and eigenvalue profiles (Figure 3). The radius of the gyration value was also below 2.15 nm. The accessible surface areas of the solvent were found to be around 180 nm^2.

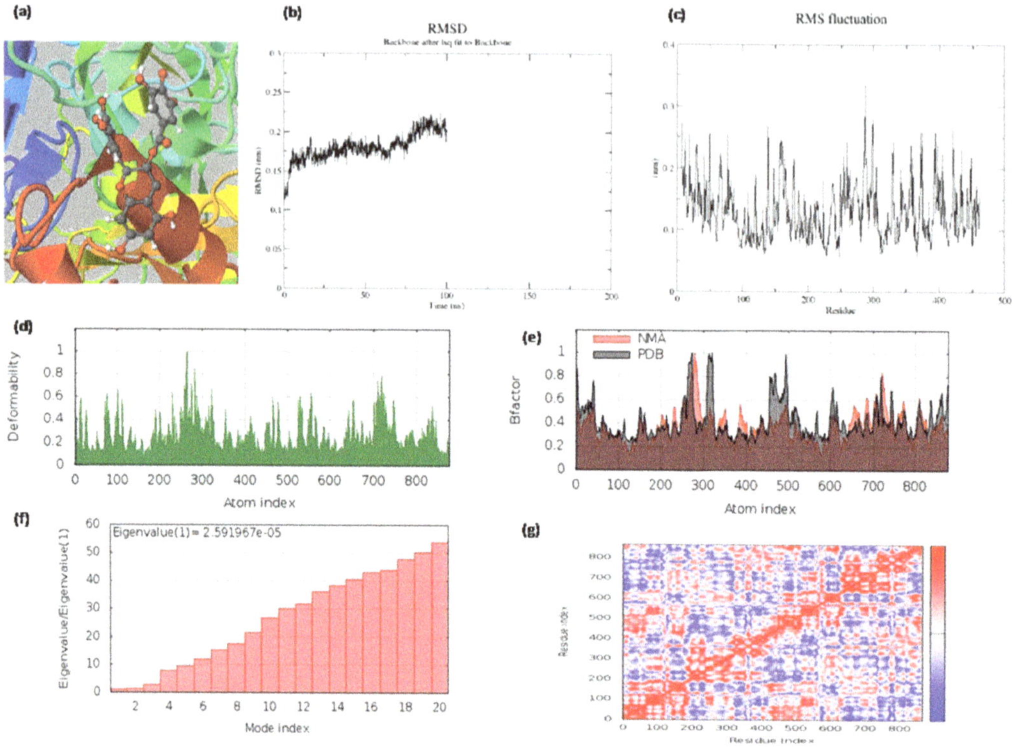

Figure 3. (**a**) The binding pocket; (**b**) the root mean square deviations (RMSD) of backbone atoms relative to the starting complexes during 100 ns MD; (**c**) protein RMSF plot (on this plot, peaks indicate areas of the protein that fluctuate the most during the simulation, and protein residues that interact with the ligand are marked with green-colored vertical bars); (**d**) deformability and (**e**) B-factor profiles; (**f**) eigenvalue and (**g**) covariance matrix of the complex for epigallocatechin gallate with 4p8c, respectively.

3.3. In Silico ADMET Studies

Cytochrome P450 (CYPs) enzymes are key enzymes responsible for various factors of metabolism. Our in silico calculated ADMET (absorption, distribution, metabolism, excretion, and toxicity) values, in the body properties evaluated, for the top three best docked hits are represented in Table 3. Three GTPs, theaflavin-3-gallate, epigallocatechin, and epigallocatechin gallate (EGCG), exhibited positive human intestinal absorption profiles, negative blood–brain barrier passage, and non-carcinogenic, non-AMES toxicity, and class IV acute oral toxicity profiles.

Table 3. In silico ADMET profiling for top 3 best docked hits against target 4P8C.

Properties	Theaflavin-3-Gallate	Epigallocatechin	Epigallocatechin Gallate (EGCG) *
CYP450 2C9 Substrate	Non-substrate	Non-substrate	Non-substrate
CYP450 2D6 Substrate	Non-substrate	Non-substrate	Non-substrate
CYP450 3A4 Substrate	Non-substrate	Non-substrate	Non-substrate
Human Ether-a-go-go-Related Gene Inhibition	Weak inhibitor	Weak inhibitor	Weak inhibitor
AMES Toxicity	Non-AMES toxicity	Non-AMES toxicity	Non-AMES toxicity
Carcinogens	None	None	None
Acute Oral Toxicity	IV	IV	IV
P-glycoprotein Inhibitor	Non-inhibitor	Non-inhibitor	Non-inhibitor
Rat Acute Toxicity (LD_{50}, mol/kg)	2.6693	1.8700	2.6643
Human Intestinal Absorption	+	+	+
AlogP	3.19	1.25	2.23
H-Bond Acceptor	16	7	11
H-Bond Donor	11	6	8
Tetrahymena pyriformis (pIGC50 (ug/L))	0.595	0.792	0.913
Blood–Brain Barrier	-	-	-

* Best docked.

4. Conclusions

In our current study, we noticed that epigallocatechin gallate (EGCG) has strong interactions with the 4p8c enzyme (docking score: −142.98 kcal/mol), and the following amino acid residues: GLN A:328; TYR A:60; GLY A:334; LYS A:418; VAL A:365; LEU A:317. The binding energy obtained for EGCG was higher when compared with the standard drug ciprofloxacin (docking score: −109.20 kcal/mol). Moreover, in silico ADMET analysis revealed that this compound has low human ether-a-go-go-related gene inhibition, no AMES toxicity, and no carcinogens. Considering a previous literature report on EGCG as an antimycobacterial, we investigated the same against the DprE1 enzyme via molecular docking analysis. Considering the benefits of GTPs, this study may provide further directions to develop more potent anti-TB compounds. Moreover, we also believe that synthetic analogues of EGCG could be tested for in vitro anti-TB potentials. The tea extract containing EGCG could also be tested in vitro for anti-TB enzymatic assessments.

Author Contributions: Conceptualization, S.N.M. and A.P.; methodology, S.N.M.; software, S.N.M.; writing—review and editing, S.N.M. and A.P.; visualization, S.N.M. and A.P.; supervision, A.P. All authors have read and agreed to the published version of the manuscript.

Funding: We wish to thank the Dept. of Pharmaceutical Sciences, Birla Institute of Technology, Mesra, India for financial assistance. Authors are thankful to the team 'Molegro virtual docker' and BIOVIA Discovery studio for their free trial licenses. S.M. is also thankful for the provision of IRF (PHD/PH/10006/20) (Ref. No. GO/Estb/Ph.D./IRF/2020-21/2484A) provided by BIT, Mesra, India.

Institutional Review Board Statement: Not applicable.

Informed Consent Statement: Not applicable.

Data Availability Statement: Not applicable.

Acknowledgments: The authors would like to thank the Head, Department of Pharmaceutical Sciences and Technology, BIT, Mesra for providing the research facilities for performing the current study.

Conflicts of Interest: The authors declare no conflict of interest.

References

1. Poro, K.E.; Hoekou, Y.; Pissang, P.; Kpabi, I.; Novidzro, K.M.; Dagnra, A.Y.; Tchacondo, T.; Batawila, K. In vitro Antimycobacterial Activity of Selected Medicinal Plants against Mycobacterium tuberculosis. *Int. J. Curr. Microbiol. App. Sci.* **2021**, *10*, 3201–3208.
2. WHO TB Factsheet. Available online: https://www.who.int/news-room/fact-sheets/detail/tuberculosis (accessed on 17 July 2021).
3. Paiva, L.; Rego, C.; Lima, E.; Marcone, M.; Baptista, J. Comparative analysis of the polyphenols, caffeine, and antioxidant activities of green tea, white tea, and flowers from Azorean Camellia sinensis varieties affected by different harvested and processing conditions. *Antioxidants* **2021**, *10*, 183. [CrossRef] [PubMed]
4. Sun, T.; Qin, B.; Gao, M.; Yin, Y.; Wang, C.; Zang, S.; Li, X.; Zhang, C.; Xin, Y.; Jiang, T. Effects of epigallocatechin gallate on the cell-wall structure of Mycobacterial smegmatis mc2155. *Nat. Prod. Res.* **2015**, *29*, 2122–2124. [CrossRef] [PubMed]
5. Anand, P.K.; Kaul, D.; Sharma, M. Green tea polyphenol inhibits Mycobacterium tuberculosis survival within human macrophages. *Int. J. Biochem. Cell Biol.* **2006**, *38*, 600–609. [CrossRef] [PubMed]
6. Narayanan, S.; Ramesh, K.V. Epigallocatechin gallate, a green tea polyphenol inhibits Mycobacterium smegmatis: In silico and in vitro studies. *Indian J. Pharm. Sci.* **2017**, *79*, 625–632. [CrossRef]
7. Bhardwaj, V.K.; Singh, R.; Sharma, J.; Rajendran, V.; Purohit, R.; Kumar, S. Identification of bioactive molecules from tea plant as SARS-CoV-2 main protease inhibitors. *J. Biomol. Struct. Dyn.* **2020**, *39*, 3449–3458. [CrossRef] [PubMed]
8. Kovacs, J.; PChacón, R. Abagyan. Predictions of Protein Flexibility: First Order Measures. PROTEINS: Structure, Function, and Bioinformatics. *Proteins* **2004**, *56*, 661–668. [CrossRef]
9. Mali, S.N.; Pandey, A. Balanced QSAR and Molecular Modelling to Identify Structural Requirements of Imidazopyridine Analogues as Anti-infective Agents against Trypanosomiases. *J. Comput. Biophys. Chem.* **2021**, *21*, 83–114. [CrossRef]
10. Hsu, K.C.; Chen, Y.F.; Lin, S.R.; Yang, J.M. iGEMDOCK: A graphical environment of enhancing GEMDOCK using pharmacological interactions and post-screening analysis. *BMC Bioinform.* **2011**, *12*, S33. [CrossRef] [PubMed]

engineering proceedings

MDPI

Proceeding Paper

A Wearable Temperature Sensor Network to Address the COVID-19 Pandemic Emergency †

Antonietta Fiore [1], Salvador Tufano [2], Giovanni De Rosa [2], Carmela Maria Napolitano [2], Carla D'Antò [2,3,*] and Nunzio Cennamo [4]

1 FLOS Srl, Via Vincenzo Calvanese 60, 80021 Afragola, Italy; flossrlwebstore@gmail.com
2 Ministero dell'Istruzione, Viale Trastevere 76/A, 00153 Roma, Italy; salvador.tufano@istruzione.it (S.T.); giovanni.derosa2@istruzione.it (G.D.R.); carmelamaria.napolitano@istruzione.it (C.M.N.)
3 Associazione Nazionale Scuola Italiana, Viale Bruno Buozzi 99, 00197 Roma, Italy
4 Department of Engineering, University of Campania Luigi Vanvitelli, Via Roma 29, 81031 Aversa, Italy; nunzio.cennamo@unicampania.it
* Correspondence: carla.danto2@posta.istruzione.it; Tel.: +39-320-457-5910
† Presented at the 2nd International Electronic Conference on Applied Sciences, 15–31 October 2021; Available online: https://asec2021.sciforum.net/.

Abstract: In the current COVID-19 emergency, to reduce the infection risk, several types of body temperature sensors, e.g., thermal imaging cameras and infrared thermometers, have been used to monitor people who access enclosed public spaces. In some buildings, where people are located for several hours, continuous monitoring could be useful. For this reason, in three schools, we have proposed and tested a body temperature sensor network based on wearable temperature sensors monitored via Bluetooth 5.0 using smartphones and/or custom gateways. The data are collected on a server via the internet, and custom software is used to control the measured temperature and to produce warnings automatically.

Keywords: wearable sensor; body temperature monitoring; security; COVID-19 pandemic emergency; school; sensor network

Citation: Fiore, A.; Tufano, S.; Rosa, G.D.; Napolitano, C.M.; D'Antò, C.; Cennamo, N. A Wearable Temperature Sensor Network to Address the COVID-19 Pandemic Emergency. *Eng. Proc.* **2021**, *11*, 32. https://doi.org/10.3390/ASEC2021-11125

Academic Editor: Nicholas Vassiliou Sarlis

Published: 15 October 2021

1. Introduction

Following the COVID-19 emergency, face-to-face teaching activities relating to the 2019/2020 school year had to be suspended throughout Italy from March 2020, activating distance learning to protect the school staff and students' health and guaranteeing the continuation of study.

The following school year continued to alternate between closure and reopening. Various procedures have been taken to ensure the safe resumption of classes in person, such as physical distancing, the obligation to wear a mask, and body temperature measurement. All these practices are used to prevent infection. In particular, body temperature detection is a parameter that can be characteristic of active infections.

To reduce the infections of the SARS-CoV-2 virus, several types of body temperature sensors are used to monitor people's access to enclosed spaces. More specifically, thermal imaging cameras and infrared thermometers are extensively used for this aim. However, these approaches cause the concentration of people at checkpoints, increasing the risk of exposure. In addition, the risk of COVID-19 in some buildings, where the people are located for several hours, e.g., schools, universities, or industries, could be reduced by continuously monitoring body temperature.

In many cases, a high-precision and wearable sensors network is the only possible solution to implement punctual (capillary) and continuous real-time monitoring of one or more quantities of interest (physical or biochemical) to prevent a cluster of infections and to safeguard health.

Wearable sensors technology is growing as researchers and developers learn of its potential benefits and possibilities [1].

Many applications, e.g., electronics or smartphone applications, encourage physical activity by measuring the user's daily step count and other parameters. Wearable sensors can come in the form of a wearable sensor network, also known as a wearable body sensor network, or even a wearable health monitoring system (WHMS) [2].

In a wearable sensor network, multiple sensors can communicate to one another or with a server to form a network, with great potential, particularly in medical and safety applications. In fact, wearable sensor networks have been successfully used to monitor body parameters, such as fluid, ph, temperature, blood pressure, etc., leading to the opportunity to monitor our health during activity, with immediate feedback on a smartphone or a PC [3].

However, this approach is typically used only in physical activities, and it is realized exploiting a personal device (e.g., smartwatches).

Numerous scientists have been bringing forward different research lines on body temperature monitoring sensors for 10 years, proving that wearable body sensor networks are a potential tool for the real-time monitoring of vital parameters. Their flexibility, constant, and low-cost operation make them suitable for applications ranging from medical diagnosis, patient monitoring, and disease treatment. Many branches of medical science benefit from a non-invasive and low-cost tool that enables the accurate measurement of body temperature. Body sensor networks have the potential to improve medical research [4].

Important examples of wearable sensor networks are also the applications of this technology for temperature monitoring to investigate psychological and emotional changes in biomedical applications [5] or for temperature measurements in different environments for several application fields, as in [6–8].

In this work, we present the exploitation of a wearable sensor network for the continuous monitoring of school staff and students' body temperature to reduce the risk of SARS-CoV-2 infection in the school and thus allow for on-site learning activities. More specifically, we have designed a sensing strategy based on a body temperature sensor network and its specific application to address the COVID-19 emergency in a school scenario. After that, we installed approximately 4500 devices in three schools, and its test is in progress. The three schools are located in the province of Naples (Italy), and it could be extended to other schools.

2. A Sensor Network for Body Temperature Detection

We have proposed and tested a wearable body temperature sensor to realize an interesting sensor network used to implement the continuous and real-time temperature monitoring of people in school buildings. The used temperature sensors measure the parameter of interest and transmit it via Bluetooth 5.0 to smartphones and/or custom gateways (with a maximum of about 150 per gateway and about 50 per smartphone). Then, the collected data are transmitted on a server via the internet.

As shown in Figure 1, the wearable sensor system is a small-size chip with an accuracy of ± 0.1 °C, with a detectable range between 32 °C and 43 °C that can be inset into a specific rubber bracelet for continuous and real-time temperature monitoring and data transmission. In detail, the used temperature sensor (model "Jw1656") falls into the "Bluetooth beacons" category. The latter includes BLE (Bluetooth low energy) class devices that are able to send, at regular intervals, a signal (for instance, hexadecimal values) that can be detected by other devices.

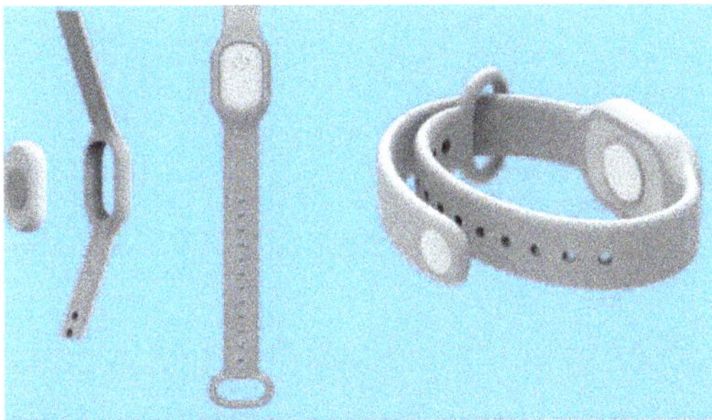

Figure 1. The packaging of the implemented wearable temperature sensor system.

The sensor system presents a communication system and other components that are integrated onboard. In its entirety, the system is composed of a frontend (Dashboard), realized through a Microsoft development environment, and a frontend (server) based on a Linux operating system (v. 8.x) with microservices logic in PHP language 7.x. The frontend permits us to handle and monitor the system. In particular, it is possible to set reference and warning temperature values, as well as alerts (text messages) to recipients; in addition, it is also possible to visualize some statistics related to temperature data acquisitions. In contrast, the backend allows us to store, handle, and process the collected data.

This sensor system chip is waterproof, and it does not provide a display to reduce battery consumption. Moreover, the smartphone app (available for IoS and Android) not only transmits the data but also displays the temperature of all the sensors connected to it.

Figure 2 show an outline of the sensor network in terms of communication architecture and capabilities, together with images of the software interface. The flow of information exchanges can be summarized as follows: the BLE sensor acquires the physical quantity (temperature) and sends the data to a smartphone. The latter, through an internet connection (TCP/IP protocol), transmits the data to the server. In detail, the data exchange between the sensor and smartphone takes place through BLE protocol (Bluetooth "2010" version 4.0). This protocol implements the master/slave principle: the master (i.e., the smartphone in our case) can handle multiple connections at the same time with different slaves (i.e., the temperature sensors), whereas each slave can only be connected to one master at a time. We have also reported more detail about the data structure in Appendix A.

The smartphone (master) keeps waiting on the channel where the sensors (slaves) send their advertising. Once the data package is received, the smartphone sends it to the server through rest calls in https with Basic Auth. About the data collection, we have chosen MySQL 8.0 (with relational architecture) as a database management system (DBMS).

To obtain a body temperature sensor network that is able to conduct constant and safe monitoring, the developed software is used to analyze the database present on the server, to control the measured temperature, and to produce warnings automatically.

The sensor network is provided with a web interface to view the data stored on the server. In particular, each school can see only the data relative to its students and staff, even if the database is the same for all to allow statistical analysis, etc. It is important to underline that the data collection is regulated by the European General Data Protection Regulation (EU) 2016/679.

More specifically, we are testing this sensor network in three schools: IPSEOA "C. Russo" in Cicciano (Naples), IPSSAR "Di Gennaro" in Vico Equense (Naples), and IS "A. Torrente" in Casoria (Naples).

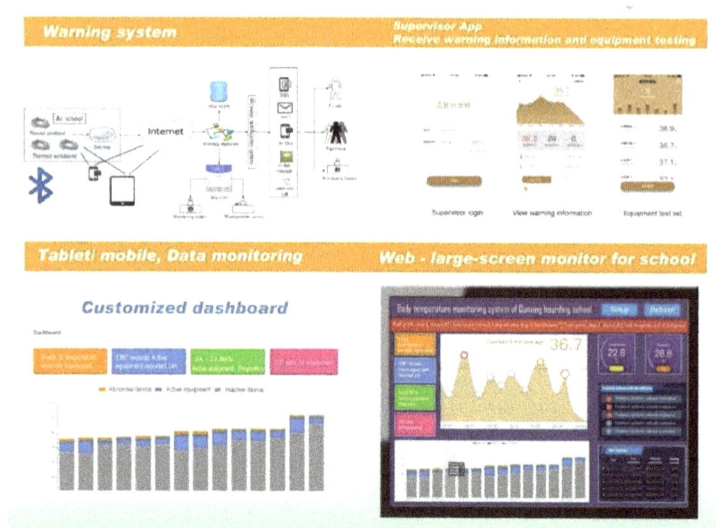

Figure 2. Outline of the sensor network and images of the software interface.

In each school, in addition to the installation of the app on smartphones and tablets, we installed a monitoring gateway covering the entire school. Moreover, we set up a background service system running in the cloud.

All students and staff members were assigned bracelets. Each wearer is a monitored object and, exploiting the smartphone app, can monitor other people connected in a range of 100 m.

3. Conclusions

We have presented an innovative strategy for continuous and real-time body temperature monitoring using a sensor network based on simple wearable sensors inserted into bracelets. This sensing approach is presently tested in three different schools. It exploits the real-time notification of the monitoring system and allows immediate isolation of the person with abnormal temperature after an alert notification.

Moreover, it could be the first paradigm for monitoring body parameters in a community exploiting a sensor network; in fact, the same sensor network system can be used to measure different parameters of interest for diagnostics and health protection. In this sense, the proposed paradigm is helpful, considering effective and far-reaching "pandemic plans". Furthermore, this technology could also monitor students' health looking for congenital pathologies and/or possible infections in progress, simply by using suitable sensors to measure the markers of interest.

Author Contributions: Conceptualization, C.D., A.F. and N.C.; methodology, C.D., A.F., N.C., G.D.R., C.M.N. and S.T.; validation, C.D., A.F., N.C., G.D.R., C.M.N. and S.T.; formal analysis, C.D., A.F., N.C., G.D.R., C.M.N. and S.T.; investigation, C.D., A.F., N.C., G.D.R., C.M.N. and S.T.; resources, G.D.R., C.M.N. and S.T.; data curation, C.D., A.F., N.C., G.D.R., C.M.N. and S.T.; writing—original draft preparation, C.D., A.F., N.C., G.D.R., C.M.N. and S.T.; writing—review and editing, C.D. and N.C.; supervision, C.D. and N.C. All authors have read and agreed to the published version of the manuscript.

Funding: This research received no external funding.

Institutional Review Board Statement: Not applicable.

Informed Consent Statement: Not applicable.

Data Availability Statement: The data is available on reasonable request from the corresponding author.

Acknowledgments: This work was supported by Ministero dell'Istruzione (Rome, Italy).

Conflicts of Interest: The authors declare no conflict of interest.

Appendix A

For the sake of completeness, in this section, we report more information regarding the data structure that the BLE sensor sends to the smartphone. The data format can be expressed as schematically represented in Figure A1, where four different parts can be distinguished. Please note that, in all parts, all numbers are expressed in hexadecimals.

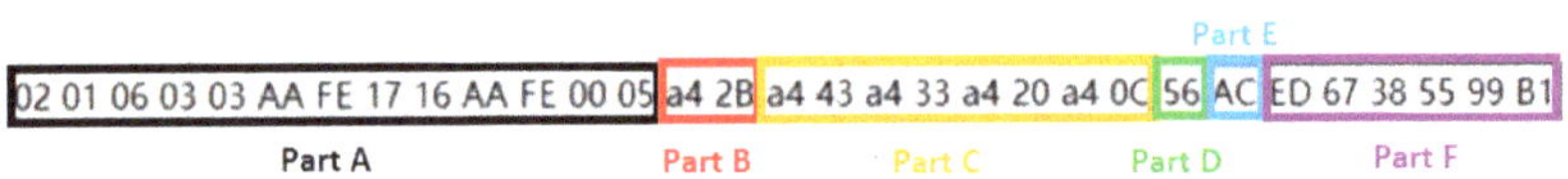

Figure A1. Schematization of the BLE sensor data structure.

In particular:

- "Part A" is reserved for "Eddystone Protocol", a specification that defines a BLE message format for proximity beacon messages;
- "Part B" deals both with the current temperature value and the time interval with which this physical quantity is acquired;
- "Part C" is relative to the history of previous temperature records;
- "Part D" is relative to the voltage;
- "Part E" is reserved as "AC" and could be used to identify data from multiple data with Bluetooth Eddystone protocol;
- "Part F" is the MAC address of the Bluetooth chip in the device.

References

1. Pantelopoulos, A.; Bourbakis, N.G. A Survey on Wearable Sensor-Based Systems for Health Monitoring and Prognosis. *IEEE Trans. Syst. Man Cybern. Part C (Appl. Rev.)* **2010**, *40*, 1–12. [CrossRef]
2. Olson, J.S.; Redkar, S. A survey of wearable sensor networks in health and entertainment. *MOJ Appl. Bionics Biomech.* **2018**, *2*, 280–287. [CrossRef]
3. Caldara, M.; Colleoni, C.; Guido, E.; Rosace, G.; Re, V.; Vitali, A. A wearable sensor platform to monitor sweat pH and skin temperature. In Proceedings of the IEEE International Conference on Body Sensor Networks, Cambridge, MA, USA, 6–9 May 2013.
4. Boano, C.A.; Lasagni, M.; Romer, K.; Lange, T. Accurate Temperature Measurements for Medical Research using Body Sensor Networks. In Proceedings of the 14th IEEE International Symposium on Object/Component/Service-Oriented Real-Time Distributed Computing Workshops, Newport Beach, CA, USA, 28–31 March 2011.
5. Li, Q.; Zhang, L.N.; Tao, X.M.; Ding, X. Review of flexible temperature sensing networks for wearable physiological monitoring. *Adv. Healthc. Mater.* **2017**, *6*, 1601371. [CrossRef] [PubMed]
6. Mikhaylov, K.; Tervonen, J.; Heikkilä, J.; Känsäkoski, J. Wireless Sensor Networks in industrial environment: Real-life evaluation results. In Proceedings of the 2nd Baltic Congress on Future Internet Communications, Vilnius, Lithuania, 25–27 April 2012; pp. 1–7.
7. Pérez, C.A.; Jiménez, M.; Soto, F.; Torres, R.; López, J.A.; Iborra, A. A system for monitoring marine environments based on Wireless Sensor Networks. In Proceedings of the OCEANS 2011 IEEE, Santander, Spain, 6–9 June 2011; pp. 1–6.
8. Low, K.S.; Win, W.N.N.; Er, M.J. Wireless Sensor Networks for Industrial Environments. In Proceedings of the International Conference on Computational Intelligence for Modelling, Control and Automation and International Conference on Intelligent Agents, Web Technologies and Internet Commerce (CIMCA-IAWTIC'06), Vienna, Austria, 28–30 November 2005; pp. 271–276.

engineering proceedings

MDPI

Proceeding Paper

A Comparative Case Analysis of Meteorological and Air Pollution Parameters between a High and Low Port Activity Period in Igoumenitsa Port †

Ioannis Logothetis [1,*], Christina Antonopoulou [1], Konstantinos Sfetsioris [2], Adamantios Mitsotakis [1] and Panagiotis Grammelis [1]

1 Centre for Research and Technology Hellas, Chemical Process and Energy Resources Institute, Thermi, GR 57001 Thessaloniki, Greece; antonopoulou@certh.gr (C.A.); adamis@certh.gr (A.M.); grammelis@certh.gr (P.G.)
2 Hellenic Association of Energy Economics, 145 62 Athens, Greece; sfetsioris@certh.gr
* Correspondence: logothetis@certh.gr

† Presented at the 2nd International Electronic Conference on Applied Sciences, 15–31 October 2021; Available online: https://asec2021.sciforum.net/.

Abstract: Among the most important trade bridges for Western Europe and the Balkans is the Port of Igoumenitsa in Western Greece, which experiences heavy traffic of goods and passengers. This case analysis primarily aims to investigate the effect of shipping and port operations on air quality in the Port of Igoumenitsa, comparing two representative short case seasons of air quality measurement campaigns. The high activity season ranges from 25 to 31 August 2018, while the low season ranges from 17 to 25 May 2018. A mobile air quality monitoring system was used to perform the analysis. To compare the air quality of the examined periods, the wind speed rose diagrams, the correlation analysis, as well as the hourly variations in concentrations of pollutants and meteorological parameters, were studied. In addition, the impact of meteorology and atmospheric circulation on local air quality were investigated. For study implementation, various types of data, obtained from NASA Worldview application, Barcelona Supercomputing Center (BSC), and Giovanni online system (NASA-GES-DISC) were included. The results indicate that port operations affect the air quality in the Port of Igoumenitsa, as the high season showed higher concentrations of air pollutants compared to the low season. A notable exception was the concentration of PMs, which was affected by an African dust transfer event during the low season. Finally, the findings indicate that climatic factors affect the pollution levels of the case analysis, and emphasize the importance of developing a green and sustainable management system within the port.

Keywords: air quality; shipping; port activity; climate; meteorology; synoptic atmospheric circulation; African dust transfer; eastern Mediterranean; green and sustainable port management

Citation: Logothetis, I.; Antonopoulou, C.; Sfetsioris, K.; Mitsotakis, A.; Grammelis, P. A Comparative Case Analysis of Meteorological and Air Pollution Parameters between a High and Low Port Activity Period in Igoumenitsa Port. *Eng. Proc.* **2021**, *11*, 33. https://doi.org/10.3390/ASEC2021-11118

Academic Editor: Nunzio Cennamo

Published: 15 October 2021

1. Introduction

The port activities and shipping are a major source of pollutants for the coastal areas, affecting not only air quality and its degradation [1,2] but also human health [3,4]. Although the port-related pollutants are not the dominant source of marine pollution, they are related to health problems and have an impact on mortality levels [5,6]. In particular, it turns out that human health problems are associated with exposure to $PM_{2.5}$ emitted by shipping [1,7]. Evidently, emissions from shipping contribute significantly to the NO_x, SO_2, and PM concentrations over port and coastal regions, while between 2007 and 2012, shipping accounted for about 1 billion tons of GHGs [8]. The IMO's 4th Greenhouse Gas Study [9] projected that the GHGs shipping emissions will increase up to 50% by 2050 compared to the 2018 levels. Consequently, increasing shipping and port activities contribute to climate change [10]. In particular, the increased concentration of pollutants in the atmosphere affects the weather and climate [11], producing positive climate forcing and enhancing the climate risk.

The atmospheric circulation is also a significant factor that affects the air quality [2,7]. The transfer and transformation of pollutants from Europe as well as the mineral dust from North Africa affect the pollution levels of the southeastern Mediterranean [12]. In particular, the peak of dust transfer is usually identified during the autumn and spring season affecting the concentration of PMs [13]. Finally, dust episodes have an impact on the climate since they modify the albedo and solar radiation on the earth's surface [12].

The port of Igoumenitsa is an important trade hub between Western and Eastern Europe, which connects the main Italian ports with Western Greece. About 1.5 million passengers and 140,000 trucks move through the port each year. Based on the lack of studies regarding the effect of shipping and atmospheric features (meteorology and synoptic atmospheric circulation) on the air quality in Western Greece, this analysis seeks to fill this gap. A comparison of the two seasons is characterized by the high and low port activity, respectively. Finally, this study highlights the importance of a larger measurement campaign in order to further improve the knowledge regarding the impact of shipping on the coastal port region of Igoumenitsa.

2. Methods and Data

Igoumenitsa (39.50615° N, 20.265534° E) is located in Western Greece (Figure 1) and it is a commercial hub for Southeastern Europe. The port of the city has high trade activity as a large number of passengers, cars, and trucks use it on an annual basis. A mobile air quality monitoring station, equipped with ambient air pollution sensors (HORIBA sensors) and in accordance with the EN regulations for certified measurements, was employed at the port of Igoumenitsa. Sensor recordings include meteorological parameters (wind direction—WDir; °, wind speed—WS; m/s, temperature—T; °C, relative humidity; RH—%, pressure—P; hPa), particle matters (PM_1, $PM_{2.5}$, PM_4, PM_{10}, $PM_{\_tot}$; μg/m^3), and gases (CO; ppm, NO; ppb, NO_2; ppb, NO_x; ppb, O_3; ppb, SO_2; ppb).

Figure 1. The location of the mobile air quality monitoring station in Igoumenitsa port (red star).

To study the impact of harbor operation in the port of Igoumenitsa during 2018, two cases were examined. The first case, with high port activity (HS), is considered the period from 25 to 31 August 2018, and the second, with low port activity (LS), is from 17 to 25 May 2018. In order to investigate the relation between the meteorology and concentration of pollutants, the Spearman correlation is employed for both studied seasons. Moreover, the hourly variation of meteorology parameters and concentration of pollutants are calculated to study the effect of hourly port activity during the two seasons. To examine the effect of synoptic atmospheric circulation on concentrations of PMs in the port, a satellite image provided by NASA Worldview application is also used (Figure 2a). Additionally, a figure provided by the DREAM8b model, operated by the Barcelona Supercomputing Center, is shown (Figure 2b). Results from MERRA-2 model version 2 (M2T1NXAER v5.12.4), available from the Giovanni online data system generated by NASA Goddard Earth Sciences (GES) Data and Information Services Center (DISC) [14], are also included in the analysis. Finally, the $PM_{2.5}/PM_{10}$ ratio (R_{PM}) is usually used to characterize the origin of pollutants (the anthropogenic or natural sources) [2,15]. In order to investigate the impact of shipping traffic and port operations (anthropogenic activities) in the port of Igoumenitsa, the R_{PM} is calculated for both case seasons (HS and LS).

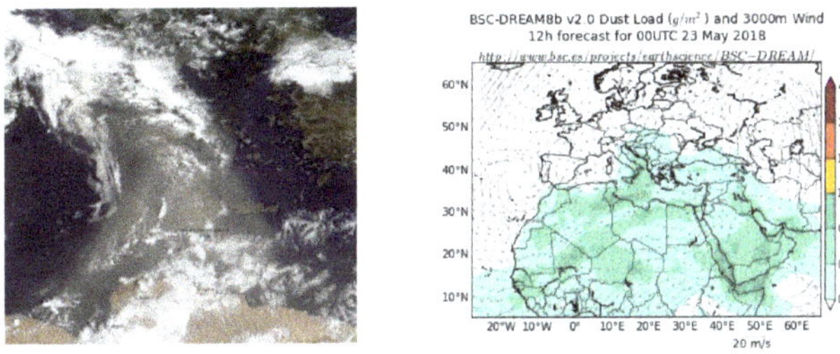

(a) (b)

Figure 2. (**a**) Image of African dust transport in Greece on 23 May 2018 (MODIS visible imagery). (**b**) Dust load and 3000 m wind speed for 23 May 2018, (BSC-DREAM8b).

3. Results

Figure 3 shows the wind-rose plot for the two studied periods, where the wind speeds are stronger in the LS (compared to HS) and blow mainly from the western sector, as well as the eastern directions for the HS and LS, respectively.

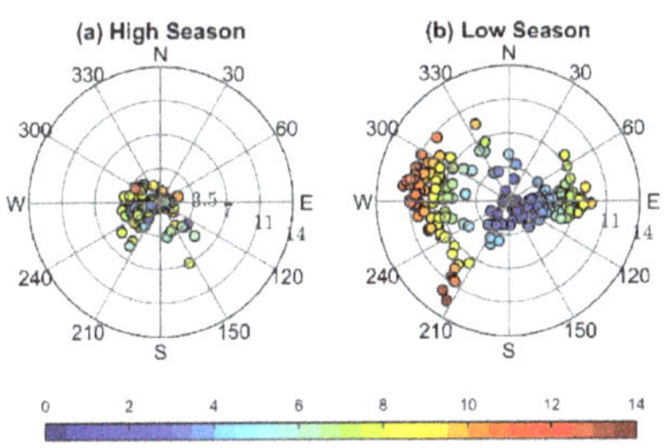

Figure 3. Wind-rose plot for (**a**) high port activity season (HS) and (**b**) low port activity season (LS).

The relation (negative correlation) of the wind speed with the concentration of pollutants indicates that higher winds tend to reduce the concentration for the majority of pollutants (PM_{10}, PM_{tot}, CO, NO_2, and NO_x). The wind speed is positively correlated with the O3 concentration since the O3 concentration is affected by photochemical reactions during the sunlight hours [11]. Additionally, the concentration of O3 is significantly negatively correlated (about -0.65) with NO_2 and NO_x. To conclude, the analysis shows that meteorological conditions significantly affect the air quality [16,17].

Figure 4 shows the hourly variation of meteorology parameters (Figure 4a–e) and the concentration of pollutants (Figure 4f–p) for both studied seasons. The analysis shows that the concentration of pollutants is maximized during the hours with high ship traffic (7:00–11:00 LT and 19:00–24:00 LT) indicating that harbor operations degrade the air quality in the port of Igoumenitsa. Comparing the two seasons studied, the O_3 and NO_x concentrations increased by about 10 ppb for HS. Additionally, the concentrations of NO_2 and SO_2 are increased ~2.5 ppb for the HS in comparison to the LS. Moreover, during the HS, the concentration of PM_1 is increased ~2.5 $\mu g/m^3$, while in contrast, the concentration of PM_4, as well as PM_{10}, is reduced ~5.5 $\mu g/m^3$. The positive difference of PM_4 and PM_{10}, between HS and LS, is explained by the synoptic atmospheric circulation. In particular, it must be stated that the Saharan dust is a major component for PM_{10} over the Mediterranean basin [18]. Previous studies have shown that the concentration of PM_{10} is maximized mainly during the transitional seasons (autumn and spring) [19]. In particular, for the

Heraklion and Athens (Greece), the concentration of PM_{10} is maximized in April during 2003–2006, which is possible due to the dust transport [12,13]. During the low port activity season, the African dust transfer event affects the eastern Mediterranean and the Western Greece and peaks on 23 May 2018 (Figure 2a,b). Furthermore, the results of the analysis of MERRA-2 model (available from the Giovanni data system) indicate that during the LS the difference of mean dust column mass density (gr/m^2) increases compared to the high season, affecting the PM concentrations (mainly PM_{10} and PM_4) over Western Greece and the port of Igoumenitsa (Figure 5).

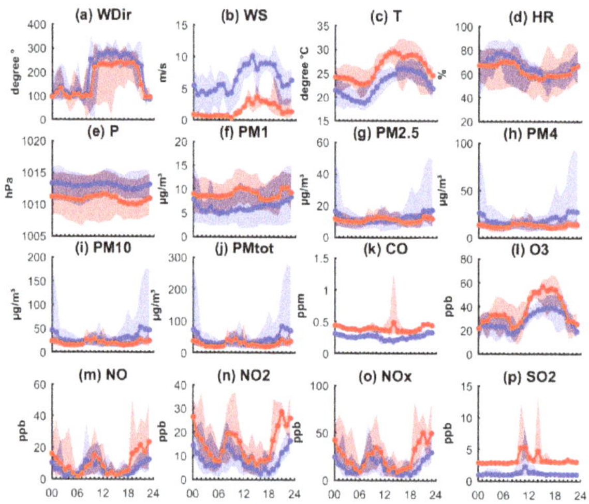

Figure 4. Mean diurnal variation of meteorology factors (**a–e**) and air quality parameters (**f–p**). The red/blue lines show the high/low port activity season. The colored regions are the range between the lower and higher hourly variability at every hour of the season.

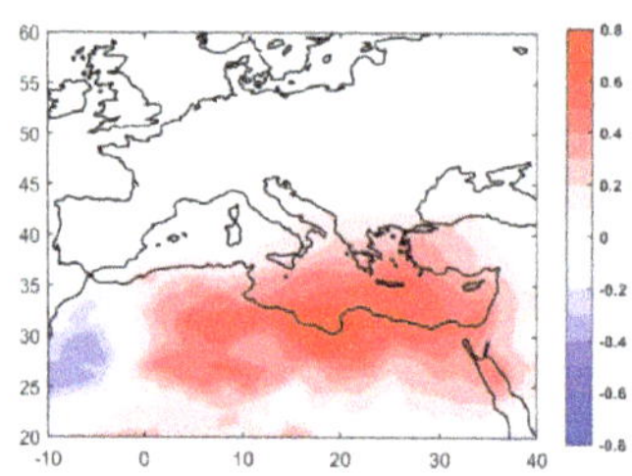

Figure 5. Composite difference of mean dust column mass density (gr/m^2) between the low port activity and high port activity season. Red/blue colors indicate an increase/decrease of the mean dust column mass density (gr/m^2).

In general, human activities and meteorological conditions are important factors that affect the ratio between $PM_{2.5}$ and PM_{10} (R_{PM}) [15]. The higher (lower) R_{PM} is more related to anthropogenic (natural) sources [15]. Additionally, the R_{PM} can be used to identify the source of the PMs, indicating whether the pollution is primarily due to anthropogenic or natural activity [2,20]. In order to investigate the impact of harbor operation on Igoumenitsa port, the R_{PM} is calculated. Figure 6 presents the variability of R_{PM} for the two studied seasons. During the HS, the R_{PM} ratio is about 0.13 higher compared to the LS (Figure 6c). Consequently, during the HS, the shipping has a more significant effect on the pollution levels of Igoumenitsa.

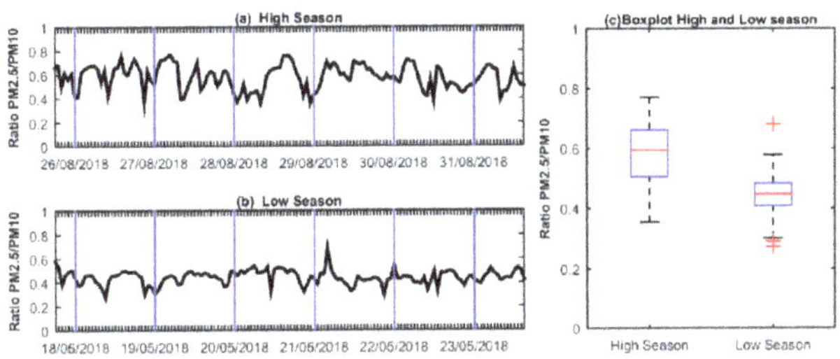

Figure 6. Timeseries of R_{PM} for the (**a**) high port activity season (HS) and (**b**) low port activity season (LS). (**c**) Boxplot of R_{PM} for the high and low port activity season.

4. Conclusions

This work examines the effect of shipping and port operation on the pollution levels of Igoumenitsa in Western Greece. The analysis is carried out through a comparison of the high (HS) and low (LS) port activity season during 2018 and the obtained findings indicate that meteorology affects the concentration of pollutants. In particular, wind speed tends to reduce the concentration of the majority of pollutants. For both seasons, the elevated concentration of pollutants occurs during the hours of high shipping traffic. It should be noted that the concentration of pollutants generally increases during the HS. The PM_4 and PM_{10} concentrations are excluded due to the effect of a Saharan dust transfer episode. The aforementioned dust episode occurs during the LS, affecting the PM concentrations over Western Greece. The higher ratio of $PM_{2.5}$ to PM_{10} (R_{PM}) during the HS indicates that the shipping contributes to the increase in the air pollution of Igoumenitsa. The analysis emphasizes that during the studied periods (HS and LS), the shipping and harbor activities affect the air quality of Igoumenitsa. Finally, this case study demonstrates that a further investigation of air quality in the port for a long-temporal period can provide important results, which can help the competent authorities in adopting energy saving practices that reduce air pollution footprint in ports.

Author Contributions: Conceptualization, I.L., A.M. and P.G.; methodology, I.L.; software, I.L.; validation, I.L.; formal analysis, I.L.; investigation, I.L.; resources, I.L.; data curation, I.L.; writing—original draft preparation, I.L.; writing—review and editing, I.L., C.A., A.M., K.S.; visualization, I.L.; supervision, A.M. and P.G.; project administration, P.G.; funding acquisition, A.M. and P.G. All authors have read and agreed to the published version of the manuscript.

Funding: This research was funded by the Connecting Europe Facility of the European Union, grant number 2014-EU-TM-0673-S and the APC was funded by Poseidon Med II Action No. 2014-EU-TM-0673-S.

Institutional Review Board Statement: Not applicable.

Informed Consent Statement: Not applicable.

Data Availability Statement: Publicly available datasets were analyzed in this study. These data can be found here: http://www.bsc.es/ess/bsc-dust-daily-forecast/ (accessed on 14 June 2021), https://worldview.earthdata.nasa.gov (accessed on 10 June 2021), and https://giovanni.gsfc.nasa.gov/giovanni/ (accessed on 3 June 2021).

Acknowledgments: The study received support from the POSEIDON MEDII project, which is co-financed by the Connecting Europe Facility (CEF) Transport Sector of the European Union. The authors would like to thank the DEPA Commercial S.A. (project coordinator) and DESFA S.A. (project technical coordinator). In addition, we would like to thank the Igoumenitsa Port Authority S.A. and personnel for providing information regarding the port activity to guarantee the representative sampling of recordings for the analysis. We acknowledge the Envirosys Ltd. Environmental Applications Technological Equipment, Athens, Greece, for the collaboration and the availability of the Horiba Ltd. Sensors. Moreover, the authors would like to acknowledge the (NMMB/BSC-Dust or BSC-DREAM8b) model, operated by the Barcelona Supercomputing Center (http://www.bsc.es/ess/bsc-dust-daily-forecast/) (accessed on 14 June 2021) for the provided image. We acknowledge the use of imagery from NASA's Worldview application (https://worldview.earthdata.nasa.gov (accessed on 10 June 2021), part of NASA's Earth Observing System Data and Information System (EOSDIS). Part of the analysis used in this study was produced with the Giovanni online data system, developed and maintained by the NASA Goddard Earth Sciences (GES) Data and Information Services Center (DISC) (accessed on 3 June 2021).

Conflicts of Interest: The authors declare no conflict of interest.

References

1. Tang, L.; Ramacher, M.O.P.; Moldanová, J.; Matthias, V.; Karl, M.; Johansson, L.; Jalkanen, J.-P.; Yaramenka, K.; Aulinger, A.; Gustafsson, M. The impact of ship emissions on air quality and human health in the Gothenburg area—Part 1: 2012 emissions. *Atmos. Chem. Phys.* **2020**, *20*, 7509–7530. [CrossRef]
2. Logothetis, I.; Antonopoulou, C.; Sfetsioris, K.; Mitsotakis, A.; Grammelis, P. Comparison analysis of the effect of high and low port-activity season on air quality in the port of Heraklion. In Proceedings of the 4th International Electronic Conference on Atmospheric Sciences, Online, 16–31 July 2021; MDPI: Basel, Switzerland, 2021. [CrossRef]
3. Gobbi, G.P.; Liberto, L.; Barnaba, F. Impact of port emissions on EU-regulated and non-regulated air quality indicators: The case 29 of Civitavecchia (Italy). *Sci. Total Environ.* **2020**, *719*, 134984. [CrossRef] [PubMed]
4. Prussi, M.; Scarlat, N.; Acciaro, M.; Kosmas, V. Potential and limiting factors in the use of alternative fuels in the Euro-pean 18 maritime sector. *J. Clean. Prod.* **2021**, *291*, 125849. [CrossRef] [PubMed]
5. Viana, M.; Fann, N.; Tobías, A.; Querol, X.; Rojas-Rueda, D.; Plaza, A.; Aynos, G.; Conde, J.A.; Fernández, L.; Fernández, C. Environmental and health benefits from designating the Marmara Sea and the Turkish Straits as an emission control area (ECA). *Environ. Sci. Technol.* **2015**, *49*, 3304–3313. [CrossRef] [PubMed]
6. Sofiev, M.; Winebrake, J.J.; Johansson, L.; Carr, E.W.; Prank, M.; Soares, J.; Vira, J.; Kouznetsov, R.; Jalkanen, J.-P.; Corbett, J.J. Cleaner fuels for ships provide public health benefits with climate tradeoffs. *Nat. Commun.* **2018**, *9*, 406. [CrossRef]
7. World Health Organization (WHO), Europe. *Health Risks of Particulate Matter from Long-Range Transboundary Air Pollution*; Joint WHO/Convention Task Force on the Health Aspects of Air Pollution, European Centre for Environment and Health Bonn Office, World Health Organization: Geneva, Switzerland, 2006.
8. Smith, T.W.P.; Jalkanen, J.P.; Anderson, B.A.; Corbett, J.J.; Faber, J.; Hanayama, S.; O'Keeffe, E.; Parker, S.; Johansson, L.; Aldous, L.; et al. *Third IMO Greenhouse Gas Study 2014*; International Maritime Organization: London, UK, 2015; 327p.
9. International Marina Organization. *Fourth IMO GHG Study 2020 Executive Summary*; International Maritime Organization: London, UK, 2021.
10. Wang, X.; Shen, Y.; Lin, Y.; Pan, J.; Zhang, Y.; Louie, P.K.K.; Li, M.; Fu, Q. Atmospheric pollution from ships and its impact on local air quality at a port site in Shanghai. *Atmos. Chem. Phys.* **2019**, *19*, 6315–6330. [CrossRef]
11. Shilenje, Z.W. Observed surface ozone trend in the year 2012 over Nairobi, Kenya. *Atmosfera* **2014**, *27*, 377–384. [CrossRef]
12. Mitsakou, C.; Kallos, G.; Papantoniou, N.; Spyrou, C.; Solomos, S.; Astitha, M.; Housiadas, C. Saharan dust levels in Greece and received inhalation doses. *Atmos. Chem. Phys.* **2008**, *8*, 7181–7192. [CrossRef]
13. Kallos, G.; Astitha, M.; Katsafados, P.; Spyrou, C. Long-range transport of anthropogenically and naturally produced particulate matter in the Mediterranean and North Atlantic: Current state of knowledge. *J. Appl. Meteorol. Clim.* **2007**, *46*, 1230–1251. [CrossRef]
14. Acker, J.G.; Leptoukh, G. Online Analysis Enhances Use of NASA Earth Science Data. *Eos Trans. AGU* **2007**, *88*, 14–17. [CrossRef]
15. Fan, H.; Zhao, C.; Yang, Y.; Yang, X. Spatio-Temporal Variations of the PM2.5/PM10 ratios and Its Application to Air Pollution Type Classification in China. *Front. Environ. Sci.* **2021**, *9*, 692440. [CrossRef]
16. Merico, E.; Gambaro, A.; Argiriou, A.; Alebic-Juretic, A.; Barbaro, E.; Cesari, D.; Chasapidis, L.; Dimopoulos, S.; Dinoi, A.; Donateo, A.; et al. Atmospheric im-pact of ship traffic in four Adriatic-Ionian port-cities: Comparison and harmonization of different approaches. *Transp. Res. Part D Transp. Environ.* **2017**, *50*, 431–445. [CrossRef]
17. Karagiannidis, A.; Poupkou, T.; Giannaros, C.; Giannaros, D.; Melas, A. Argiriou The air quality of a Mediterranean urban environment area and its relation to major meteorological parameters. *Water Air Soil Pollut.* **2015**, *226*, 2239. [CrossRef]
18. Pey, J.; Querol, X.; Alastuey, A.; Forastiere, F.; Stafoggia, M. African dust outbreaks over the Mediterranean Basin during 2001-2011: PM10 concentrations, phenomenology and trends, and its relation with synoptic and mesoscale meteorology. *Atmospher. Chem. Phys.* **2013**, *13*, 1395–1410. [CrossRef]
19. Gerasopoulos, E.; Kouvarakis, G.; Babasakalis, P.; Vrekoussis, M.; Putaud, J.P.; Mihalopoulos, N. Origin and variability of particulate matter (PM10) mass concentrations over the Eastern Mediterrane. *Atmos. Environ.* **2006**, *40*, 4679–4690. [CrossRef]
20. Zhao, C.; Li, Y.N.; Zhang, F.; Sun, Y.L.; Wang, P.C. Growth rates of fine aerosol particles at a site near Beijing in June 2013. *Adv. Atmos. Sci.* **2018**, *35*, 209–217. [CrossRef]

engineering proceedings

MDPI

Proceeding Paper

Propelling the Penetration of Electric Vehicles in Pakistan by Optimal Placement of Charging Stations †

Hafiz Owais Ahmed Khan, Faisal Saeed * and Naveed Arshad

SBA School of Science and Engineering, Lahore University of Management Sciences (LUMS), Lahore 54792, Pakistan; 19060025@lums.edu.pk (H.O.A.K.); naveedarshad@lums.edu.pk (N.A.)

* Correspondence: 19060005@lums.edu.pk

† Presented at the 2nd International Electronic Conference on Applied Sciences, 15–31 October 2021; Available online: https://asec2021.sciforum.net/.

Abstract: The world is rapidly advancing towards the electrification of mobility owing to the substantial benefits of emission reduction. Adhering to international trends and environmental obligations, the Government of Pakistan (GOP) also intended to adopt 30@30 plug-in-electric vehicles (PEVs) across the country, which implies 30 percent of new sales will be of PEVs until 2030. Despite the policy guidelines introduced by the GOP as well as incentives for vehicle fleet electrification and indigenization, the foremost challenge is the lack of a PEV charging infrastructure placement plan for the country. In this regard, an optimal locality map for level-3 or direct current fast charging (DCFC) stations' installation is proposed, considering traffic volume, service area, and local grid facility while ensuring the availability of charging stations across all major networks of the country. The area of focus for this is National Highway 5, known as N5, and the Motorway-2 (M2) Network. The paper also provides insights into the techno-economic analysis of the proposed charging station installation spots. The results are extremely encouraging and reveal the proposed PEV charging stations under observation on the highways from Lahore to Islamabad consumed an electricity share of 3 MW–0.13 MW based on minimum to maximum traffic volume scenarios, respectively. The study is impactful and ultimately paves a way forward for the aggravation of the EV market share by considering the initial investment and a payback period of 7 months. With the help of this study, better planning in terms of EV penetration size and its requirement for public DCFC stations can be implemented, and the exact recipe for the growth of the supportive industry with the pace of PEVs' perforation can be executed.

Keywords: charging stations; PEVs; optimal location; DCFC; route node coverage; techno-economic analysis

Citation: Khan, H.O.A.; Saeed, F.; Arshad, N. Propelling the Penetration of Electric Vehicles in Pakistan by Optimal Placement of Charging Stations. *Eng. Proc.* **2021**, *11*, 34. https://doi.org/10.3390/ASEC2021-11189

Academic Editor: Nunzio Cennamo

Published: 15 October 2021

1. Introduction

Fossil-fuel-based internal combustion engines (ICEs) are one of the key factors which account for 50% of environmental pollution [1]. Developing countries suffer more because of old and inefficient engines used in their transportation network which are the cause of transport-generated pollution, particularly in Asia, Africa, and the Middle East, ranging from 12 to 70% [2,3]. The challenge of transportation pollution can only be overcome by changing the transport fleet from ICE to plug-in electric vehicles (PEVs) [3]. To encourage maximum PEV penetration, there must be a coordinated network of fast charging stations available publicly with private parties involved to also enable the rapid market penetration of PEVs. In recent years, many researchers have focused on the optimal placement of charging stations by continuing to study areas such as the environment, commerce, self-sustainability, etc. [4–7].

Presently, Pakistan lacks a PEV charging infrastructure plan to facilitate the adoption of PEVs on a wide scale in the country. To solve this problem robustly, a similar approach as discussed in [7] is adopted with slight improvements in a model for the optimal placement of direct current fast charging (DCFC) stations based on the flow calculation by using the dataset provided by the National Highways and Motorways authority. The considered networks for this contextual analysis are the Motorway 2 (M2) and National Highway 5 (N5) networks from Lahore to Islamabad. These routes are more active traffic routes than the rest of the road networks in the country, and also, the region covering these routes is among the most densely populated areas of the country. Moreover, the study is focused on proposing an optimal PEV charging station plan for intercity routes to ensure long-range, anxiety-free traveling in the future.

2. Electrical Charging Stations Locality Deployment Model

To maximize PEVs' market share, a coordinated charging station (CS) network along highways and motorways is suggested. In this study, all vehicles were considered as cars, and heavy-duty traffic was not considered. Charging time was assumed to be 30 min for standardization, and the charger electricity consumption was 50 KW. The PEV charging port and CS charging port adopted the same type of standardization for the convenience of installation purposes. The tariff was assumed to be 35 rupees for dedicated load EV charging by the distribution companies and an annual 10% rupee devaluation. As the charging process interrupts the journey, only DCFC chargers were considered. To determine CS sites, we only considered rest-places with basic rest-place facilities as candidate sites. These facilities are available on the candidate site and also no farther than 250 m from it and are categorized as: (i) *basic facility location:* parking, small shops, and prayer provision (ii) *medium facility location:* supermarket, dining court, and minimum rest-place facility (iii) *superior facility location:* High-end rest and accommodation facility, food courts, and additional facilities such as a pharmacy, etc. By considering these facilities, the potential location of CSs could be selected based on the re-defined equation detailed in [7] for each nominated site, and the process is illustrated in Figure 1.

$$PL_i = a_1x_{1,i} + a_2x_{2,i} + a_3x_{3,i} + a_4x_{4,i} + a_5x_{5,i} \tag{1}$$

where PL_i = potential location of candidate site, 'i', $x_{1,i}$ = security level on nearby roads at the candidate site, 'i', $x_{2,i}$ = evaluation value of traffic volume on nearby roads at the candidate site, 'i', $x_{3,i}$ = evaluation value of service level of the candidate site, 'i', $x_{4,i}$ = evaluation value of the distance between two candidate sites, 'i', $x_{5,i}$ = electricity availability at the candidate site 'i', while a_1, a_2, a_3, a_4, and a_5 are the weights of variables.

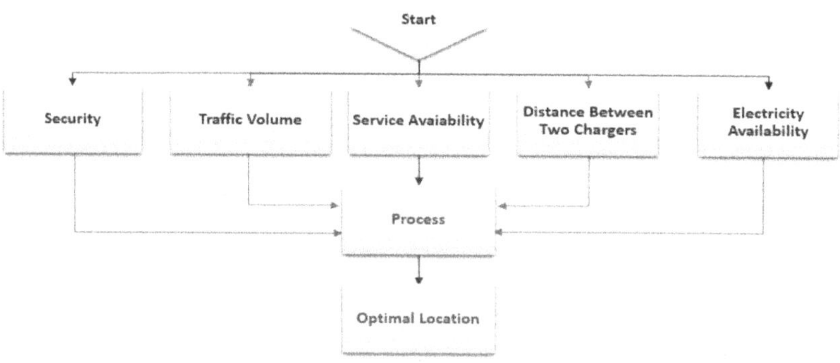

Figure 1. Algorithm for optimal location determination for installation of PEV charging stations.

The parameters in (1) require exploration for the precise determination of optimal CS spots. In (1) $x_{1,i}$ is the security factor for the CSs as well as for the nearby roads. The value of '$x_{2,i}$' is the sum of average daily traffic volume that passes through national highways and motorways within the range of 20–100 km (Km) from the CS (N_i) (vehicle/day) location. We considered the traffic volume of the directions from where the rest-place is accessible. The value of x_1 is calculated according to the equation below [7]:

$$x_{1,i} = \begin{cases} 0, & if\ N_i \leq fmin \\ \frac{N_i - fmin}{fmax} 0.5, & if\ fmin < N_i < fmax \\ 1, & if\ N_i \geq fmax \end{cases} \quad (2)$$

where N_i = number of vehicle flow, and $fmax$ = maximum vehicle flow $fmin$ = minimum vehicle flow. We defined the limit values according to the calculations by using the dataset. In terms of service level, $x_{3,i}$, a basic service facility is ranked as 1, medium is ranked as 2, while a superior service level at CS locations is given a rank of 3. $a_4 x_{4,i}$ is assumed to be constant as the distance between two candidate sites on the motorway network is fixed (service areas also have a fixed location), while on the N5 network, a supposition is made that there must be a charging station after every 40 km. Additionally, $x_{5,i}$ factor ensures the availability of national power grids, transmission, and distribution networks for PEV CS integration at each candidate site.

3. Results and Discussions

To determine the optimal charging station locations based on the dataset, vehicle flow was calculated at N5 north, from Lahore to Islamabad, and at motorway M2 from Islamabad to Lahore. The dataset consisted of data of vehicle flow for April 2019 as depicted in Figure 2a,b, and for March 11 to the April 14 of the year 2020, respectively, as shown in Figure 2c. This particular dataset is important because it covered the pre-COVID-19 (2019) as well as the post-COVID-19 (2020) period. So, in this way, we gained the regular maximum vehicle flow data as well as the minimum vehicle flow data. Due to the availability of minimum vehicle flow data, different case scenarios could be developed, and we also learned the minimum amount of the traffic that would flow in any bad scenario.

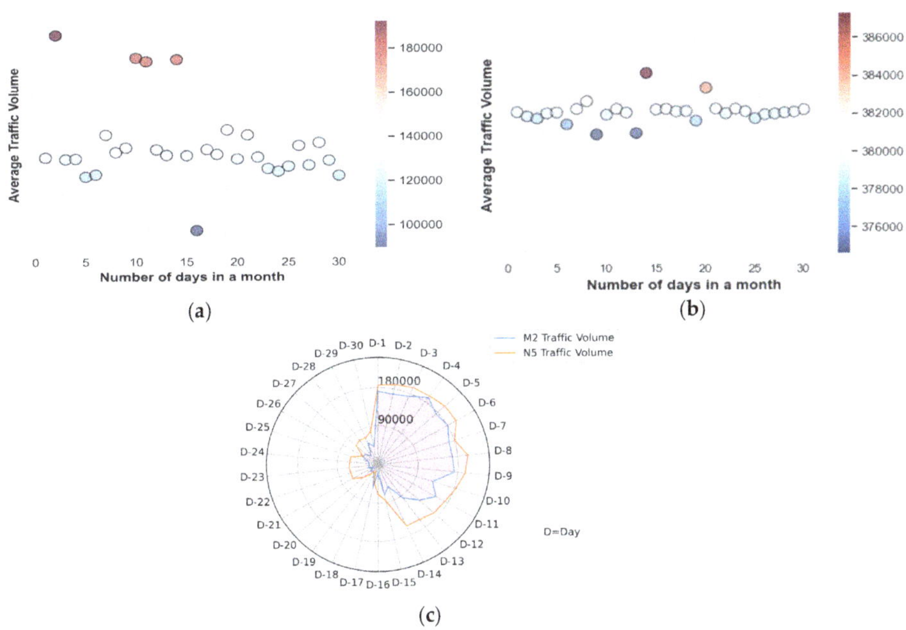

Figure 2. Vehicle flow data pre-COVID scenario on (**a**) M2; (**b**) N5; and (**c**) post-COVID scenario of M2 and N5.

Considering the provision of facilities, the study area was divided into different zones on M2 and N5, as shown in Figure 3. The zones were divided according to the traffic data and the nature of the facilities available. The PEV population was distributed between these zones. Considering the zones, proposed locations with distances in-between the two CSs are enlisted in Table 1. Further, battery size and the mileage range of different models of cars were also considered for this investigation (see Table 2 [8]). From the dataset, the average vehicle flow was calculated in the normal period as well as during the COVID-19 period (see Tables 3 and 4) by adopting approach detailed in [9,10]. By assuming the differences ranged from 1% to 10%, we developed the scenarios as listed in Tables 5 and 6 for the N5 and M2 highways, respectively. In this way, the goal of the research effort to establish a certain number of priority CSs was accomplished by maximizing the service of charging stations. It is to be noted that when calculating the distance from the demand point to the candidate point, the mathematical model mentioned in (1) and (2) and the after-mentioned principles were adhered to for the optimal placement of PEV CSs. The finalized scenario including transmission network infrastructure and the proposed potential charging station candidates for the M2 and N5 routes are depicted in Figure 4.

Table 1. Proposed Charging Station Locations and Distance.

Sr. No.	City	Longitude, Latitude	Distance
A	**N5 network**		
1.	Rawat	32.4805288, 72.687214	15.7
2.	Mandra toll plaza	32.4277489, 72.40935	59
3.	Deena	33.0285967, 73.598110	20
4.	Sarae Alamgir	32.907611040865, 73.730340115134	8
5.	Kharian	32.8830768, 73.7785187	46
6.	Gujrat exit	32.5959182, 74.0378919	9
7.	Gujrat	32.4797136, 74.091663	10.3
8.	WazirAbad	32.4026824, 74.1224383,	13.5
9.	Gakhar	32.317067, 74.143303	13.6
10.	Gujranwala		17
11.	MoreAimanAbad	32.0488448, 74.2085573	11.4
12.	Kamoke	31.995427755973, 74.218015463522	11
13.	Sadhoke	31.862328, 74.24472	12.7
12.	Muridke	31.6420817, 74.2032471	11.4
15.	Kala Shah Kaku	31.7338073, 74.2655666	46.2
B	**M2 network**		
16.	Chakri	33.3203856, 72.7829902	45
17.	KalarKahar	32.869405, 72.65204	73.5
28.	Bhera	32.453259538502, 72.886018340599	46.5
19.	Sial Mor	31.9680162, 73.1120396	77.5
20.	Sukheki	31.906767165001, 73.56816594500	48.6

Table 2. Travelling Range of Different Electrical Vehicle Cars [8].

Manufacturer	Range (Km)	Battery Size (KWh)
Tesla	483	60–100
Xpeng (China)	706	60–100
Chevrolet	355	60–100
Nissan	270	22–40
BMW	246	22–40
Kia	160	22–40
Volkswagen	130	12–20
TopSun	**300**	50

Table 3. Percentage of Electric Vehicle Flow in Normal Days.

Percentage of EVs from Total Vehicle	Number of EVs	Percentage of Different Models (50%, 30, 20%)	Number of Charger at Each Location
10%	100	50, 30, 20	100
5%	50	25, 15, 10	50
2.5%	25	13, 7, 5	25
1%	10	5, 3, 2	10

Table 4. Percentage of Electric Vehicle Flow in COVID-19.

Percentage of EVs from total Vehicle	Number of EVs	Percentage of Different Models (50%, 30, 20%)	Number of Charger at Each Location
10%	6	3, 2, 1	6
5%	3	2, 1, 0	3
2.5%	3	1, 1, 1	3
1%	3	1, 1, 1	3

Table 5. Power consumption at 15 Stations of N5.

Period	Demand	Power Usage (MW)
Normal period	Maximum	3
Normal period	Minimum	0.3
COVID-19	Maximum	1.98
COVID-19	Minimum	0.18

Table 6. Power consumption at 5 stations of M2.

Period	Demand	Power Usage (MW)
Normal period	Maximum	1.6
Normal period	Minimum	0.16
COVID-19	Maximum	1.36
COVID-19	Minimum	0.13

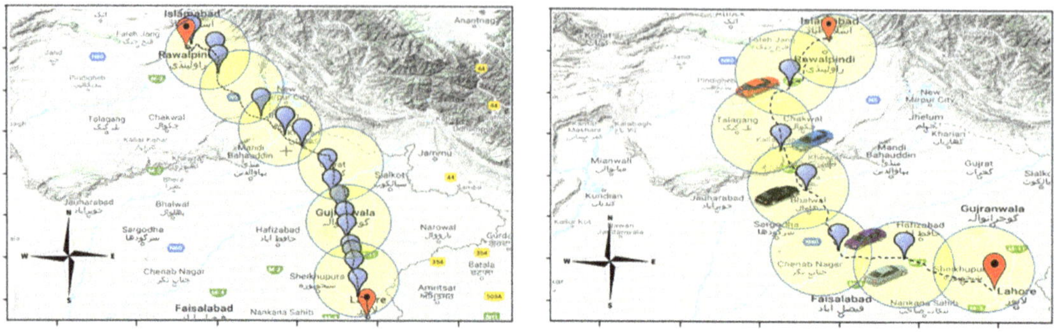

Figure 3. Zones to be covered for proposed PEV charging installation on (**left**) N5 north; and (**right**) M2 from Lahore to Islamabad. It is to be noted that figures are not according to the scale and only indicate the approximate zone areas.

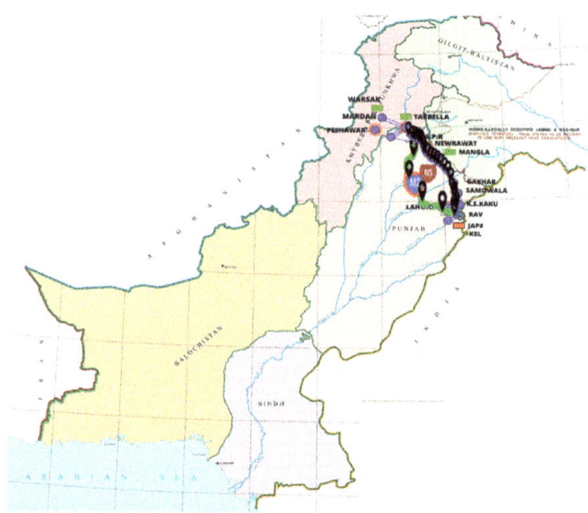

Figure 4. Proposed candidate sites for PEV charging stations on M2 motorway and N5 with transmission network.

Further, economic analysis about the investment and payback period was also taken into account for a feasibility analysis of the proposed model. For this, we considered the initial cost of investment, variable cost, operational cost, etc., as listed in Table 7.

Table 7. Economic analysis parameters for the installation of PEV CSs.

Parameters	Cost (USD)
Charger	20,000
Installation	1000
New Connection	2500
Operation and Maintenance	10% of overall
Electricity Tariff per kWh	0.142
Electricity Taxes per kWh	0.11
Rupee Devaluation	10% annually
Miscellaneous	1000
Total cost (excluding electric cost and taxes)	29,400

The charger mentioned above is the DCFC, with two ports for charging at one time. The installation cost included the labor cost, material cost, and other such parameters; the new connection cost was the cost of the regulator, and in the case of the transformer, there was a minimum cost both for the regulator and transformer. The operational and maintenance cost was taken as the 10% annual cost. The electricity and taxes costs were obtained from the provider, while we had to consider the rupee devaluation for investment and some miscellaneous charges, as this is the new technology, and there will inevitably be some unknown annual charges. Even during the period of strict lockdown during the COVID-19 pandemic, the minimum EV flow was 3 at each point in 1 h. So, at least two chargers are needed at one optimal location point. If the charging cost is assumed to be 0.31 USD/KWh and the installed charger worked for 24 h, then:

Total 1 day selling cost = 0.31 * (24 * 2) * 50 KW = 744 USD/KWh; Total 30 days selling cost = 22,320 USD/KWh, while:

Total 30 days actual electric cost is = 18,144 USD/KWh. Profit for 30 days = 4176 USDKWh.

Total investment recovery time = 29,400/4176 = 7 months.

So, in almost 7 months, the total investment will be recovered, even when the devaluation (or, if we remain in dollars, considering the interest rate at 10%) is also considered.

4. Conclusions

In the implementation process, a N5 road and motorway map was derived, and the results are presented in the above section. To address the problem of location selection during electric vehicle charging station planning, this paper proposed a location method based on regional information and future predicted demand. According to the battery life of an electric vehicle, we determined the service range of a charging station. Based on the cost constraints, we determined the number of CSs to determine the optimal location for a PEV CSs. The method proposed in this paper can obtain an ideal charging station planning scheme that meets requirements and provides a guiding significance and application value for the location and constant volume of an electric vehicle charging station.

Author Contributions: Conceptualization, H.O.A.K., N.A.; methodology, H.O.A.K., F.S., N.A.; validation, H.O.A.K., F.S., N.A.; formal analysis, H.O.A.K., F.S., N.A.; investigation, H.O.A.K., F.S., N.A.; resources, H.O.A.K., N.A.; data curation, H.O.A.K., N.A.; writing—original draft preparation, H.O.A.K., F.S., N.A.; writing—review and editing, H.O.A.K., F.S., N.A.; visualization, H.O.A.K., F.S., N.A.; supervision, N.A.; project administration, N.A.; All authors have read and agreed to the published version of the manuscript.

Funding: This research received no external funding.

Institutional Review Board Statement: Not applicable.

Informed Consent Statement: Not applicable.

Data Availability Statement: Not applicable.

Acknowledgments: The authors would like to acknowledge the various support of the National Highway Authority (NHA) of Pakistan on the provision of traffic flow data for the successful completion of this research study.

Conflicts of Interest: The authors declare no conflict of interest.

References

1. Sheykhi, M.; Chahartaghi, M.; Safaei Pirooz, A.A.; Flay, R.G.J. Investigation of the effects of operating parameters of an internal combustion engine on the performance and fuel consumption of a CCHP system. *Energy* **2020**, *211*, 119041. [CrossRef]
2. Tauqeer, H.A.; Saeed, F.; Yousuf, M.H.; Ahmed, H.; Idrees, A.; Khan, M.H.; Gelani, H.E. Proposed model of sustainable resource management for smart grid utilization. *World Electr. Veh. J.* **2021**, *12*, 70. [CrossRef]
3. Hamidi, A.; Ramavandi, B. Evaluation and scientometric analysis of researches on air pollution in developing countries from 1952 to 2018. *Air Qual. Atmos. Health* **2020**, *13*, 797–806. [CrossRef]
4. Elma, O. A dynamic charging strategy with hybrid fast-charging station for electric vehicles. *Energy* **2020**, *202*, 117680. [CrossRef]
5. Szumska, E.M.; Jurecki, R.S. Parameters influencing on electric vehicle range. *Energies* **2021**, *14*, 4821. [CrossRef]

6. Arif, S.M.; Lie, T.T.; Seet, B.C.; Ayyadi, S.; Jensen, K. Review of electric vehicle technologies, charging methods, standards and optimization techniques. *Electronics* **2021**, *10*, 1910. [CrossRef]
7. Csonka, B.; Csiszár, C. Determination of charging infrastructure location for electric vehicles. *Proc. Transp. Res. Procedia* **2017**, *27*, 768–775. [CrossRef]
8. Saeed, F.; Waris, M.D.; Rehman, T.U.; Khan, M.A.; Khan, M.H.; Gelani, H.E. A Comparative Study of Grid-Tied PV Systems Employing CIGS and Crystalline Solar Modules. In Proceedings of the 2021 Mohammad Ali Jinnah University International Conference on Computing (MAJICC), Karachi, Pakistan, 15–17 July 2021; pp. 1–7.
9. Milev, G.; Hastings, A.; Al-Habaibeh, A. The environmental and financial implications of expanding the use of electric cars—A Case study of Scotland. *Energy Built Environ.* **2021**, *2*, 204–213. [CrossRef]
10. Parker, N.; Breetz, H.L.; Salon, D.; Conway, M.W.; Williams, J.; Patterson, M. Who saves money buying electric vehicles? Heterogeneity in total cost of ownership. *Transp. Res. Part D Transp. Environ.* **2021**, *96*, 102893. [CrossRef]

engineering proceedings

MDPI

Proceeding Paper

Quantification of Losses in a Photovoltaic System: A Review †

Faisal Saeed [1,*] and Abdullah Zohaib [2]

1 SBA School of Science and Engineering, Lahore University of Management Sciences (LUMS), Lahore 54792, Punjab, Pakistan
2 Department of Electrical Engineering, University of Engineering and Technology Lahore, Lahore 54890, Punjab, Pakistan; abdullahzohaib3@gmail.com
* Correspondence: 19060005@lums.edu.pk

† Presented at the 2nd International Electronic Conference on Applied Sciences, 15–31 October 2021.

Abstract: In this paper, we characterized and reviewed the emergence of fundamental and extended losses that limit the efficiency of a photovoltaic (PV) system. Although there is an upper theoretical bound to the power conversion efficiency of solar cells, i.e., the Shockley Queisser limit, in a practical environment, the consideration of inevitable losses in a whole PV system is imperative to optimally harvest solar energy. In this regard, this study quantifies the losses from a PV cell level to the whole PV system. It was perceived that reported losses on the PV cell level included the low energy bandgap, thermalization, recombination (surface and bulk recombination), optical absorption, space charge region, finite thickness, and metal contact loss, and it was determined that cutting techniques mainly constrained the power conversion efficiency of the solar cell. Furthermore, the detailed PV array losses were classified as mismatch power losses, dust accumulation losses, temperature effects, material quality losses, and ohmic wiring losses. The unavoidable system losses were quantified as inverter losses, maximum power point tracking losses, battery losses, and polarization losses. The study also provides insights into potential approaches to combat these losses and can become a useful guide to better visualize the overall phenomenology of a PV System.

Keywords: PV cell; PV modules; losses; quantification

Citation: Saeed, F.; Zohaib, A. Quantification of Losses in a Photovoltaic System: A Review. *Eng. Proc.* **2021**, *11*, 35. https://doi.org/10.3390/ASEC2021-11200

Academic Editor: Nicholas Vassiliou Sarlis

Published: 19 October 2021

1. Introduction

In the last few years, photovoltaics (PV) have emerged as a pioneer technology to meet the energy demands of small-scale consumers to those of the commercial sector and provide a cost-beneficial solar power generation system that can be used to offset the electricity costs from utility providers as well as alleviate the burden on the national electricity grid. Another major advantage of PV systems is the emission reduction benefits [1,2]. Presently, the installed PV capacity is around 109 GW_p, and this could cross 149 GW_p by 2022 according to the International Energy Agency (IEA), France [3]. This trend certainly demonstrates unparalleled progress in efficiency enhancement in the area of photovoltaics combined with power electronic-aided hybrid converters as well as cutting edge cost benefits, yet the emergence of losses in real environmental conditions is inevitable, as these losses cannot be eliminated beyond fundamental limits [4–6].

PV cells harvest solar energy to yield photogenerated power. The performance of solar cells depends on the available solar insolation and the spectral distribution of incident wavelengths over the surface of the PV system. The output of the solar cell is generally measured in standard testing conditions (STC); irradiance 1000 W/m^2, temperature 25 °C, and standard earth spectrum AM 1.5 G, where G stands for global and includes both direct and diffuse radiation [7–9]. The solar cell performance is characterized based on parameters including open-circuit voltage (V_{oc}), the voltage at the maximum power point (V_{mp}), the short circuit current (I_{sc}), current at the maximum power point (I_{mp}), and the maximum power point (P_{mp}), which can be extracted from the current–voltage (I–V) characteristics shown in Figure 1 [5]. The efficiency (η) of the solar cell is the ratio of available solar energy to the converted electrical energy, which can be calculated using the percentage of the maximum power point and the surface area of the solar cell (A) into irradiance (I_r), which is provided by Equation (1).

$$\eta = \frac{P_{mp}}{A \times I_r} \ (\%) \tag{1}$$

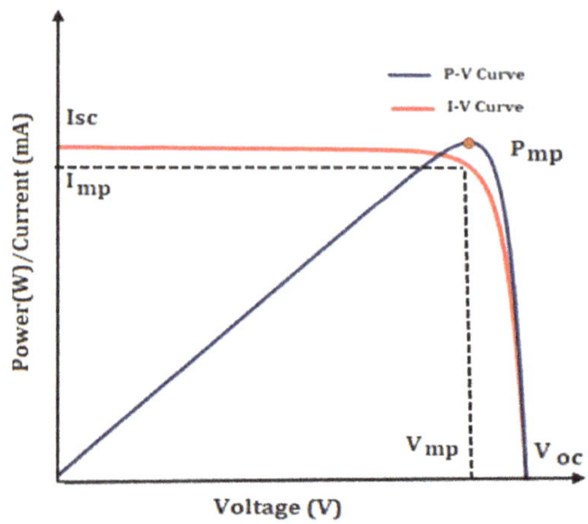

Figure 1. Generic current–voltage (I–V) and power–voltage (P–V) characteristics of a photovoltaic cell [5].

In real environmental conditions, several factors affect the performance of PV cells. Herein, we first reviewed the major losses from PV cells to the overall PV system and subsequently characterized and presented the losses in a pictorial form for better visualization and understanding for the reader.

2. Quantification of Losses in a Photovoltaic System

2.1. Losses in a Photovoltaic Cell

The loss mechanisms in a PV cell are initiated by the fundamental inability of the solar absorber-layer material (silicon, gallium arsenide, perovskite, copper indium gallium selenide (CIGS), among others) to potentially absorb all incident light wavelengths [10]. Incident light wavelengths with a photon energy (E_{ph}) less than the energy bandgap (E_g) of the absorber layer are unable to be absorbed. Such losses are the *below energy band gap losses* and are shown mathematically by Equation (2) [11].

$$Below\ Eg\ Loss = \int_0^{Eg} E{\cdot}GP(E, \Omega_A, T_S, \mu = 0)dE \tag{2}$$

The photons with energy $E_{ph} > E_g$ generate electron–hole pairs. However, the carriers with high kinetic energy sometimes decay to the band edges quickly from their initial excited states to reach their thermal equilibrium states, releasing their excess energy upon interaction with the crystal lattice. Such losses can be categorized under *thermalization loss*, and the mathematical relationship is given in Equation (3) [12–14].

$$Thermalization\ loss = \frac{Eg \int_0^{\lambda g} \Phi(\lambda)d\lambda}{\int_0^{\lambda g} \Phi(\lambda)\frac{hc}{\lambda}d\lambda} \tag{3}$$

Thermodynamic studies on a PV cell demonstrated that at temperature > 0 K, a voltage drop is associated with the PV cell, which is termed as *etendue loss* [15]. Moreover, *Fermi level losses*; losses associated with the displacement of the V_{oc} and E_g relationship and *electron kinetic losses; and* losses underlying the inefficacious use of the carriers' kinetic energy during the thermalization process are among the major thermodynamic losses that limit the efficiency of solar cells [14–16]. Besides this, operating solar cells at P_{mp} could also result in reduced output performance because of series and shunt resistance effects and is referred to as *fill factor loss* [17].

In practical scenarios, part of the incident light that falls on the surface of a solar cell is reflected or transmitted instead of being absorbed. Such losses are referred to as *optical losses* [18]. The reflected portion of the incident light is also separately named the *reflection loss* [13,18]. The reflection losses directly reduce the I_{sc} of solar cells. Similarly, the finite thickness or geometry of the solar cell contributes to *transmission losses* in a PV cell [13,18]. In a wafer-based solar cell, the part of the cell that makes contact with the front side of the cell (from where light enters) is made of a finger and bus bar. These metal contacts shadow some light, which can be up to 10% [16–18]. Such losses tend to create *area losses/losses due to metal coverage*.

The photons on the solar cell generate electron–hole pairs, and these generated carriers need to be separated in order to reach their respective metal contacts before they recombine. The recombination of the carriers can be attributed to *recombination losses* in a solar cell. Recombination losses can be further classified as (i) surface recombination; (ii) bulk recombination; (iii) depletion region recombination; and (iv) recombination at the metal contacts [19].

2.2. Photovoltaic Array Losses

Under same environmental conditions/STC, identical PV cell/module/arrays sometimes exhibit un-identical P_{mp} values because of manufacturing errors that can be attributed as *mismatch power loss* [20]. It is to be noted that under heterogeneous irradiation conditions (partial shading), mismatch power loss is modeled separately due to variation in the module performance/physical environments.

The accumulation of dust over the surface of the PV module results in reduced photogenerated power and also affects the angle of incidence reaching the absorber layer of the solar cell. Such losses are referred to as *dust accumulation losses* [21]. Besides these varied irradiance values that accumulate over time, irradiance losses and temperature impacts (hot spot issues), temperature losses, and DC wiring ohmic losses seriously affect the power conversion efficiency of PV modules [22,23].

2.3. System-Level Losses

On a system level, the *inverter losses, batter losses, maximum power point tracking* (*MPPT*) *topology losses*, and potential-induced degradation or polarization losses are among the major types of PV system losses that result in reduced PV system performance over time [24,25].

For better understanding, the above-mentioned PV cell system losses have been shown pictorially in Figure 2.

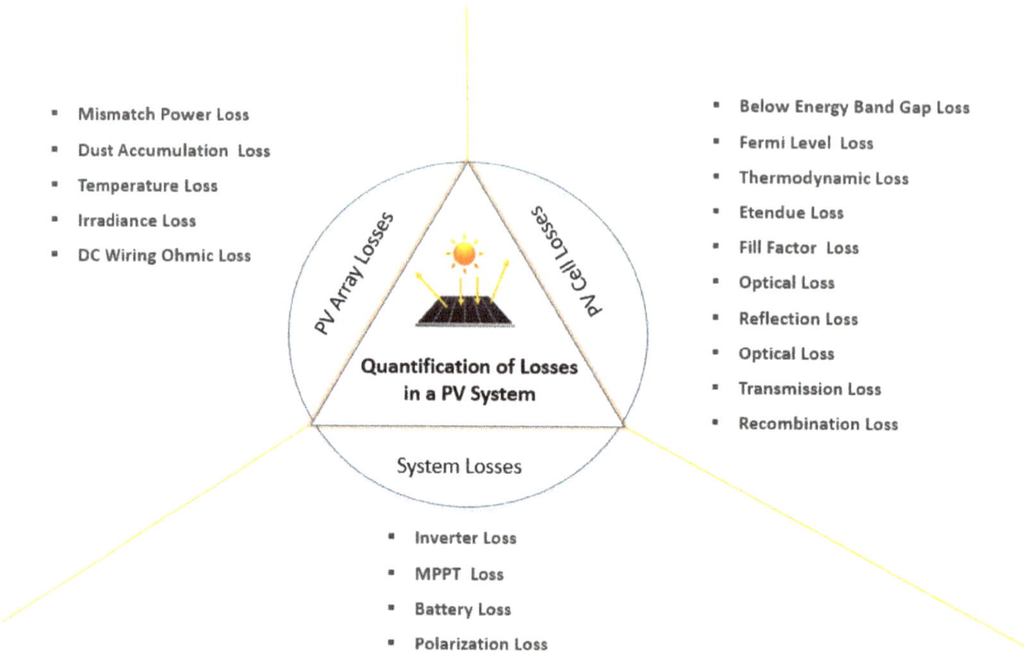

Figure 2. Characterization of losses in a photovoltaic system: cell to system level.

3. Possible Ways to Combat Losses

3.1. Addressing Photovoltaic Cell-Level Losses

The *below energy band gap, thermalization, Fermi level losses,* and *etendue losses* can be addressed by employing an absorber layer material with low E_g or multi-junction approaches. In emerging PV technology, tuning the energy bandgap of organic/inorganic absorber layer properties can be useful to combat the above-mentioned issues. The *optical* and *reflection losses* can be addressed by using surface texturing and anti-reflective coatings (the material should have good transmittance). The *transmission losses* can be addressed by employing an appropriate wafer geometry and thickness to absorb the maximum amount of incident light wavelengths. The *area losses* can be mitigated by reducing the widths of the finger over the top surface while expanding the contact size of the back metal. The *surface recombination losses* can be reduced by passivating the surface to reduce dangling bonds or by adopting a window layer to limit the path of the minority charge carriers at the maximum amount. *Depletion region recombination losses* are not the most prominent type of loss. *Bulk recombination losses* can be addressed by using a pure semi-conductor material while rear surface passivation approaches could aid in combating metal contact recombination sites [11–19].

3.2. Addressing Photovoltaic Array Losses

The *mismatch power losses* can be addressed via the application of by-pass/blocking diodes or cell-cutting approaches. The *dust accumulation losses* can be addressed by properly cleaning the PV module with demineralized water or with an electro-static cleaning system. The *temperature losses* can be addressed by considering appropriate module technology (crystalline, crystalline PERC, thin-film), while DC wiring losses can be mitigated by using wires with good conductance and a minimum number of connections [20–25].

3.3. System-Level Losses

With the employment of efficient power electronic-aided topologies, inverter, MPPT, and polarization losses can be addressed [25,26]. Proper battery sizing, advancement towards dry batteries rather than lead–acid Batteries, and moderate temperature, battery dispatch strategies can aid in mitigating *battery losses* in PV systems [27].

4. Conclusions

Depending on the nature of the losses experienced in a PV system reported in the literature, we broadly and briefly classified the major types of losses that are responsible for the reduced efficacy of whole PV systems at the PV cell level, array level, and system level and presented them in a pictorial form. Further, we discussed potential solutions to overcome fundamental and extended losses in PV systems. This illustration may become a brief and useful guide to create awareness of issues that may occur at the PV cell fabrication level and how they affect the whole PV system.

Author Contributions: Conceptualization, F.S., A.Z.; methodology, F.S., A.Z.; validation, F.S., A.Z.; formal analysis, F.S., A.Z.; investigation, F.S., A.Z.; writing—original draft preparation, F.S., A.Z.; writing—review and editing, F.S., A.Z.; visualization, F.S., A.Z. All authors have read and agreed to the published version of the manuscript.

Funding: This research received no external funding.

Data Availability Statement: Not applicable.

Conflicts of Interest: The authors declare no conflict of interest.

References

1. Anvari-Moghaddam, A.; Vahidinasab, V.; Mohammadi-Ivatloo, B.; Razzaghi, R.; Mohammadi, F. Emerging technologies for the energy systems of the future. *Inventions* **2021**, *6*, 23. [CrossRef]
2. McKuin, B.; Zumkehr, A.; Ta, J.; Bales, R.; Viers, J.H.; Pathak, T.; Campbell, J.E. Energy and water co-benefits from covering canals with solar panels. *Nat. Sustain.* **2021**, *4*, 609–617. [CrossRef]
3. Tauqeer, H.A.; Saeed, F.; Yousuf, M.H.; Ahmed, H.; Idrees, A.; Khan, M.H.; Gelani, H.E. Proposed model of sustainable resource management for smart grid utilization. *World Electr. Veh. J.* **2021**, *12*, 70. [CrossRef]
4. Olczak, P.; Olek, M.; Matuszewska, D.; Dyczko, A.; Mania, T. Monofacial and bifacial micro pv installation as element of energy transition—The case of poland. *Energies* **2021**, *14*, 499. [CrossRef]
5. Saeed, F.; Yousuf, M.H.; Tauqeer, H.A.; Akhtar, M.R.; Abbas, Z.A.; Khan, M.H. Performance benchmark of multi-layer neural network based solar MPPT for PV applications. In Proceedings of the 2021 International Conference on Emerging Power Technologies, ICEPT 2021, Topi, Pakistan, 10–11 April 2021.
6. Ahmed, W.; Sheikh, J.A.; Farjana, S.H.; Mahmud, M.A.P. Defects impact on pv system ghg mitigation potential and climate change. *Sustainability* **2021**, *13*, 7793. [CrossRef]
7. Saeed, F.; Waris, M.D.; Rehman, T.U.; Khan, M.A.; Khan, M.H.; Gelani, H.E. A Comparative Study of Grid-Tied PV Systems Employing CIGS and Crystalline Solar Modules. In Proceedings of the 2021 IEEE Mohammad Ali Jinnah University International Conference on Computing (MAJICC), Karachi, Pakistan, 15–17 July 2021; pp. 1–7.
8. Saeed, F.; Tauqeer, H.A.; Idrees, A.; Ali, M.Z.; Raza, A.; Khan, M.A. Buffer Layered PbS Colloidal Quantum Dot Solar Cell with Enhanced Efficiency. In Proceedings of the 2021 4th International Conference on Energy Conservation and Efficiency, ICECE 2021—Proceedings, Lahore, Pakistan, 16–17 March 2021.
9. Saeed, F.; Abbas, Z.A.; Akhtar, M.R.; Yousuf, M.H.; Idrees, A.; Tauqeer, H.A. Intelligent Hybrid Energy Resource Connected Demand Side Load Management System-Case of Pakistan. In Proceedings of the 2021 4th International Conference on Energy Conservation and Efficiency, ICECE 2021—Proceedings, Lahore, Pakistan, 16–17 March 2021.
10. Kapsalis, V.; Kyriakopoulos, G.; Zamparas, M.; Tolis, A. Investigation of the photon to charge conversion and its implication on photovoltaic cell efficient operation. *Energies* **2021**, *14*, 3022. [CrossRef]
11. Nayak, P.K.; Mahesh, S.; Snaith, H.J.; Cahen, D. Photovoltaic solar cell technologies: Analysing the state of the art. *Nat. Rev. Mater.* **2019**, *4*, 269–285. [CrossRef]
12. Amin, N.; Karim, M.R.; Alothman, Z.A. Optical losses of frontal layers in superstrate cds/cdte solar cells using opal2. *Coatings* **2021**, *11*, 943. [CrossRef]
13. Luo, D.; Su, R.; Zhang, W.; Gong, Q.; Zhu, R. Minimizing non-radiative recombination losses in perovskite solar cells. *Nat. Rev. Mater.* **2020**, *5*, 44–60. [CrossRef]
14. López, E.; Martí, A.; Antolín, E.; Luque, A. On the potential of silicon intermediate band solar cells. *Energies* **2020**, *13*, 3044. [CrossRef]

15. Vasiliev, M.; Nur-E-Alam, M.; Alameh, K. Recent developments in solar energy-harvesting technologies for building integration and distributed energy generation. *Energies* **2019**, *12*, 1080. [CrossRef]
16. Yadav, P.; Prochowicz, D.; Saliba, M.; Boix, P.P.; Zakeeruddin, S.M.; Grätzel, M. Interfacial kinetics of efficient perovskite solar cells. *Crystals* **2017**, *7*, 252. [CrossRef]
17. Min, K.H.; Min, K.H.; Kim, T.; Kang, M.G.; Song, H.E.; Kang, Y.; Lee, H.S.; Kim, D.; Park, S.; Lee, S.H. An analysis of fill factor loss depending on the temperature for the industrial silicon solar cells. *Energies* **2020**, *13*, 2931. [CrossRef]
18. Kosyachenko, L.A.; Mathew, X.; Paulson, P.D.; Lytvynenko, V.Y.; Maslyanchuk, O.L. Optical and recombination losses in thin-film Cu(In,Ga)Se_2 solar cells. *Sol. Energy Mater. Sol. Cells* **2014**, *130*, 291–302. [CrossRef]
19. Bai, Q.; Yang, H.; Cheng, X.; Wang, H. Recombination parameters of the diffusion region and depletion region for crystalline silicon solar cells under different injection levels. *Appl. Sci.* **2020**, *10*, 4887. [CrossRef]
20. Bosman, L.B.; Leon-Salas, W.D.; Hutzel, W.; Soto, E.A. PV system predictive maintenance: Challenges, current approaches, and opportunities. *Energies* **2020**, *16*, 1398. [CrossRef]
21. Altıntaş, M.; Arslan, S. The study of dust removal using electrostatic cleaning system for solar panels. *Sustainability* **2021**, *13*, 9454. [CrossRef]
22. Vieira, R.G.; de Araújo, F.M.U.; Dhimish, M.; Guerra, M.I.S. A comprehensive review on bypass diode application on photovoltaic modules. *Energies* **2020**, *13*, 2472. [CrossRef]
23. Bai, J.; Zong, X. Global solar radiation transfer and its loss in the atmosphere. *Appl. Sci.* **2021**, *11*, 2651. [CrossRef]
24. Derbeli, M.; Barambones, O.; Silaa, M.Y.; Napole, C. Real-time implementation of a new MPPT control method for a DC-DC boost converter used in a PEM fuel cell power system. *Actuators* **2020**, *9*, 105. [CrossRef]
25. Zhang, S.; Peng, J.; Qian, H.; Shen, H.; Wei, Q.; Lian, W.; Ni, Z.; Jie, J.; Zhang, X.; Xie, L. The impact of thermal treatment on light-induced degradation of multicrystalline silicon PERC solar cell. *Energies* **2019**, *12*, 416. [CrossRef]
26. Maxim, A.A.; Sadyk, S.N.; Aidarkhanov, D.; Surya, C.; Ng, A.; Hwang, Y.H.; Atabaev, T.S.; Jumabekov, A.N. PMMA thin film with embedded carbon quantum dots for post-fabrication improvement of light harvesting in perovskite solar cells. *Nanomaterials* **2020**, *10*, 291. [CrossRef] [PubMed]
27. Dufo-López, R.; Cortés-Arcos, T.; Artal-Sevil, J.S.; Bernal-Agustín, J.L. Comparison of lead-acid and li-ion batteries lifetime prediction models in stand-alone photovoltaic systems. *Appl. Sci.* **2021**, *11*, 1099. [CrossRef]

engineering proceedings

MDPI

Proceeding Paper

A Novel PDMS-Based Microfeature-Size Fabrication Method for Biocompatible and Flexible Devices †

Fatemeh Mashayekhi *, Faezeh Shanehsazzadeh and Mehdi Fardmanesh

Department of Electrical Engineering, Sharif University of Technology, Tehran 1458889694, Iran; f.shanehsazzadeh@alum.sharif.edu (F.S.); fardmanesh@sharif.edu (M.F.)

* Correspondence: fatemeh.mashayekhi@ee.sharif.edu

† Presented at the 2nd International Electronic Conference on Applied Sciences, 15–31 October 2021; Available online: https://asec2021.sciforum.net/.

Abstract: This article proposes a novel cost-effective method to achieve microfeature-sized patterns on Polydimethylsiloxane (PDMS) substrates. As a biocompatible, flexible, economical, and easy-to-use polymer benefiting the trait of mechanical impedance close to that of soft tissues, PDMS is the best candidate to be used where we need communication between the electrical circuits and soft tissues. Additionally, PDMS can be matched with tissue's different shapes and doesn't cause any trauma. The proposed approach eliminates complex and high-cost manufacturing methods of microfeature-sized patterns on PDMS, such as conventional microfabrication methods. Our technique takes advantage of not requiring standard photolithography processes, making it simple and cost-effective. This manner can be used for various purposes, such as micro-fluidic chip fabrication, bio-sensing applications, neuroscience research and neural prosthetics such as electrocorticogram (ECoG) and, in general, where microfeature-size patterning on PDMS is required. To prove the method's functionality, we fabricated a test sample. Firstly, the scaffold was fabricated using a conventional laser engraver and Poly(methylmethacrylate) (PMMA). Then, a mold was made using this scaffold from PDMS. In the last step, a typical commercial photoresist was applied as an anti-adhesion layer between the PDMS mold and the sample to make the sample peel off the mold surface easily. The final sample indicated that the pattern's feature size was around 200 micrometers and that the required patterns were very close to the desired form possible.

Keywords: microfabrication; Polydimethylsiloxane (PDMS); flexible substrate; biocompatible substrate; biosensor; microfluidics

Citation: Mashayekhi, F.; Shanehsazzadeh, F.; Fardmanesh, M. A Novel PDMS-Based Microfeature-Size Fabrication Method for Biocompatible and Flexible Devices. *Eng. Proc.* **2021**, *11*, 36. https://doi.org/10.3390/ASEC2021-11132

Academic Editor: Roger Narayan

Published: 15 October 2021

1. Introduction

Flexible, elastic, and durable soft materials lead the path for future electronics applications in diagnostics and personal healthcare [1]. As a biocompatible, flexible, simple processing, optically transparent, and cost-effective polymer, PDMS (polydimethylsiloxane) is one of the most frequently applied substrate layers [2,3] in various applications including wearable sensors [4,5], epiretinal prosthetics [6–8], electronic textiles [9,10], stretchable conductors [11,12], etc. PDMS can be made in different shapes to fit biological tissues. Moreover, its mechanical impedance matching property to soft tissues such as the spinal cord and the oxygen-permeability make it one of the best MEA/neural tissue interfaces [13–15]. Other flexible substrate options include parylene-c [16,17], Ecoflex 00-30 [18], silicon elastomers like RTV-2 [19], and so on. PDMS Young's modulus is reported in the range of 0.4–1.0 MPa [20] while Young's moduli of parylene and polyimide, as other popular materials, are respectively in the range of 4–4.5 and 2.3–2.8 GPa [21,22]. These make PDMS a suitable substrate material for communication between an electrical circuit and soft tissues [23,24]. For instance, PMDS is employed in neural stimulation devices to activate targeted neurons accurately [25,26]. We desired a sample that was biocompatible, flexible, and long-lasting, thus PDMS was the ideal material pick for the sample. Another PDMS's benefit is its simplicity in fabrication. Whereas, for example, the parylene-c fabrication method necessitates specialized equipment, making it expensive [16]. Another option is Ecoflex 00-30, which has excellent stretchability but should be assessed for biocompatibility for each application [27]. RTV-2 is a low-cost silicone rubber that may be utilized as a protection layer for electronics systems since it remains flexible over a large temperature range of −80 °C to +250 °C. RTV-2 has a low surface tension, which allows it to replicate surface detail and makes it a good choice for molding applications [19]. Consequently, PDMS was the ideal material pick for the sample in many applications because of its biocompatibility, flexibility, long-lasting, simple fabrication process, ets. Which convinces us to focus on PDMS as samples' body material. Following this, since we need a flexible mold in the proposed approach, we tested Ecoflex00-30, RTV-2, and PDMS as mold materials to choose the best one.

Besides material features, the processing technique also plays an important role. Micropatterning is one of the most critical processes in a device fabrication procedure. PDMS is typically patterned using either the conventional photolithography method (by adding photosensitive composites) [28,29] or molding techniques [30]. The conventional photolithography method is incredibly accurate and can manufacture micro-feature size patterns. Not only this approach is complicated, but also the procedure must be performed in a particular environment, such as a clean and yellow room, with special instruments, and by professionals. The alternative choice would be to use the 3D-printing technique to design and fabricate molds. A peeling-off operation is required when using typical 3D-printing materials, however micro-feature size patterns on micrometer thickness substrates cannot be achieved and this is where peeling off might be challenging. Although the peeling-off procedure can be omitted by employing solvable 3D-printing materials, these molds are not reusable.

This article provided an easy method to micropattern PDMS and the resulting mold can be reused several times. Using this approach, Ecoflex00-30, RTV-2, and PDMS were used to create distinct molds. Comparing different molds materials, PDMS was eventually chosen as the final mold material. An anti-adhesion layer was required to separate the PDMS mold from the sample. Separation was evaluated using silicone spray, water and sugar solution, and a typical commercial photoresist as a low-cost photoresist layer. The commercial photoresist was then chosen. The resulting sample's patterns were quite similar to the desired ones, with a feature size of roughly 200 micrometers which is desired for many applications.

2. Materials and Methods

To fabricate the device firstly, we designed the desired pattern using COREL DRAW. After that, a conventional laser engraver was used to make patterns on the PMMA. PMMA was chosen as scaffold material due to its availability and the fact that high accuracy can be easily achieved with conventional laser. Lasers' power and delay time were optimized to obtain the best result on the PMMA scaffold. The chosen design contained two parallel lines with a thickness of 200 micrometers and a distance of 1000 micrometers from each other, as shown in Figure 1.

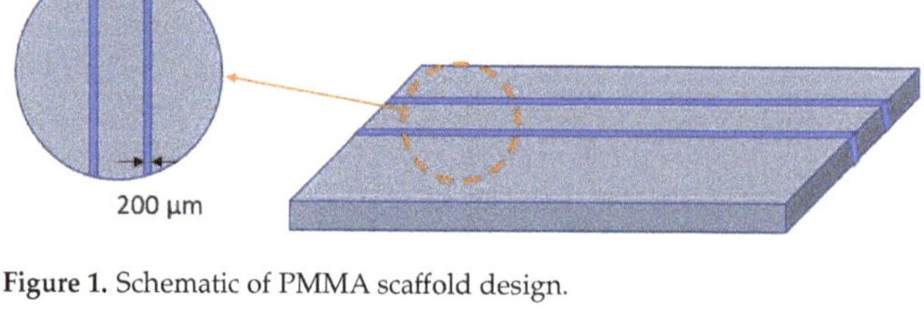

Figure 1. Schematic of PMMA scaffold design.

After that, for mold fabrication, RTV-2 was chosen for the first test. Its mixture was made by adding the second part of RTV-2 and diluent liquid, 4% and 5% of the mass, respectively, and then adequately mixed with a glass stirring rod. Then the mixture was placed in a desiccator to eliminate the bubbles for less than 5 min. After that, it was poured on the scaffold and cured for 24 h on a hotplate at 120 °C. The resulted mold was so sticky and unsuitable for our needs.

Ecoflex00-30 was the second material tested for mold fabrication. Its two parts were homogenized in 1:1 ratio and bubbles were eliminated in the same manner previously described. The resulting compound was poured on the scaffold and allowed to cure for 1 h in an oven at 60 °C. After the curing phase, it was clear that Ecoflex00-30 was stuck to the PMMA scaffold and couldn't be peeled off properly.

The PDMS substance was the third to be tried. SYLGARD 184 was the product that we utilized. The curing agent and base elastomer were combined in a 1:10 ratio to achieve our desired mechanical characteristics. After that, a desiccator was used for around 20 min to clear all of the bubbles. The mixture was then poured over the scaffold and left to cure for one hour in the oven at 85 °C. The PDMS mold simply separated from the PMMA surface due to the poor adhesion between PMMA and PDMS. Patterns on the PDMS mold were evaluated and found to be in good condition. The final PDMS mold and PMMA scaffold are shown in Figure 2.

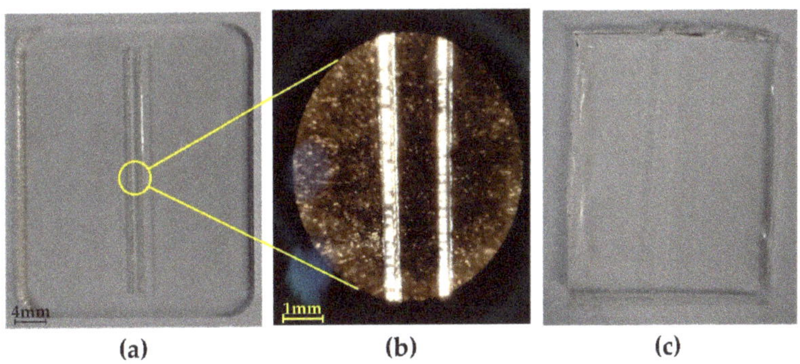

Figure 2. (**a**) Engraved PMMA based layer to be used as the scaffold; (**b**) Magnification of engraved patterns on PMMA scaffold; (**c**) Fabricated mold from PDMS.

Since both the mold and sample were made from PDMS, an anti-adhesion layer was essential to make two layers separation feasible. After selecting the best material for mold fabrication, several anti-adhesion coatings such as silicon spray, water and sugar solvent, and a typical commercial photoresist were investigated. Peeling off was feasible with the applying a silicon spray layer, although patterns were not in the ideal condition, and peeling off thin sample layers was challenging. The water and sugar solvent with different concentrations prevent two layers from adhering, but patterns were not adequate because of the considerable sugar particles size compared to micropatterns. As the last tested anti-adhesion layer, a low-cost commercial photoresist, was spin-coated on the mold and baked at 90 °C for around 10 min. The desired photoresist should only have the appropriate density to achieve a layer with a thickness in the range of micrometers after spin coating. Additionaly, it should be detached from the PDMS surface without any damage to the surface. As a result, it could be any commercial photoresist without any critical consideration. Here, after spin coating, a layer with a thickness of about 2 micrometers was obtained, which was thin enough compared to patterns. Using photoresist, the sample was separated from the mold effortlessly and patterns on the sample were in very good shape.

For the sample layer, PDMS was made as explained before and the mixture was spin-coated on the photoresist. Spin-coating was started at 400 rpm/s and continued in 500 rpm/s to obtain a 300 micrometers thickness uniform layer which was then cured for

1 h at 85 °C in the oven. After curing was completed, two layers were separated pretty readily and patterns were in the best shape. Figure 3 represents the entire procedure and Figure 4 depicts the final sample results. Different tested substrate materials and results are summarized in Table 1.

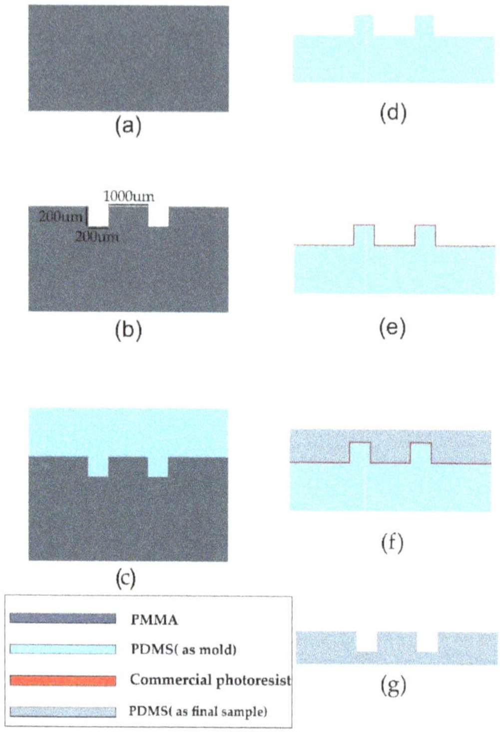

Figure 3. Fabrication procedure. (**a**) PMMA plate; (**b**) Engraved PMMA based scaffold; (**c**) Spin coated PDMS on the scaffold to fabricate mold; (**d**) Fabricated PDMS based mold; (**e**) Spin coated anti-adhesion layer on the mold; (**f**) Spin coated PDMS on the anti-adhesion layer; (**g**) Final sample after separation from mold and eliminating anti-adhesion layer.

Figure 4. Final fabricated sample. pattern's feature size is 200 micrometers.

Table 1. Different substrate materials and their tested results.

Mold Materials	PMMA Peeling Off Feasibility	Silicon Spray Anti-Adhesion Layer	Water and Sugar Solution Anti-Adhesion Layer	Commercial Photoresist Anti-Adhesion Layer
RTV-2	☑	☒	☒	☒
Ecoflex00-30	☒	☒	☒	☒
PDMS	☑	☑	☑	☑

3. Discussion

In this work, a new fabrication method of micropatterning on PDMS substrate was proposed. The presented fabrication approach doesn't include the standard photolithography process so it doesn't necessitate expensive material, equipment, and professionalists making it a simple cost-effective patterning method. Compared to other methods [28,29], we didn't add any extra materials to PDMS to make it photosensitive in our novel patterning approach, either. This is important since any other additives would change PDMS characteristics, such as biocompatibility and flexibility. When benzophenone is added to PDMS, for example, live cells would inevitably die [31]. The use of photoPDMS (PDMS with photosensitive particles inserted) may hinder bioapplications, necessitating the use of an extraction method. In the extraction step, several different chemical materials such as n-pentane, xylene, ethanol (200 proof) are employed to unbound oligomers and enhance polymer cross-linking [31]. Although the extraction process would promote PDMS biocompatibility, it alters material properties and this might be undesired, thus the new form of PDMS should be characterized like the way it's done in [32]. Added materials are costly, as well as, this process might take several days to be completed.

The fabrication of 3D-printed-based molds can be done in two ways. The traditional technique of 3D printing employs unsolved materials, demanding a peeling-off procedure. Due to the unavoidable peeling off step, micropatterns on micro thickness substrates can not be achievable. In comparison with the solvable 3D-printing-based technique, our method has the advantage of being reusable. If 3D-printed solvable molds are employed to achieve micropatterns, the built mold can only be used once. As a result, the entire patterning process takes a lot more time and money. However, PDMS-based molds are used in this paper, making the procedure repeatable. Table 2 shows a summary of techniques to produce micropatterns on PDMS.

Table 2. Micropattern fabrication techniques on PDMS.

Micropatterning Technique	Advantages	Fabrication Challenges
photolithography methods	• Tunable and nanoscale feature size	• complex fabrication process • requiring high-cost photoresist • high-cost additional materials • needing extraction process in bioapplications • requiring characterization process
Molding techniques	• achievable microfeature size patterns • relatively cost-effective	• not repeatable • high-cost materials • slow process
PDMS-based-molding technique	• achievable microfeature size patterns • low cost • repeatable	• requiring optimization for spin coating speed • limitation on fabrication of 3D structures

4. Conclusions

In this article, we proposed an innovative and cost-effective method for creating micropatterns on PDMS substrate. There was a biocompatible, micrometer thickness, and the sample substrate desired, leading to the choice of PDMS as the material for the sample substrate. We optimized a conventional laser engraver to create our desired pre-designed pattern on a PMMA sheet, then used it as the scaffold. Following that, Ecoflex 00-30, RTV-2, and PDMS were evaluated to determine the best mold material candidate. PDMS was found to be the best option, benefiting its flexibility property making it possible to be peeled off from scaffold without causing any damage. Choosing PDMS for both mold and sample material required an anti-adhesion layer to make mold and sample separation feasible. Silicon spray, water and sugar mixture in different concentrations, and a low-cost photoresist were tested to hinder mold and sample adhesion. The low-cost photoresist was the final choice because of its desired density and thickness, and also it could be spin-coated on mold leading to a uniform layer. Using this photoresist, the sample was separated from the mold easily and patterns were in good shape indicating that microfeature-size patterns on micro thickness substrate layers are achievable employing the presented method.

Author Contributions: Conceptualization, F.M. and F.S.; methodology, F.M. and F.S.; investigation, F.M.; resources, M.F.; data curation, F.M. and F.S.; writing—original draft preparation, F.M.; writing—review and editing, F.S.; visualization, F.M. and F.S.; supervision, F.S. and M.F.; project administration, F.S. All authors have read and agreed to the published version of the manuscript.

Funding: This research received no external funding.

Institutional Review Board Statement: Not applicable.

Informed Consent Statement: Not applicable.

Conflicts of Interest: The authors declare no conflict of interest.

References

1. Heath, J.R.; Ratner, M.A. *Molecular Electronics*; American Institute of Physics: College Park, MD, USA, 2003.
2. Li, L.; Chin, W.S. Rapid fabrication of a flexible and transparent Ag Nanocubes@ PDMS Film as a SERS substrate with high performance. *ACS Appl. Mater. Interfaces* **2020**, *12*, 37538–37548. [CrossRef] [PubMed]
3. Qi, D.; Zhang, K.; Tian, G.; Jiang, B.; Huang, Y. Stretchable electronics based on PDMS substrates. *Adv. Mater.* **2021**, *33*, 2003155. [CrossRef] [PubMed]
4. Yu, Y.; Peng, S.; Blanloeuil, P.; Wu, S.; Wang, C.H. Wearable Temperature Sensors with Enhanced Sensitivity by Engineering Microcrack Morphology in PEDOT: PSS–PDMS Sensors. *ACS Appl. Mater. Interfaces* **2020**, *12*, 36578–36588. [CrossRef] [PubMed]
5. Chen, J.; Zhu, Y.; Jiang, W. A stretchable and transparent strain sensor based on sandwich-like PDMS/CNTs/PDMS composite containing an ultrathin conductive CNT layer. *Compos. Sci. Technol.* **2020**, *186*, 107938. [CrossRef]
6. Ko, H.C.; Stoykovich, M.P.; Song, J.; Malyarchuk, V.; Choi, W.M.; Yu, C.J.; Geddes Iii, J.B.; Xiao, J.; Wang, S.; Huang, Y.; et al. A hemispherical electronic eye camera based on compressible silicon optoelectronics. *Nature* **2008**, *454*, 748–753. [CrossRef] [PubMed]
7. Seo, H.W.; Kim, N.; Kim, S. Fabrication of subretinal 3D microelectrodes with hexagonal arrangement. *Micromachines* **2020**, *11*, 467. [CrossRef]
8. Zhou, M.; Kang, D.H.; Kim, J.; Weiland, J.D. Shape morphable hydrogel/elastomer bilayer for implanted retinal electronics. *Micromachines* **2020**, *11*, 392. [CrossRef]
9. Choi, M.C.; Kim, Y.; Ha, C.S. Polymers for flexible displays: From material selection to device applications. *Prog. Polym. Sci.* **2008**, *33*, 581–630. [CrossRef]
10. Ojuroye, O.; Torah, R.; Beeby, S. Modified PDMS packaging of sensory e-textile circuit microsystems for improved robustness with washing. *Microsyst. Technol.* **2019**, 1–18. [CrossRef]
11. Choi, S.; Han, S.I.; Kim, D.; Hyeon, T.; Kim, D.H. High-performance stretchable conductive nanocomposites: Materials, processes, and device applications. *Chem. Soc. Rev.* **2019**, *48*, 1566–1595. [CrossRef]
12. Kim, J.; Lee, M.; Shim, H.J.; Ghaffari, R.; Cho, H.R.; Son, D.; Jung, Y.H.; Soh, M.; Choi, C.; Jung, S.; et al. Stretchable silicon nanoribbon electronics for skin prosthesis. *Nat. Commun.* **2014**, *5*, 5747. [CrossRef] [PubMed]
13. Shanehsazzadeh, F.; Rouhi, S.; Ahmadvand, T.; Namazi, M.; Kiani, S.; Fardmanesh, M. A Novel, Low Cost and Versatile Fabrication Method of Flexible Multi-electrode Array for Spinal Cord Stimulation. In Proceedings of the 2020 27th National and 5th International Iranian Conference on Biomedical Engineering (ICBME), Tehran, Iran, 26–27 November 2020; pp. 138–142.
14. Hasan, M.N.; Radwan, A.N.; Kim, M.; Kucukal, E.; Maji, D.; Pashaei, V.; Chung, C.Y.; Kakkar, A.; Gurkan, U.A. Emerging micro and nanotechnologies in neuroscience: Devices, fabrication methods, and implementation in monitoring of neural activity and drug delivery. *Technology* **2019**, *7*, 57–83. [CrossRef]
15. Ahmadvand, T.; Mirsadeghi, S.; Shanehsazzadeh, F.; Kiani, S.; Fardmanesh, M. A Novel Low-Cost Method for Fabrication of 2D Multi-Electrode Array (MEA) to Evaluate Functionality of Neuronal Cells. *Proceedings* **2020**, *60*, 51.
16. Ortigoza-Diaz, J.; Scholten, K.; Larson, C.; Cobo, A.; Hudson, T.; Yoo, J.; Baldwin, A.; Weltman Hirschberg, A.; Meng, E. Techniques and considerations in the microfabrication of Parylene C microelectromechanical systems. *Micromachines* **2018**, *9*, 422. [CrossRef] [PubMed]

17. Sanzari, I.; Callisti, M.; De Grazia, A.; Evans, D.J.; Polcar, T.; Prodromakis, T. Parylene C topographic micropattern as a template for patterning PDMS and Polyacrylamide hydrogel. *Sci. Rep.* **2017**, *7*, 5764. [CrossRef] [PubMed]
18. Kim, Y.S.; Mahmood, M.; Lee, Y.; Kim, N.K.; Kwon, S.; Herbert, R.; Kim, D.; Cho, H.C.; Yeo, W.H. All-in-One, wireless, stretchable hybrid electronics for smart, connected, and ambulatory physiological monitoring. *Adv. Sci.* **2019**, *6*, 1900939. [CrossRef] [PubMed]
19. RTV-2 Silicones; The Durability, Stability and High Temperature Resistance of RTV-2 Silicone Rubber. Available online: https://www.elkem.com/silicones/technologies/elastomers/rtv-2/ (accessed on 28 February 2022).
20. Armani, D.; Liu, C.; Aluru, N. Re-configurable fluid circuits by PDMS elastomer micromachining. In Proceedings of the Technical Digest. IEEE International MEMS 99 Conference. Twelfth IEEE International Conference on Micro Electro Mechanical Systems (Cat. No. 99CH36291), Orlando, FL, USA, 21 January 1999.
21. Suzuki, Y. Micromachined high-aspect-ratio parylene beam and its application to low-frequency seismometer Micro Electro Mechanical Systems. In Proceedings of the Sixteenth Annual International Conference on Micro Electro Mechanical Systems, MEMS-03 Kyoto, Kyoto, Japan, 23 January 2003.
22. Rousche, P.J.; Pellinen, D.S.; Pivin, D.P.; Williams, J.C.; Vetter, R.J.; Kipke, D.R. Flexible polyimide-based intracortical electrode arrays with bioactive capability. *IEEE Trans. Biomed. Eng.* **2001**, *48*, 361–371. [CrossRef] [PubMed]
23. Terkan, K.; Zurita, F.; Jamal Khalaf, T.; Rinklin, P.; Teshima, T.; Kohl, T.; Wolfrum, B. Soft peripheral nerve interface made from carbon nanotubes embedded in silicone. *APL Mater.* **2020**, *8*, 101111. [CrossRef]
24. Guo, L.; DeWeerth, S.P. Implementation of integratable PDMS-based conformable microelectrode arrays using a multilayer wiring interconnect technology. In Proceedings of the Annual International Conference of the IEEE Engineering in Medicine and Biology Society, Minneapolis, MN, USA, 3–6 September 2009.
25. Branner, A.; Stein, R.B.; Normann, R.A. Selective stimulation of cat sciatic nerve using an array of varying-length microelectrodes. *J. Neurophysiol.* **2001**, *85*, 1585–1594. [CrossRef]
26. McDonnall, D.; Clark, G.A.; Normann, R.A. Interleaved, multisite electrical stimulation of cat sciatic nerve produces fatigue-resistant, ripple-free motor responses. *IEEE Trans. Neural Syst. Rehabil. Eng.* **2004**, *12*, 208–215. [CrossRef]
27. Ecoflex ™ 00-30. Available online: https://www.smooth-on.com/products/ecoflex-00-30/ (accessed on 28 February 2022).
28. Cho, D.; Park, J.; Kim, T.; Jeon, S. Recent advances in lithographic fabrication of micro-/nanostructured polydimethylsiloxanes and their soft electronic applications. *J. Semicond.* **2019**, *40*, 111605. [CrossRef]
29. Cong, H.; Pan, T. Photopatternable conductive PDMS materials for microfabrication. *Adv. Funct. Mater.* **2008**, *18*, 1912–1921. [CrossRef]
30. Chen, Q.; Zhao, J.; Ren, J.; Rong, L.; Cao, P.F.; Advincula, R.C. 3D printed multifunctional, hyperelastic silicone rubber foam. *Adv. Funct. Mater.* **2019**, *29*, 1900469. [CrossRef]
31. Millet, L.J.; Jain, A.; Gillette, M.U. Less is More: Oligomer extraction and hydrothermal annealing increase PDMS bonding forces for new microfluidics assembly and for biological studies. *bioRxiv* **2017**, 150953. [CrossRef]
32. Lee, J.N.; Park, C.; Whitesides, G.M. Solvent compatibility of poly (dimethylsiloxane)-based microfluidic devices. *Anal. Chem.* **2003**, *75*, 6544–6554. [CrossRef] [PubMed]

engineering proceedings

MDPI

Proceeding Paper

Enhanced Reconstruction of Spatially Incoherent Digital Holograms Using Synthetic Point Spread Holograms †

Vijayakumar Anand [1,2,*], Joseph Rosen [3], Soon Hock Ng [1], Tomas Katkus [1], Denver P. Linklater [4], Elena P. Ivanova [4] and Saulius Juodkazis [1,5]

1 Optical Sciences Centre and ARC Training Centre in Surface Engineering for Advanced Materials (SEAM), School of Science, Swinburne University of Technology, Hawthorn, VIC 3122, Australia; soonhockng@swin.edu.au (S.H.N.); tkatkus@swin.edu.au (T.K.); sjuodkazis@swin.edu.au (S.J.)
2 Institute of Physics, University of Tartu, 50411 Tartu, Estonia
3 School of Electrical and Computer Engineering, Ben-Gurion University of the Negev, P.O. Box 653, Beer-Sheva 8410501, Israel; rosenj@bgu.ac.il
4 School of Science, RMIT University, Melbourne, VIC 3000, Australia; denver.linklater@rmit.edu.au (D.P.L.); elena.ivanova@rmit.edu.au (E.P.I.)
5 Tokyo Tech World Research Hub Initiative (WRHI), School of Materials and Chemical Technology, Tokyo Institute of Technology, 2-12-1, Ookayama, Meguro-ku, Tokyo 152-8550, Japan
* Correspondence: vanand@swin.edu.au

† Presented at the 2nd International Electronic Conference on Applied Sciences, 15–31 October 2021; Available online: https://asec2021.sciforum.net/.

Abstract: Coded aperture imaging (CAI) methods offer multidimensional and multispectral imaging capabilities with minimal resources than what is needed in a lens-based direct imager. In the CAI method, the light diffracted from an object is modulated by a coded mask, and the resulting intensity distribution is recorded. Most of the CAI techniques involve two steps: the recording of the point spread function (PSF) and object intensity under identical conditions and with the same coded mask. The image of the object is reconstructed by computationally processing the PSF and object intensity. The above recording and reconstruction procedure precludes the introduction of special beam characteristics in imaging, such as a direct imager. In this study, a postprocessing approach is developed, where synthetic PSFs capable of introducing special beam characteristics when processed with the object intensity are generated using an iterative algorithm. The method is applied to generate edge-enhanced images in both CAI as well as Fresnel incoherent correlation holography methods.

Keywords: edge enhancement; coded aperture imaging; Fresnel incoherent correlation holography; phase-retrieval algorithm; holography; incoherent imaging; high-speed imaging

Citation: Anand, V.; Rosen, J.; Ng, S.H.; Katkus, T.; Linklater, D.P.; Ivanova, E.P.; Juodkazis, S. Enhanced Reconstruction of Spatially Incoherent Digital Holograms Using Synthetic Point Spread Holograms. *Eng. Proc.* **2021**, *11*, 37. https://doi.org/10.3390/ASEC2021-11162

Academic Editor: Nunzio Cennamo

Published: 15 October 2021

1. Introduction

The coded aperture imaging (CAI) technique is a widely used computational optical method that has efficiently replaced the need for high-quality optical components in direct lens-based imagers with computational methods [1,2]. In direct imaging methods, the image of an object is directly formed on the image sensor. In CAI, two steps are necessary for imaging. In the first step, a point object is mounted in the object plane, and the light from it is modulated by a coded mask (CM), and the resulting intensity distribution—point spread function (I_{PSF})—is recorded. In the next step, an object is mounted at the same location as the point object, and with the same CM and identical conditions, a second intensity distribution is recorded. The two intensity distributions are processed in a computer to reconstruct the object information. In a linear, shift-invariant system, the object intensity (I_O) can be expressed as a convolution of the object function O with the PSF, $I_O = O \otimes I_{PSF}$, where '$\otimes$' is a 2D convolutional operator. The image reconstruction is carried out by a cross-correlation given as $I_R = I_O * I_{PSF}$, where '*' is a 2D correlational operator.

The above principle of recording and reconstruction precludes the introduction of special beam characteristics in the imaging system. In direct imaging methods and well-established holography methods such as Fresnel incoherent correlation holography (FINCH), the introduction of beam characteristics is easy and straightforward. Let us consider the case of edge enhancement which is a useful technique in many applications [3,4]. In direct imaging, edge enhancement is achieved by modulating light using a vortex filter [5]. In FINCH, the object hologram is usually formed by interfering two object waves with different quadratic phase modulations and reconstructed by numerically propagating the recorded hologram to one of the image planes [6,7]. In FINCH, edge enhancement is introduced by a modulo-2π phase addition of a vortex filter to one of the quadratic phase masks used to modulate the object wave [8]. The resulting hologram, when propagated to one of the image planes, yields an edge-enhanced image of the object. A similar approach was attempted in FINCH with the reconstruction method of CAI in a simulative study, and no edge enhancement was noticed [9,10]. In FINCH, the original reconstruction mechanism is independent of the modulating function, and whether the phase mask is quadratic or quadratic with a vortex filter, the hologram is numerically propagated to one of the image planes. In FINCH, when the CAI reconstruction method is implemented, i.e., instead of numerical backpropagation, a cross-correlation with the PSF is carried out, and the scenario becomes different as the reconstructing function, i.e., PSF, is dependent upon the modulation function. The PSF and the object holograms are recorded under identical conditions and using the same modulation function of the vortex filter. Therefore, in this case, during reconstruction by cross-correlation, the edge-enhancing characteristics of the vortex filter are not expressed. In this study, an iterative algorithm is developed which can synthesize special PSFs from the recorded PSFs. The special PSFs when processed with the recorded object intensity distributions can produce enhanced images of the object. While the proposed approach can be used for many applications, in this manuscript, the edge enhancement is demonstrated.

2. Materials and Methods

The optical configuration of the generalized imaging system is shown in Figure 1a. The light from a point object was modulated by optical modulators consisting of lenses, and CMs and I_{PSF} were recorded and provided as input into an iterative algorithm shown in Figure 1b. In this case, the only requirement was that the imaging system had to be a linear shift-invariant system and, therefore, the optical configuration could either be as simple with a single optical modulator or multiple optical elements and components. The iteration occurred between two planes of interest P_1 and P_2 as shown in Figure 1a. The ground truth image was the required output I_D by cross-correlation between the recorded I_{PSF} and synthetic I_{PSF}. This output was transferred to every recorded object point when the synthetic I_{PSF} was cross-correlated with the object intensity distribution. The recorded I_{PSF} was Fourier transformed, and the complex conjugate was calculated ($\widetilde{I}_{PSF}{}^{*}$). The Fourier transform of the initial guess synthetic PSF was assumed to be a random phase-only function, which was multiplied by $\widetilde{I}_{PSF}{}^{*}$, and the result was Fourier transformed. The resulting complex amplitude's magnitude was replaced by the far-field diffraction pattern of a vortex filter [11], but its phase was retained. The resulting complex amplitude was inverse Fourier transformed, and the result was multiplied by $\widetilde{I}_{PSF}{}^{-1*}$ and I^{-1}. This process was iterated until an optimal solution was obtained [12]. Similar to most iterative algorithms, the optimal solution was achieved using an error function that could be the root mean square error between the ground truth and the result obtained from the algorithm. The resulting solution was correlated with the recorded object intensity distribution using a nonlinear filter to reconstruct the edge-enhanced image of the object [13].

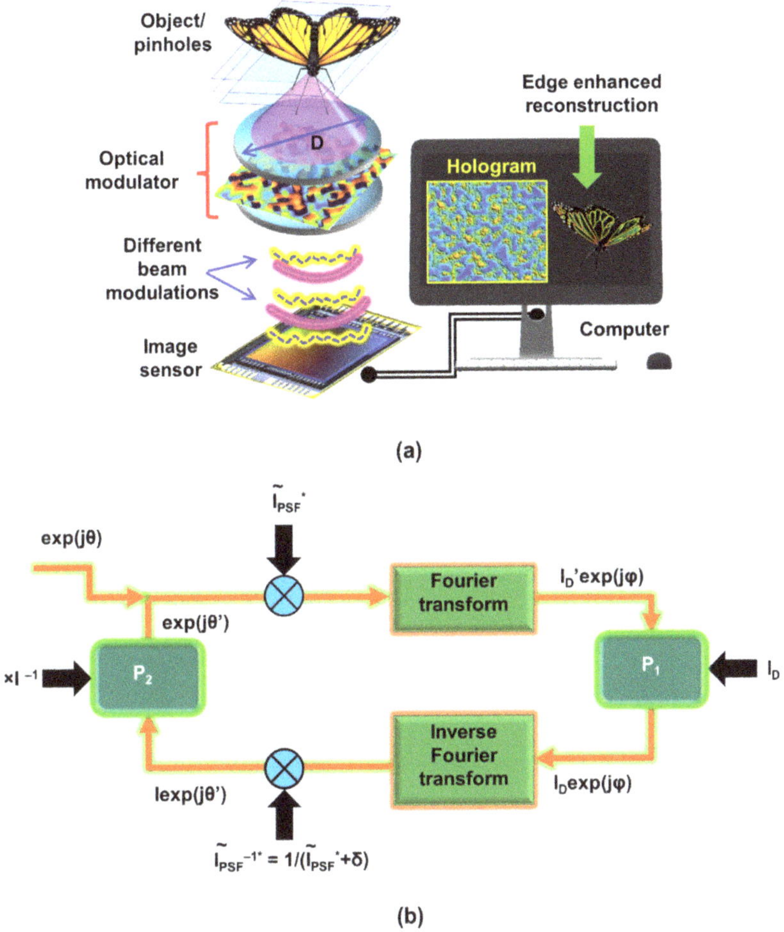

Figure 1. (**a**) Optical configuration of a generalized imaging system. (**b**) Schematic of the iterative algorithm for generating the synthetic PSFs. The symbol (~) above I_{PSF} represents a Fourier transform operation, ⊗ represents a multiplication operation, and * represents a complex conjugate. The far-field diffraction of a vortex filter input was I_D in P_1. The inverse of the matrix I was multiplied at P_2 to obtain a phase-only output.

3. Results

The proposed method, being a completely postprocessing one, neither required a vortex filter as in the conventional method, nor did it affect the temporal resolution of the system. The method was applied to two cases: FINCH and CAI.

The FINCH experiment was carried out using randomly multiplexed diffractive lenses mounted between the object and the image sensor [9,10]. Two FINCH holograms were recorded: PSF and object hologram as shown in Figure 2a,b, respectively. The reconstructed image is shown in Figure 2c. The phase of the Fourier transform of the synthesized PSF is shown in Figure 2d, and the edge-enhanced reconstruction is shown in Figure 2e. In the CAI experiment, a PSF was recorded (Figure 2f) by modulating the light diffracted from a point object by a mask consisting of a random array of pinholes. In the next step, a spark was generated and recorded, as shown in Figure 2g [2,14,15]. The recorded intensity distribution was processed with the PSF, and the reconstructed image is shown in Figure 2h.

The phase of the Fourier transform of the synthesized PSF is shown in Figure 2i, and the edge-enhanced reconstruction is shown in Figure 2j.

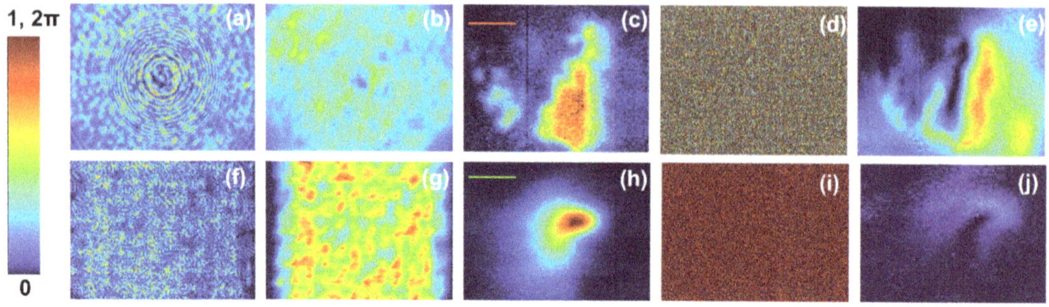

Figure 2. Images of the (**a**) PSF, (**b**) object intensity distributions, (**c**) reconstructed image of Fungi sample with recorded PSF, (**d**) phase of the Fourier transform of synthetic PSF, and (**e**) edge-enhanced reconstruction for FINCH system. Images of the (**f**) PSF, (**g**) object intensity distributions, (**h**) reconstructed image of a spark with recorded PSF, (**i**) phase of the Fourier transform of synthetic PSF, and (**j**) edge-enhanced reconstruction for CAI system. The red scale bar is 150 μm, and the green scale bar is 1 mm.

4. Conclusions

A completely computational enhancement method was developed for CAI methods utilizing the indirect reconstruction method of the cross-correlation of the object intensity distribution with the PSF. The developed method could be implemented offline, and so it did not affect the temporal resolution of the imaging system. Furthermore, the method did not require any additional optical component such as vortex filters, and so it is low-cost in comparison to existing edge-enhancement methods. Secondly, the edge enhancement in FINCH required at least three camera shots for every object, i.e., for the m objects, 3m camera shots were needed. The proposed method required only a single camera shot, and the PSF recording was conducted only once and did not need to be repeated for every new object. Therefore, for the m objects, the number of camera shots needed was only m + 1. Therefore, when m was large, the impact was clear. The preliminary results were promising when implemented for the fungi sample with well-defined boundaries and a spark event involving a gradual variation in intensity. A strong edge-enhancement was seen in the regions with rapid changes in intensity while a mild edge enhancement is noticed in the regions with a slow variation in intensity in the case of the spark image as expected. The method could be directly extended for implementing many operations such as blurring, sharpening, etc., of reconstructed images without the need for additional optical experiments in CAI methods. Future studies will involve applying the developed method for different image enhancement tasks and quantitative comparison studies with existing methods.

Author Contributions: Conceptualization, J.R. and V.A.; methodology, J.R.; algorithm, J.R.; software, J.R. and V.A.; validation, V.A., S.H.N., T.K., D.P.L. and S.J.; resources, S.J.; DOE fabrication, V.A. and T.K.; bio samples preparation, D.P.L. and E.P.I.; writing—original draft preparation, V.A., J.R. and S.J.; writing—review and editing, V.A., J.R. and S.J.; visualization, J.R.; supervision, J.R., E.P.I. and S.J.; project administration, S.J.; funding acquisition, S.J. All authors have read and agreed to the published version of the manuscript.

Funding: NATO grant no. SPS-985048 and Australian Research Council (LP190100505). J.R. was supported by The Israel Science Foundation (ISF) (1669/16). Funded by European Union's Horizon 2020 research and innovation programme under grant agreement No. 857627 (CIPHR).

Institutional Review Board Statement: Not applicable.

Informed Consent Statement: Not applicable.

Data Availability Statement: All data generated or analyzed during this study are included in this published article.

Acknowledgments: The Nano Lab fabrication facility at Swinburne University of Technology is acknowledged.

Conflicts of Interest: The authors declare no conflict of interest.

References

1. Rosen, J.; Vijayakumar, A.; Kumar, M.; Rai, M.R.; Kelner, R.; Kashter, Y.; Bulbul, A.; Mukherjee, S. Recent advances in self-interference incoherent digital holography. *Adv. Opt. Photonics* **2019**, *11*, 1–66. [CrossRef]
2. Anand, V.; Ng, S.H.; Maksimovic, J.; Linklater, D.; Katkus, T.; Ivanova, E.P.; Juodkazis, S. Single shot multispectral multidimensional imaging using chaotic waves. *Sci. Rep.* **2020**, *10*, 13902. [CrossRef] [PubMed]
3. Monnin, P.; Bulling, S.; Hoszowska, J.; Valley, J.F.; Meuli, R.; Verdun, F.R. Quantitative characterization of edge enhancement in phase contrast X-ray imaging. *Med. Phys.* **2004**, *31*, 1372–1383. [CrossRef]
4. Yue, Y.; Croitoru, M.M.; Bidani, A.; Zwischenberger, J.B.; Clark, J.W. Nonlinear multiscale wavelet diffusion for speckle suppression and edge enhancement in ultrasound images. *IEEE Trans. Med. Imaging* **2006**, *25*, 297–311.
5. Tian, N.; Fu, L.; Gu, M. Resolution and contrast enhancement of subtractive second harmonic generation microscopy with a circularly polarized vortex beam. *Sci. Rep.* **2015**, *5*, 13580. [CrossRef] [PubMed]
6. Rosen, J.; Brooker, G. Digital spatially incoherent Fresnel holography. *Opt. Lett.* **2007**, *32*, 912–914. [CrossRef] [PubMed]
7. Rosen, J.; Brooker, G. Non-Scanning Motionless Fluorescence Three-Dimensional Holographic Microscopy. *Nat. Photonics* **2008**, *2*, 190–195. [CrossRef]
8. Bouchal, P.; Bouchal, Z. Selective edge enhancement in three-dimensional vortex imaging with incoherent light. *Opt. Lett.* **2012**, *37*, 2949–2951. [CrossRef] [PubMed]
9. Anand, V.; Katkus, T.; Lundgaard, S.; Linklater, D.P.; Ivanova, E.P.; Ng, S.H.; Juodkazis, S. Fresnel incoherent correlation holography with single camera shot. *Opto-Electron. Adv.* **2020**, *3*, 200004.
10. Anand, V.; Katkus, T.; Ng, S.H.; Juodkazis, S. Review of Fresnel incoherent correlation holography with linear and non-linear correlations. *Chin. Opt. Lett.* **2021**, *19*, 020501. [CrossRef]
11. Vijayakumar, A.; Bhattacharya, S. *Design and Fabrication of Diffractive Optical Elements with MATLAB*; SPIE Press Book: Bellingham, WA, USA, 2017.
12. Anand, V.; Rosen, J.; Ng, S.H.; Katkus, T.; Linklater, D.P.; Ivanova, E.P.; Juodkazis, S. Edge and Contrast Enhancement Using Spatially Incoherent Correlation Holography Techniques. *Photonics* **2021**, *8*, 224. [CrossRef]
13. Rai, M.R.; Vijayakumar, A.; Ogura, Y.; Rosen, J. Resolution enhancement in nonlinear interferenceless COACH with point response of subdiffraction limit patterns. *Opt. Express* **2019**, *27*, 391–403. [CrossRef] [PubMed]
14. Vijayakumar, A.; Ng, S.H.; Katkus, T.; Juodkazis, S. Spatio-spectral-temporal imaging of fast transient phenomena using a random array of pinholes. *Adv. Photonics Res.* **2021**, *2*, 2000032.
15. Ng, S.H.; Anand, V.; Katkus, T.; Juodkazis, S. Invasive and Non-Invasive Observation of Occluded Fast Transient Events: Computational Tools. *Photonics* **2021**, *8*, 253. [CrossRef]

engineering proceedings

MDPI

Proceeding Paper

Formulation of Effervescent Compact Detergent Tablets with Unique Chemical Compositions †

Rohan Mestri [1], Suraj N. Mali [2,*] and Amit Pratap [1,*]

1 Department of Oils, Oleochemicals and Surfactants Technology, Institute of Chemical Technology (University under Section 3 of UGC Act 1956; Formerly UDCT/UICT), Nathalal Parekh Road, Matunga (East), Mumbai 400 019, India; amitprat2001@gmail.com

2 Department of Pharmaceutical Sciences and Technology, Institute of Chemical Technology (University under Section 3 of UGC Act 1956; Formerly UDCT/UICT), Nathalal Parekh Road, Matunga (East), Mumbai 400 019, India

* Correspondence: mali.suraj1695@gmail.com (S.N.M.); ap.pratap@ictmumbai.edu.in (A.P.)

† Presented at the 2nd International Electronic Conference on Applied Sciences, 15–31 October 2021; Available online: https://asec2021.sciforum.net/.

Abstract: The tablet form of detergent powder is one of the new delivery systems of detergent. It is a compact form of detergent powder with highly active ingredients. The tablet form of detergent reduces the volume of the powder. Due to its compact format, it affects the transportation and packing cost. Thus, we aimed to formulate the unique detergent powder with the chemical combinations of various surface-active agents. The detergent formulation thus contains linear alkyl benzene sulphonate (LABSA), alfa sulfo methyl esters, sodium tripolyphosphate (STPP), sodium hydroxide (NaOH), sodium silicate, sodium sulphate, etc. In our study, the detergent powder is mixed with various disintegrating agents such as corn starch, sodium carboxymethyl cellulose (sodium CMC), silicic acid, sodium carbonate and citric acid. Our compact detergent powder showed better detergency properties. This helps to instantly disintegrate and disperse when contacted with water. Detergent powder composition in our currently formulated tablets caused effervescences and disintegrated within 30 s at room temperature in water. The tablet detergent showed better performance than market detergents.

Keywords: detergent powder; disintegrating agents; tablet; delivery system

Citation: Mestri, R.; Mali, S.N.; Pratap, A. Formulation of Effervescent Compact Detergent Tablets with Unique Chemical Compositions. *Eng. Proc.* **2021**, *11*, 38. https://doi.org/10.3390/ASEC2021-11186

Academic Editor: Nunzio Cennamo

Published: 15 October 2021

1. Introduction

Cleaning products usually consist of ingredients or materials, when treated with water, helps to remove dirt or foreign matter from the surface [1]. The key ingredients in detergent powder are surfactants and builders as they perform the main role in washing processes, and they directly impact detergency performance [2]. The tablet form of the detergent powder is one of the novel formats of detergents having a compact form with highly active ingredients. One or two tablets are enough for the washing instead of one scoop of detergent powder. The life of the tablet detergent is, however, in the hands of the ultimate user, as the properties of the product do not deteriorate significantly before use. Herein, we demonstrate the formulation of a unique composition-based tablet detergent compact for detergency applications.

2. Materials and Methods

Linear alkyl benzene sulphonic acid (LABSA) was obtained from M/s Godrej Industries Pvt. Ltd. Mumbai, Maharashtra, India. as a gift Sample. Sodium tri polyphosphate (STPP), sodium sulphate, sodium silicate, carbon black, and lauric acid were procured from M/s Hi Media Laboratory Pvt. Ltd. Mumbai, Maharashtra, India. Caustic soda and starch were procured from M/s Finar India Pvt. Ltd. Mumbai, Maharashtra, India.

Disintegrating agents silicic acid, sodium bicarbonate and citric acid were obtained from M/s Loba Chemical Pvt. Ltd. Mumbai, Maharashtra, India. Ready for Dying cotton (RFD) was obtained from the Textile Department, Institute of Chemical Technology, Mumbai, India. Coconut oil was procured from the local market of Mumbai, India.

2.1. Formulation of Detergent Tablets

The detergent tablet manufacturing process consists of (i) neutralization of LABSA (ii) formulation of detergent and (iii) compressing to tablet form along with disintegrating agents (Figure 1).

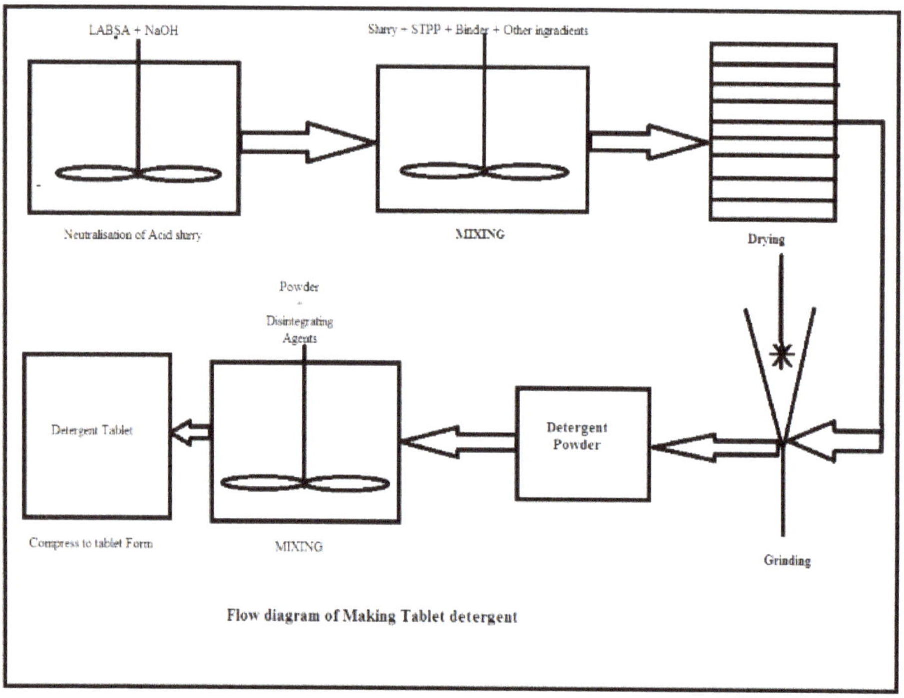

Figure 1. Flow diagram of making tablet detergent compact.

(i) Neutralization of Acid Slurry of LABSA LABSA has an acid value of 225 $\frac{mg\ of\ KOH}{gm\ sample}$, which was neutralized by NaOH solution. As per stoichiometry, LABSA (100 g) was neutralized with 16 g NaOH (Table 1).

Table 1. The acid value of LABSA.

Sr. No.	Sample	Weight of Sample	B.R.	Acid Value
1	LABSA	2.25 g	90 mL	225 $\frac{mg\ of\ KOH}{gm\ sample}$
2	LABSA after neutralization	2.5 g	0.0	0.0 $\frac{mg\ of\ KOH}{gm\ Sample}$

(ii) Formulation of Detergent Neutralized LABSA was then mixed with sodium silicate, sodium sulphate, water and STPP at 55 °C using an overhead stirrer. The quantity of sodium silicate and STPP depended upon active matter and hardness of water [3–6]. This mixture of surfactant, binder and other ingredients were dried using a vacuum oven at 70 °C under a pressure of 350 mm for 4 to 5 h. The dried mixture was converted in fine powder form with the help of a mixer or grinder or ball mill.

(iii) Compressing to Tablet Form The fine powder was then mixed with various disintegrating agents such as starch, silicic acid, citric acid, and sodium bicarbonate. This mixture of detergent powder and disintegrating agents were compressed with the help of the tablet machine for converting detergents to tablet form. When this tablet was contacted with enough water, the CO_2 was released from disintegrating agents, which facilitated dispersion of the tablet in water.

2.2. Detergency Test

2.2.1. Fabric Soiling

The mixture of carbon black, mineral oil and lauric acid (28.4:17.9:17.9 w/w) was mixed with 35.8 g coconut oil to form a slurry. This slurry was further mixed with 500 mL carbon tetrachloride. The cotton fabric (100% RFD cotton) having a size 10 × 10 cm was soaked in the soiling medium for 15 to 20 min. These soiled fabrics were dried at 80 °C in the oven for 3 h.

2.2.2. Washing

The soiled fabric was washed using Terg—O—Tometer (M/s Wadegati Pvt. Ltd., Mumbai, Maharashtra, India). The soiled fabric was further finally washed with a detergent solution of 1000 mL at 100 rpm and 50 °C for 20 min followed by a rinsing time of 10 min. The process was repeated for various concentrations of detergents (0.1, 0.25 and 0.5%) in tap water. After washing, the detergency (%) was calculated using Lambert and Sanders formula using reflectance of washed fabric (Rw), soiled fabric before washing (Rs) and unsoiled fabric (Ro) [7].

3. Results and Discussion

3.1. Disintegration Time

Effervescences of gases (mostly CO_2 and O_2) were released immediately after contact with enough water. It helped to penetrate water in the compact form of detergent and to disintegrate the tablet in water. The disintegration of the tablet was dependent upon the % of the binder (starch) and gas removing material (citric acid and silicic acid) used. A minimum of 15% starch was required along with 20% citric acid and sodium bicarbonate or silicic acid to disintegrate the detergent from of the tablet (Table 2).

Table 2. Disintegration time of various formulated tablets.

Sr. No.	Detergent Powder (%)	Disintegrating Agents (g)				Volume of Water (mL)	Disintegration Time
		Starch	Silicic Acid	Citric Acid	Sodium Bicarbonate		
1	100	-	-	-	-	100	78 h
2	70	30	-	-	-	100	50 h
3	70	15	15	-	-	100	2 h
4	60	15	25	-	-	100	2 h
5	60	15	5	8	7	100	3 min
6	70	-	-	15	15	100	2 h
7	60	15	-	15	10	100	3 min
8	60	10		18	12	100	1 min
9	60	10	10	10	10	100	3.5 min
10	60	15	5	10	10	100	1 min
11	50	15	5	15	15	100	0.5 min

3.2. Cleaning Performance of Detergents and Their Tablets

The cleansing properties of detergent and its tablet with marketed detergent were studied by a % detergency test. The effect of disintegrating agents was also studied at various concentrations (0.1%, 0.5% and 1%) and room temperature using distilled water. The cotton used for detergency was 100% RFD cotton white colour fabric. It was observed that the detergency of the tablet having a disintegrating time of 0.5 min. gave better stain removal than marketed powder and other detergent tablets (Table 3 and Figure 2).

Table 3. Detergency (soil removal) of liquid detergency for soiled cotton fabric.

Sample	Concentration	% Detergency
Detergent Powder	0.1 0.25 0.5	63.19 64.8 72.59
Detergent tablet having disintegrating time 0.5 min	0.1 0.25 0.5	62.45 65.12 72.43
Detergent tablet having disintegrating time 2.0 min	0.1 0.25 0.5	59.32 61.29 63.84
Commercial detergent powder	0.1 0.25 0.5	44.19 48.32 53.11

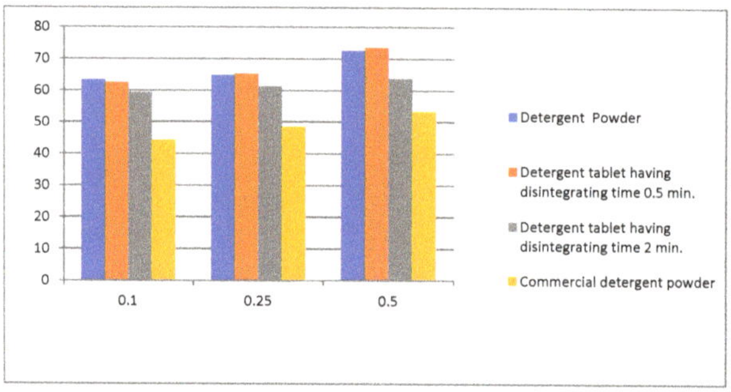

Figure 2. Graph of % detergency vs. concentrations.

3.3. Foam Stability and Height

Foaming stability and height are important aesthetic properties of detergent. Foam is nothing but trapping of gas in the liquid. The foaming height and stability were observed in Ross & Mill apparatus having a volume of 3 L. The detergent and its tablet foam were constant at all concentrations; it was also noticed that the foam height and stability was the same as that of the market detergent sample (Table 4).

Table 4. Foam stability and height of detergent tablet and powder.

Sample	Concentration	Time (min)					
		0	5	10	15	20	25
		Height (cm)					
Detergent Powder	0.1	26.5	26.5	26.5	26.4	25.6	24.2
	0.25	26.5	26.5	26.5	26.1	24.8	24.0
	0.5	26.5	26.5	26.5	26.3	24.6	23.8
Detergent tablet having disintegrating time 0.5 min	0.1	26.5	26.5	26.5	25.6	25.0	23.8
	0.25	26.5	26.5	26.5	25.8	24.2	23.4
	0.5	26.5	26.5	26.5	24.5	24.0	22.8
Detergent tablet having disintegrating time 2.0 min	0.1	26.5	26.5	26.5	24.5	24.5	24
	0.25	26.5	26.5	26.5	25.4	24	24
	0.5	26.5	26.5	26.5	25	25	24.5
Commercial detergent powder	0.1	23.2	23.2	22.4	22	22	21
	0.25	23.2	23.2	22.3	22	22	21.5
	0.5	23.0	23.0	22.3	22	22	21.5

3.4. Wetting Ability

The wetting properties of detergent and its tablet were observed using a 100% RFD cotton disk having size 0.5 cm radius, which was dipped in detergent solutions of various concentrations and at ambient temperature. The wetting property of detergent depended upon the concentration of detergent. The detergent tablet had good wetting properties due to the effervescence of gases (CO_2 and O_2) from the disintegrating agent's citric acid, silicic acid, and sodium bicarbonate. The wetting time of detergent in 1.5% solution was 38 s, hile their tablet has a wetting time of 32 s (Table 5).

Table 5. Wetting ability of samples.

Sr. No.	Batch	Concentration (%)		
		0.5	1	1.5
1	Detergent Powder	55 s	38 s	32 s
	Detergent tablet	63 s	41 s	38 s
2	Commercial detergent powder	24 s	25 s	23 s

3.5. Tablet Friability

The friability test is the method used to find the loss of product during transportation. The test was carried out using a tablet friability machine, which is commonly used in pharmaceuticals tablet testing. All 10 tablets having the same weight (650 mg/tablet) were selected. The drum was rotated 100 times. Tablets were removed and weighed. The % friability was calculated by the following formula. The % friability of detergent tablets having the disintegrating time of 2 min was better than the detergent tablet having the disintegrating time of 30 s. (Table 6). The tablet disintegration pattern or time was dependent on the composition of tablet composition and pressure. The composition and size of a tablet and the quantity of water played an important role in disintegration and dispersion. A minimum of 15% starch was required along with 25% citric acid and sodium bicarbonate or silicic acid to disintegrate the tablet rapidly. The hardness of the tablet was due to the pressure applied during the compression of the detergent powder. Tablet hardness should be greater than 4.5 kg/(cm)^2. The hardness of the tablet directly affected the friability test, which measured the loss during the transportation. The % loss during the transportation was less than 2%. Due to the compact form of powder, its volume was reduced up to 50%, thus reducing packing and transportation costs. The rate of disintegration of the tablet was less than 30 s without appreciable energy. In the compact form, the powder particles are tightly bound, and the minimum part of the powder is contacted with the environment. Due to this, the powder does not catch moisture easily.

Table 6. % Friability of tablets.

Sr. No.	Batch No.	% Friability of Tablet
1	Detergent tablet having disintegrating time 2 min	26.91
2	Detergent tablet having disintegrating time 30 s	2

4. Conclusions

Detergent tablets can be successfully used for the washing of fabric. They consist of of active ingredients and disintegrating agents, which releases gases such as carbon dioxide, oxygen, and hydrogen after contacting with enough water. One or two tablets are enough for the washing instead of one scoop. Effervescent tablets are convenient to use and handle and are preferred over existing forms. The detergent powder composition in the effervescent tablets contains 95% useful matter and disintegrates within 30 s at room temperature in water. The tablet detergent showed better performance than various market detergents. Thus, compact detergent tablets have an added benefit of ease of convenience.

Author Contributions: Conceptualization, A.P.; methodology, A.P. and R.M.; writing—review and editing, S.N.M. All authors have read and agreed to the published version of the manuscript.

Funding: Not applicable.

Institutional Review Board Statement: Not applicable.

Informed Consent Statement: Not applicable.

Data Availability Statement: Not applicable.

Acknowledgments: The authors would like to thank the ICT, Mumbai, India for their support.

Conflicts of Interest: The authors declare no conflict of interest.

References

1. Singh, A.; Sharma, A.; Bansal, S.; Sharma, P. Comparative interaction study of amylase and surfactants for potential detergent formulation. *J. Mol. Liq.* **2018**, *261*, 397–401. [CrossRef]
2. Cheng, K.C.; Khoo, Z.S.; Lo, N.W.; Tan, W.J.; Chemmangattuvalappil, N.G. Design and performance optimisation of detergent product containing a binary mixture of anionic-nonionic surfactants. *Heliyon* **2020**, *6*, e03861. [CrossRef] [PubMed]
3. Chateau, M.E.; Galet, L.; Soudais, Y.; Fages, J. Processing a detergent powder formulation: Direct compression, and high shear wet granulation followed by compression. *Powder Technol.* **2005**, *157*, 191–198. [CrossRef]
4. Ahmadian, H.; Ghadiri, M. Analysis of enzyme dust formation in detergent manufacturing plants. *Adv. Powder Technol.* **2007**, *18*, 53–67. [CrossRef]
5. Tai, X.M.; Song, J.Y.; Du, Z.P.; Liu, X.; Wang, T.; Wang, G. The performance test of fatty acid methyl ester sulfonates and application in the dishwashing liquid detergent. *J. Dispers. Sci. Technol.* **2018**, *39*, 1422–1426. [CrossRef]
6. Ponnusamy, T.; Dubal, S.A.; Momin, S.A. Studies in detergency: Influence of different factors for removing motor oil stain from the cotton fabric. *J. Dispers. Sci. Technol.* **2008**, *29*, 1123–1128. [CrossRef]
7. Kogawa, A.C.; Cernic, B.G.; do Couto, L.G.D.; Salgado, H.R.N. Synthetic detergents: 100 years of history. *Saudi Pharm. J.* **2017**, *25*, 934–938. [CrossRef] [PubMed]

engineering proceedings

MDPI

Proceeding Paper

Surface Plasmon Resonance Sensor Based on Inkjet 3D Printing †

Lorena Saitta [1,*], Nunzio Cennamo [2], Claudio Tosto [1], Francesco Arcadio [2], Maria Elena Fragalà [3,4], Luigi Zeni [2] and Gianluca Cicala [1,3]

[1] Department of Civil Engineering and Architecture, University of Catania, 95125 Catania, Italy; claudio.tosto@unict.it (C.T.); gianluca.cicala@unict.it (G.C.)
[2] Department of Engineering, University of Campania Luigi Vanvitelli, 81031 Aversa, Italy; nunzio.cennamo@unicampania.it (N.C.); francesco.arcadio@unicampania.it (F.A.); luigi.zeni@unicampania.it (L.Z.)
[3] INSTM Udr Catania, 95125 Catania, Italy; me.fragala@unict.it
[4] Department of Chemical Sciences, University of Catania, 95125 Catania, Italy
* Correspondence: lorena.saitta@phd.unict.it; Tel.: +39-3896891258

† Presented at the 2nd International Electronic Conference on Applied Sciences, 15–31 October 2021; Available online: https://asec2021.sciforum.net/.

Abstract: A novel surface plasmon (SPR) sensor was designed, manufactured and experimentally tested. A novel approach was followed to fabricate the sensor. It is based on a combination of both the inkjet 3D printing process and the use of optical adhesives, which were used as an alternative solution to the use of plastic optical fibers (POFs). The obtained experimental results showed good performances, at least in terms of figure of merit (FOM), for the 3D-printed sensor, which were quite similar to those gained by an SPR–POF configuration. Next, through a cost analysis, the possibility of manufacturing the SPR sensor at a low cost was demonstrated, thus being economically advantageous towards conventional sensors.

Keywords: 3D printing; additive manufacturing; photocurable resin; plasmonic sensor

Citation: Saitta, L.; Cennamo, N.; Tosto, C.; Arcadio, F.; Fragalà, M.E.; Zeni, L.; Cicala, G. Surface Plasmon Resonance Sensor Based on Inkjet 3D Printing. *Eng. Proc.* **2021**, *11*, 39. https://doi.org/10.3390/ASEC2021-11127

Academic Editor: Saulius Juodkazis

Published: 15 October 2021

1. Introduction

In the last few years, the continuous demand for sensitive sensors operating in diverse application fields has led to developing innovative platforms based on different working principles [1–3]. Among them, the surface plasmon resonance (SPR) technique is commonly used as a detection method [4]. In particular, the SPR working principle relies on the refractive index discrepancy at the interface between a dielectric medium and a metallic nanofilm. This family of sensors can be used to analyze different substances, such as pollutants, pesticides, toxic metals, viruses and other molecules. As an alternative to silicon-based technologies, SPR sensors could be fabricated by means of a novel technique relying on inkjet 3D printing. One of the most remarkable advantages of using 3D printing technologies resides in the possibility to realize more complex geometries, different from basic cylindrical fibers [5–7], and obtain a freedom design approach. An additional benefit of developing organic optoelectronic devices is their low cost compared to silicon-based ones [8] as clean rooms are not needed, unlike microelectronics industries.

With regard to 3D-printed plasmonic sensors, despite several approaches having recently been presented [9], one of the common downsides is represented by the necessity of polishing the printed surfaces before gold sputtering in order to obtain the required SPR performance [10]. For complex designs with restricted access to all surfaces, this strategy can be a limiting factor. In addition, several strategies involve the use of expensive resin [5], which increases the total cost of the developed sensor.

In order to overcome the above-mentioned issues, in this work, a novel SPR sensor has been designed and manufactured via an inkjet 3D printing process combined with optical adhesive use. The developed SPR sensor is easy to fabricate since no further surface modifications (i.e., lapping procedures) are required. This approach was previously adopted as a substitute to plastic optical fibers (POFs) [11] but with no reference to SPR phenomena. The numerical and experimental results have been presented as well. Eventually, through a cost analysis, it has been demonstrated that the 3D-printed sensor is economically advantageous.

2. 3D-Printed Surface Plasmon Resonance Sensor

2.1. SPR Sensor Design and Fabrication

The SPR sensor has been designed disassembled and is composed of four different parts by using Autodesk® Fusion 360 (Figure 1), and then the STL (Standard Triangle Language) files were generated. The latter is a de facto standard file that describes the external closed surfaces of the original CAD model and forms the basis for the slicing procedure. Next, the G-Code instructions for the 3D printer were realized via the software Objet Studio™. Finally, the sensor construction was performed by using the PolyJet 3D printer Stratasys Objet260 Connex 1 (Stratasys, Los Angeles, CA, USA). The used material was a liquid photopolymer ink (VeroClear RGD810). Once the SPR sensor parts construction was completed (Figure 2), the waveguide core of the 3D-printed optical device was fabricated. Thus, the UV photopolymer adhesive (NOA88, Edmund Optics, Nether Poppleton, York, UK) was microinjected into the sensor channel and cured for 10 min by means of a lamp bulb with UVA emission at 365 nm, as shown in Figure 3.

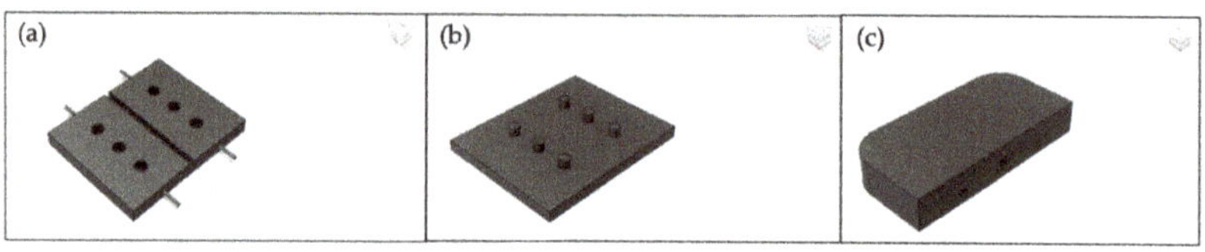

Figure 1. Disassembled parts of the surface plasmon resonance (SPR) sensor designed on Autodesk® Fusion 360. (**a**) Substrate having the functionality of cladding for the waveguide core; (**b**) cover as cladding for the upper part of the waveguide core; (**c**) support for fitting with 1 mm POF waveguides.

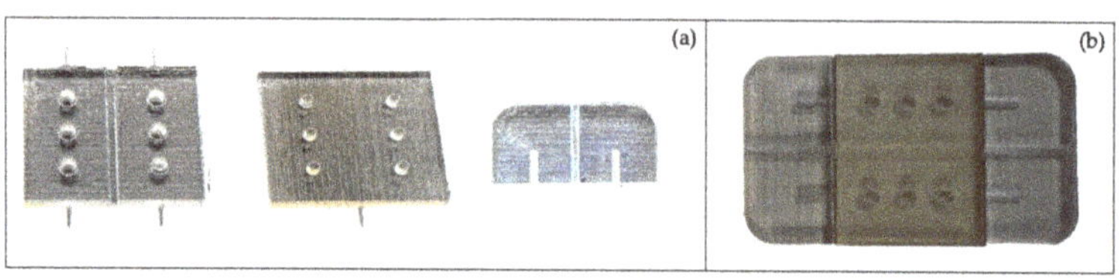

Figure 2. (**a**) SPR sensor's 3D-printed disassembled parts. (**b**) Assembled SPR sensor.

Next, to generate the SPR phenomenon, the cured core, without any further surface modification, was gold-sputtered with a coater (Bal-Tec SCD 500, Schalksmühle, Germany) in such a way as to present a noble metal nanofilm. The thickness of the sputtered gold was about 60 nm.

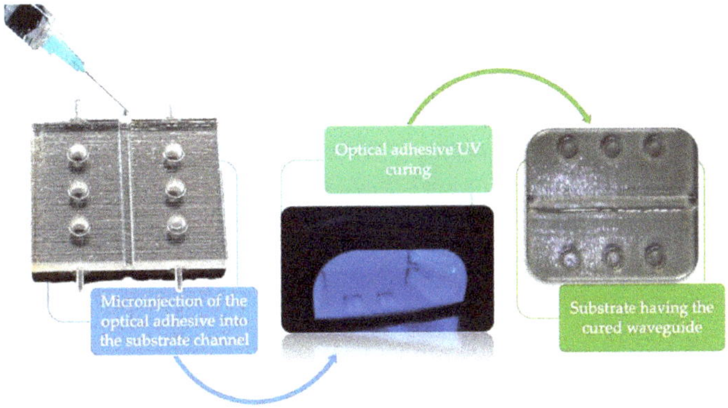

Figure 3. Waveguide core made of cured optical adhesive (NOA88) fabrication.

2.2. Experimental Setup

To monitor the developed plasmonic sensor, a simple experimental setup has been used, as shown in Figure 4, to carry out a low-cost sensor system. In particular, it comprises a halogen lamp, used as a white light source (HL-2000LL, Ocean Optics, Dunedin, FL, USA), two POF patches (1 mm total diameter) to couple and collect the light into the 3D-printed plasmonic sensor and a spectrometer (FLAME-S-VIS-NIR-ES, Ocean Optics, Dunedin, FL, USA) with a detection range from 350 nm to 1023 nm.

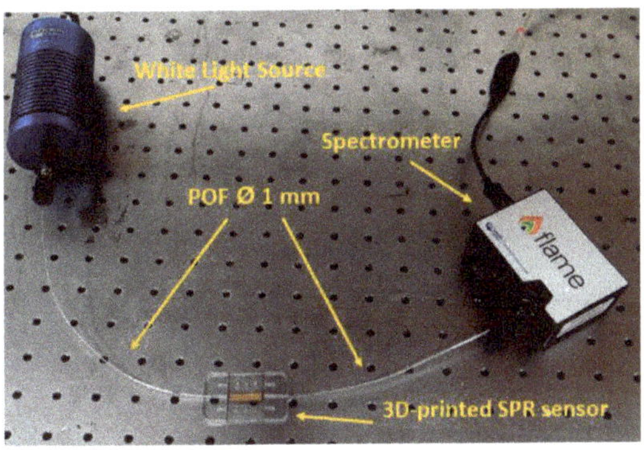

Figure 4. Experimental setup used to test the developed 3D-printed SPR sensor.

3. Results

3.1. Experimental Results

The experimental measurements have been obtained by exploiting the experimental setup reported in Figure 4. In particular, several water–glycerin solutions, whose refractive index ranges from 1.332 to 1.382, have been used to test the sensor performances. Figure 5 reports the normalized SPR transmitted spectra obtained using these water–glycerin solutions in contact with the gold sensing surface.

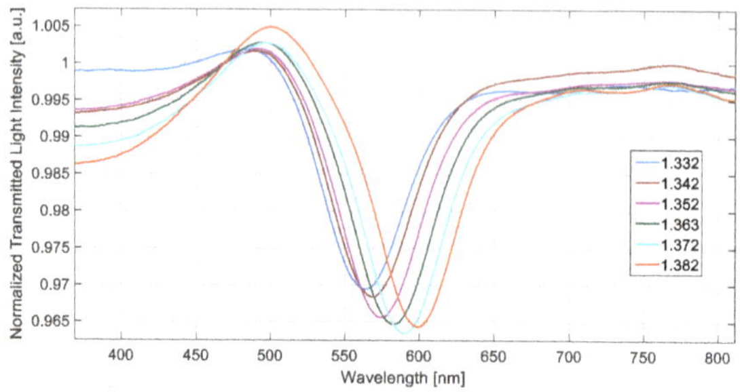

Figure 5. SPR spectra obtained at different refractive indices (from 1.332 to 1.382).

For the proposed 3D-printed SPR sensor, the obtained results have shown a good sensitivity, equal to about 710 nm/RIU (in the considered refractive index range). This value has been calculated by considering a linear sensor response [12]. Moreover, the 3D-printed SPR sensor has also denoted a figure of merit (FOM) equal to 13.6 RIU^{-1} [12], and this value is very similar to the one obtained with another low-cost SPR sensor based on D-shaped POFs [13]. The best improvement with respect to Ref. [13] is related to an approximate 40% improvement in the signal to noise ratio (SNR) [12,13].

3.2. Cost Analysis

By categorizing the cost parameters as process, material and machine, the cost needed to fabricate the SPR sensor was modeled. Focusing on the raw material cost, it was modeled by considering the raw material cost (393.11 €/kg for the model material VeroClear RGD810, 126.74 €/kg for the support material FullCure705, 2.50 €/ml for the optical adhesive NOA88) and the quantity needed to manufacture the model (i.e., 0.017 kg of VeroClear RGD810, 0.006 kg of FullCure705 and 1 ml of NOA88). Next, with the purchase, installation and maintenance costs of the 3D printing machine used being known, the cost model even considered the depreciation of the machine itself. To complete this operation, input parameters were used for the model both in terms of the depreciation cost (i.e., 10 €/h) and the printing time needed to manufacture the sensor (equal to 0.47 h in this case). In conclusion, the last input parameter considered was related to the power cost. In this case, the power requirements for the instrument used during the 3D printing process were considered. In detail, in the model, the power cost (0.10 €/kWh) and the printing time (0.47 h), which correspond to the usage of the machine in terms of time, were taken into account.

The resulting cost allocation is shown in Figure 6. The raw materials cost (model material VeroClear RGD810 393.11 €/kg, support material FullCure705 126.74 €/kg, optical adhesive NOA88 2.50 €/mL) had the greatest impact (equal to 66%) since the 3D printer employed only uses proprietary materials. As a result, the determined price for one sensor was ∼15 €, which resulted to be much cheaper than a traditional sensor. The costs can be further reduced in the future by using new vat-photopolymerization printers that are being developed and that use more low-price materials (i.e., 50 €/kg).

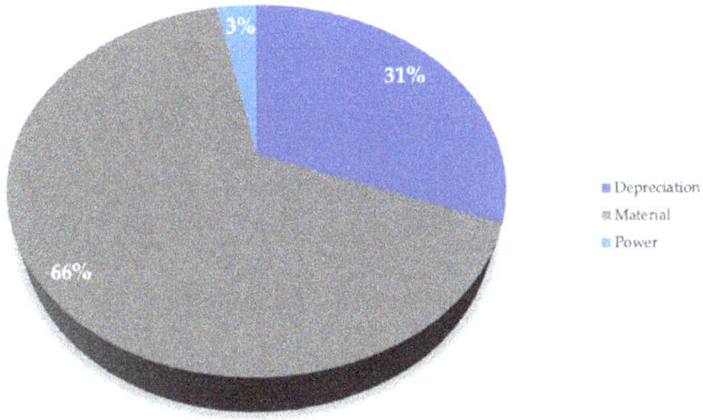

Figure 6. Cost allocation pie chart.

4. Conclusions

Once a CAD model was accomplished, by using the inkjet 3D printing technology, a cheap SPR sensor was manufactured. It is an innovative approach to obtain sensors in a fast way for mass production. The manufacturing cost resulted to be a very low-price (∼15 €), making the proposed approach very cost-effective. Moreover, the total cost for this device could be further decreased by using cheaper resins through the LCD printing.

The experimental analysis performed showed good performances for the SPR sensor fabricated. Indeed, the test run showed a figure of merit quite similar to the POF-based SPR sensor, while the sensitivity resulted to be somewhat minor. It is important to underline that, in order to improve the sensor's performances, other photocurable resins could be used in the fabrication process. In fact, by changing the waveguide core material, i.e., its refractive index, it will be possible to tune the SPR sensitivity, making it possible to obtain an improved sensor configuration.

For all these reasons, the fabricated sensor could represent the starting point for developing a new class of plasmonic biochemical sensors for several applications where high sensitivity and real-time, label-free detection are strictly required. Moreover, with the versatility of the proposed sensing approach, it could also be possible to include these kinds of sensors in a "smart city" environment, for instance, to monitor air pollutants and water quality, in view of the so-called "internet of things".

Author Contributions: Conceptualization, N.C. and G.C.; methodology, N.C. and G.C.; validation, L.S., F.A. and C.T.; investigation, L.S., F.A., C.T., N.C. and G.C.; resources, N.C. and G.C.; writing—original draft preparation, M.E.F., L.Z., L.S. and F.A.; writing—review and editing, G.C. and N.C.; supervision, N.C. and G.C. All authors have read and agreed to the published version of the manuscript.

Funding: This research received no external funding.

Institutional Review Board Statement: Not applicable.

Informed Consent Statement: Not applicable.

Data Availability Statement: The data are available on reasonable request from the corresponding author.

Acknowledgments: Gianluca Cicala acknowledges the funding received for this project from Università degli Studi di Catania under the Grant Scheme PIACERI with the project MAF-moF "Materiali multifunzionali per dispositive micro-optofluidici", Project Coordinator Maria Elena Fragalà. Gianluca Cicala also acknowledges Italian MIUR grant number 20179SWLKA Project Title Multiple Advanced Materials Manufactured by Additive technologies (MAMMA), under the PRIN funding Scheme, Project Coordinator G.C.

Conflicts of Interest: The authors declare no conflict of interest.

References

1. Barshilia, D.; Chau, L.-K.; Chang, G.-E. Low-cost planar waveguide-based optofluidic sensor for real-time refractive index sensing. *Opt. Express* **2020**, *28*, 27337–27345. [CrossRef] [PubMed]
2. Lončar, M.; Doll, T.; Vučković, J.; Scherer, A. Design and Fabrication of Silicon Photonic Crystal Optical Waveguides. *J. Light. Technol.* **2000**, *18*, 1402. [CrossRef]
3. De Kerf, T.; Pipintakos, G.; Zahiri, Z.; Vanlanduit, S.; Scheunders, P. Identification of Corrosion Minerals Using Shortwave Infrared Hyperspectral Imaging. *Sensors* **2022**, *22*, 407. [CrossRef]
4. Tong, L.; Wei, H.; Zhang, S.; Xu, H. Recent Advances in476 Plasmonic Sensors. *Sensors* **2014**, *14*, 7959–7973. [CrossRef] [PubMed]
5. Haring, A.P.; Khan, A.U.; Liu, G.; Johnson, B.N. 3D Printed Functionally Graded Plasmonic Constructs. *Adv. Opt. Mater.* **2017**, *5*, 1700367. [CrossRef]
6. Garcìa-Astrain, C.; Lenzi, E.; De Aberasturi, D.J.; Henriksen-Lacey, M.; Binelli, M.R.; Liz-Marzaàn, L.M. 3D-Printed Biocompatible Scaffolds with Built-In Nanoplasmonic Sensors. *Adv. Funct. Mater.* **2020**, *30*, 2005407. [CrossRef]
7. Xu, Y.; Wu, X.; Guo, X.; Kong, B.; Zhang, M.; Qian, X.; Mi, S.; Sun, W. The Boom in 3D-Printed Sensor Technology. *Sensors* **2017**, *17*, 1166. [CrossRef] [PubMed]
8. Forrest, S. The path to ubiquitous and low-cost organic electronic appliances on plastic. *Nature* **2004**, *428*, 911–918. [CrossRef] [PubMed]
9. Lambert, A.; Valiulis, S.; Cheng, Q. Advances in Optical Sensing and Bioanalysis Enabled by 3D Printing. *ACS Sens.* **2018**, *3*, 2475–2491. [CrossRef] [PubMed]
10. Hinman, S.S.; McKeating, K.S.; Cheng, Q. Plasmonic Sensing with 3D Printed Optics. *Anal. Chem.* **2017**, *89*, 12626–12630. [CrossRef] [PubMed]
11. Prajzler, V.; Kulha, P.; Knietel, M.; Enser, H. Large core plastic planar optical splitter fabricated by 3D printing technology. *Opt. Commun.* **2017**, *400*, 38–42. [CrossRef]
12. Cennamo, N.; Saitta, L.; Tosto, C.; Arcadio, F.; Zeni, L.; Fragalá, M.E.; Cicala, G. Microstructured Surface Plasmon Resonance Sensor Based on Inkjet 3D Printing Using Photocurable Resins with Tailored Refractive Index. *Polymers* **2021**, *13*, 2518. [CrossRef] [PubMed]
13. Cennamo, N.; Massarotti, D.; Conte, L.; Zeni, L. Low Cost Sensors Based on SPR in a Plastic Optical Fiber for Biosensor Implementation. *Sensors* **2011**, *11*, 11752–11760. [CrossRef] [PubMed]

engineering proceedings

MDPI

Proceeding Paper

Partial Differential Equations of Motion for a Single-Link Flexible Manipulator †

Mohammed Bouanane [1,*], Rachad Oulad Ben Zarouala [2] and Abdellatif Khamlichi [2]

1 Systems of Communication and Detection Laboratory, Department of Physics, Faculty of Sciences, Abdelmalek Essaâdi University, Tetouan 93000, Morocco
2 Industrial and Civil Engineering Department, National School of Applied Sciences, Abdelmalek Essaâdi University, Tetouan 93000, Morocco; rachad.oulad@gmail.com (R.O.B.Z.); khamlichi7@yahoo.es (A.K.)
* Correspondence: bouanane.mohammed-etu@uae.ac.ma
† This paper is an extended version of our paper published in presented at the 2nd International Electronic Conference on Applied Sciences, 15–31 October 2021; Available online: https://asec2021.sciforum.net/.

Abstract: Robot manipulators have played an enormous role in the industry during the twenty-first century. Due to the advances in materials science, lightweight manipulators have emerged with low energy consumption and positive economic aspect regardless of their complex mechanical model and control techniques problems. This paper presents a dynamic model of a single link flexible robot manipulator with a payload at its free end based on the Euler–Bernoulli beam theory with a complete second-order deformation field that generates a complete second-order elastic rotation matrix. The beam experiences an axial stretching, horizontal and vertical deflections, and a torsional deformation ignoring the shear due to bending, warping due to torsion, and viscous air friction. The deformation and its derivatives are assumed to be small. The application of the extended Hamilton principle while taking into account the viscoelastic internal damping based on the Kelvin–Voigt model expressed by the Rayleigh dissipation function yields both the boundary conditions and the coupled partial differential equations of motion that can be decoupled when the manipulator rotates with a constant angular velocity. Equations of motion solutions are still under research, as it is required to study the behavior of flexible manipulators and develop novel ways and methods for controlling their complex movements.

Keywords: flexible manipulator; Euler-Bernoulli beam; Viscoelasticity; Kelvin-Voigt model; Rayleigh dissipation function; extended Hamilton principle; partial differential equations

Citation: Bouanane, M.; Zarouala, R.O.B.; Khamlichi, A. Partial Differential Equations of Motion for a Single-Link Flexible Manipulator. *Eng. Proc.* **2021**, *11*, 40. https://doi.org/10.3390/ASEC2021-11120

Academic Editor: Nunzio Cennamo

Published: 15 October 2021

1. Introduction

The focus of robotics research in the last decade has been on building lightweight manipulators due to their low energy consumption despite their complex mechanical models and control systems. Lightweight manipulators are considered flexible manipulators that suffer from flexural effects, which leads to growing interest toward modeling and control architecture of such systems. In general, the research is restricted to single-link flexible manipulator [1] due to the intricacy of serial link flexible manipulators. In the literature, the single link is usually modeled by one deformation parameter [2], and the kinematics of the Euler–Bernoulli beam is usually approached by the assumed traditional deformation field that cannot allow having an orthogonal elastic rotation matrix to the second-order. For this article, the deformations and their partial derivatives are assumed to be small. The kinematic model described in Section 2.1 is based on the complete second-order deformation field [3]. Section 2.2 presents the dynamics model that includes the kinetic energy and potential energy of the system that is composed of gravitational and strain potential energies due to gravity and elasticity. Section 2.3 takes into account the Rayleigh dissipation function due to motor friction and the viscoelastic internal damping based on the Kelvin–Voigt model. Section 2.4 gives the motion equations using the extended Hamilton principle that yields four partial differential equations satisfied by the deformation variables and seven boundary conditions. Lastly, Section 3 deals with the decoupling of partial differential equations in a particular case which allows small simplifications of the equations.

2. Mechanical Modeling

The system consists of a base subjected to an applied torque T_{mot} by a motor, a flexible link modeled as an Euler–Bernoulli beam with a circular cross-section with radius R and length L, and a payload with mass m_p and inertia matrix I_p at the free end of the link. The beam is subjected to an axial stretching $u(x,t)$, a horizontal deflection $v(x,t)$, a vertical deflection $w(x,t)$ and a torsional deformation $\phi(x,t)$, as shown in (Figure 1a,b) where the axis Z_1 is perpendicular to the ground. The beam deformations and their partial derivatives are assumed to be small, and shear due to bending, warping due to torsion, and air viscous friction are neglected. To simplify the notation, $u(x,t)$, $v(x,t)$, $w(x,t)$, $\phi(x,t)$, $\frac{d}{dt}(.)$, $\frac{d}{dx}(.)$ are denoted by u, v, w, ϕ, $(\dot{.})$ and $(.)'$ respectively.

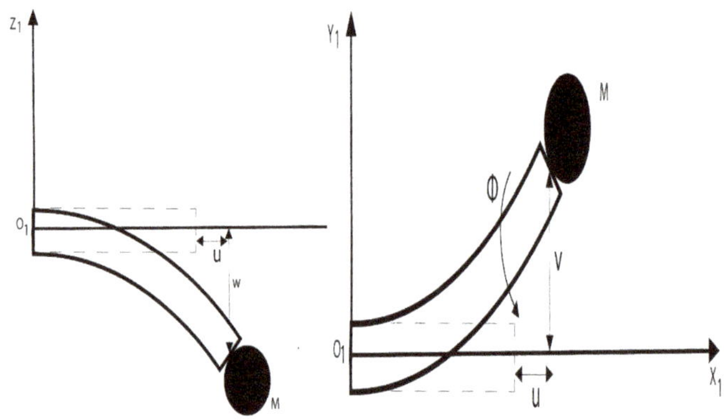

(**a**) Front view (**b**) Top view

Figure 1. Flexible beam with payload.

2.1. Kinematics

Let $\mathcal{R}_0$ be an inertial frame with origin O_0, $\mathcal{R}_1$ a frame attached to the motor with origin O_1 that coincides with O_0, and $\mathcal{R}_{dm}$ a frame attached to the cross-section of mass dm whose axes are parallel to those of $\mathcal{R}_1$ before deformation and whose origin O_{dm} is the center of the cross-section that is at a distance x from O_1 along the neutral axis of the link before deformation. The rotation matrix of $\mathcal{R}_1$ relative to $\mathcal{R}_0$ [4] is ${}^0R_1 = R_{Z_0,\theta}$ which means the frame R_1 rotates by an angle θ of about Z_0.

The position of O_{dm} relative to $\mathcal{R}_1$ expressed in $\mathcal{R}_1$ after deformation [3] expressed by:

$$^1\overrightarrow{O_1O_{dm}} = [x + u - \frac{1}{2}\int_0^x (v'^2 + w'^2)ds, v, w]^T \tag{1}$$

The rotation matrix of $\mathcal{R}_{dm}$ relative to $\mathcal{R}_1$ after deformation [3] is:

$$^1R_{dm} = \begin{bmatrix} 1 - \frac{1}{2}(v'^2 + w'^2) & -v' + u'v' - w'\phi & -w' + u'w' + v'\phi \\ v' - u'v' & 1 - \frac{1}{2}(v'^2 + \phi^2) & -\phi - \frac{1}{2}v'w' \\ w' - u'w' & \phi - \frac{1}{2}v'w' & 1 - \frac{1}{2}(w'^2 + \phi^2) \end{bmatrix} \tag{2}$$

${}^1R_{dm}$ is verified to be orthogonal to the second-order of Taylor expansion in the deformation variables. Let P be a point of the cross-section with (x, y, z) its coordinates relative to $\mathcal{R}_1$ before deformation. The position of P relative to $\mathcal{R}_1$ expressed in $\mathcal{R}_1$ after deformation [4] is

$$ {}^1\overrightarrow{O_1P} = {}^1\overrightarrow{O_1O_{dm}} + {}^1R_{dm}\, {}^{dm}\overrightarrow{O_{dm}P} $$

where ${}^{dm}\overrightarrow{O_{dm}P} = [0, y, z]^T$ and ${}^0\overrightarrow{O_0P} = {}^0R_1\, {}^1\overrightarrow{O_1P}$.

Let $\mathcal{R}_2$ be a frame attached to the free end of the link whose origin is O_2 and obtained from $\mathcal{R}_{dm}$ by replacing x by L (for example v(x,t) at x=L becomes $v(L,t)$, shortened v_L). If the position of the center of mass C of the payload relative to $\mathcal{R}_2$ expressed in $\mathcal{R}_2$ is ${}^2\overrightarrow{O_2C} = [c, 0, 0]^T$, then the position of C relative to $\mathcal{R}_1$ expressed in $\mathcal{R}_1$ is given by:

$$ {}^1\overrightarrow{O_1C} = [L + u_L - \frac{1}{2}\int_0^L (v'^2 + w'^2)ds + c(1 - \frac{1}{2}(v_L'^2 + w_L'^2)), v_L + c(v_L' - u_L'v_L'), w_L + c(w_L' - u_L'w_L')]^T \quad (3) $$

Since ${}^1\overrightarrow{O_1C} = {}^1\overrightarrow{O_1O_2} + {}^1R_2\, {}^2\overrightarrow{O_2C}$, and ${}^1R_2$ is deduced from $1R_{dm}$ by replacing x by L; hence, ${}^0\overrightarrow{O_0C} = {}^0R_1\, {}^1\overrightarrow{O_1C}$. The angular velocity of $\mathcal{R}_1$ relative to $\mathcal{R}_0$ expressed in $\mathcal{R}_0$ is ${}^0\overrightarrow{\Omega_{1/0}} = [0, 0, \dot{\theta}]^T$. The angular velocity of $\mathcal{R}_{dm}$ relative to $\mathcal{R}_1$ expressed in $\mathcal{R}_1$ [4] is found from the following matrix

$$ S = {}^1\dot{R}_{dm}\, {}^1R_{dm}^T = \begin{bmatrix} 0 & -\omega_z & \omega_y \\ \omega_z & 0 & -\omega_x \\ -\omega_y & \omega_x & 0 \end{bmatrix} $$

Hence,

$$ {}^1\overrightarrow{\Omega_{dm/1}} = [\omega_x, \omega_y, \omega_z]^T $$

The Taylor expansion of ${}^1\overrightarrow{\Omega_{dm/1}}$ to the second-order in the deformation variables and after simplification gives:

$$ \omega_x \approx \dot{\phi} + \frac{1}{2}(v'\dot{w}' - \dot{v}'w') \qquad \omega_y \approx -\dot{w}' + \dot{u}'w' + u'\dot{w}' + v'\dot{\phi} \qquad \omega_z \approx \dot{v}' - \dot{u}'v' - u'\dot{v}' + \dot{\phi}w' \quad (4) $$

Hence, the angular velocity of $\mathcal{R}_{dm}$ relative to $\mathcal{R}_0$ expressed in $\mathcal{R}_0$ is given by:

$$ {}^0\overrightarrow{\Omega_{dm/0}} = {}^0\overrightarrow{\Omega_{1/0}} + {}^0R_1\, {}^1\overrightarrow{\Omega_{dm/1}} $$

The gravity vector is represented in $\mathcal{R}_0$ by: ${}^0\overrightarrow{g} = [0, 0, -g]^T$.

2.2. Dynamics

2.2.1. Kinetic Energy

The kinetic energy T of the system is the sum of kinetic energies: T_B of the base, T_l of the flexible link and T_p of the payload. Where $T_B = \frac{1}{2}I_B\dot{\theta}^2$, with I_B is the base inertia about the Z_0 axis. The kinetic energy of the link [5] is given by:

$$ T_l = \frac{1}{2}\iiint_V v(P/0)^2 dm = \frac{1}{2}\int_{z=-R}^{R}\int_{y=-\sqrt{R^2-z^2}}^{\sqrt{R^2-z^2}}\int_{x=0}^{L} \rho\, v(P/0)^2 dxdydz \quad (5) $$

Since the beam cross-section is circular, $y^2 + z^2 = r^2,\ r \in [0, R]$ and the last triple integral is written [6] as:

$$T_l = \frac{1}{2}\int_{r=0}^{R}\int_{\gamma=0}^{2\pi}\int_{x=0}^{L} \rho v(P/0)^2 r dr d\gamma dx \tag{6}$$

where $y = rcos(\gamma)$, $z = rsin(\gamma)$. Therefore, the kinetic energy of the link linearized to the second-order and after simplifications is given by:

$$\begin{aligned} T_l = \frac{\rho}{2}\Big\{ & \pi R^2 \int_0^L (\dot{u}^2 + \dot{v}^2 + \dot{w}^2)dx + \frac{1}{4}\pi R^4 \int_0^L (\dot{v}'^2 + \dot{w}'^2 + 2\dot{\phi}^2)dx + \dot{\theta}^2\Big[\frac{1}{3}\pi R^2 L^3 + \frac{1}{4}\pi R^4 L + \pi R^2 \int_0^L (u^2+v^2)dx + \\ & \frac{1}{4}\pi R^4 \int_0^L w'^2 dx + 2\pi R^2 \int_0^L xu dx - \frac{1}{2}\pi R^2 \int_0^L (L^2 - x^2)(v'^2 + w'^2)dx\Big] + 2\dot{\theta}\Big[\pi R^2 \int_0^L x\dot{v}dx - \\ & \frac{1}{4}\pi R^4 \int_0^L (-\dot{v}' + \dot{u}'v' + u'\dot{v}' - 2w'\dot{\phi})dx + \pi R^2 \int_0^L (u\dot{v} - \dot{u}v)dx\Big]\Big\} \end{aligned} \tag{7}$$

The kinetic energy of the payload [7] is expressed by :

$$T_p = \frac{1}{2}\overrightarrow{\Omega_{p/0}}.I_p\overrightarrow{\Omega_{p/0}} + \frac{1}{2}m_p v(C/0)^2 \tag{8}$$

where $I_p = \begin{bmatrix} I_1 & I_4 & I_5 \\ I_4 & I_2 & I_6 \\ I_5 & I_6 & I_3 \end{bmatrix}$ and $\overrightarrow{\Omega_{p/0}}$ is obtained from $\overrightarrow{\Omega_{dm/0}}$ by replacing x by L; hence, the expression of T_p linearized to the second-order in the deformation variables is given by:

$$\begin{aligned} T_p = \frac{1}{2}\Big[& I_1\Big(\dot{\phi}_L^2 cos(\theta)^2 + \dot{w}_L'^2 sin(\theta)^2 + 2\dot{\phi}_L\dot{w}_L' cos(\theta)sin(\theta)\Big) + I_2\Big(\dot{\phi}_L^2 sin(\theta)^2 + \dot{w}_L'^2 cos(\theta)^2 - 2\dot{\phi}_L\dot{w}_L' cos(\theta)sin(\theta)\Big) \\ & + I_3\Big(\dot{\theta}^2 + \dot{v}_L'^2 + 2\dot{\theta}(\dot{v}_L' - \dot{u}_L'v_L' - u_L'\dot{v}_L' + \dot{\phi}_L w_L')\Big) + 2I_4\Big((\dot{\phi}_L^2 - \dot{w}_L'^2)cos(\theta)sin(\theta) - \dot{\phi}_L\dot{w}_L'(2cos(\theta)^2 - 1)\Big) + \\ & 2I_5\Big(\dot{\theta}\Big((\dot{\phi}_L + \frac{1}{2}(v_L'\dot{w}_L' - \dot{v}_L'w_L'))cos(\theta) - (-\dot{w}_L' + \dot{u}_L'w_L' + u_L'\dot{w}_L' + v_L'\dot{\phi}_L)\sin(\theta)\Big) + \dot{v}_L'\dot{\phi}_L cos(\theta) + \dot{v}_L'\dot{w}_L' sin(\theta)\Big) \\ & + 2I_6\Big(\dot{\theta}\Big((\dot{\phi}_L + \frac{1}{2}(v_L'\dot{w}_L' - \dot{v}_L'w_L'))sin(\theta) + (-\dot{w}_L' + \dot{u}_L'w_L' + u_L'\dot{w}_L' + v_L'\dot{\phi}_L)\cos(\theta)\Big) + \dot{v}_L'\dot{\phi}_L sin(\theta) - \dot{v}_L'\dot{w}_L' cos(\theta)\Big)\Big] + \\ & \frac{1}{2}m_p\Big[\dot{u}_L^2 + \dot{v}_L^2 + \dot{w}_L^2 + c^2(\dot{v}_L'^2 + \dot{w}_L'^2) + 2c(\dot{v}_L\dot{v}_L' + \dot{w}_L\dot{w}_L') + \dot{\theta}^2\Big(L^2 + u_L^2 + v_L^2 + c^2(1 - w_L'^2) + 2L[u_L - \frac{1}{2}\int_0^L (v'^2 + w'^2)ds + \\ & c(1 - \frac{1}{2}(v_L'^2 + w_L'^2))] + 2c(u_L + v_L v_L') - c\int_0^L (v'^2 + w'^2)ds\Big) + 2\dot{\theta}\Big((L+c)(\dot{v}_L + c(\dot{v}_L' - \dot{u}_L'v_L' - u_L'\dot{v}_L')) + u_L(\dot{v}_L + c\dot{v}_L') - \\ & \dot{u}_L(v_L + cv_L')\Big)\Big] \end{aligned} \tag{9}$$

2.2.2. Potential Energy

The potential energy V of the system is the sum of potential energies: V_B of the base , V_l of the flexible link and V_p of the payload. The potential energy V_B of the base which is its gravitational potential energy equals a constant C_B because its mass center is fixed in the inertial frame $\mathcal{R}_0$ whose origin level is taken as reference $V_B = C_B$.The potential energy of the link is the sum of its gravitational potential energy and its strain energy:

$$V_l = V_{gravit} + V_{str} \tag{10}$$

V_{gravit} is the gravitational potential energy of the link [7] that equals:

$$V_{gravit} = -\int_{r=0}^{R}\int_{\gamma=0}^{2\pi}\int_{x=0}^{L} \vec{g}\,\overrightarrow{O_0P}\rho r dr d\gamma dx = \rho g \pi R^2 \int_{x=0}^{L} w dx \tag{11}$$

V_{str} is the strain energy of the link [8] and it is the sum of strain energies due to different strains:

$$V_{str} = V_u + V_v + V_w + V_\phi \tag{12}$$

The expressions of different strain energies [9] are:

$$V_u = \frac{1}{2}\iiint_V Eu'^2 dV = \frac{1}{2}\pi R^2 E \int_0^L u'^2 dx \qquad V_v = \frac{1}{2}\iiint_V Ev''^2 y^2 dV = \frac{1}{8}\pi R^4 E \int_0^L v''^2 dx \qquad V_w = \frac{1}{2}\iiint_V Ew''^2 z^2 dV = \frac{1}{8}\pi R^4 E \int_0^L w''^2 dx$$

$$V_\phi = \frac{1}{2}\iiint_V Gr^2 \phi'^2 dV = \frac{1}{4}\pi R^4 G \int_0^L \phi'^2 dx$$

where E and G are the young modulus and the shear modulus of the beam material respectively. The potential energy of the payload is its gravitational potential energy that equals:

$$V_p = -m_p \vec{g}\,\overrightarrow{O_0 C} = m_p g\Big(w_L + c(w'_L - u'_L w'_L)\Big) \tag{13}$$

2.3. Rayleigh Dissipation Function

Rayleigh dissipation function $\mathscr{R}$ expresses the energy dissipated due to motor friction and internal damping effect of each deformation (u, v, w, ϕ), the dissipation is based on the Kelvin–Voigt model [10], and can be expressed [11] as follows:

$$\mathscr{R} = \mathscr{R}_{mot} + \mathscr{R}_u + \mathscr{R}_v + \mathscr{R}_w + \mathscr{R}_\phi \tag{14}$$

where

$$\mathscr{R}_u = \frac{1}{2}\iiint_V \sigma_u^d \dot{\epsilon}_u dV = \frac{1}{2}\pi R^2 C_X \int_{x=0}^L \dot{u}'^2 dx \qquad \mathscr{R}_v = \frac{1}{2}\iiint_V \sigma_v^d \dot{\epsilon}_v dV = \frac{1}{8}\pi R^4 C_Y \int_{x=0}^L \dot{v}''^2 dx \qquad \mathscr{R}_w = \frac{1}{2}\iiint_V \sigma_w^d \dot{\epsilon}_w dV = \frac{1}{8}\pi R^4 C_Z \int_{x=0}^L \dot{w}''^2 dx$$

$$\mathscr{R}_\phi = \frac{1}{2}\iiint_V \tau_\phi^d \dot{\gamma}_\phi dV = \frac{1}{4}\pi R^4 C_\Phi \int_{x=0}^L \dot{\phi}'^2 dx \qquad \mathscr{R}_{mot} = \frac{1}{2} b_m \dot{\theta}^2$$

Since

$$|\epsilon_u| = |u'| \;,\; \sigma_u^d = C_X \dot{\epsilon}_u, |\epsilon_v| = |yv''| = |rcos(\gamma)v''| \;,\; \sigma_v^d = C_Y \dot{\epsilon}_v,$$

$$|\epsilon_w| = |zw''| = |rsin(\gamma)w''| \;,\; \sigma_w^d = C_Z \dot{\epsilon}_w, |\gamma_\phi| = |r\phi'| \;\; and \;\; \tau_\phi^d = C_\Phi \dot{\gamma}_\phi.$$

2.4. Motion Equations

The extended Hamilton principle [12] is used to get motion equations and boundary conditions: $0 = \int_{t_1}^{t_2} (\delta T - \delta V + T_{mot}\,\delta\theta + \delta\zeta)dt$ where $\delta\zeta$ is the variation of work done by the dissipative forces, its expression is derived from Rayleigh dissipation function as follows: If the expression of Rayleigh dissipation function is given by: $\mathscr{R} = \frac{1}{2}\iiint_V \sigma^d\,\dot{\epsilon}\,dV$, then the expression of work variation $\delta\zeta$ done by dissipative forces is: $\delta\zeta = -\iiint_V \sigma^d\,\delta\epsilon\,dV$. Hence, using the fact that the beam is clamped at the joint i.e., $u(0,t) = v(0,t) = w(0,t) = \phi(0,t) = 0, v'(0,t) = w'(0,t) = 0$

The dynamic equation associated with θ is given by:

$$
\begin{aligned}
T_{mot} = {} & b_m\dot{\theta} + \frac{1}{2}I_B\ddot{\theta} - \Bigg[I_1\Big(cos(\theta)sin(\theta)(\dot{w}_L'^2 - \dot{\phi}_L^2) + \dot{\phi}_L\dot{w}_L'(2cos(\theta)^2 - 1)\Big) + I_2\Big(cos(\theta)sin(\theta)(\dot{\phi}_L^2 - \dot{w}_L'^2) - \dot{\phi}_L\dot{w}_L'(2cos(\theta)^2 - 1)\Big) \\
& + I_4\Big((\dot{\phi}_L^2 - \dot{w}_L'^2)(2cos(\theta)^2 - 1) + 4\dot{\phi}_L\dot{w}_L' cos(\theta)sin(\theta)\Big) + \dot{\theta}\big[(\dot{\phi}_L + \frac{1}{2}(v_L'\dot{w}_L' - \dot{v}_L'w_L'))(-I_5 sin(\theta) + I_6 cos(\theta)) + \\
& (-\dot{w}_L' + \dot{u}_L'w_L' + u_L'\dot{w}_L' + v_L'\dot{\phi}_L)(-I_5 cos(\theta) - I_6 sin(\theta))\big] + \dot{v}_L'\big[\dot{\phi}_L(-I_5 sin(\theta) + I_6 cos(\theta)) + \dot{w}_L'(I_5 cos(\theta) + I_6 sin(\theta))\big] - \\
& \frac{\partial}{\partial t}\Big(I_3(\dot{\theta} + \dot{v}_L' - \dot{u}_L'v_L' - u_L'\dot{v}_L' + \dot{\phi}_L w_L') + (\dot{\phi}_L + \frac{1}{2}(v_L'\dot{w}_L' - \dot{v}_L'w_L'))(I_5 cos(\theta) + I_6 sin(\theta)) + \\
& (-\dot{w}_L' + \dot{u}_L'w_L' + u_L'\dot{w}_L' + v_L'\dot{\phi}_L)(-I_5 sin(\theta) + I_6 cos(\theta)) + m_p\Big\{\dot{\theta}\Big(L^2 + u_L^2 + v_L^2 + c^2(1 - w_L'^2) + 2L\big[u_L - \frac{1}{2}\int_0^L (v'^2 + w'^2)ds + \\
& c(1 - \frac{1}{2}(v_L'^2 + w_L'^2))\big] + 2c(u_L + v_L v_L') - c\int_0^L (v'^2 + w'^2)ds\Big) + (L + c)(\dot{v}_L + c(\dot{v}_L' - \dot{u}_L'v_L' - u_L'\dot{v}_L')) + u_L(\dot{v}_L + c\dot{v}_L') - \\
& \dot{u}_L(v_L + cv_L')\Big\}\Big)\Bigg] + \frac{\rho}{2}\Big\{\frac{\partial}{\partial t}\Big(2\dot{\theta}\Big[\frac{1}{3}\pi R^2 L^3 + \frac{1}{4}\pi R^4 L + \pi R^2\int_0^L (u^2 + v^2)dx + \frac{1}{4}\pi R^4\int_0^L w'^2 dx + 2\pi R^2\int_0^L xudx - \\
& \frac{1}{2}\pi R^2\int_0^L (L^2 - x^2)(v'^2 + w'^2)dx\Big]\Big) + \\
& 2\frac{\partial}{\partial t}\Big(\pi R^2\int_0^L x\dot{v}dx - \frac{1}{4}\pi R^4\int_0^L (-\dot{v}' + \dot{u}'v' + u'\dot{v}' - 2w'\dot{\phi})dx + \pi R^2\int_0^L (u\dot{v} - \dot{u}v)dx\Big)\Big\}
\end{aligned}
\tag{15}
$$

- ⋆ The equation satisfied by u:

$$
0 = \frac{\rho}{2}\Big(-2\pi R^2\ddot{u} + 2\pi R^2\dot{\theta}^2 u + 2\pi R^2\dot{\theta}^2 x + \pi R^2(4\dot{\theta}\dot{v} + 2\ddot{\theta}v) - \frac{1}{2}\pi R^4\ddot{\theta}v''\Big) + \pi R^2 C_X\dot{u}'' + \pi R^2 E u'' \tag{16}
$$

- ⋆ The equation satisfied by v:

$$
\begin{aligned}
0 = {} & \frac{\rho}{2}\Big(-2\pi R^2\ddot{v} + \frac{1}{2}\pi R^4\ddot{v}'' + 2\pi R^2\dot{\theta}^2 v - \pi R^2\dot{\theta}^2(2xv' + (x^2 - L^2)v'') - \pi R^2(4\dot{\theta}\dot{u} + 2\ddot{\theta}u + 2x\ddot{\theta}) - \frac{1}{2}\pi R^4\ddot{\theta}u''\Big) \\
& - \frac{1}{4}\pi R^4 C_Y\dot{v}'''' - \frac{1}{4}\pi R^4 E v'''' + m_p\dot{\theta}^2(L + c)v''
\end{aligned}
\tag{17}
$$

- ⋆ The equation satisfied by w:

$$
\begin{aligned}
0 = {} & \frac{\rho}{2}\Big(-2\pi R^2\ddot{w} + \frac{1}{2}\pi R^4\ddot{w}'' - \frac{1}{2}\pi R^4\dot{\theta}^2 w'' - \pi R^2\dot{\theta}^2(2xw' + (x^2 - L^2)w'') - \pi R^4\dot{\theta}\dot{\phi}'\Big) - \frac{1}{4}\pi R^4 C_Z\dot{w}'''' + m_p\dot{\theta}^2(L + c)w'' \\
& - \frac{1}{4}\pi R^4 E w'''' - \rho g\pi R^2
\end{aligned}
\tag{18}
$$

- ⋆ The equation satisfied by ϕ:

$$
0 = \frac{\rho}{2}\Big(-\pi R^4\ddot{\phi} - \pi R^4(\ddot{\theta}w' + \dot{\theta}\dot{w}')\Big) + \frac{1}{2}\pi R^4 C_\Phi\dot{\phi}'' + \frac{1}{2}\pi R^4 G\phi'' \tag{19}
$$

- ⋆ Since the free end of the beam is at $x = L$, the following quantities δu_L, $\delta u_L'$, δv_L, $\delta v_L'$, δw_L, $\delta w_L'$, and $\delta\phi_L$ are arbitrary, therefore the final equations of boundary conditions are:

$$
0 = -\pi R^2 C_X\dot{u}_L' + \frac{\rho}{4}\pi R\ddot{\theta}v_L' - \pi R^2 E u_L' - \frac{\partial}{\partial t}\Big[m_p\Big(\dot{u}_L - \dot{\theta}(v_L + cv_L')\Big)\Big] + m_p\big[\dot{\theta}^2(u_L + L + c) + \dot{\theta}(\dot{v}_L + c\dot{v}_L')\big] \tag{20}
$$

$$
\begin{aligned}
0 = {} & -\frac{\partial}{\partial t}\Big[-I_3\dot{\theta}v_L' + w_L'(-I_5\dot{\theta}sin(\theta) + I_6\dot{\theta}cos(\theta)) - m_p\dot{\theta}c(L + c)v_L'\Big] - I_3\dot{\theta}\dot{v}_L' - I_5\dot{\theta}\dot{w}_L' sin(\theta) + I_6\dot{\theta}cos(\theta)\dot{w}_L' - m_p\dot{\theta}c(L + c)\dot{v}_L' \\
& + m_p g c w_L'
\end{aligned}
\tag{21}
$$

$$0 = \frac{1}{4}\pi R^4 C_Y \dot{v}_L''' - \frac{\rho}{4}\pi R^4 \ddot{v}_L' + \frac{\rho}{4}\pi R^4 \ddot{\theta}(u_L' - 1) + \frac{1}{4}\pi R^4 E v_L''' - \frac{\partial}{\partial t}\left[m_p\left(\dot{v}_L + c\dot{v}_L' + \dot{\theta}(L + c + u_L)\right)\right] + m_p\left(\dot{\theta}^2(v_L + cv_L') - \dot{\theta}\dot{u}_L - \dot{\theta}^2(L+c)v_L'\right) \tag{22}$$

$$0 = -\frac{1}{4}\pi R^4 C_Y \dot{v}_L'' - \frac{1}{4}\pi R^4 E v_L'' - \frac{\partial}{\partial t}\Big[I_3\left(\dot{v}_L' + \dot{\theta}(1 - u_L')\right) + I_5\left(-\frac{1}{2}\dot{\theta} w_L' cos(\theta) + \dot{\phi}_L cos(\theta) + \dot{w}_L' sin(\theta)\right) + I_6\left(-\frac{1}{2}\dot{\theta} w_L' sin(\theta) + \dot{\phi}_L sin(\theta) - \dot{w}_L' cos(\theta)\right) + m_p\left(c^2\dot{v}_L' + c\dot{v}_L + \dot{\theta}c(L+c)(1-u_L') + \dot{\theta}cu_L\right)\Big] - I_3\dot{\theta}\dot{u}_L' + I_5\left(\frac{1}{2}\dot{\theta}\dot{w}_L' cos(\theta) - \dot{\theta}\dot{\phi}_L sin(\theta)\right) + I_6\left(\frac{1}{2}\dot{\theta}\dot{w}_L' sin(\theta) + \dot{\theta}\dot{\phi}_L cos(\theta)\right) + m_p\left(\dot{\theta}^2(-Lcv_L' + cv_L) - \dot{\theta}c(L+c)\dot{u}_L' - \dot{\theta}c\dot{u}_L\right) \tag{23}$$

$$0 = \frac{1}{4}\pi R^4 C_Z \dot{w}_L''' + \frac{1}{4}\pi R^4 E w_L''' - m_p g - \frac{\rho}{4}\pi R^4 \ddot{w}_L' + \frac{\rho}{4}\pi R^4 \dot{\theta}^2 w_L' + \frac{\rho}{2}\pi R^4 \dot{\theta}\dot{\phi}_L - m_p(\ddot{w}_L + c\ddot{w}_L') - m_p\dot{\theta}^2(L+c)w_L' \tag{24}$$

$$0 = -\frac{1}{4}\pi R^4 C_Z \dot{w}_L'' - \frac{1}{4}\pi R^4 E w_L'' - m_p g c(1 - u_L') - \frac{\partial}{\partial t}\Big[I_1\left(sin(\theta)^2\dot{w}_L' + cos(\theta)sin(\theta)\dot{\phi}_L\right) + I_2\left(cos(\theta)^2\dot{w}_L' - cos(\theta)sin(\theta)\dot{\phi}_L\right) + I_4\left(-2\dot{w}_L' cos(\theta)sin(\theta) - \dot{\phi}_L(2cos(\theta)^2 - 1)\right) + I_5\left(\frac{1}{2}\dot{\theta}v_L' cos(\theta) + sin(\theta)(\dot{v}_L' - \dot{\theta}(u_L' - 1))\right) + I_6\left(\frac{1}{2}\dot{\theta}v_L' sin(\theta) + cos(\theta)(-\dot{v}_L' + \dot{\theta}(u_L' - 1))\right) + m_p\left(c^2\dot{w}_L' + c\dot{w}_L\right)\Big] + I_3\dot{\theta}\dot{\phi}_L + I_5\left(-\frac{1}{2}\dot{\theta}\dot{v}_L' cos(\theta) - \dot{\theta}\dot{u}_L' sin(\theta)\right) + I_6\left(-\frac{1}{2}\dot{\theta}\dot{v}_L' sin(\theta) + \dot{\theta}\dot{u}_L' cos(\theta)\right) - m_p\dot{\theta}^2 c(L+c)w_L' \tag{25}$$

$$0 = -\frac{1}{2}\pi R^4 C_\Phi \dot{\phi}_L' - \frac{1}{2}\pi R^4 G \phi_L' - \frac{\partial}{\partial t}\Big[I_1\left(cos(\theta)^2\dot{\phi}_L + cos(\theta)sin(\theta)\dot{w}_L'\right) + I_2\left(sin(\theta)^2\dot{\phi}_L - cos(\theta)sin(\theta)\dot{w}_L'\right) + I_4\left(2cos(\theta)sin(\theta)\dot{\phi}_L - \dot{w}_L'(2cos(\theta)^2 - 1)\right) + I_5\left((\dot{\theta} + \dot{v}_L')cos(\theta) - \dot{\theta}v_L' sin(\theta)\right) + I_6\left((\dot{\theta} + \dot{v}_L')sin(\theta) + \dot{\theta}v_L' cos(\theta)\right)\Big] \tag{26}$$

u, v, ϕ must also satisfy these conditions: $u(x,0) = \lim_{t\to\infty} u(x,t) = 0$, $v(x,0) = \lim_{t\to\infty} v(x,t) = 0, \phi(x,0) = \lim_{t\to\infty}\phi(x,t) = 0$ and w must satisfy $w(x,0) = \lim_{t\to\infty} w(x,t) = \tilde{w}(x)$ whose expression [13] is given by: $\tilde{w}'(x) = tan\left(\frac{x(2a-x)}{2b}\right)$, since $\tilde{w}(0) = 0$, then $\tilde{w}(x) = \int_0^x tan\left(\frac{l(2a-l)}{2b}\right)dl$, where $a = L - \delta, b = \frac{EI}{F}$, the expression of the foreshortening δ term due to beam bending [14] is given by: $\delta = -\frac{1}{2}\int_0^L \tilde{w}'^2(x)dx$, where payload weight F equals $m_p g$, and beam area second moment I equals:$I = \iint y^2 dydz = \frac{\pi R^4}{4}$.

3. Discussion

Considering the reference of angle θ is zero when the manipulator is at rest ($t = 0$) and the angular velocity is constant ($\dot{\theta} = \Omega$), then θ and $\dot{\theta}$ are replaced by Ωt and Ω respectively in the equations of the previous section. Equation (16) yields $\dot{u} = L_1(v)$, taking the time derivative of Equation (15) and using the last expression yields $L_2(v) = 0$. Equation (17) yields $\dot{\phi}' = L_3(w) + c$, taking both time and spatial derivatives of Equation (18) and using the last expression yields $L_4(w) = 0$, where c is a constant and L_1, L_2, L_3, L_4 are linear operators. Hence, the motions equations are decoupled but the boundary conditions are still coupled. The goal of future work is to develop a numerical method for solving previous partial differential equations with coupled boundary conditions while ensuring the stability of the solutions. Once the solutions are found, the mechanical modeling will be generalized to flexible manipulators with serial links where the payload attached to each link is the rest of the chain.

4. Conclusions

Modeling the single-link flexible manipulator as an Euler–Bernoulli beam with a payload at its free end subjected to small deformations, and using a rotation matrix orthogonal to the second-order of Taylor expansion in the deformations variables, the extended Hamilton principle is applied to get both the motion equations and boundary conditions. The motion partial differential equations are decoupled when the angular velocity is constant. Once the solutions are available, it will help to study more accurately the movements of flexible manipulators and to find new techniques for robust control of such systems.

Author Contributions: Conceptualization, M.B. and R.O.B.Z.; methodology, M.B.; investigation, A.K. All authors have read and agreed to the published version of the manuscript.

Funding: This research did not receive any specific grant from funding agencies in the public, commercial, or not-for-profit sectors.

Institutional Review Board Statement: Not applicable.

Informed Consent Statement: Not applicable.

Data Availability Statement: Not applicable.

Conflicts of Interest: The authors declare no conflict of interest.

Nomenclature

T_{mot}	Torque applied by the motor
R	Beam cross section radius
L	Beam length
m_p	Payload mass
I_p	Payload inertia matrix
$u(x,t)$	Axial stretching
$v(x,t)$	Horizontal deflection
$w(x,t)$	Vertical deflection
$\phi(x,t)$	Torsional deformation
R_0	Inertial frame
R_1	Motor frame
R_{dm}	Cross section frame
R_2	Beam free end frame
O_0,O_1,O_{dm},O_2	Origins of frames R_0,R_1,R_{dm},R_2 respectively
C	Payload center of mass
${}^{i}\overrightarrow{v}$	Vector v expressed in frame R_i
${}^{0}\overrightarrow{\Omega_{1/0}}$	Rotation vector of frame R_1 relative to R_0 expressed in R_0
${}^{0}\overrightarrow{\Omega_{dm/0}}$	Rotation vector of frame R_{dm} relative to R_0 expressed in R_0
${}^{1}\overrightarrow{\Omega_{dm/1}}$	Rotation vector of frame R_{dm} relative to R_1 expressed in R_1
w_x,w_y,w_z	${}^{1}\overrightarrow{\Omega_{dm/1}}$ coordinates in frame R_1
T,T_B,T_l,T_p	Kinetic energy of the system,base,link,payload
I_1,I_2,I_3,I_4,I_5,I_6	Inertia matrix coefficients
ρ	Beam mass density
θ	Rotation angle of frame R_1 relative R_0
V,V_B,V_l,V_p	Potential energy of the system,base,link,payload
V_{gravit},V_{str}	link gravitational and strain potential energy
V_u,V_v,V_w,V_ϕ	Strain energies due to u, v, w, ϕ
$\mathscr{R}_{mot}$,$\mathscr{R}_u$,$\mathscr{R}_v$,$\mathscr{R}_w$,$\mathscr{R}_\phi$	Rayleigh dissipation function due to motor,u,v,w,ϕ
E, G	Beam young moduls and shear modulus
$\epsilon_u, \epsilon_v, \epsilon_w$	strains due to u, v, w

$\sigma_u^d, \sigma_v^d, \sigma_w^d, \tau_\phi^d$	stress due to internal damping caused by u,v,w and
C_X, C_Y, C_Z	Internal damping coefficient along x_1,y_1,z_1 axis
γ_ϕ, τ_ϕ^d	Shear strain and shear stress due to torsional damping
ζ, F	Work done by dissipative forces and payload weight
δ, b_m	Foreshortening term,motor viscous friction coefficient
C_ϕ, I	Torsional deformation coefficient, area second moment

References

1. Tokhi, M.; Azad, A. *Flexible Robot Manipulators: Modelling, Simulation and Control*; IET: London, UK, 2008.
2. Saad, M. Modeling of a one flexible link manipulator. In *Robot Manipulators: Trends and Development*; IntechOpen: London, UK, 2010; p. 73.
3. Shi, P.; McPhee, J.; Heppler, G. A deformation field for Euler–Bernoulli beams with applications to flexible multibody dynamics. *Multibody Syst. Dyn.* **2001**, *5*, 79–104. [CrossRef]
4. Jazar, R. *Theory of Applied Robotics: Kinematics, Dynamics, and Control*; Springer: Berlin/Heidelberg, Germany, 2010. [CrossRef]
5. Shabana, A. *Dynamics of Multibody Systems*; Cambridge University Press: Cambridge, UK, 2014.
6. Stewart, J. *Essential Calculus: Early Transcendentals*; Cengage Learning: Boston, MA, USA, 2012.
7. Díaz, E.; Díaz, O. *Ditzinger 3D Motion of Rigid Bodies*; Springer: Berlin/Heidelberg, Germany, 2019. [CrossRef]
8. Low, K.; Vidyasagar, M. A Lagrangian Formulation of the Dynamic Model for Flexible Manipulator Systems. *J. Dyn. Syst. Meas. Control* **1988**, *110*, 175–181. [CrossRef]
9. Gavin, H. *Strain Energy in Linear Elastic Solids*; Department of Civil and Environmental Engineering, Duke University: Durham, NC, USA, 2015.
10. Reddy, J. *An Introduction to Continuum Mechanics*; Cambridge University Press: Cambridge, UK, 2007.
11. Chen, W. Bending vibration of axially loaded Timoshenko beams with locally distributed Kelvin–Voigt damping. *J. Sound Vib.* **2011**, *330*, 3040–3056. [CrossRef]
12. Meirovitch, L. *Analytical Methods in Vibrations*; Pearson: London, UK, 1967.
13. Carrillo, E. The cantilevered beam: An analytical solution for general deflections of linear-elastic materials. *Eur. J. Phys.* **2006**, *27*, 1437. [CrossRef]
14. Li, Q.; Wang, T.; Ma, X. A note on the foreshortening effect of a flexible beam under oblique excitation. *Multibody Syst. Dyn.* **2010**, *23*, 209–225. [CrossRef]

engineering proceedings

MDPI

Abstract

Seed Priming with Pectic-Oligosaccharides Improved Seed Germination and Growth of Chili †

Wascharin Udchumpisai *, Yuree Wandee, Ditpon Kotatha and Dudsadee Uttapap

Division of Biochemical Technology, School of Bioresources and Technology, King Mongkut's University of Technology Thonburi, Bangkhuntien, Bangkok 10150, Thailand; yuree.wan@kmutt.ac.th (Y.W.); ditpon.kot@kmutt.ac.th (D.K.); dudsadee.utt@kmutt.ac.th (D.U.)

* Correspondence: udchumpisai.w@outlook.co.th; Tel.: +66-871666446

† Presented at the 2nd International Electronic Conference on Applied Sciences, 15–31 October 2021; Available online: https://asec2021.sciforum.net/.

Abstract: The aim of this study was to assess the impact of pectic-oligosaccharides (POSs) obtained from oxidative degradation of pomelo peel with H_2O_2 under alkaline conditions on seed germination and growth of chili using the seed priming technique. Two types of POSs (POS-I and POS-II) with different size distributions were prepared. Chili seeds were soaked in 500 ppm of POS solutions for 16 h at two temperatures, 30 and 50 °C, with moderate shaking, and then air-dried. The primed seeds were planted on wet filter paper in a Petri dish at 30 °C for 9 d and the effects of priming on germination and growth were observed. Priming of seeds with POS at 30 °C increased the germination percentage and vigor index at 9 d after sowing by 16.7–20.5% and 16.0–25.5%, respectively, whereas root and shoot length did not differ from the hydropriming. However, there were no significant differences in all growth parameters between POS-I and POS-II treatments. Seedling length and the vigor index of seeds primed with POS at 30 °C (29.3–31.0 mm, 2693.9–2914.0) were also significantly higher than those of non-primed seeds (22.9 mm, 2062). Priming the seeds at 50 °C had an adverse effect on seed germination, i.e., the germination percentage was reduced by 24.4–31.9% compared to seeds primed at 30 °C. It also resulted in a significant reduction in chili seed growth, possibly due to the inactivation of some enzymes in chili seeds.

Keywords: chili; germination; growth; oxidative degradation; pectic-oligosaccharides; pomelo; seed priming

Citation: Udchumpisai, W.; Wandee, Y.; Kotatha, D.; Uttapap, D. Seed Priming with Pectic-Oligosaccharides Improved Seed Germination and Growth of Chili. *Eng. Proc.* **2021**, *11*, 41. https://doi.org/10.3390/ASEC2021-11159

Academic Editor: Roger Narayan

Published: 15 October 2021

1. Introduction

Poor seedling growth and seedling vigor lead to poor planting and yield. In various crops, different seed treatment techniques are used to improve the germination and seedling vigor after harvest [1]. Seed priming is the conventional strategy for improving crop production and minimizing the negative effects against stress. Partial seed pre-hydration promotes membrane protein hydration and activates various metabolism processes, as well as early germination events in seeds. These activities stop after the re-drying of seeds [2,3]. Several priming technologies such as osmopriming, solid matrix priming, hormopriming, biopriming and chemical priming have been developed recently to increase the efficiency of germination and strength of seeds [2].

Plant biostimulators are natural compounds that can improve plant life processes without an adverse impact on principal natural plant pathways such as the regulation of carbon and nitrogen metabolism, the control of secondary metabolism, and the activation of plant enzyme production related to growth and detoxification, etc., [4,5]. Hu et al. [6] reported that soaking maize seeds in alginate-derived oligosaccharide solution for 15 h can increase root and shoot growth on day 7 up to 18% and 46%, compared to seeds soaked in water. Chitosan microparticles can enhance the germination and vigor index of tomato seeds and also improve the root and stem development [7]. The study of Nandhini and Somasundaram [8] showed that priming maize seeds with lipo-chito oligosaccharide improved the tolerance of seeds to salinity, as indicated by the increasing germination stress tolerance index.

Pectic-oligosaccharides (POS) are oligosaccharides produced from the partial depolymerization of pectin using different techniques including chemical, hydrothermal, and enzymatic degradation. POS products include oligo-galacturonides (OGA), galacto-oligosaccharides (GalOS), rhamnogalacturonan-oligosaccharides (RGOS), and others [9–11]. Although there has been research on POS obtained from a variety of raw materials, such as orange and apple pomaces, there have been few studies on POS derived from pomelo peel. Moreover, the bioavailability of POS is mostly examined in food applications, such as prebiotics properties and antioxidant activity, etc., with less work being done on their usage in agriculture. In recent years, many researchers have found that OGAs affect the morphogenesis and organogenesis of various plants, such as inducing root formation, stimulating growth and improving robustness of the in vitro cultivated plants [12]. This study aimed to assess the impact of POSs derived from pomelo peel on chili seed germination and seedling growth by using the priming technique.

2. Materials and Methods

2.1. Materials

Chili seeds were prepared from dried chili purchased from a local market. Fruit peels of pomelo (*Citrus maxima Merr.*) cultivar Khao-Yai were collected from local vendor in Nakhon Pathom Province, Thailand. Galacturonic acid (GalA), digalacturonic acid and trigalacturonic acid were purchased from Sigma Chemical Co. (St. Louis, MO, USA). The H_2O_2 and other chemical reagents used were of analytical reagent grade.

2.2. Preparation of POS

The albedo part of pomelo peels was washed, chopped into small pieces and dried at 40 °C before being pulverized to powder with a particle size less than 200 µm. POSs were produced from the dried pomelo peel powder by using H_2O_2 under alkaline condition, as described by Wandee et al. [13], with minor modifications. Briefly, dried pomelo peel powder was hydrolyzed with 5% (*v*/*v*) H_2O_2 in 50 mM NaOH at 80 °C for 5 h with continuous stirring. The mixture was collected and centrifuged at 6136× *g* for 10 min. Absolute ethanol was added to the supernatant at a ratio of 1:1 (*v*/*v*) and kept overnight at 4 °C. The suspension was centrifuged at 6136× *g* for 20 min. The pellet was collected and washed 3–4 times with 95% ethanol to obtain POS-I, whereas another 2 vol of absolute ethanol was added to the supernatant and kept overnight at 4 °C. The precipitate was collected to obtain POS-II. The POS precipitates were dried in a hot air oven at 40 °C for 24 h. The dried POS was ground and sifted through a 200 µm sieve to obtain fine powders.

For the POS size distribution analysis, 20 µL of POS solution was injected into a Shimadzu HPLC system consisting of LC-20AD pump, RID-10A detector, two serially linked columns (both Shodex OHpak SB-802.5 HQ, 8 mm × 300 mm, Showa Denko K.K., Tokyo, Japan) with a specified guard column, and a computer with a data processing software program (CLASS-VP). An isocratic elution with 0.1 M $NaNO_3$ containing 0.01 M NaN_3 was carried out at 50 °C and a flow rate of 0.8 mL·min^{-1}. Galacturonic acid, digalacturonic acid and trigalacturonic acid were used as standards.

2.3. Imbibition Curve

Seeds were disinfected by soaking in 1% (*v*/*v*) sodium hypochlorite containing 0.2% (*v*/*v*) Tween 80 for 30 min and rinsed three times with sterilized water. The seeds were dried overnight in an oven at 37 °C. Five grams of sterilized seeds were soaked in 25 mL of sterilized water (or 500 ppm POS solution) for 72 h with shaking at 30 °C. At specified times, 0.3 g of soaked seeds were collected, weighed, and dried at 105 °C for 48 h. Three replicates of each treatment were conducted, and the water content of seeds was calculated using the following equation:

$$\text{Water content } (\%) = \left(\frac{W_s - W_d}{W_d}\right) \times 100 \quad (1)$$

where W_s and W_d are the weights of soaked seeds and dried seeds, respectively.

2.4. Seed Priming Treatment

Five grams of disinfected seeds were fully immersed in 25 mL of priming media (water and 500 ppm POS solution) with shaking at two temperatures, 30 and 50 °C. Priming time was based on the early time of the stationary phase (phase II) of the imbibition curve. After priming, seeds were washed with sterile water, removed of excess water, and dried in an oven at 37 °C for 48 h. The dried primed seeds were kept at room temperature until use.

2.5. Determinations of Seed Germination and Seedling Characteristics

Seeds were germinated on a filter paper wetted with 5 mL of distilled water in Petri dishes and kept in the dark at 30 ± 2 °C. The seed growth was conducted over a nine-day period. Each seed soaking treatment was replicated three times with 50 seeds each. The approach developed by the Association of Official Seed Analysts (AOSA) [14] was used to count seed germination on a daily basis. A seed was deemed germinated when the radical length was at least 2 mm long. Other seed growth characteristics including percentage of germination and vigor index [15] were evaluated using the following equations.

$$\text{Percentage of germination } (\%\text{G}) = \frac{\text{Total no. of germinated seeds}}{\text{Total no. of initial seeds}} \times 100 \quad (2)$$

$$\text{Vigor index} = \%\text{G} \times \text{seedling length (cm)} \quad (3)$$

2.6. Statistical Analyses

The experimental data were analyzed using analysis of variance (ANOVA) and expressed as mean values ± standard deviations. A Tukey test was conducted to examine significant differences among experimental mean values ($p \leq 0.05$).

3. Results and Discussion

3.1. Size Distribution of POS

HPSEC chromatograms of POS-I and POS-II, and relative amount of POS species contained in each sample (% weight basis) are shown in Figure 1. The lower retention time indicates the larger molecular size of POS species. Although the same molecular size range was shown in both POS fractions, the relative area under the peaks of POS-I and POS-II was clearly different. A higher relative area under area #1 and a lower relative area under area #2 − #6 of POS-I indicated that POS-I contained a higher proportion of large-sized POS species than POS-II. The percent masses of POS species having a size larger than > DP5 of POS-I and POS-II were 81.9 and 73.8%, respectively.

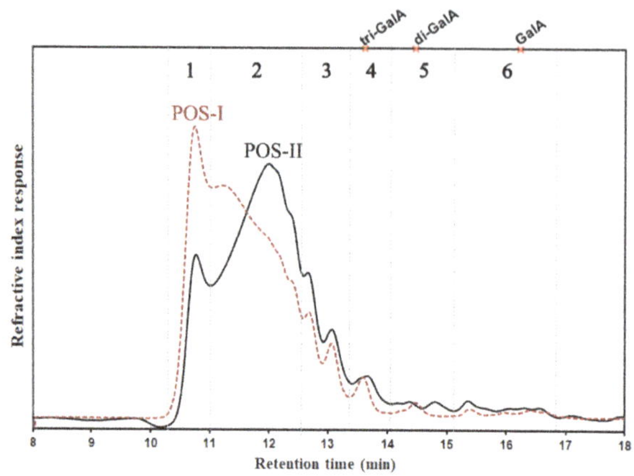

Area No.	Relative area (%)					
	1	2	3	4	5	6
POS I	24.0	57.9	10.2	2.8	2.0	3.1
POS II	12.4	61.4	14.0	4.3	3.4	4.5

Figure 1. HPSEC chromatograms of POSs and relative amount of POS species in POS-I (dot line) and POS-II (solid line).

3.2. Imbibition Curve

Imbibition curves of chili seeds when soaked in water and POS solutions for 48 h are shown in Figure 2. Generally, the imbibition curve of the seeds can be divided into three stages; imbibition, metabolism activation and elongation [16]. For the imbibition stage, the seed imbibes water rapidly to activate enzyme activity. Water contents of chili seeds in all treatments jumped from 1.2% to 39.8–42.2% during the first 2 h.

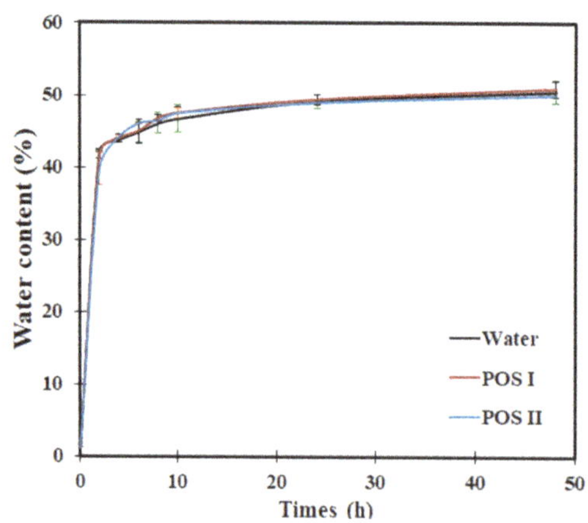

Figure 2. Imbibition curves of chili seeds when soaked in water, and 500 ppm of POS-I and POS-II solutions.

After 2 h, the seeds still further imbibed water but the rate of water imbibition decreased markedly. Water contents of seeds soaked in water and POS solutions were in the range of 42.2–50.5% and 39.8–50.9%, respectively. The metabolic processes of seeds were activated during this phase to prepare seeds for germination. However, the germination was not observed in during this period.

3.3. Germination and Seed Growth Parameters

Germination curves of chili seeds primed with different treatments are shown in Figure 3. The germination curve of chili seeds primed at 30 °C indicated that germination rates of seeds primed with POS solutions were significantly higher than that with hydropriming, especially after 6 d of incubation. Seeds primed with POS-I showed slightly higher germination percentage than those with POS-II. When compared to the non-primed seeds, seeds primed with POS solutions showed observably higher germination. Priming at 50 °C had adverse effect on seed germination, i.e., the germination percentage was reduced by 24.4–31.9% compared to seeds primed at 30 °C.

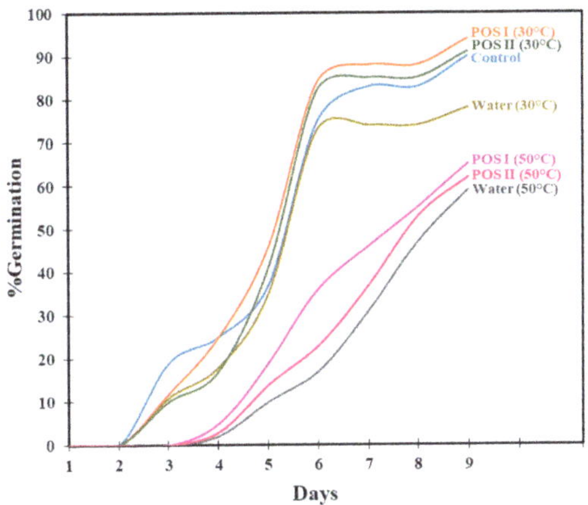

Figure 3. Germination curve (cumulative germination percentage vs. time, in d) of chili seeds after priming with different treatments.

The growth parameters of chili seedlings after incubating in Petri dish for 9 d are summarized in Table 1. The non-primed and hydroprimed seeds developed shoots 11.6 and 15.7 mm in length on average, respectively. Under the same cultivation environment, the seeds primed with POS-I and POS-II solutions demonstrated significantly longer shoots, 17.7 and 18.1 mm, respectively, while the root length in all treatments was not significantly different. POSs also increased the chili seedling length and vigor index by 27.9–35.4% and 30.6–41.3%, respectively, when compared to non-priming treatment. However, there were no significant differences in all growth parameters between POS-I and POS-II treatments. Priming the seeds at 50 °C resulted in a significant reduction in chili seed growth, possibly due to inactivation of some enzymes in chili seeds [17].

Table 1. Root length, shoot length, seedling length and vigor index of chili seeds primed with water, and 500 ppm of POS-I and POS-II solutions at 9 d after sowing compared to non-primed seeds (control).

Treatments	Root Length (mm)	Shoot Length (mm)	Seedling Length (mm)	Vigor Index
Non-priming (Control)	11.2 ^{BC}	11.6 ^{B}	22.9 ^{B}	2062.0 ^{C}
Priming at 30 °C				
Water	14.1 ^{A}	15.7 ^{A}	29.8 ^{A}	2322.0 ^{BC}
POS-I	13.3 ^{AB}	17.7 ^{A}	31.0 ^{A}	2914.0 ^{A}
POS-II	11.2 ^{BC}	18.1 ^{A}	29.3 ^{A}	2693.9 ^{AB}
Priming at 50 °C				
Water	6.0 ^{E}	6.2 ^{C}	12.2 ^{D}	720.0 ^{E}
POS-I	7.4 ^{DE}	6.8 ^{C}	14.2 ^{CD}	925.7 ^{DE}
POS-II	9.6 ^{CD}	9.4 ^{BC}	19.0 ^{BD}	1178.0 ^{D}

Means with different letters (^{A}, ^{B}, …) within columns are significantly different at $p \leq 0.05$.

4. Conclusions

Pectic-oligosaccharides (POS) derived from pomelo peel can promote chili seed development, as evidenced by the increase in the germination percentage and vigor index of seeds after hydropriming treatment compared to non-primed seeds. Although the percentage of germination of seeds primed with POS-I was slightly higher than that of those primed with POS-II, there was no significant difference in chili seed growth parameters. Priming of chili seeds at a high temperature (50 °C) had adverse effects on seed germination and seedling growth.

Author Contributions: Conceptualization, D.U.; Methodology, W.U. and Y.W.; Validation, W.U.; Formal analysis, W.U. and D.K.; Investigation, W.U. and Y.W.; Resources, D.U.; Data curation, W.U.; Writing—original draft preparation, W.U.; Writing—review and editing, W.U., Y.W. and D.U.; Visualization, W.U. and D.K.; Supervision, D.U.; Project administration, D.U.; Funding Acquisition, D.U. All authors have read and agreed to the published version of the manuscript.

Funding: This research has been funded by King Mongkut's University of Technology Thonburi, Thailand through the Post-doctoral Fellowship, and the National Research Council of Thailand through the NRCT Senior Research Scholar Program (Contract No.814-2020).

Institutional Review Board Statement: Not applicable.

Informed Consent Statement: Not applicable.

Data Availability Statement: The data presented in this study are available on request from the corresponding author.

Conflicts of Interest: The authors declare no conflict of interest.

References

1. Taylor, A.G.; Allen, P.S.; Bennett, M.A.; Bradford, K.J.; Burris, J.S.; Misra, M.K. Seed enhancements. *Seed Sci. Res.* **1998**, *8*, 245–256. [CrossRef]
2. Lutts, S.; Benincasa, P.; Wojtyla, L.; Kubala, S.S.; Pace, R.; Lechowska, K.; Quinet, M.; Garnczarska, M. Seed Priming: New Comprehensive Approaches for an Old Empirical Technique. In *New Challenges in Seed Biology—Basic and Translational Research Driving Seed Technology*; Araujo, S., Balestrazzi, A., Eds.; IntechOpen: London, UK, 2016.
3. Maiti, R.; Rajkumar, D.; Jagan, M.; Pramanik, K.; Vidyasagar, P. Effect of Seed Priming on Seedling Vigour and Yield of Tomato and Chilli. *Int. J. Bio-Resour. Stress Manag.* **2013**, *4*, 119–125.
4. Thakur, M.; Sohal, B.S. Role of Elicitors in Inducing Resistance in Plants against Pathogen Infection: A Review. *ISRN Biochem.* **2013**, *2013*, 762412. [CrossRef] [PubMed]
5. Posmyk, M.M.; Szafranska, K. Biostimulators: A New Trend towards Solving an Old Problem. *Front. Plant Sci.* **2016**, *7*, 748. [CrossRef] [PubMed]
6. Hu, X.; Jiang, X.; Hwang, H.; Liu, S.; Guan, H. Promotive effects of alginate-derived oligosaccharide on maize seed germination. *J. Appl. Phycol.* **2004**, *16*, 73–76. [CrossRef]
7. Colman, S.L.; Salcedo, M.F.; Mansilla, A.Y.; Iglesiasa, M.J.; Fiol, D.F.; Saldañac, S.M.; Chevalier, A.A.; Casalongué, C.A.; Alvarez, A.A. Chitosan microparticles improve tomato seedling biomass and modulate hormonal, redox and defense pathways. *Plant Physiol. Biochem.* **2019**, *143*, 203–211. [CrossRef] [PubMed]
8. Nandhini, D.U.; Somasundaram, E. Lipo-Chito Oligosaccharides Enhances Germination Tolerance of Maize to Salinity Stress. *Int. J. Curr. Microbiol. App. Sci.* **2017**, *6*, 437–443. [CrossRef]
9. Babbar, N.; Dejonghe, W.; Sforza, S.; Elst, K. Enzymatic pectic oligosaccharides (POS) production from sugar beet pulp using response surface methodology. *J. Food Sci.* **2017**, *54*, 3707–3715. [CrossRef] [PubMed]
10. Beatriz, M.; Belén, G.; Patricia, G.; Beatriz, G.; Alonso, J.L. Pectic Oligosaccharides and Other Emerging Prebiotics. In *Probiotics and Prebiotics in Human Nutrition and Health*; Rao, V., Ed.; IntechOpen: London, UK, 2016.
11. Gullón, B.; Gómez, B.; Martínez-Sabajanes, M.; Yánez, R.; Parajó, J.C.; Alonso, J.L. Pectic oligosaccharides: Manufacture and functional properties. *Trends Food Sci. Technol.* **2013**, *30*, 153–161. [CrossRef]
12. Cabrera, J.C.; Wégria, G.; Onderwater, R.C.A.; González, G.; Nápoles, M.C.; Falcón-Rodríguez, A.B.; Costales, D.; Rogers, H.J.; Diosdado, E.; González, S.; et al. Practical use of oligosaccharins in agriculture. *Acta Hortic.* **2013**, *1009*, 195–212. [CrossRef]
13. Wandee, Y.; Uttapap, D.; Mischnick, P.; Rungsardthong, V. Production of pectic-oligosaccharides from pomelo peel pectin by oxidative degradation with hydrogen peroxide. *Food Chem.* **2021**, *348*, 129078. [CrossRef] [PubMed]
14. Association of Official Seed Analysts (AOSA). Rules for testing seeds. *J. Seed Technol.* **1990**, *12*, 1–112.
15. Amnuaysin, N.; Korakotchakorn, H.; Chittapun, S.; Poolyarat, N. Seed germination and seedling growth of rice in response to atmospheric air dielectric-barrier discharge plasma. *Songklanakarin J. Sci. Technol.* **2018**, *40*, 819–823.

16. Bewley, J.D.; Black, M. *Seed: Physiology of Development and Germination*, 2nd ed.; Springer: New York, NY, USA, 1994; pp. 1–445.
17. Qi, F.; Zhang, F. Cell Cycle Regulation in the Plant Response to Stress. *Front. Plant Sci.* **2019**, *10*, 1765. [CrossRef] [PubMed]

engineering proceedings

Proceeding Paper

High-Resolution Distributed Liquid Level Sensor Based on a Self-Heating Approach †

Ester Catalano, Agnese Coscetta, Luigi Zeni and Aldo Minardo *

Department of Engineering, Università della Campania Luigi Vanvitelli, Via Roma 29, 81031 Aversa, Italy; ester.catalano@unicampania.it (E.C.); agnese.coscetta@unicampania.it (A.C.); luigi.zeni@unicampania.it (L.Z.)
* Correspondence: aldo.minardo@unicampania.it; Tel.: +39-081-501-0435

† Presented at the 2nd International Electronic Conference on Applied Sciences, 15–31 October 2021; Available online: https://asec2021.sciforum.net/.

Abstract: In this work, we propose a high-resolution distributed liquid level sensor based on a Cobalt-based, high-attenuation fiber (HAF), and a high-spatial resolution (5 mm) Brillouin Optical Frequency-Domain Analysis (BOFDA) sensor. In our method, the interrogating laser has a dual role: on one side, it excites the acoustic wave involved in the scattering phenomenon; on the other side, it heats up the fiber in a manner dependent on the surrounding medium (air or liquid). The proposed method has the potentiality of determining the liquid level with high spatial resolution, without requiring any additional component compared to a conventional BOFDA sensor.

Keywords: distributed optical fiber sensors; liquid detection

Citation: Catalano, E.; Coscetta, A.; Zeni, L.; Minardo, A. High-Resolution Distributed Liquid Level Sensor Based on a Self-Heating Approach. *Eng. Proc.* **2021**, *11*, 42. https://doi.org/10.3390/ASEC2021-11131

Academic Editor: Saulius Juodkazis

Published: 15 October 2021

1. Introduction

High-resolution liquid level sensing is relevant to industry monitoring. Compared to conventional electrical liquid-level sensors, optical-fiber sensors provide key advantages such as immunity to electromagnetic interference, good corrosion resistance, and high sensitivity. Actively heated fibers can be used for liquid level sensing, in which some heat source is employed to raise the fiber temperature. The temperature reached by the fiber, or the time required to recover the original temperature after heat source removal, may be employed to gain information about the medium surrounding the fiber, such as its phase (liquid or gas) [1], its speed [2,3], or its thermal conductivity [4]. Heat can be conveyed either electrically by Joule effect, or optically by use of specialty fibers. The latter solution is preferrable in those cases in which one should avoid the use of electrical currents (e.g., in hazardous environments). Specialty fibers whose core is doped with transition metal ions (namely Co^{2+} or Vn^{+}) can be employed, where absorption of light at specific wavelengths produces heat due to nonradiative relaxation. While conventional approaches recover the temperature using fiber Bragg gratings (FBGs) inscribed into the specialty fiber, a more convenient approach relies on the use of fully distributed sensing techniques. Chen et al. employed Optical Frequency-Domain Reflectometry (OFDR) in a Co^{2+}-doped fiber, realizing liquid level sensing at cryogenic temperatures [5]. In their work, the authors used an optical source at 1550 nm for heating, and a wavelength scanning laser in the C-band for sensing. More recently, a dual wavelength approach has been demonstrated by our group, in which a 1550 nm laser source was used for heating, while an 850 nm Brillouin setup was used for distributed temperature sensing [6]. The dual wavelength approach exploits the wavelength-selective absorption of the Cobalt-doped fiber, so that a wavelength falling into the low absorption band of the specialty fiber can be chosen for sensing. However, this also complicates the setup, due to the necessity to use separate laser sources and wavelength-division multiplexing (WDM) components.

In this work, we propose the use of a Brillouin Optical Frequency-Domain Analysis configuration featuring a spatial resolution of 5 mm, in order to perform liquid level sensing along a Co^{2+}-doped fiber. Differently from Ref. [6], a single wavelength approach is followed here, where the 1550 nm wavelength is adopted both for sensing and heating. The amount of self-heating is controlled by acting on the optical intensity of the pump (and probe) light used for BOFDA sensing. We will show experimentally that, the air/liquid interface can be easily recognized from the acquired temperature profile. In fact, the heat transfer in liquids is much more efficient than in air, leading to an abrupt decrease of the temperature along the optically heated fiber along the segment immersed into water.

2. Experimental Results

The BOFDA method consists in the acquisition of the amplitude and phase of the modulation impressed on a c.w. probe beam, due to the stimulated Brillouin scattering interaction with a counterpropagating pump beam, for a range of modulation frequencies. The two beams are separated by a frequency offset in the range of 10–11 GHz, corresponding to the so-called Brillouin Frequency Shift (BFS) of the fiber. The method is usually adopted for determining the BFS distribution along the fiber, with a spatial resolution not usually achievable with common time-domain reflectometric techniques [7]. As the BFS is linearly proportional to the temperature, the method is here applied to determine the temperature change along the Co^{2+}-doped fiber resulting from self-heating.

The experimental scheme adopted for the measurements is shown in Figure 1. The light from an external cavity laser, with a linewidth less than 100 kHz, is split into distinct branches for pump/probe generation. In the upper (probe) branch, the laser beam is double-sideband (DSB) modulated by means of an intensity electro-optic modulator (IM1) driven by an RF synthesizer. The upper sideband is filtered out by means of a narrowband fiber Bragg grating (FBG), while the lower sideband is first amplified by an Erbium-Doped Fiber Amplifier (EDFA1), then launched into one end of the fiber under test (FUT). In the lower (pump) branch, the laser beam is modulated by another electro-optic modulator (IM2) biased at its quadrature point and driven by the RF output of the vector network analyzer (VNA). The modulated pump passes through a polarization switch, used to suppress any polarization dependency from the measurements. Finally, the probe beam is amplified by EDFA2 and launched into the opposite end of the FUT. The backscattered light from the pump is fed into a high-bandwidth photodetector. The VNA covers the range from 300 kHz to 20 GHz, and thus permits us to investigate the Brillouin response at a minimum spatial resolution of 5 mm.

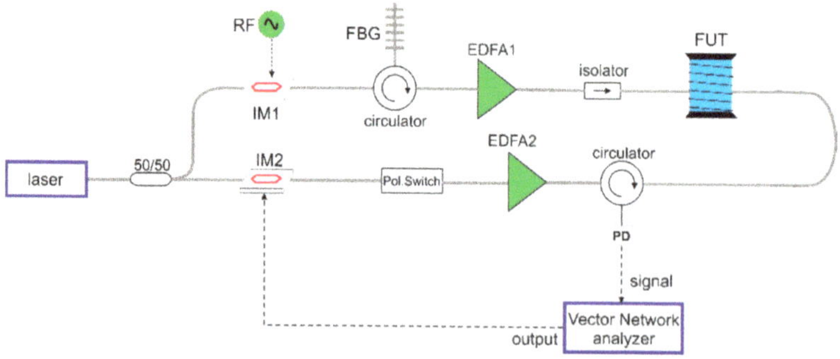

Figure 1. Experimental setup. IM, intensity modulator; FBG, fiber-Bragg grating; EDFA, erbium-doped fiber amplifier; Pol. Switch, polarization switch; PD, photodetector; FUT, fiber under test.

As a preliminary test, we prepared a FUT composed by a strand of 3 m of SMF-28 fiber, followed by a 9 cm length of high-attenuation fiber (HAF) with 40 dB/m attenuation at 1550 nm, and another strand of 3 m of SMF-28 fiber. We note that the attenuation along the HAF was ≈3 dB. We report in Figure 2 the cross-correlation between the Brillouin Gain Spectrum (BGS) at the far section of the HAF (i.e., at the section closest to the probe injection point), as acquired by the BOFDA sensor at a probe power of 16 dBm, with the BFS acquired by the same setup at slightly higher probe powers. Note that all measurements were done with the HAF surrounded by air. While the results shown in Figure 2 are quite noisy (mainly because of the low SBS efficiency of the HAF), a positive trend of the cross-correlation peak with the injected power can be appreciated. In fact, by performing a quadratic fitting around each peak, a shift of the cross-correlation peak of ≈28 MHz can be estimated when increasing the probe power from 16 dBm to 19 dBm. This corresponds to an approximate variation of 28 °C of the inner temperature of the fiber as a result of a 3-dB variation of the injected probe power (from 40 mW to 80 mW). This preliminary result confirms the occurrence of self-heating on the HAF during BOFDA measurements.

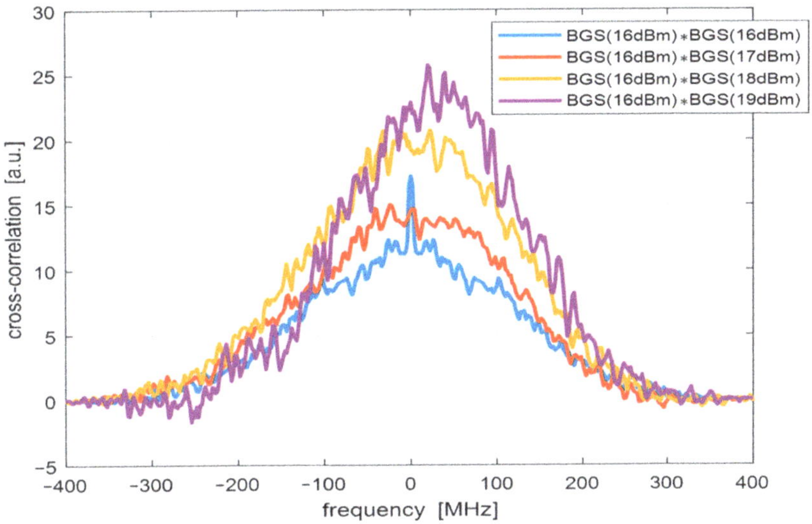

Figure 2. Cross-correlation between the BGS at the final section of the HAF, as acquired at various probe powers. The symbol * in the legend represents the cross-correlation operator.

As a next step, the BFS distributions along the FUT were measured upon immersion of the cobalt doped fiber into water at various levels. The temperature profiles, acquired by immerging progressively the HAF into the water at a step of 5 mm, are shown in Figure 3. The low Brillouin gain along the HAF is responsible of the irregularities exhibited by the temperature profiles. Nonetheless, each temperature profile shows an abrupt change in correspondence of the water/air interface crossing point. The position where the temperature drops abruptly can be used as an indicator of the liquid position. In particular, we set a threshold equal to 45 °C, so that the level position was identified as the threshold crossing position. The only exceptions are the profiles acquired with the fiber totally outside the water (for which the temperature was always higher than the preset threshold), or totally immersed into water (for which the temperature was always lower than the preset threshold). For these two special cases, the measured level was set to 0 cm and 9 cm, respectively.

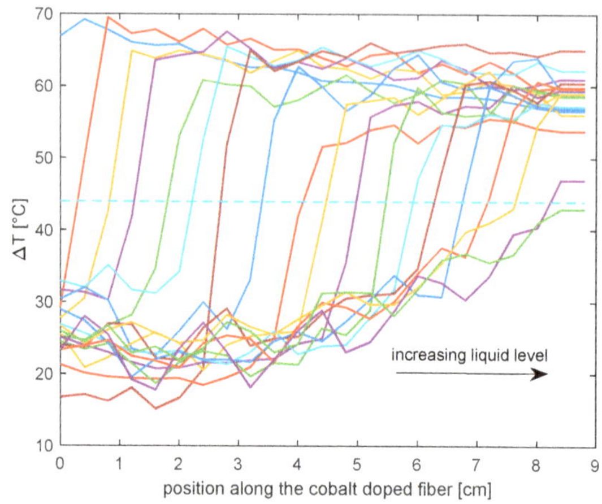

Figure 3. Temperature rise profiles acquired along the cobalt doped fiber, for L_W ranging from 0 cm to 9 cm at a step of 5 mm. The dashed line represents the threshold used for liquid level position measurements (From [8]).

In order to improve the resolution of the level measurement in the intermediate cases, we have performed a linear interpolation of the temperature profiles, acquired at a spatial resolution of 8 mm, using an interpolated step of 1 mm. In Figure 4, we report the liquid level obtained by localizing the threshold-crossing position in the interpolated temperature profiles, compared to the level set by experiments. The reported graph shows that a good linear relationship exists between the experimentally determined liquid level and the nominal level. In particular, the acquired data indicate a maximum deviation between the measured level position and the linear fitting curve equal to 2.3 mm, while the mean square error is 1.1 mm. Note that, while the reported experiments were performed using water as liquid medium, any other liquid could be used as well. In fact, our system is not sensitive to the refractive index of the liquid, as the optical field remains well confined in the HAF core. The only physical parameter of the liquid medium coming into play is its heat transfer coefficient, which is usually much larger than the corresponding coefficient in air, leading to the abrupt change in temperature distribution observed experimentally.

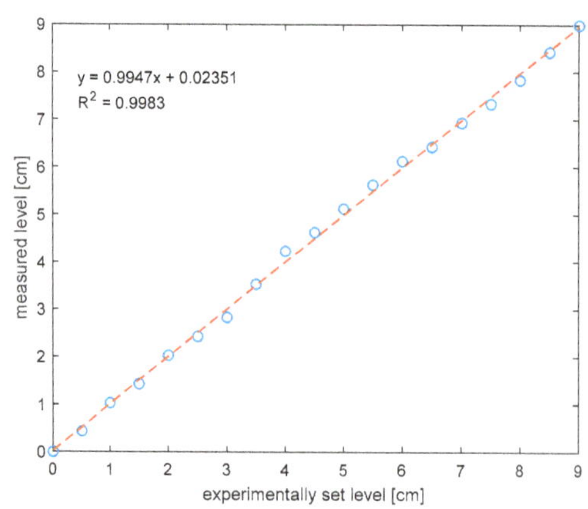

Figure 4. Measured level versus experimentally set level. Circles represents the experimental data, while the solid line represents the linear fitting curve (From [8]).

In conclusion, we have demonstrated the use of a self-heating approach for liquid level sensing measurements in cobalt-doped optical fibers. The main novelty of our proposal is the absence of any dedicated laser source for optical heating. In fact, we have shown that the strong c.w. components of the pump and probe waves, involved in the BOFDA configurations, are sufficient to heat the HAF at a sufficient temperature, allowing the localization of the air/liquid interface with a resolution dictated by the spatial resolution of the BOFDA setup. Future works will be devoted to enlarging the sensing range so as to cover a larger number of applications.

Author Contributions: Conceptualization, A.M.; validation, E.C. and A.C.; writing, A.M.; funding acquisition, A.M. and L.Z. All authors have read and agreed to the published version of the manuscript.

Funding: This research was funded by Università della Campania Luigi Vanvitelli, Programma VALERE.

Institutional Review Board Statement: Not applicable.

Informed Consent Statement: Not applicable.

Data Availability Statement: Data are available upon reasonable request.

Conflicts of Interest: The authors declare no conflict of interest.

References

1. Chen, K.P.; McMillen, B.; Buric, M.; Jewart, C.; Xu, W. Self-heated fiber Bragg grating sensors. *Appl. Phys. Lett.* **2005**, *86*, 143502. [CrossRef]
2. Sayde, C.; Thomas, C.K.; Wagner, J.; Selker, J. High-resolution wind speed measurements using actively heated fiber optics. *Geophys. Res. Lett.* **2015**, *42*, 10064–10073. [CrossRef]
3. Garcia-Ruiz, A.; Dominguez-Lopez, A.; Pastor-Graells, J.; Martins, H.F.; Martin-Lopez, S.; Gonzalez-Herraez, M. Long-range distributed optical fibre hot-wire anemometer based on chirped-pulse Φ-OTDR. *Opt. Express* **2018**, *26*, 463–476. [CrossRef] [PubMed]
4. Agliata, R.; Bogaard, T.A.; Greco, R.; Minardo, A.; Mollo, L.; Steele-Dunne, S.C. Non-invasive water content estimation in a tuff wall by DTS. *Constr. Build. Mater.* **2019**, *197*, 821–829. [CrossRef]
5. Chen, T.; Wang, Q.; Chen, R.; Zhang, B.; Lin, Y.; Chen, K.P. Distributed liquid level sensors using self-heated optical fibers for cryogenic liquid management. *Appl. Opt.* **2012**, *51*, 6282–6289. [CrossRef] [PubMed]
6. Coscetta, A.; Catalano, E.; Cerri, E.; Zeni, L.; Minardo, A. A Dual-Wavelength Scheme for Brillouin Temperature Sensing in Optically Heated Co2+-Doped Fibers. *IEEE Sens. J.* **2020**, *20*, 1349–1354. [CrossRef]
7. Bernini, R.; Minardo, A.; Zeni, L. Distributed Sensing at Centimeter-Scale Spatial Resolution by BOFDA: Measurements and Signal Processing. *IEEE Photonics J.* **2012**, *4*, 48–56. [CrossRef]
8. Catalano, E.; Vallifuoco, R.; Zeni, L.; Minardo, A. Distributed liquid level sensor based on Brillouin Optical Frequency-Domain Analysis. *IEEE Sens. J.* **2022**. [CrossRef]

engineering proceedings

MDPI

Proceeding Paper

The 3D-Printed Low-Cost Delta Robot *Óscar*: Technology Overview and Benchmarking †

César M. A. Vasques [1,*] and Fernando A. V. Figueiredo [2,3,*]

1 ProMetheus, Escola Superior de Tecnologia e Gestão, Instituto Politécnico de Viana do Castelo, Rua Escola Industrial e Comercial de Nun'Álvares, 4900-347 Viana do Castelo, Portugal
2 Smile.Tech—Robótica, 4430-699 Vila Nova de Gaia, Portugal
3 Instituto Superior Politécnico Gaya (ISPGaya), 4400-103 Vila Nova de Gaia, Portugal
* Correspondence: cmavasques@gmail.com (C.M.A.V.); ff@smlt.pt (F.A.V.F.)
† Presented at the 2nd International Electronic Conference on Applied Sciences, 15–31 October 2021; Available online: https://asec2021.sciforum.net/.

Abstract: Robotics is undoubtedly one of the most influential fields of modern technology in changing the very nature of our society. Parallel Delta robots have for a long time been mainly focused on a niche market; however, compared to serial anthropomorphic robots they present several simplicity and improved dynamics features. Additive manufacturing (AM) and 3D-printing technologies are enabling rapid changes in robotic engineering as we classically know it, allowing for greater creativity and freedom in mechatronics design and innovation. The effective benefits of far-reaching design freedom in terms of geometry, materials, and manufacturing accessibility are now starting to become apparent, answering many complex technical questions and scientific uncertainties that go beyond basic design and functional knowledge and that require engineering skills and scientific analysis. The Delta robot, as one of the most significant industrialized parallel robots due to its simplicity, is considered in this work, which provides an overview of the multidisciplinary aspects of the new Smile.Tech's 3D-printed and low-cost Delta robot, the *Óscar* family. We provide a concise analysis of the current state of the art and use of Delta robots, as well as a discussion of the Delta architecture, interface software, and virtual operation environments. The article concludes with a market analysis, a summary of the major manufacturers and currently available Delta models as well as a benchmarking study of their major operating and technical features.

Keywords: robotics; Delta; *Robótica* platform; *Óscar*; 3D-printing; low-cost; benchmarking

Citation: Vasques, C.M.A.; Figueiredo, F.A.V. The 3D-Printed Low-Cost Delta Robot *Óscar*: Technology Overview and Benchmarking. *Eng. Proc.* **2021**, *11*, 43. https://doi.org/10.3390/ASEC2021-11173

Academic Editor: Saulius Juodkazis and Nicholas Vassiliou Sarlis

Published: 15 October 2021

1. Introduction

Recent advancements in artificial intelligence and 5G telecommunications services have resulted in robotic technology influencing an increasing number of aspects of our lives at work and at home. As a result of easy access to robots and the more interesting interactions between humans and robots, our daily routines and work practices are being changed and automated in a safe way, which is encouraging research and the spread of cutting-edge robotic technology and collaborative work environments.

It is well known that, in robotics, there are essentially two kinds of kinematics—serial and parallel—and a combination thereof. Serial robots typically consist of a sequence of linkage arms sequentially connected through their joints until the moving tool carrier, whereas parallel robots take effect on the moving tool carrier directly with several simultaneously actuated linkage arms, the archetypal serial and parallel robots being the anthropomorphic arm and the 3D spatial moving platform used in ludic and professional motion simulators.

In contrast to serial kinematics, in parallel kinematics, not all joints carry drive engines, which are usually located on the base platform and, therefore, the moving mass is smaller and high-dynamic tasks can be more easily addressed. The resulting low moving masses of parallel kinematics enable both high-dynamics and very precise mechanics.

It is widely claimed that parallel robots are intrinsically more accurate than serial robots because their errors are averaged instead of added cumulatively, an assertion that has been somewhat confirmed [1]. Engineering experience and scientific reasoning agree on the strengths and weaknesses of serial and parallel architectures: small footprint, large workspace, simple modeling, but large movable masses, for serial manipulators; low movable masses, high dynamic capabilities, but large footprint and small workspaces, for parallel manipulators [2]. These enhanced mechanical features make parallel robots better suited for handling and assembly tasks.

The so-called *Delta robot* is a type of parallel robot whose fundamental concept is based on parallelograms (Figure 1). A parallelogram enables a fixed orientation of an output link relative to an input link. By employing three such parallelograms, the orientation of the mobile platform is completely constrained, leaving only three purely translational degrees of freedom. The three parallelograms' input links are mounted on rotating levers via revolute joints, which can be actuated in two ways: by rotational servomotors motors (DC or AC) or by linear actuators—the former being the most frequent.

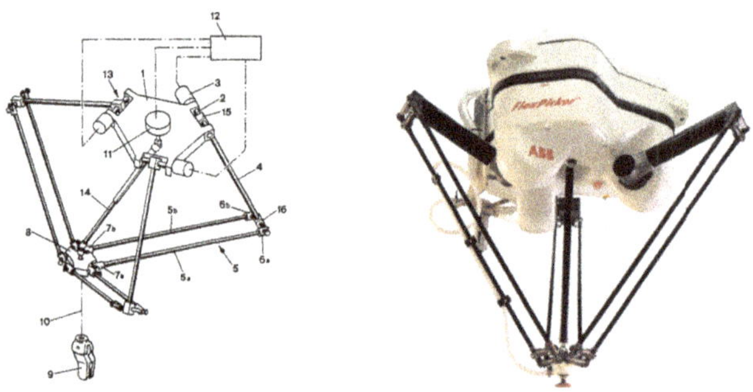

Figure 1. Schematic of the original Clavel's Delta robot patented in 1990 [3] and the first ABB Flexible Automation's Delta robot IRB 340 FlexPicker launched in 1999. Source: ABB Robotics.

The Delta robot, as one of the most significant industrialized parallel robots due to its simplicity, is considered in this work, which provides an overview of the multidisciplinary aspects of the new Smile.Tech's 3D-printed and low-cost Delta robot, the *Óscar* family. It addresses a concise and overview analysis of the current state of the art and use of Delta robots, as well as discussion of the Delta architecture, interface software, and virtual operation environments. The article concludes with a market analysis, a summary of the major manufacturers and currently available Delta models, and a benchmarking study of their major technical and operating features.

2. State of the Art and Use of Delta-like Robots

The Delta robot is a type of parallel robot that has established itself as one of the most successful parallel robot designs, certainly with over several hundred thousand active robots worldwide today [4,5]. It was first invented by Clavel in the 1980s [3,6,7] at École Polytechnique Fédérale de Lausanne (EPFL) in Switzerland with the objective to develop an industrial robot for automating a monotonous manual packaging process involving the manipulation of very light chocolates (of a few grams) at a very high speed (a few transfers per second).

Based on a patent license, ABB launched in 1999 the very successful and well-known commercial Delta robot *IRB 340 FlexPicker* (Figure 1). Originally developed with 3 translational degrees of freedom (DoF), it is typically constituted by three control arms driven by servomotors, which are mounted to a base plate; an optional telescopic drive shaft is used to transmit rotary motion from the base to an end-effector mounted on the mobile platform adding one additional rotational DoF.

The key design feature is the use of parallelograms in the arms, which maintains the orientation of the end-effector (moving tool carrier) and therefore ensures parallel motion, by contrast to the well-know *Stewart platform* that can change the orientation of its end-effector [8]. Allowing three spatial translations, the manipulator utilizes its symmetry and rigidity to create a simplistic system that can perform pick-and-place on small objects and assembly operations at high speeds with accuracy.

Alternatively, the Delta robot can also be driven by linear actuators, instead of rotary ones, in a Delta design often referred to as *Linapod* or *linear Delta* [9,10]; while in theory there is this design possibility, rotary actuation is probably the most interesting one implemented into commercial products and applications; for further details on linear Deltas, the reader is referred to [11,12].

Assembly and pick-and-place applications usually require at least one additional rotational DoF, mostly around the axis perpendicular to the mobile platform plane. Such motions, with three translational and one rotational DoF, are generally called Schoenflies (or SCARA) motions [13], resembling the kinematics of the well known SCARA robots that have been around for decades. Delta robots, in comparison to these, are generally considered to be faster (i.e., with a shorter transfer time) and with a lower maximum payload. In this vein, the so-called *hybrid Delta robots* comprise an additional orientation mechanism with one, two, or three DoF, mounted in series to the three DoF position mechanism, forming a *hybrid parallel-serial manipulator*, or mounted in parallel, forming a *cooperating machine* [14]. The basic idea behind a hybrid manipulator is to split the task of manipulation into two parts: position and orientation. The position mechanism controls the end effector's position and the orientation mechanism controls its orientation.

Regarding conventional hybrid manipulators, FANUC created an holder assembly resembling a wrist having three additional rotational DoF, mounted in series to the parallel 3 DoF Delta position mechanism, to which a tool is attached and orientated [15]. Coordinating the motions of the position and orientation mechanisms enables for example 6-axis machining of a workpiece. This strategy has a number of significant advantages. To begin, the parallel construction enables the realization of a high stiffness, low inertia, and high speed machine tool. Secondly, its direct and inverse kinematic solutions are available in closed-form, which simplifies control and path planning problems significantly. Thirdly, in comparison to fully parallel platform manipulators, it has a relatively large workspace. Fourthly, its position and orientation kinematics are completely decoupled. Lastly, it makes extensive use of revolute joints, which can be precisely manufactured at a low cost. Overcoming the shortcomings of expensive and imprecise spherical joints as used within six-limbed 6 DoF Stewart platform architectures, and introducing a new generation of hybrid manipulators comprising independent position and orientation mechanisms, each with three complementary DoF, hybrid manipulators were also developed at the University of Maryland [14,16–18] employing only revolute joints.

Thus, a possible solution to implement additional orientation DoF is the already mentioned mechanism rotationally actuated by a motor that is usually fixed on the base platform and connected through an intermediary telescopic drive shaft and universal joints to a mechanism pivotally mounted on the end-effector (element 14 in the left-hand side of Figure 1); other variants may consider the end-effector manipulated by a flexible cable, a small motor mounted directly or the pull of a cable wound on a drum with a torsion spring. Other designs, as the one created by FANUC, utilize the basic Delta structure in conjunction with a three DoF rotating head or serial robotic wrist positioned on the end-effector powered by three independent motors mounted on the frame.

Based on Delta technology, Pierrot [19] suggested a new distinctive family of four DoF parallel Delta robots considering an articulated (and not rigid) movable platform with embedded joints—the *H4-family*—with four independent kinematic chains with an H-shaped end-effector, with the hope that such designs would improve the maximum range of motion in rotation. The most frequent mechanism in this family is the H4 robot where, in contrast to the original Delta robot concept, four kinematic chains connected to the traveling plate through revolute joints are operated by four angular motors. The gripper may rotate due to an extra gear-based amplification mechanism. Extensive research and testing of this and likewise model I4 have shown that none of these designs is optimum for extremely rapid pick-and-place operations and that problems may arise from greatly varying Jacobian condition number, risk of singular configurations, self-collisions, and short service-life of prismatic joints for the I4.

Later, these designs were reviewed and improved, and a prototype was developed with a superior actuator arrangement and only revolute joints on its articulated traveling plate—the Par4 [2]. The first commercially available version of the high-speed pick-and-place parallel manipulator employing the Par4 architecture (Adept Quattro), was free of any singular postures within its workspace and the preliminary prototype succeeded in achieving high speeds (more than 4 m/s) and high accelerations (more than 15 g).

Serial-parallel hybrid and fully-parallel designs with Schoenflies motion are discussed more in detail in [2,12]. Regarding further research and development activities, during the last decades, extensive research on dimensional synthesis, workspace and singular configurations, kinematics, dynamics, control, vibrations, position accuracy, calibration, stiffness, optimization, gravity compensation, and mechanical design of Delta robots has been conducted. Refer, for example to [20–26] for an overview of the main contributions.

As mentioned, Delta robots are typically used in applications where the robot picks up products in groups and arranges them in a container or assembly pattern. Deltas can now distinguish and select from a variety of size, color, and shape options, as well as pick and place objects based on a programmed pattern. The packaging industry, as well as the medical and pharmaceutical industries, benefit from Delta robots' high speed. It is also used in surgery due to its stiffness. Additionally, high-precision assembly operations in a clean room for electronic components are possible. A Delta robot's structure can also be used to create haptic controllers and, recently, the technology was adapted for use with 3D printers. These printers are faster, can be built for less than $200, and perform well in comparison to traditional Cartesian printers [27]. 3D printing challenges and the analysis of the contribution of key process parameters in printing results and quality is discussed in [28,29]; there are a wide variety of uses for 3D printing, ranging from mechanical components and human implants to musical instruments and personal protection equipment, such as COVID-19 protection [30–34]. Numerous other broader application areas for Delta robots include micro robotics, visual control, dynamic balancing, medical haptic devices, and redundancy [4,5,22,35,36]. These are merely a handful of the many possible application scenarios with many more in development.

3. The *Óscar* Delta Robot Family Development

Automated systems have become increasingly prevalent in recent years in smaller and smaller businesses, offices, schools, and households, owing to the advent of mass production, 3D printing, and open source robotic solutions. It is relatively easy to find examples of this in any field. Automated software completion, scheduled reminders on our agenda, automated data collection, self-driving vehicles, automated access control, and even automated vacuum cleaning and kitchen robots are all possible these days. While the statements above are accurate, despite their exponential growth in industrial settings, non-cartesian robotic manipulators have not yet been widely adopted. Our assessment indicates that there are three significant obstacles maintaining the existing state of affairs:

1. A good match between arm length and payload results in a high moment of inertia, which makes direct drive actuators extremely heavy and economically inefficient. With the addition of a reduction system to the output shaft, smaller actuators may be used and, in this case, any backlash results in a significant positioning error; although backlash-free reducers are commercially available, they are quite expensive.

2. Safety! Long arms necessitate the use of powerful actuators. When these machines are operating in close proximity to humans, they must protect the latter in the event of contact. As a result, it is necessary to distinguish between the torques required to maneuver a payload and the torques generated by a collision, which is not always easy to do with proven guaranteed success.
3. Difficulty of use; the programming and commissioning of robotic systems is more complex than it appears at first sight, namely with regard to safety.

With the above in mind, the iterative design procedure for the *Óscar* Delta robot included many stages and a long time since its first version was set back in 2016 (see Figure 2). The market for collaborative robotics applications is widely perceived to be developing to satisfy the demands of consumers or clients across a variety of sectors and applications with varying specifications. This was also discovered through market research and the company's expertise identifying the demands of future consumers, indicating that expanding into a library of off-the-shelf integrating parts and family of Delta solutions was the way to go.

The first step in developing the design was to ascertain the true nature of the problem, which was accomplished through analysis. This is a critical stage because incorrectly defining the problem can result in time being wasted on designs that do not meet the requirements. Following the analysis, it was possible to create and document a specification of the requirements, with some of the primary requirements being an excellent cost-performance ratio and safety suitability for collaborative and effective broad-scope use. Various concepts were developed throughout the conceptual stage, followed by preliminary engineering analyses of the most promising solutions, as illustrated in Figure 2.

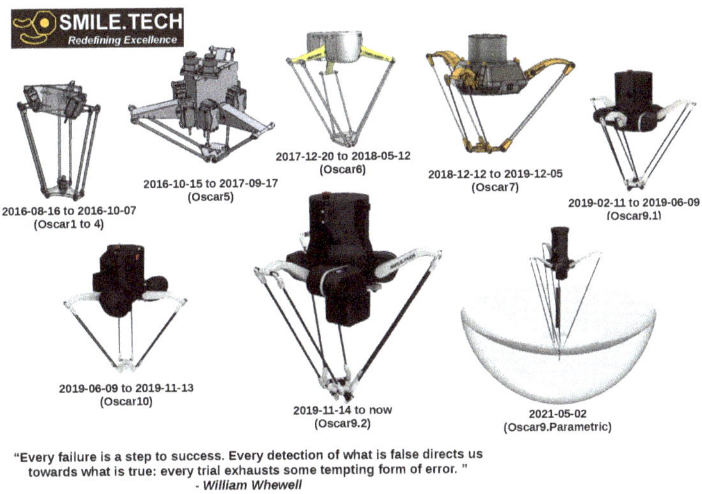

Figure 2. History of the iterative design, successive Delta robot model versions and *Óscar* family.

Outline solutions were developed and detailed enough to indicate the methods for achieving each of the required functions, e.g., approximate sizes, shapes, materials, costs, and performance (whether by test or analysis). Additionally, this entails determining what has been done previously to address similar issues; there is no point in reinventing the wheel and subsequent versions have been built upon previous knowledge about the former versions. The various solutions were weighed, and the most appropriate ones were chosen. Oftentimes, evaluation involved modeling the system and then simulating it to determine how it might react to various inputs and selecting the most suitable solution. The details of the chosen design have now been worked out, including the creation of prototypes or mock-ups based on 3D printing of selected components in order to ascertain the optimal design details. The chosen designs were then translated into working drawings, circuit diagrams, and so on, so that the items can be promptly manufactured and dispatched as part of a continuous technology improvement process and market entrance.

4. The Smile.Tech's SMLT *Robótica* Platform

The Smile.Tech's SMLT *Robótica* Platform (STRP) is an in-house dedicated cost-effective robotic platform that outperforms competitors in terms of efficiency, accuracy, reliability, and safety. It makes use of stepper motors due to their simplicity of position control and, more significantly, their simplicity of velocity restriction. These motors are connected to low-backlash mechanical reducers that incorporate absolute position feedback on the output shaft. The torque created by this assembly can be simply computed from its elastic deformation, allowing the system to calculate the magnitude and direction of force at all times.

As depicted in Figure 3, the platform is composed of multiple components connected in a variety of ways. The PDR-20 (*Óscar* 9.2 model version) is a Delta robot that is built upon the STRP and comprises four major key components, where these key components can be arranged to build any kind of manipulation robot kinematics:

- The programming and simulation software (Virtu3D, ReTRoControl and ReTroControlEnvisioned);
- The *Real Time Robot Controller* (ReTRoC);
- The *Robotic Stepper Controller* (RoSteC);
- The *Belt Elastic Actuator for Robotics* (BEAR).

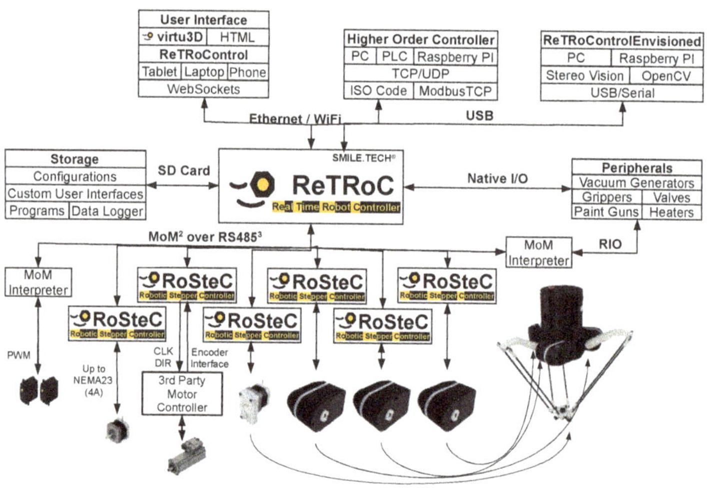

Figure 3. Overview of the Smile.Tech's SMLT Robótica Platform (STRP) and product breakdown structure.

The STRP is a comprehensive system built on three interconnected pillars: the BEAR, the ReTRoC and the Virtu3D (Figure 4). By substituting elasticity for backlash in the transmission, the BEAR eliminates positioning uncertainty and provides mechanical compliance. It is a low-cost actuator based on the popular NEMA 17 stepper motor series. It includes instrumentation and a controller called RoSteC, which calculates torque based on transmission deformation and is capable of compensating for, or virtually increasing compliance.

All robotic actuators (BEARs) are connected via an RS485 communication network and are managed by the ReTRoC. This controller is capable of interpreting commands sent by higher-order controllers or pre-programmed sequences stored on an SD card. Additionally, it manages motion and trajectory, converts coordinates, and calculates forces and their directions using data from the actuators' torque sensors. Lastly, Virtu3D is an online tool that works on any computer, tablet, or smart-phone, that can be used to program and simulate robotic systems and that is capable of connecting to the robot controller via Ethernet or WiFi (Figure 5).

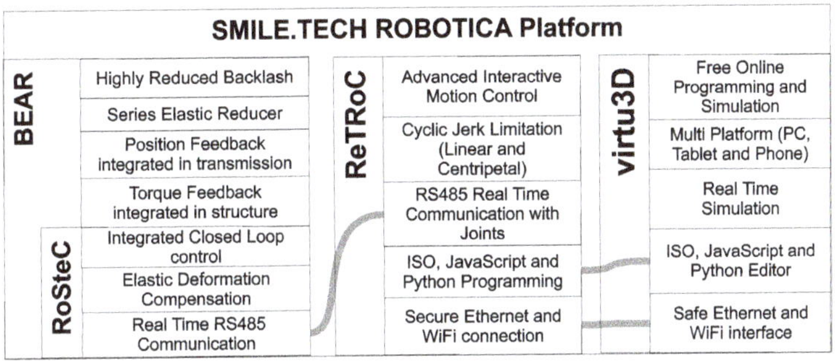

Figure 4. Pillars of the Smile.Tech's SMLT *Robótica* Platform (STRP) and features diagram.

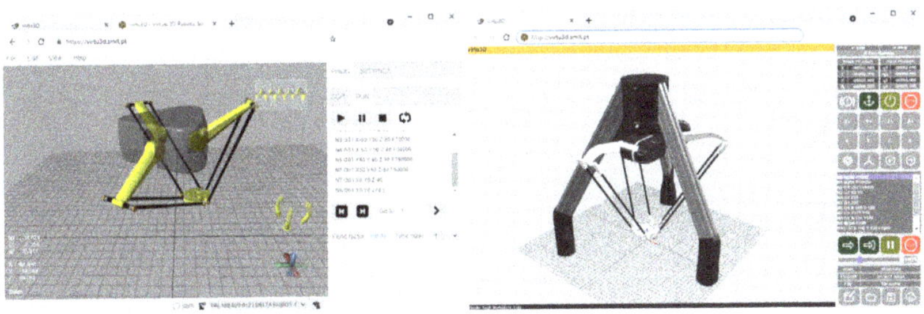

Figure 5. Virtu3D, a web-based application that can be used on any computer, tablet, or smartphone to program and simulate robotic systems and is capable of connecting to the robot controller via Ethernet or WiFi. On-line kinematics simulator available at https://virtu3d.smlt.pt; current (**left**, accessed on 11 March 2022) and future (**right**) versions.

5. Market, Manufacturers, and Benchmarking

Currently, the robotics industry mainly serves two distinct markets—industrial and professional services—commonly lumped together. The emergence of new service robots has been noticed by many organizations throughout the world, and steps have been taken to support the new developments [37]. However, industrial robots and professional service robots have different operational requirements and different costs and have recently been shown to have very different market growth rates [38].

Industrial robots have been around since the 1970s—the archetypal industrial robot being a mechanical anthropomorphic arm with varying number of DoF, found in factories around the world. In the manufacturing industry, the biggest users of industrial robots by descending order are, in general, the automotive, electrical/electronics, metal and machinery, plastics and chemicals, and food and beverage sectors. In contrast to industrial robots, professional service robots are more recent, taking off within the last decade, and are typically used outside of manufacturing lines to assist humans rather than replace them. They can have wheels to make them mobile, some are anthropomorphic, but they are not intended for the kinds of heavy tasks that most industrial robots tackle.

Thus far, professional service robots have been most popular in the retail, hospitality, health care, and logistics (in warehouse or fulfillment settings) industries, although some have also started to be used in space and defense, agriculture, and demolition. In addition to the industrial and professional service robots used by enterprises, there are two other large and quickly growing consumer robot markets—consumer service and entertainment robots; the former designed for tasks, such as vacuuming, mowing the lawn, and washing windows and the latter consisting mainly of toys, some of which are fairly sophisticated and mainly made in Asia.

The foregoing analysis was detailed in a recent market study [39] that indicated that nearly 1 million robots were expected to be sold for enterprise use in 2020, and over half of them were expected to be professional service robots, generating more than $16 billion in revenue, 30% more than in 2019. The market for professional service robots is growing much faster than that for industrial robots, where professional service robots are on the verge of passing industrial robots in terms of units and revenue.

On the other hand, although 97% of all of the robots sold in 2019 were consumer service and toy robots, they represent only 14% of robotics industry revenue, where industrial robots lead with 49% of revenues; but the growth drivers of 5G, artificial intelligence chips, and affordability of robotic technology are also likely to have a strong influence on consumer robots growth in the future and certainly contribute to a rapid market change.

Regarding collaborative robots, another study [40] suggested that the market size is expected to reach a value of $1.09 billion at a compound annual growth rate (CAGR) of 16.08% with accelerating momentum during 2021–2025; 34% of the growth will originate from the Asia-Pacific region and key countries are presently US, China, and Germany. Refer also to [41] for a recent tracing of the evolution of service robotics.

For the Delta's, still another market study [42] predicted that the global Delta robots market size will grow by $242 million during 2019–2023 with a CAGR of almost 9%, also accelerating, with 50% coming from the Asia-Pacific region; the market is fairly fragmented, where one of the key trends for the period is the development of vision-integrated Delta robots, bringing up new application possibilities.

Delta robots are reported to have been typically designed and optimized to meet the requirements of extremely fast product handling, low cost, and easy disassembly for cleaning in the food (e.g., dough cutting or pancake stacking) and packaging industries (e.g., top-loading, feed placement, and assortment placement). Additionally, the medical field, the electronics industry, pick-and-place tasks for cells and wafers in the photovoltaic industry, laser cutting, high-speed milling, drilling, wood tooling, and use as a 3D haptic device with force feedback in the gaming industry are possible application highlights. Additionally, agricultural applications have been reported to include a Delta-based prototype for manipulating soft vegetables, an autonomous field robot platform equipped with a Delta robot for treating individual plants, and a non-chemical vegetable weed controller. See also [4,12] and the references therein for further details on Delta applications and an overview of the historical developments of the Delta robot on the market and academia.

According to [5,12], the license for the Delta robot was purchased in 1987 from EPFL to a Swiss company, Demaurex, which started the industrial development process and began manufacturing Delta robots for the packaging industry at that time. In 1996, Demaurex bought the license and merged with Sigpack Systems to increase competitiveness and internationalization to sell in the world market. The IRB340 FlexPicker Delta robot was also launched in 1999 by ABB Flexible Automation that purchased also a license. Later, in 2004, the merged Demaurex and Sigpack Systems were incorporated by the Bosch Packaging Technology division. Since then, and fostered also by the expiration of the Clavel's original patent, parallel robots with Delta-like architecture have attracted the interest by industry and researchers all over the world. The global Delta robot market has become moderately fragmented since then also due to the expiraton of the original Clavel's patent and progressively others, with various manufacturers located all over the world. However, several top-rated companies, such as ABB (Swedish–Swiss), FANUC (Japan), Kawasaki Heavy Industries (Japan), Adept/Omrom (USA/Japan), Yaskawa Electric (Japan), Codian Robotics (Netherlands), Penta Robotics (Netherlands), Estun Automation (China), and Bosch (Germany), are apparently leading the way.

As such, Delta robots have nowadays popular usage in picking and packaging in factories because they can be quite fast, some executing up to 300 picks per minute. For its stiffness it is also used for surgery and other applications include high precision assembly operations in a clean room for electronic components. The structure of a Delta robot can also be used to create haptic controllers [43–45]. Agriculture apps [46] and others used in mechanical machining processes [47,48]. Surveys on parallel robots with Delta-like architecture can be found e.g., in [5,12].

A list of commercially available Delta-robots along with their main features is presented in Table 1. As a benchmarking study, the features of the *Óscar* 9 SMLT Delta robot PDR-20 are also presented. The criteria for the benchmarking analysis was to select consolidated manufacturers and Delta models with a payload in the same range of the SMLT PRD-20, i.e., <5 kg, and with a comparable architecture and number of DoF. As can be seen, the prices and weight differ significantly at the cost of sacrificing speed, for equivalent (or, at least, not significant for the majority of applications) position accuracy. The top two models represent two consolidated and high-performance models from ABB and FANUC; the middle model is an unusually featured model from Festo; and the fourth model is the most comparable model from Igus, targeted at a similar clientele as the SMLT PDR-20.

Together, they provide a streamlined overview of the market and competing products, allowing an accurate and concise benchmarking for Delta robots selection. The SMLT PDR-20 (*Óscar* 9.2 model version) Delta robot, whose features were mainly determined by analysis of design and engineering judgment, is shown to be the most affordable one and to cover a range of operating requirements and cost-benefit trade-off not yet fully covered by the most well-known models available on the market.

Table 1. Benchmarking of technical specifications of selected Delta robots in the same class of the PDR-20 of Smile.Tech's *Óscar* family.

Model Reference	Weight /kg	Workspace ($r \times z$) /mm^2	Payload /kg	Speed (max) /m·s^{-1}	Accel. (max) /m·s^{-2}	Power (Drive) /W	DoF #	Repetition Accuracy /µm	Price (Reference) €
ABB IRB 360-3/1130	120	565 × 250	3	10	100	477	4	±100	25,000
FANUC M-1iA/1HL	18	210 × 150	1	N/A	N/A	200	3	±30	14,000
Festo EXPT 95	62	200 × 150	5	7	110	N/A	3	±100	20,000
Igus DLE-DR-0003	15	150 × 40	5	3	60	72	3	±500	7000
SMLT PDR-20	6	320 × 100	2	0.35	50	200	3	±500	5000

N/A: Not available; r and z: radius and height of the cylindrical robot workspace envelope.

6. Conclusions

Almost 40 years after the original ideas, Delta robots continue to serve a niche market for high-speed pick-and-place applications, and a huge number of new additional applications are anticipated and currently under development. During the past decade, the patents that have expired and the surge of new areas of application have resulted in an increase in research and development. Also, the growing scientific emphasis on extended architectures with more rotational DoF has resulted in a variety of serial-parallel hybrid and completely parallel systems able to respond to commercial demands for new and more sophisticated handling jobs requiring also increased payload capacity and non-contaminating designs.

This article describes the process and major technical developments and features of a platform for developing safe, easy-to-use, precise, and reliable robotic manipulators that are also affordable for general public use—the Smile.Tech's *SMLT Robótica Platform* (STRP) and the *SMLT Delta Robot Óscar* family. The STRP project aims to leverage technologies that, despite their longevity, are not yet widely adopted. Two examples include additive manufacturing and substituting transmission elasticity by transmission backlash. Another significant feature of the STRP is that its success is determined not only by the achievement of the proposed technical features but also by the acceptance and satisfaction of its derived products by customers. Thus, during the design and integration processes performed by Smile.Tech, several aspects of parallel robots were covered, ranging from modeling (geometric, kinematic, dynamic, elasticity, etc.) to control, while taking into account singularity analysis, repetition accuracy, calibration, design optimization, durability, and a variety of other scientific and technology development issues. In this vein, the article presents a focused analysis of the current state of the art and use of Delta robots as well as a discussion of the Delta architecture, interface software, virtual operation environments, and the technologies involved. Moreover, the article presents a market analysis, a summary of the major manufacturers and currently available Delta models, and a benchmarking study of their major operating and technical features.

Overall, this work is expected to contribute to and enable a sustained selection and identification of the primary technical characteristics of leading Delta robots in the same class, as well as the advantages of the Smile.Tech's Delta robot model SMLT PDR-20 and its success in being an appellation to the collaborative robot market due to its affordable price, high level of safety, and exceptional cost–benefit ratio.

Author Contributions: Conceptualization, C.M.A.V. and F.A.V.F.; methodology, C.M.A.V. and F.A.V.F.; software, F.A.V.F.; formal analysis, C.M.A.V. and F.A.V.F.; investigation, C.M.A.V.; resources, F.A.V.F.; data curation, F.A.V.F.; writing—original draft preparation, C.M.A.V. and F.A.V.F.; writing—review and editing, C.M.A.V.; funding acquisition, C.M.A.V. and F.A.V.F. All authors have read and agreed to the published version of the manuscript.

Funding: The authors gratefully acknowledge the support provided by the Foundation for Science and Technology (FCT) of Portugal, within the scope of the project of the Research Unit on Materials, Energy and Environment for Sustainability (proMetheus), Ref. UID/05975/2020, financed by national funds through the FCT/MCTES.

Conflicts of Interest: The authors declare no conflict of interest.

References

1. Briot, S.; Bonev, I.A. Are parallel robots more accurate than serial robots? *Trans. Can. Soc. Mech. Eng.* **2007**, *31*, 445–455. [CrossRef]
2. Pierrot, F.; Nabat, V.; Company, O.; Krut, S.; Poignet, P. Optimal design of a 4-DOF parallel manipulator: From academia to industry. *IEEE Trans. Robot.* **2009**, *25*, 213–224. [CrossRef]
3. Clavel, R. Device for the Movement and Positioning of an Element in Space. U.S. Patent 4,976,582, 11 December 1990.
4. Bonev, I. Delta Parallel Robot—The Story of Success. 2001. Available online: http://www.parallemic.org/Reviews/Review002.html (accessed on 13 July 2021).
5. Poppeova, V.; Uricek, J.; Bulej, V.; Sindler, P. Delta robots—Robots for high speed manipulation. *Teh.-Vjesn.-Tech. Gaz.* **2011**, *18*, 435–445.
6. Clavel, R. DELTA: A fast robot with parallel geometry. In Proceedings of the 18th International Symposium on Industrial Robots, Sydney, Australia, 26–28 April 1988; pp. 91–100.
7. Clavel, R. Conception d'un Robot Parallle Rapide à 4 Degrés de Liberté. Ph.D. Thesis, École Polytechnique Fédérale de Lausanne, Lausanne, Switzerland, 1991. [CrossRef]
8. Stewart, D. A Platform with six degrees of freedom. *Proc. Inst. Mech. Eng.* **1965**, *180*, 371–386. [CrossRef]
9. Wurst, K.H. LINAPOD—Machine Tools as Parallel Link Systems Based on a Modular Design. In *Parallel Kinematic Machines*; Boër, C.R., Molinari-Tosatti, L., Smith, K.S., Eds.; Springer: London, UK, 1999; pp. 377–394. [CrossRef]
10. Merlet, J.P. *Parallel Robots*, 2nd ed.; Kluwer Academic Publishers: Dordrecht, The Netherlands, 2006.
11. Bouri, M.; Clavel, R. The Linear Delta: Developments and applications. In Proceedings of the 41st International Symposium on Robotics (ISR 2010) and 6th German Conference on Robotics (ROBOTIK 2010), Munich, Germany, 7–9 June 2010; p. 8.
12. Brinker, J.; Corves, B. A survey on parallel robots with Delta-like architecture. In Proceedings of the 14th IFToMM World Congress, Taipei, Taiwan, 25–30 October 2015; pp. 407–414. [CrossRef]
13. Company, O.; Pierrot, F.; Nabat, V.; de la O Rodriguez, M. Schoenflies motion generator: A new non redundant parallel manipulator with unlimited rotation capability. In Proceedings of the 2005 IEEE International Conference on Robotics and Automation, Barcelona, Spain, 18–22 April 2005; pp. 3250–3255. [CrossRef]
14. Tsai, L.W. Multi-Degree-of-Freedom Mechanisms for Machine Tools and the Like. U.S. Patent 5,656,905, 12 August 1997.
15. Kinoshita, S.; Yamashiro, H. Parallel Robot Provided with Wrist Section Having Three Degrees of Freedom. U.S. Patent 8,893,578 B2, 25 November 2014.

16. Tsai, L.W.; Joshi, S. Kinematic analysis of 3-DOF position mechanisms for use in hybrid kinematic machines. *J. Mech. Des.* **2002**, *124*, 245–253. [CrossRef]
17. Tsai, L.W.; Stamper, R. A parallel manipulator with only translational degrees of freedom. In *Proceedings of the Volume 2B: 24th Biennial Mechanisms Conference*; American Society of Mechanical Engineers: New York, NY, USA, 1996. [CrossRef]
18. Tsai, L.W.; Walsh, G.; Stamper, R. Kinematics of a novel three DOF translational platform. In Proceedings of the IEEE International Conference on Robotics and Automation, Minneapolis, MN, USA, 22–28 April 1996; Volume 4, pp. 3446–3451. [CrossRef]
19. Pierrot, F.; Company, O. H4: A new family of 4-DOF parallel robots. In Proceedings of the 1999 IEEE/ASME International Conference on Advanced Intelligent Mechatronics, Atlanta, GA, USA, 19–23 September 1999; pp. 508–513. [CrossRef]
20. Gogu, G. *Structural Synthesis of Parallel Robots, Part 2: Translational Topologies with Two and Three Degrees of Freedom*; Springer: Dordrecht, The Netherlands, 2009. [CrossRef]
21. Zhao, Y. Singularity, isotropy, and velocity transmission evaluation of a three translational degrees-of-freedom parallel robot. *Robotica* **2012**, *31*, 193–202. [CrossRef]
22. Brinker, J.; Corves, B.; Wahle, M. A comparative study of inverse dynamics based on Clavel's Delta robot. In Proceedings of the 14th world congress in mechanism and machine science. Taipei, Taiwan, 25–30 October 2015; pp. 89–98. [CrossRef]
23. Patil, M.J.; Deshmukh, M.P. A review paper on introduction of parallel manipulator and control system. *Int. J. Eng. Res. Technol.* **2015**, *4*, 959–963.
24. Brinker, J.; Funk, N.; Ingenlath, P.; Takeda, Y.; Corves, B. Comparative study of serial-parallel Delta robots with full orientation capabilities. *IEEE Robot. Autom. Lett.* **2017**, *2*, 920–926. [CrossRef]
25. Brinker, J.; Corves, B.; Takeda, Y. On the motion/force transmissibility and constrainability of Delta parallel robots. In *Computational Kinematics*; Mechanisms and Machine Science; Springer: Berlin/Heidelberg, Germany, 2018; pp. 340–348. [CrossRef]
26. Brinker, J.; Schmitz, M.; Takeda, Y.; Corves, B. Dynamic modeling of functionally extended Delta-like parallel robots with virtual tree structures. In *ROMANSY 22—Robot Design, Dynamics and Control*; CISM International Centre for Mechanical Sciences; Springer: Berlin/Heidelberg, Germany, 2019; pp. 171–179. [CrossRef]
27. Wikipedia. Delta Robot—Wikipedia, The Free Encyclopedia. 2021. Available online: http://en.wikipedia.org/w/index.php?title=Delta%20robot&oldid=1030986049 (accessed on 6 September 2021).
28. Kantaros, A.; Piromalis, D. Employing a low-cost desktop 3D printer: Challenges, and how to overcome them by tuning key process parameters. *Int. J. Mech. Appl.* **2021**, *10*, 11–19. [CrossRef]
29. Vasques, C.M.A.; Resende, P.M.; Cavadas, A.M.; Abrantes, J.C. Quality analysis of a demonstration 3D-printed maraging steel part. In Proceedings of the 2nd International Electronic Conference on Applied Sciences, Online, 15–31 October 2021. [CrossRef]
30. Kantaros, A.; Diegel, O. 3D printing technology in musical instrument research: Reviewing the potential. *Rapid Prototyp. J.* **2018**, *24*, 1511–1523. [CrossRef]
31. Coimbra, M.R.C.; Barbosa, T.P.; Vasques, C.M.A. Preliminary design and validation of a 3D-printed continuously variable transmission for an electric vehicle prototype. *Eng. Proc.* **2021**, *11*, 11. [CrossRef]
32. Kantaros, A.; Laskaris, N.; Piromalis, D.; Ganetsos, T. Manufacturing zero-waste COVID-19 personal protection equipment: A case study of utilizing 3D printing while employing waste material recycling. *Circ. Econ. Sustain.* **2021**, *1*, 851–869. [CrossRef] [PubMed]
33. Kantaros, A.; Piromalis, D. Fabricating lattice structures via 3D printing: The case of porous bio-engineered scaffolds. *Appl. Mech.* **2021**, *2*, 289–302. [CrossRef]
34. Vasques, C.M.A.; Gonçalves, F.C.; Cavadas, A.M. Manufacturing and testing of 3D-printed polymer isogrid lattice cylindrical shell structures. In Proceedings of the 2nd International Electronic Conference on Applied Sciences, Online, 15–31 October 2021. [CrossRef]
35. Bloss, R. Review of new robot designs and the rise of remote programming in manufacturing technology. *Ind. Robot. Int. J.* **2013**, *40*, 213–217. [CrossRef]
36. Singh, B.; Sellappan, N.; Kumaradhas, P. Evolution of industrial robots and their applications. *Int. J. Emerg. Technol. Adv. Eng.* **2013**, *3*, 763–768.
37. Virk, G.S.; Moon, S.; Gelin, R. ISO Standards for Service Robots. In Proceedings of the Eleventh International Conference on Climbing and Walking Robots and the Support Technologies for Mobile Machines, Coimbra, Portugal, 8–10 September 2008; World Scientific Press: Singapore; pp. 133–138. [CrossRef]
38. Global Delta Robots Market. Market.us. Available online: https://market.us/report/delta-robots-market (accessed on 13 July 2021).
39. Stewart, D.; Wigginton, C.; Casey, M. Robots on the move: Professional service robots set for double-digit growth. In *Deloitte Insights—Technology, Media, and Telecommunications Predictions 2020*; Deloitte Development LLC: Hermitage, TN, USA, 2019; pp. 18–29.
40. Collaborative Robots Market by Application and Geography—Forecast and Analysis 2021–2025. Technavio. Available online: https://www.technavio.com/report/collaborative-robots-market-industry-analysis (accessed on 17 July 2021).
41. Ott, I.; Savin, I.; Konop, C. Tracing the evolution of service robotics: Insights from a topic modeling approach. In *Kiel Working Paper No. 2180*; Kiel Institute for the World Economy (IfW): Kiel, Germany, 2021; 47p.
42. Global Delta Robots Market 2019–2023. Technavio. Available online: https://www.technavio.com/report/global-delta-robots-market-industry-analysis (accessed on 17 July 2021).

43. Grange, S.; Conti, F.; Helmer, P.; Rouiller, P.; Baur, C. Delta haptic device as a nanomanipulator. In *Microrobotics and Microassembly III*; Nelson, B.J., Breguet, J.M., Eds.; SPIE: Bellingham, WA, USA, 2001. [CrossRef]
44. Mitsantisuk, C.; Ohishi, K. Haptic human-robot collaboration system based on Delta robot with gravity compensation. In Proceedings of the IECON 2016—42nd Annual Conference of the IEEE Industrial Electronics Society, Florence, Italy, 23–26 October 2016. [CrossRef]
45. Liu, G.; Chen, Y.; Xie, Z.; Geng, X. GA\SQP optimization for the dimensional synthesis of a Delta mechanism based haptic device design. *Robot.-Comput.-Integr. Manuf.* **2018**, *51*, 73–84. [CrossRef]
46. Hussmann, S.; Knoll, F.J.; Meissner, A.; Holtorf, T. Development and evaluation of a low-cost Delta robot system for weed control applications in organic farming. In Proceedings of the 2019 IEEE International Instrumentation and Measurement Technology Conference (I2MTC), Auckland, New Zealand, 20–23 May 2019. [CrossRef]
47. Ayyıldız, E.A.; Ayyildiz, M. Development of a 3-axis Parallel Kinematic Machine for Milling Wood Material—Part 1: Design. *Bioresources* **2017**, *12*, 9326–9337.
48. Ayyıldız, E.A.; Ayyıldız, M.; Kara, F. Optimization of surface roughness in drilling medium-density fiberboard with a parallel robot. *Adv. Mater. Sci. Eng.* **2021**, *2021*, 6658968. [CrossRef]

engineering proceedings

MDPI

Proceeding Paper

Ultra-Short-Pulse Lasers—Materials—Applications †

Molong Han [1], Daniel Smith [1], Soon Hock Ng [1,*], Vijayakumar Anand [1], Tomas Katkus [1] and Saulius Juodkazis [1,2,*]

1 Optical Sciences Centre and ARC Training Centre in Surface Engineering for Advanced Materials (SEAM), School of Science, Computing and Engineering Technologies, Swinburne University of Technology, Hawthorn, VIC 3122, Australia; molonghan@swin.edu.au (M.H.); danielsmith@swin.edu.au (D.S.); vanand@swin.edu.au (V.A.); tkatkus@swin.edu.au (T.K.)

2 World Research Hub Initiative (WRHI), School of Materials and Chemical Technology, Tokyo Institute of Technology, 2-12-1, Ookayama, Meguro-ku, Tokyo 152-8550, Japan

* Correspondence: soonhockng@swin.edu.au (S.H.N); saulius.juodkazis@gmail.com (S.J.)

† Presented at the 2nd International Electronic Conference on Applied Sciences, 15–31 October 2021; Available online: https://asec2021.sciforum.net/.

Abstract: We overview recent developments of $3D^{\pm}$ (additive/subtractive) manufacturing/printing from the point of view of laser development, beam delivery tools, applications, and materials. The average power of ultra-short-pulsed lasers has followed a Moore's scaling trajectory, doubling every two years, for the past 20 years. This requires fast beam scanning solutions and beam delivery control for larger-area applications. New material synthesis with high spatial resolution is provided at the high intensity TW/cm^2-PW/cm^2 exposure site. Net-shape manufacturing with a reduced number of post-processing steps is a practical trait of $3D^{\pm}$ printing. With computer numerical control (CNC) optimised using artificial intelligence (AI), the future of $3D^{\pm}$ manufacturing is discussed.

Keywords: 3D printing; ablation; light–matter interaction; femtosecond lasers; nanoscale

Citation: Han, M.; Smith, D.; Ng, S.J.; Anand, V.; Katkus, T.; Juodkazis, S. Ultra-Short-Pulse Lasers—Materials—Applications. *Eng. Proc.* **2021**, *11*, 44. https://doi.org/10.3390/ASEC2021-11143

Academic Editor: Nunzio Cennamo

Published: 15 October 2021

1. Laser Source and Beam Delivery

1.1. Ultra-Short-Pulse Laser Evolution

Laser, as a non-contact energy delivery tool, has a unique capability harnessed for fundamental research in the inertial confinement fusion (ICF), which recently became a step closer by reaching the burning plasma condition [1]. A laser intensity increase over the years after its invention in 1960 is a constant trend important for the basic science of light–matter/plasma interactions, as outlined in a roadmap review [2]. Matter at extreme conditions at pressures above 1 megabar (10^{11} Pa) is currently one of the most active fields of research [3].

Since the year 2000, the average laser power of ultra-short (sub-1 ps) pulsed lasers has increased as $Power = 2^{N/2}$, with N being the number of years from the beginning of the trend, which parallels Moore's law for the number of transistors in an integrated circuit. This conclusion is achieved following the evolution of ultra-short-pulsed laser amplitude produced over the last 20 years, presented recently [4]. Initially based on the chirped pulse amplification (CPA), which was awarded the Nobel prize in 2018, more recent approaches exploit different cavity geometries as well as amplification via the divided pulse and coherent beam combination. These strategies further increase the extracted power from solid-state and fibre laser systems and make them more compact. Ultra-short lasers with powers in the sub-1 kW range, $\sim$1 mJ pulse energies and at the repetition rates up to $\sim$1 MHz have become available.

New modes of laser operation bring the capability of combining ultra-short pulses into MHz–GHz bursts with a controlled number of pulses per burst [5]. It was shown that this burst mode of operation delivers ablation rates for metal and dental tissue on the order of 3 mm^3/min. This is the rate that reaches that of current Electrical Discharge Machining/Grinding (EDM/G) computer numerical control (CNC) tools. This parity between material removal rate by discharge spark and laser beam was achieved in 2016. The burst mode advantage is in the possibility to fine tune material removal to the most efficient fluence [J/cm^2] [6], which is empirically determined to be $e^2 = 7.4$ times larger than the ablation threshold for the given material [7]. Fine tuning the optimum ablation rate is achieved by changing the pulse number per irradiation spot, using beam scanning [8], and control over the number of pulses per burst. For comparison of different fabrication conditions, the volume [mm^3] ablated per 1 W average power per time 1 min, $V_a \sim mm^3/W/min \sim mm^3/(W.s) \sim mm^3/J$, is used. This is the ablated volume-per-energy delivered by the laser for subtractive machining ($3D^{(-)}$ printing). Interestingly, we show here that the volumetric energy density $Energy/Volume \sim J/mm^3$ is the right measure for the additive mode of $3D^{(+)}$ printing by ultra-short laser pulses [9]. It is not surprising that accounting for the energy deposition in the volume of light–matter interaction is the essential measure for the both additive and subtractive $3D^{(+)}$ and $3D^{(-)}$ modes of 3D fabrication.

1.2. Use of High-Average-Power Laser Beam

High-average-power sub-kW laser systems are targeting industrial applications. With the exponential $2^{N/2}$ increase in laser power indicated above, the most efficient use of this photon budget is required. To handle high laser power, new beam delivery systems are developed for the distribution of energy in a very well-controlled and precise way over the workpiece. Photonic crystal fibres (holy-fibres), flexible delivery units and polygon scanners with beam travel rates up to 1 km/s are readily available; interestingly, polygon scanners now used for the fastest beam delivery became available from mid-1980 and are on a similar growth trajectory to fs lasers. Galvano and polygon scanners further contribute to the compactness, versatility and safety of high-power handling. It is noteworthy that the scanning of the laser beam in cash-counter machines is an example of an application where speed and safety are delivered simultaneously. This is especially important for open-space and field-deployable applications, e.g., surface texturing by ablation ripples for the creation of hydrophobic, anti-icing and biocidal surfaces [10]. These applications are particularly suitable for fast beam scanning techniques. One of the most demanding applications for surface treatment is in the solar cells industry. Anti-reflection coatings and packaging for 20+ years continuous performance in open air have to be delivered. With the promise of increasing the efficiency of Si solar cells from the current 18% (for mass-produced cells) to one closer to the theoretical Shockley–Queisser limit of ∼31%, the use of photonic crystal patterns on Si surfaces is an invitation to use fast laser scanning for laser texturing [11]. Scanning of large (cm-scale) areas without stitching errors and maintaining sub-wavelength precision of laser patterning by combined sample and beam scan was recently introduced for 3D polymerisation [12]. This approach is inherently scalable to larger (meter-scale) patterning in atmospheric (room) conditions, required for patterning surfaces for injection moulding die surfaces, texturing steel and fibre composites for anti-frosting and water repelling properties in the aviation industry, and potentially for solar cells in the future.

2. Materials

Materials are a major and critical part for the $3D^{\pm}$ manufacturing ecosystem (Figure 1). New polymerisable mixtures of colloidal particles and standard photo-polymerisable resists/resins can be tailored for the required material composition. Calcination of the polymerised composites can be transferred into a glass, polycrystalline or ceramic state with feature sizes down to the nanoscale [13]. Cutting and drilling of dielectrics, e.g., dicing of sapphire substrates in the light emitting diode (LED) industry, and metal/composite processing with high precision and minimal heat-affected zone (HAZ) for complex 3D geometries can be carried out most efficiently with ultra-short laser pulses [14]. This versatility in terms of material processing stems from well-controlled energy delivery in space and time. Even small energy pulses have high intensities—TW/cm^2 and above—and can turn non-absorbing dielectrics into ionised plasma with strong energy deposition. Internal modification of the interior volume of dielectrics becomes feasible with these energies. It was demonstrated that high-pressure and high-temperature phases of materials can be created and retained down to room ambience due to ultra-fast thermal quenching of a small modified volume [15,16]. Internally confined micro-explosions occurring in the high-Young-modulus dielectrics create conditions similar to the centre of the Earth—hence, warm dense matter (WDM). The micro-explosion hydrodynamics follows the established and tested macroscopic versions [17]. New and metastable phases of materials, e.g., amorphous sapphire, can be produced by tightly focused fs laser pulses [18].

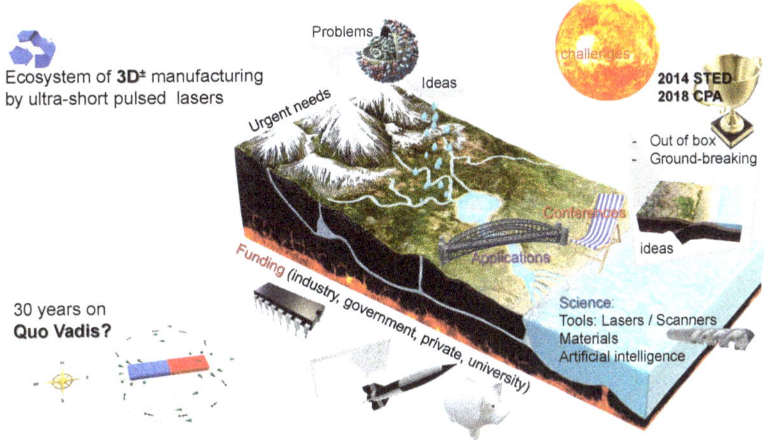

Figure 1. Ecosystem of $3D^{\pm}$ manufacturing based on development of lasers, beam/stage scanners, computer numerical control (CNC), artificial intelligence (AI). Increasing field of applications in material processing and creation of new materials is developing via different funding sources.

The mass production of colloidal nanoparticles of different materials in water with fs laser pulses scanned at speeds exceeding that of bubble formation is already an industrial process. The benefits of such nanoparticles are that surfaces are free from surfactants used in chemical synthesis. The size distribution of these colloids can be controlled via interaction with simultaneously generated coherent white light continuum (WLC) [19].

A large impact on the development of material processing by ultra-short laser pulses was driven by the quest for higher resolution—ultimately, super-resolution—which can deliver the fabrication of 3D objects with sub-diffraction λ/NA and sub-wavelength resolution; NA is the numerical aperture of the optics used, and λ is the wavelength. The method of stimulated emission depletion (STED) microscopy, demonstrated in 2000 and awarded the Nobel prize in 2014, influenced the community of fs laser users who widely relied on table-top microscopes used for the polymerisation of nano-micro-structures and optical memory. Due to the threshold effect of material modification, tens-of-nm resolution in 3D can be achieved by direct fs laser write via the fine tuning of the pulse energy. This is even without critical point drying (CPD) equipment, which is typically used to avoid deformations made by surface tension during the wet development stage; a 30 nm 3D feature size was obtained using the threshold effect in common SU8 [20].

3. Applications

Beyond material processing, ultra-short laser pulses are used in an ever-increasing range of applications, especially due to the available high power and dramatic reductions in size. Ultra-short laser pulses in the vis-IR spectral range have potential for data communication, especially in non-scattering ambience, e.g., for space applications due to high frequency—hence, large bandwidth is required for fast data communications. It is a recognisable trend in wireless and mobile communications.

Direct energy deposition applications already range from defence to 3D$^+$ printing (e.g., powder sintering). In the practical, high-fluence/-intensity application of laser cutting, the use of linearly shaped focal regions, e.g., Gauss–Bessel beams, is proving to be a viable solution [21,22].

In multi-dimensional optical memory, the usual 3D positioning of memory bits for laser writing and readout by luminescence or scattering [23,24] is augmented by a polarisation degree of freedom due to nano-gratings, which form two extra dimensions via form birefringence. Fs-inscribed optical memory bits withstand 1100 °C temperatures [25]. Optical memory is of significant interest due to its thermal stability and durability.

Coming full circle, for high-spatial-resolution studies with single fs laser pulses and interference patterns [26–29], the most recent development of high-precision direct write shows the possibility of fabricating nanoscale grooves down to 20 nm width on a solid-state dielectric film (equivalent of a resist) [30]. Precise energy control by the orientation of linear polarisation allows the patterning of single nanoscale features: bumps, voids and grooves [31,32].

For the commercial viability of any technical solution, it is necessary for it to deliver a bridging solution in product manufacturing that is unique: better before cheaper. Based on the commercial success of a particular implementation, other areas as well as more fundamental research are funded (Figure 1). It is increasingly difficult to make improvements to production line processes as a new project due to complications of a fast-moving industry cycle (<1 year), in contrast to academic research, which is a multi-year endeavour, e.g., can be measured in duration of PhD projects (∼3–4 years). Due to this complexity and lengthy project review (∼0.5 years), the entry point between academia and industry is most efficient for small-scale proof of the principle applications. Rapid prototyping, which is the key advantage of 3D$^\pm$ printing by ultra-short laser pulses, is the most promising pathway for industry–academia engagement. The trend for using artificial intelligence (AI) in the CNC control of processes is rapidly evolving. Recently, predictions of the optical properties of complex 3D multilayered structures of different materials for specific spectral functions were AI generated with convincing fidelity [33].

Author Contributions: Conceptualisation, S.J., S.H.N.; investigation, M.H., D.S., V.A., T.K.; writing—original draft preparation, S.H.N., S.J.; review and editing, all authors. All authors have read and agreed to the published version of the manuscript.

Funding: This research was funded by the Australian Research Council, grant number LP190100505. S.J. is grateful for startup funding from the Nanotechnology Facility at Swinburne and to the Workshop-on-Photonics for the technology transfer project, which installed the first industrial-grade microfabrication setup in Australia.

Institutional Review Board Statement: Not applicable.

Informed Consent Statement: Not applicable.

Data Availability Statement: The data presented in this study are available on request from the corresponding author.

Conflicts of Interest: The authors declare no conflict of interest.

References

1. Zylstra, A.B.; Hurricane, O.A.; Callahan, D.A.; Kritcher, A.L.; Ralph, J.E.; Robey, H.F.; Ross, J.S.; Young, C.V.; Baker, K.L.; Casey, D.T.; et al. Burning plasma achieved in inertial fusion. *Nature* **2022**, *601*, 542–548. [CrossRef] [PubMed]
2. Mourou, G.A.; Tajima, T.; Bulanov, S.V. Optics in the relativistic regime. *Rev. Mod. Phys.* **2006**, *78*, 309–371. [CrossRef]
3. Batani, K.; Batani, D.; He, X.; Shigemori, K. Recent progress in matter in extreme states created by laser. *Matter Radiat. Extrem.* **2022**, *7*, 013001. [CrossRef]
4. Mottay, E. Femtosecond Lasers for High Throughput Surface Engineering. In *SPIE Photonics West Industry Events*; SPIE: Bellingham, WA, USA, 2021; Volume 11768, p. 1176807.

5. Kerse, C.; Kalaycıoğlu, H.; Elahi, P.; Çetin, B.; Kesim, D.; Akçaalan, O.; Yavaş, S.; Aşık, M.; Öktem, B.; Hoogland, H.; et al. Ablation-cooled material removal with ultrafast bursts of pulses. *Nature* **2016**, *537*, 84–88. [CrossRef] [PubMed]
6. Förster, D.; Jäggi, B.; Michalowski, A.; Neuenschwander, B. Review on Experimental and Theoretical Investigations of Ultra-Short Pulsed Laser Ablation of Metals with Burst Pulses. *Materials* **2021**, *14*, 3331. [CrossRef] [PubMed]
7. Hodgson, N.; Steinkopff, A.; Heming, S.; Allegre, H.; Haloui, H.; Lee, T.; Laha, M.; VanNunen, J. Ultrafast laser machining: process optimization and applications. In *SPIE Laser Applications in Microelectronic and Optoelectronic Manufacturing (LAMOM) XXVI*; SPIE: Bellingham, WA, USA, 2021; Volume 11673, p. 1167308.
8. Vanagas, E.; Kawai, J.; Tuzilin, D.; Kudryashov, I.; Mizuyama, A.; Nakamura, K.G.; Kondo, K.I.; Koshihara, S.Y.; Takesada, M.; Matsuda, K.; et al. Glass cutting by femtosecond pulsed irradiation. *J. Microlith. Microfab. Microsyst.* **2004**, *3*, 358–363. [CrossRef]
9. Skliutas, E.; Lebedevaite, M.; Kabouraki, E.; Baldacchini, T.; Ostrauskaite, J.; Vamvakaki, M.; Farsari, M.; Juodkazis, S.; Malinauskas, M. Polymerization mechanisms initiated by spatio-temporally confined light. *Nanophotonics* **2021**, *10*, 1211–1242. [CrossRef]
10. Bonaccurso, E. Laser-treated superhydrophobic surfaces to reduce ice build-up in aeronautical applications. In *SPIE Photonics West Industry*; SPIE: Bellingham, WA, USA, 2021; Volume 11768, p. 117680M.
11. Maksimovic, J.; Hu, J.; Ng, S.; Katkus, T.; Seniutinas, G.; Nishijima, Y.; John, S.; Juodkazis, S. Large area $\gg 1$ cm^2 light trapping patterns for Si solar cells. In *SPIE Advanced Fabrication Technologies for Micro/Nano Optics and Photonics XIV*; SPIE: Bellingham, WA, USA, 2021; Volume 11696, p. 116960T.
12. Jonušauskas, L.; Gailevičius, D.; Rekštytė, S.; Baldacchini, T.; Juodkazis, S.; Malinauskas, M. Mesoscale laser 3D printing. *Opt. Express* **2019**, *27*, 15205–15221. [CrossRef]
13. Gailevičius, D.; Padolskytė, V.; Mikoliūnaitė, L.; Šakirzanovas, S.; Juodkazis, S.; Malinauskas, M. Additive-manufacturing of 3D glass-ceramics down to nanoscale resolution. *Nanoscale Horiz.* **2019**, *4*, 647–651. [CrossRef]
14. Jonušauskas, L.; Gailevičius, D.; Mikoliūnaitė, L.; Sakalauskas, D.; Šakirzanovas, S.; Juodkazis, S.; Malinauskas, M. Optically Clear and Resilient Free-Form µm-Optics 3D-Printed via Ultrafast Laser Lithography. *Materials* **2017**, *10*, 12. [CrossRef]
15. Juodkazis, S.; Nishimura, K.; Tanaka, S.; Misawa, H.; Gamaly, E.E.; Luther-Davies, B.; Hallo, L.; Nicolai, P.; Tikhonchuk, V. Laser-Induced Microexplosion Confined in the Bulk of a Sapphire Crystal: Evidence of Multimegabar Pressures. *Phys. Rev. Lett.* **2006**, *96*, 166101. [CrossRef]
16. Ivanova, E.P.; Hasan, J.; Webb, H.K.; Gervinskas, G.; Juodkazis, S.; Truong, V.K.; Wu, A.H.F.; Lamb, R.N.; Baulin, V.; Watson, G.S.; et al. Bactericidal activity of nanostructured black silicon. *Nat. Commun.* **2013**, *4*, 2838. [CrossRef] [PubMed]
17. Juodkazis, S.; Misawa, H.; Gamaly, E.G.; Luther-Davis, B.; Hallo, L.; Nicolai, P.; Tikhonchuk, V. Is the nano-explosion really microscopic? *J. Non-Crystall. Solids* **2009**, *355*, 1160–1162. [CrossRef]
18. Juodkazis, S.; Nishimura, K.; Misawa, H.; Ebisui, T.; Waki, R.; Matsuo, S.; Okada, T. Control over the State of Crystallinity: Sapphire. *Adv. Mat.* **2006**, *18*, 1361–1364. [CrossRef]
19. Kubiliūtė, R.; Maximova, K.; Lajevardipour, A.; Yong, J.; Hartley, J.S.; Mohsin, A.S.M.; Blandin, P.; Chon, J.W.M.; Clayton, A.H.A.; Sentis, M.; et al. Ultra-pure, water-dispersed Au nanoparticles produced by femtosecond laser ablation and fragmentation. *Int. J. Nanomed.* **2013**, *8*, 2601–2611.
20. Juodkazis, S.; Mizeikis, V.; Seet, K.K.; Miwa, M.; Misawa, H. Two-photon lithography of nanorods in SU-8 photoresist. *Nanotechnology* **2005**, *16*, 846–849. [CrossRef]
21. Marcinkevicius, A.; Juodkazis, S.; Matsuo, S.; Mizeikis, V.; Misawa, H. Application of Bessel Beams for Microfabrication of Dielectrics by Femtosecond Laser. *Jpn. J. Appl. Phys.* **2001**, *40*, L1197–L1199. [CrossRef]
22. Mikutis, M.; Kudrius, T.; Šlekys, G.; Paipulas, D.; Juodkazis, S. High 90% efficiency Bragg gratings formed in fused silica by femtosecond Gauss-Bessel laser beams. *Opt. Mat. Express* **2013**, *3*, 1862–1871. [CrossRef]
23. Juodkazis, S.; Rode, A.V.; Gamaly, E.G.; Matsuo, S.; Misawa, H. Recording and reading of three-dimensional optical memory in glasses. *Appl. Phys. B* **2003**, *77*, 361–368. [CrossRef]
24. Watanabe, M.; Juodkazis, S.; Sun, H.B.; Matsuo, S.; Misawa, H.; Miwa, M.; Kaneko, R. Transmission and Photoluminescence Images of Three-Dimensional Memory in Vitreous Silica. *Appl. Phys. Lett.* **1999**, *74*, 3957–3959. [CrossRef]
25. Watanabe, M.; Sun, H.B.; Juodkazis, S.; Takahashi, T.; Matsuo, S.; Suzuki, Y.; Nishii, J.; Misawa, H. Three-Dimensional Optical Data Storage in Vitreous Silica. *Jpn. J. Appl. Phys.* **1998**, *27*, L1527–L1530. [CrossRef]
26. Juodkazis, S.; Matsuo, S.; Misawa, H.; Mizeikis, V.; Marcinkevicius, A.; Sun, H.B.; Tokuda, Y.; Takahashi, M.; Yoko, T.; Nishii, J. Application of femtosecond laser pulses for microfabrication of transparent media. *Appl. Surf. Sci.* **2002**, *197–198*, 705–709. [CrossRef]
27. Efimov, O.; Juodkazis, S.; Misawa, H. Intrinsic single and multiple pulse laser-induced damage in silicate glasses in the femtosecond-to-nanosecond region. *Phys. Rev. A* **2004**, *69*, 042903. [CrossRef]
28. Seet, K.K.; Mizeikis, V.; Juodkazis, S.; Misawa, H. Three-Dimensional Horizontal Circular Spirals Photonic Crystals with stop gaps below 1 µm. *Appl. Phys. Lett.* **2006**, *88*, 221101. [CrossRef]
29. Kondo, T.; Juodkazis, S.; Misawa, H. Reduction of Capillary Force for High-Aspect Ratio Nanofabrication. *Appl. Phys. A* **2005**, *81*, 1583–1586. [CrossRef]
30. Li, Z.Z.; Wang, L.; Fan, H.; Yu, Y.H.; Chen, Q.D.; Juodkazis, S.; Sun, H.B. O-FIB: far-field-induced near-field breakdown for direct nanowriting in an atmospheric environment. *Light Sci. Appl.* **2020**, *9*, 41. [CrossRef]

31. Buividas, R.; Rekštytė, S.; Malinauskas, M.; Juodkazis, S. Nano-groove and 3D fabrication by controlled avalanche using femtosecond laser pulses. *Opt. Mat. Express* **2013**, *3*, 1674–1686. [CrossRef]
32. Vanagas, E.; Kudryashov, I.; Tuzhilin, D.; Juodkazis, S.; Matsuo, S.; Misawa, H. Surface nanostructuring of borosilicate glass by femtosecond nJ energy pulses. *Appl. Phys. Lett.* **2003**, *82*, 2901–2903. [CrossRef]
33. So, S.; Yang, Y.; Lee, T.; Rho, J. On-demand design of spectrally sensitive multiband absorbers using an artificial neural network. *Photonics Res.* **2021**, *9*, B153–B158. [CrossRef]

engineering proceedings

MDPI

Proceeding Paper

Comparing two Fitting Algorithms for Determining the Cole–Cole Parameters in Blood Glucose Problems †

Roberto Dima *, Giovanni Buonanno and Raffaele Solimene

Department of Engineering, University of Campania, 81031 Aversa, Italy; giovanni.buonanno@unicampania.it (G.B.); raffaele.solimene@unicampania.it (R.S.)

* Correspondence: roberto.dima@unicampania.it

† Presented at the 2nd International Electronic Conference on Applied Sciences, 15–31 October 2021 ; Available online: https://asec2021.sciforum.net/.

Abstract: This paper addresses the nonlinear inverse problem of estimating the parameters of the Cole–Cole model used to describe the behavior of the complex permittivity of blood samples. Such a model provides an efficient and accurate representation of biological tissues in the entire frequency band considered and reduces the complexity of the experimental data to a few parameters. In this way, it is possible to extract a "synthetic view" of the dielectric properties of tissues in such a way that more information on the glucose concentration can be derived, in addition to the resonance peak or phase shift. In order to perform the fitting of the Cole–Cole model, two different algorithms were used and compared: the Levenberg–Marquardt and the variable projection algorithms. Synthetic data present in the literature were used to evaluate the performances obtainable with these methods. In particular, Monte Carlo analysis was used in order to evaluate the accuracy and the precision that these two methods provide in the process of estimating the parameters involved, with respect to the starting points of the parameters. The results obtained showed that the variable projection algorithm always outperformed the Levenberg–Marquardt one, although the former has a greater computational burden than the latter.

Keywords: glucose measurement; Cole–Cole model; Levenberg–Marquardt algorithm; variable projection algorithm; blood dielectric properties; nonlinear fitting problem

Citation: Dima, R.; Buonanno, G.; Solimene, R. Comparing two Fitting Algorithms for Determining the Cole–Cole Parameters in Blood Glucose Problems. *Eng. Proc.* **2021**, *11*, 45. https://doi.org/10.3390/ASEC2021-11188

Academic Editor: Roger Narayan

Published: 15 October 2021

1. Introduction

Diabetes is a metabolic disorder that afflicts millions of people in the world [1]. It degrades the cell's ability to absorb glucose from the bloodstream because of the improper regulation of the insulin hormone. For this reason, great efforts have been dedicated to the development of non-invasive glucose monitoring devices, which may considerably improve the quality of life for diabetics [2].

This work is particularly concerned with microwave sensor technology that relies on the change in the dielectric and conductivity properties of blood plasma as a function of the glucose concentration in order to track such a change.

In this framework, developing accurate and precise fitting methods for blood models, at different glucose concentrations, is essential for the development of robust electromagnetic (EM)-based techniques that could be employed for non-invasive, continuous glucose monitoring. Indeed, accurate electromagnetic tissue modeling is of paramount importance since it affects the simulation stage required for sensor design [3]. Moreover, extracting a "synthetic view" (in terms of a few parameters) of the sensor response data is essential for analyzing patterns and possibly extracting more information, besides the resonance peak or phase shift, about glucose concentration.

In this paper, the aim was just to address the fitting problem. More in detail, starting from the dielectric spectrum, which is assumed known over a certain number of frequencies, we aimed at estimating the parameters of a single-pole Cole–Cole model. As is well known, this entails solving a nonlinear inverse problem, which here was addressed by two different methods: the classical Levenberg–Marquardt method [4,5] and the variable projection algorithm [6]. We evaluated how sensitive the two methods are with respect to the starting points of the parameters and with what accuracy and precision these parameters can be estimated. In order to check the two methods, we first generated synthetic relative permittivities by employing a single-pole Cole–Cole model, using data from the literature [7] as true values for its parameters, and then solved an inverse problem in order to trace these values by resorting to the two aforesaid methods.

Although higher-order models could perform better, we considered a first-order model to perform the comparison in the simplest possible case.

2. Methods

2.1. Cole–Cole Model

The Cole–Cole model [8] is widely used to describe the complex relative permittivity of biological tissues, $\varepsilon_{\rm r}(\omega) = \varepsilon(\omega)/\varepsilon_0$, and its equation is:

$$\varepsilon_{\rm r}(\omega) = \varepsilon_\infty + \sum_{n=1}^{N} \frac{\varepsilon_{{\rm s}n} - \varepsilon_\infty}{1 + (j\omega\tau_n)^{1-\alpha_n}} - \frac{\sigma}{j\omega\varepsilon_0} \tag{1}$$

in which N is the number of poles and thus the order of the model, $\varepsilon_\infty = \lim_{\omega\to\infty} \varepsilon_{\rm r}(\omega)$ is the permittivity at high frequencies, σ is the static ionic conductivity and $\varepsilon_{{\rm s}n} = \lim_{\omega\to 0} \varepsilon_{\rm r}(\omega)$, and τ_n, and α_n are the static permittivity, the relaxation time constant, and the so-called distribution parameter of the n-th addend of the summation, respectively. Such a model incorporates the Debye model [9]. Indeed, the main difference between the Debye and the Cole–Cole models is that the latter includes the exponent $1-\alpha$, with $0 \le \alpha \le 1$. When the exponent becomes smaller, the relaxation time distribution becomes broader, i.e., the transition between low- and high-frequency values becomes wider and the peak on the imaginary part of the spectrum also becomes wider.

The complexity of both the structure and composition of biological material is such that the dispersion region of each pole may be broadened by multiple contributions to it. The broadening of the dispersion could be empirically accounted for by using the Cole–Cole model [10], which is expected to give more accurate dielectric spectrum curve fitting.

2.2. Curve Fitting Algorithms

Let be $\boldsymbol{x}$ the vector of model parameters and P its length and M the number of frequency points for which the measures are taken. We define the data vector ($\top$ stands for transposition) $\boldsymbol{y} = [y(\omega_1)\dots y(\omega_m)\dots y(\omega_M)]^\top$ in which the m-th component of the vector $\boldsymbol{y}$ is the observed value $y(\omega_m)$. Let also $\boldsymbol{\varepsilon}_{\rm r} = [\varepsilon_{\rm r}(\omega_1;\boldsymbol{x})\dots\varepsilon_{\rm r}(\omega_m;\boldsymbol{x})\dots\varepsilon_{\rm r}(\omega_M;\boldsymbol{x})]^\top$ be the model vector, here given by Equation (1)), with $\varepsilon_{\rm r}(\omega_m;\boldsymbol{x})$ being the estimation at ω_m.

Solving the least-squares problem means finding $\hat{\boldsymbol{x}}$ such that:

$$\hat{\boldsymbol{x}} = \arg\min_{\boldsymbol{x}\in\mathbb{R}^P} \left\{ \frac{1}{2} \|\boldsymbol{\varepsilon}_{\rm r}(\boldsymbol{x}) - \boldsymbol{y}\|_2^2 \right\} \tag{2}$$

in which the function to minimize, $\Psi = \frac{1}{2}\|\boldsymbol{\varepsilon}_{\rm r}(\boldsymbol{x}) - \boldsymbol{y}\|_2^2$, is the ℓ_2 quadratic norm of the misfit $\boldsymbol{r} = \boldsymbol{\varepsilon}_{\rm r}(\boldsymbol{x}) - \boldsymbol{y}$, which is a nonlinear function such that $\boldsymbol{r} : \mathbb{R}^P \mapsto \mathbb{R}^M$ with $P \ll M$.

We addressed the nonlinear fitting problem with two methods: the Levenberg–Marquardt algorithm (LMA) and the variable projection algorithm (VPA).

The Levenberg–Marquardt algorithm [4,5] acts more as a gradient-descent method when the parameters are far from their optimal value and acts more as the Gauss–Newton method when the parameters are close to their optimal value [11]. The equation for the step $\boldsymbol{h}$ at the kth iteration is:

$$\left(J(\boldsymbol{x}_k)^\top J(\boldsymbol{x}_k) + \lambda_k I\right)\boldsymbol{h} = -J(\boldsymbol{x}_k)^\top \boldsymbol{f}(\boldsymbol{x}_k) \tag{3}$$

where J is the Jacobian of f and λ_k is the damping parameter. It controls both the magnitude and direction of $\boldsymbol{h}$, and it is chosen at each iteration. It can be shown [5] that, at each iteration, Equation (3) solves the minimization problem over a reduced set of admissible solutions, i.e., those that satisfy $\|\boldsymbol{h}\| \leq R(\lambda)$, limiting the correction step to within a region near $\boldsymbol{x}_k$. The radius of the trust region $R = R(\lambda)$ is a strictly decreasing function with $\lim_{\lambda \to \infty} R(\lambda) = 0$. When $\lambda_k = 0$, the step $\boldsymbol{h}$ is identical to that of the Gauss–Newton method, i.e., the same direction and maximum magnitude. As $\lambda \to \infty$, $\boldsymbol{h}$ tends towards the steepest descent direction, with the magnitude tending towards 0.

Based on the above, we inferred the qualitative update rule for λ_{k+1}: If $\Psi(\boldsymbol{x}_k + \boldsymbol{h}) < \Psi(\boldsymbol{x}_k)$, then the quadratic approximation works well, and we can extend the trust region, i.e., it will be $\lambda_{k+1} < \lambda_k$. Otherwise, the step is unsuccessful and we reduce the trust region, i.e., it will be $\lambda_{k+1} > \lambda_k$; in this way, the next step tends towards the negative gradient method, and a lower value of Ψ is more likely to be found.

The MATLAB implementation was used, in particular the `lscurvefit` function with the Levenberg–Marquardt option.

The variable projection algorithm [6] is a method used to solve separable nonlinear least-squares problems. The least-squares problem is said to be separable when the model parameters can be separated into two sets of parameters, one that enters linearly into the model, $\boldsymbol{c} = [c_1, \ldots, c_k]$, and another set of parameters that enter the model nonlinearly, $\boldsymbol{a} = [a_1, \ldots, a_l]$, so that $\boldsymbol{x} = [\boldsymbol{c}, \boldsymbol{a}]$. For each observation y_m of a separable nonlinear least-squares problem, the model is a linear combination of nonlinear functions that depend on nonlinear parameters, and the model function can be written as $\varepsilon_r(\omega) = \sum_{j=1}^{k} c_j \, \phi_j(\omega; \boldsymbol{a})$.

The functional Ψ is written in terms of residual vector $\boldsymbol{r}$ as:

$$\Psi(\boldsymbol{a}, \boldsymbol{c}) = \frac{1}{2} \|\boldsymbol{y} - \boldsymbol{\Phi}(\boldsymbol{a})\boldsymbol{c}\|^2 \tag{4}$$

in which the columns of the matrix $\boldsymbol{\Phi}$ are the nonlinear functions $\phi_j(\omega; \boldsymbol{a})$. The linear parameters $\boldsymbol{c}$ could be obtained from the knowledge of $\boldsymbol{a}$, by solving the linear least-squares problem:

$$\boldsymbol{c} = \boldsymbol{\Phi}(\boldsymbol{a})^{\dagger} \boldsymbol{y} \tag{5}$$

which stands for the minimum-norm solution of the linear least-squares problem for fixed $\boldsymbol{a}$, where $\boldsymbol{\Phi}(\boldsymbol{a})^{\dagger}$ is the Moore–Penrose generalized inverse of $\boldsymbol{\Phi}(\boldsymbol{a})$. By replacing this in Equation (4), we obtain the variable projection functional:

$$\Psi_{\mathrm{VP}}(\boldsymbol{a}) = \frac{1}{2} \left\| \boldsymbol{y} - \boldsymbol{\Phi}(\boldsymbol{a}) \boldsymbol{\Phi}(\boldsymbol{a})^{\dagger} \boldsymbol{y} \right\|^2 \tag{6}$$

The variable projection algorithm consists of two steps: first minimizing Equation (6) with an iterative nonlinear method and then using the optimal value found for $\boldsymbol{a}$ to solve for $\boldsymbol{c}$ in Equation (5) [12]. The principal advantage is that the iterative nonlinear algorithm used to solve the first minimization problem works in a reduced space and less initial guesses are necessary. A robust implementation in MATLAB, called `VARPRO` [13], was adapted and used to deal with complex-valued problems, choosing the Levenberg–Marquardt option for the solution of Equation (6).

2.3. Numerical Simulations

The generation of the synthetic complex relative permittivity of blood plasma relied on quadratic fits to glucose-dependent Cole–Cole parameters reported in [7]; in particular, we considered two different concentrations, 100 mg/dL and 250 mg/dL, one normal and the other of severe diabetes, respectively, in accordance with the diagnostic criteria in [1]. The data vector consisted of $M = 1000$ points in the frequency range 500 MHz–20 GHz.

In gradient-like algorithms (such as those used in this paper), the choice of the initial point is a crucial factor for the convergence of the procedure. For the single-pole model case, it is fairly easy to exploit the physical meaning of the parameters to infer an initial estimate. However, since the noise can invalidate the initial estimate, we propose to study the robustness of the two algorithms with respect to the initial point. To this end, a Monte Carlo analysis was performed, iteratively evaluating the deterministic algorithms using a set of N = 1000 uniformly distributed random initial points arranged in a 5D hypercube of the parameter space in order to statistically characterize the results. Each side of the hypercube represents an interval containing the range of variation of each parameter for the glucose concentrations considered.

We ran simulations on a machine with Intel i9-10850K (10 physical cores), 32 GB RAM, and Ubuntu 21.04, and we took advantage of the Parallel Computing Toolbox, using the `parfor` loop for running the 1000 simulations.

The intervals for generating the random initial value for each parameter (of the Cole–Cole model) were chosen from the data tabulated in [7]. In particular, the widths of these intervals were the same for each glucose concentration and were: $[1,5]$ for ε_∞, $[1,150]$ for ε_s, $[1\times10^{-14}, 1\times10^{-11}]$ for τ, $[0.1-1\times10^{-9}, 0.1+1\times10^{-9}]$ for α, and $[0,5]$ for σ. These intervals are relatively large compared to the values taken from [7] in order to test the two algorithms in sufficiently stressful situations. Only the range of variation of α is extremely small because the model used in [7] practically fixes it a priori to 0.1. Obviously, it must be taken into account that the VPA requires only generating the values for τ and α.

For an initial bland qualitative assessment, we established evaluation intervals (the same for generations) for the estimated parameters so that we could assert that a reconstruction is "good" if it falls within these ranges, and "wrong" otherwise.

Now, let $\hat{\mathbf{x}}^{(i)} = [\widehat{\varepsilon}_\infty^{(i)}, \widehat{\varepsilon}_s^{(i)}, \widehat{\tau}^{(i)}, \widehat{\alpha}^{(i)}, \widehat{\sigma}^{(i)}]$ be the vector of the parameter estimates returned by the two algorithms at the i-th simulation, and let $\hat{x}^{(i)}$ denote one of its five elements. Moreover, let $\langle x\rangle = (1/N_{\mathrm{sim}})\sum_{i=1}^{N_{\mathrm{sim}}}\hat{x}_i$ and $\sigma_x = \sqrt{[1/(N_{\mathrm{sim}}-1)]\sum_{i=1}^{N_{\mathrm{sim}}}|\hat{x}_i - \langle x\rangle|^2}$ be the sample mean and standard deviation, respectively, calculated for each parameter.

For a quantitative evaluation of the performance of the two algorithms, we then define multiple figures of merit for statistically characterizing the results of Monte Carlo analysis. For each parameter, Equations (7) and (8) define measures of accuracy and precision, respectively, defined over the entire set of reconstructions. However, such measures can be greatly affected by estimates that are very far from the true value, x_{true}, where the latter represents one of the five elements of the vector of reference values $\hat{\mathbf{x}}_{true} = [\widehat{\varepsilon}_{\infty_{true}}, \widehat{\varepsilon}_{s_{true}}, \widehat{\tau}_{true}, \widehat{\alpha}_{true}, \widehat{\sigma}_{true}]$. For this reason, we also introduced $\mathcal{A}_{cut}$ and $\mathcal{P}_{cut}$ in Equations (9) and (10) in order to define the accuracy and precision measures, respectively, which instead dampen the effect of the above isolated events. The subscript *cut* means that they were calculated on a subset obtained by eliminating $\zeta\%$ of reconstructions with lower values and $\zeta\%$ of reconstructions with higher values, in which $0 < \zeta < 50$.

$$\mathcal{A} = \left|\frac{x_{true} - \langle x\rangle}{x_{true}}\right| \times 100\% \tag{7}$$

$$\mathcal{P} = \frac{\sigma_x}{\langle x\rangle} \times 100\% \tag{8}$$

$$\mathcal{A}_{cut} = \left|\frac{x_{true} - \langle x\rangle_{cut}}{x_{true}}\right| \times 100\% \tag{9}$$

$$\mathcal{P}_{cut} = \frac{\sigma_{x_{cut}}}{\langle x\rangle_{cut}} \times 100\% \tag{10}$$

3. Results

We conducted many numerical simulations by widening more and more the generation intervals. In this paper, we report the case where the generation intervals are very large except for the α interval, due to the above explanation. In all these experiments, following the qualitative criterion mentioned above, the VPA always provided good estimations, while the same was not true for the LMA. In particular, in the case considered, the LMA provided the wrong estimates in about 260 simulations out of 1000, for each of the two glucose concentrations, while no wrong estimate was returned by the VPA. Here, by the "wrong" estimate, we mean that at least one component of the parameter vector x had a value that was outside its generation range. Graphical representations of those qualitative results are provided in Figure 1.

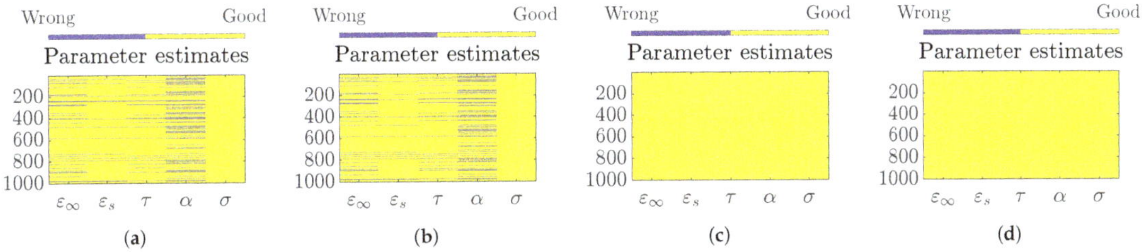

Figure 1. Graphical representations of the convergence of the LMA and VPA for 1000 simulations and 2 different glucose concentrations: (**a**) the LMA for 100 mg/dL, (**b**) the LMA for 250 mg/dL, (**c**) the VPA for 100 mg/dL, and (**d**) the VPA for 250 mg/dL.

Consistent with the qualitative results, considering the whole set of 1000 estimates, the VPA exhibited excellent accuracy and precision, while this was not the case for LMA, as can be observed in Tables 1 and 2. For this reason, for the VPA only, the results calculated by means of Equations (7) and (8) are reported, whilst for the LMA, the results derived from Equations (9) and (10), obtained by cutting 25% of the lowest values and 25% of the highest values, were also considered.

Regarding execution times, the LMA took 7 s to finish 1000 simulations, while the VPA took around 170 s.

Table 1. Figures of merit (in %) of the two algorithm in which the glucose concentration is 100 mg/dL.

		ε_∞	ε_s	τ (s)	α	σ (S/m)
	x_{true}	**2.305**	**73.28**	8.723×10^{-12}	**0.1**	**1.986**
LMA	$\mathcal{A}$	2.043×10^5	5.34×10^4	1.383×10^5	4.161×10^5	1.921
	$\mathcal{A}_{cut,25\%}$	2.257×10^{-7}	1.799×10^{-10}	1.041×10^{-9}	3.237×10^{-9}	6.762×10^{-10}
	$\mathcal{P}$	1490	2668	3188	2086	6.834
	$\mathcal{P}_{cut,25\%}$	6.285×10^{-7}	5.054×10^{-10}	2.243×10^{-9}	6.742×10^{-9}	1.881×10^{-9}
VPA	$\mathcal{A}$	1.305×10^{-7}	4.024×10^{-10}	6.201×10^{-9}	2.129×10^{-8}	4.876×10^{-10}
	$\mathcal{P}$	3.569×10^{-7}	1.13×10^{-9}	1.694×10^{-8}	5.877×10^{-8}	1.357×10^{-9}

Table 2. Figures of merit (in %) of the two algorithm in which the glucose concentration is 250 mg/dL.

		ε_∞	ε_s	τ (s)	α	σ (S/m)
	x_{true}	**2.312**	**73.26**	8.758×10^{-12}	**0.1**	**1.965**
LMA	$\mathcal{A}$	1.344×10^5	3.58×10^6	2.692×10^8	2.468×10^7	1.995
	$\mathcal{A}_{cut,25\%}$	2.248×10^{-7}	1.699×10^{-10}	1.101×10^{-9}	3.316×10^{-9}	7.198×10^{-10}
	$\mathcal{P}$	6908	3130	3162	3017	6.874
	$\mathcal{P}_{cut,25\%}$	6.199×10^{-7}	4.859×10^{-10}	2.479×10^{-9}	7.333×10^{-9}	1.985×10^{-9}
VPA	$\mathcal{A}$	1.322×10^{-7}	4.107×10^{-10}	6.306×10^{-9}	2.17×10^{-8}	5.029×10^{-10}
	$\mathcal{P}$	3.595×10^{-7}	1.153×10^{-9}	1.712×10^{-8}	5.965×10^{-8}	1.396×10^{-9}

4. Discussion

In this paper, we faced the problem of fitting the dielectric spectrum of a blood sample in order to estimate the parameters of the single-pole Cole–Cole model. In particular, we compared the performance of two different algorithms, the LMA and VPA, in terms of the accuracy and precision with respect to the starting points of the parameters.

For the parameter range considered, the VPA outperformed the LMA in robustness with respect to the initial point of the algorithm. However, analyzing the figures of merit related to the LMA, it became clear that there were erroneous reconstructions so far from the true value such that they heavily deteriorated the (standard) accuracy and precision, while the cut versions did not suffer from this problem. In fact, once the erroneous ones were removed, in all other simulations, the algorithm converged to the true values.

On the other hand, the VPA gave these good results because less initial guesses were necessary and because the iterative nonlinear algorithm used to solve the first minimization problem works in a reduced space. The big disadvantage of the VPA, or at least of the implementation used in this paper, was the execution time. This was certainly due to the numerous singular-value decompositions (SVDs) that the algorithm calculates in its runtime.

The results are promising, and the research will continue by evaluating the algorithms in increasingly realistic scenarios, including adding noise to the synthetic data.

Author Contributions: Conceptualization, R.S.; methodology, R.D. and G.B.; software, R.D.; validation, R.D. and G.B.; formal analysis, R.S., R.D. and G.B.; investigation, R.D. and G.B.; resources, R.S.; data curation, R.D.; writing—original draft preparation, R.D.; writing—review and editing, G.B. and R.D.; visualization, R.D. and G.B.; supervision, R.S. and G.B.; project administration, R.S.; funding acquisition, R.S. All authors have read and agreed to the published version of the manuscript.

Funding: This research was funded by MIUR, Project Grant PRIN 2017 "Microwave Biosensors: Enhanced Non-Invasive Methodology for Blood Glucose Monitoring". The APC was funded by the above Project Grant PRIN 2017.

Data Availability Statement: The data presented in this study are available upon request from the corresponding author.

Conflicts of Interest: The authors declare no conflict of interest.

References

1. *IDF Diabetes Atlas*, 9th ed.; Technical Report; International Diabetes Federation: Brussels, Belgium, 2019.
2. Lin, T. Non-Invasive Glucose Monitoring: A Review of Challenges and Recent Advances. *Curr. Trends Biomed. Eng. Biosci.* **2017**, *6*, 1–8. [CrossRef]
3. Costanzo, S.; Cioffi, V. Dielectric Models for the Accurate Design of Wearable Diabetes Sensors. In Proceedings of the 2019 23rd International Conference on Applied Electromagnetics and Communications (ICECOM), Dubrovnik, Croatia, 30 September–2 October 2019; pp. 1–3. [CrossRef]
4. Levenberg, K. A Method for the Solution of Certain Non-Linear Problems in Least Squares. *Q. Appl. Math.* **1944**, *2*, 164–168. [CrossRef]
5. Marquardt, D.W. An Algorithm for Least-Squares Estimation of Nonlinear Parameters. *J. Soc. Ind. Appl. Math.* **1963**, *11*, 431–441. [CrossRef]
6. Golub, G.H.; Pereyra, V. The Differentiation of Pseudo-Inverses and Nonlinear Least Squares Problems Whose Variables Separate. *SIAM J. Numer. Anal.* **1973**, *10*, 413–432. [CrossRef]
7. Karacolak, T.; Moreland, E.C.; Topsakal, E. Cole–Cole Model for Glucose-Dependent Dielectric Properties of Blood Plasma for Continuous Glucose Monitoring. *Microw. Opt. Technol. Lett.* **2013**, *55*, 1160–1164. [CrossRef]
8. Cole, K.S.; Cole, R.H. Dispersion and Absorption in Dielectrics I. Alternating Current Characteristics. *J. Chem. Phys.* **1941**, *9*, 341–351. [CrossRef]
9. Debye, P. *Polar Molecules*; The Chemical Catalog Company, Inc.: New York, NY, USA, 1929.
10. Gabriel, S.; Lau, R.W.; Gabriel, C. The Dielectric Properties of Biological Tissues: III. Parametric Models for the Dielectric Spectrum of Tissues. *Phys. Med. Biol.* **1996**, *41*, 2271–2293. [CrossRef] [PubMed]
11. Gavin, H.P. *The Levenberg–Marquardt Algorithm for Nonlinear Least Squares Curve-Fitting Problems*; Duke University: Durham, NC, USA, 2019; p. 19.
12. Golub, G.; Pereyra, V. Separable Nonlinear Least Squares: The Variable Projection Method and Its Applications. *Inverse Probl.* **2003**, *19*, R1–R26. [CrossRef]
13. O'Leary, D.P.; Rust, B.W. Variable Projection for Nonlinear Least Squares Problems. *Comput. Optim. Appl.* **2013**, *54*, 579–593. [CrossRef]

engineering proceedings

MDPI

Proceeding Paper

Drone Polariscopy—Towards Remote Sensing Applications †

Soon Hock Ng [1,*], Blake Allan [2], Daniel Ierodiaconou [2], Vijayakumar Anand [1], Alexander Babanin [3] and Saulius Juodkazis [1,4,*]

1 Optical Sciences Centre and ARC Training Centre in Surface Engineering for Advanced Materials (SEAM), School of Science, Computing and Engineering Technologies, Swinburne University of Technology, Hawthorn, VIC 3122, Australia; vanand@swin.edu.au
2 Faculty of Science Engineering & Built Environment, School of Life and Environmental Sciences, Deakin University, Princes Highway, Warrnambool, VIC 3280, Australia; b.allan@deakin.edu.au (B.A.); daniel.ierodiaconou@deakin.edu.au (D.I.)
3 Department of Infrastructure Engineering, University of Melbourne, Parkville, VIC 3010, Australia; a.babanin@unimelb.edu.au
4 World Research Hub Initiative (WRHI), School of Materials and Chemical Technology, Tokyo Institute of Technology, 2-12-1, Ookayama, Meguro-ku, Tokyo 152-8550, Japan
* Correspondence: soonhockng@swin.edu.au (S.H.N); saulius.juodkazis@gmail.com (S.J.)
† Presented at the 2nd International Electronic Conference on Applied Sciences, 15–31 October 2021; Available online: https://asec2021.sciforum.

Abstract: Remote sensing is critical for a wide range of applications, including ocean and wave monitoring, planetary exploration, agriculture, and astronomy. We demonstrate a polariscopy concept that is able to determine orientation of patterns below the optical resolution limit of a system. This technique relies on measuring at least four different polarisation angles and calculating the orientation from this set of intensity information. It was initially demonstrated on the Infrared Microspectroscopy Beamline at the Australian Synchrotron using IR light in transmission. Using a monochrome polarising camera mounted onto a drone as a remote sensing platform analogue, orientation information was extracted from 3D-printed targets in reflection. The images were taken at an altitude where conventional imaging could not resolve the test patterns. The system had a 3.33 mm ground resolution. Patterns consisting of 0.5 mm lines spaced 0.5 mm apart were detected using the method, demonstrating the capability of detecting features over six times smaller than the resolution limit. In the interest of moving towards high-speed data acquisition and processing, two methods for processing the image are compared—an analytical and a curve fitting method.

Keywords: polariscopy; remote sensing; drone; image processing

Citation: Ng, S.H.; Allan, B.; Ierodiaconou, D.; Anand, V.; Babanin, A.; Juodkazis, S. Drone Polariscopy—Towards Remote Sensing Applications. *Eng. Proc.* **2021**, *11*, 46. https://doi.org/10.3390/ASEC2021-11161

Academic Editors: Nunzio Cennamo and Nicholas Vassiliou Sarlis

Published: 15 October 2021

1. Introduction

The ability to detect periodic features below the resolution limit of a system is valuable to all forms of imaging. This is especially important in areas such as remote sensing, where the cost of launching a satellite is dependent on its mass. Polarisation information adds an additional dimension to image data and has shown value in machine vision, where it is utilised for analysis of reflections [1]. In remote sensing, the orientation detection of waves is studied using synthetic aperture radar (SAR). The polarisation orientation has been used on an existing resolved image [2,3]. These satellites need large, heavy antennas to achieve their resolutions. However, in such a contexts, polarisation has not been used for diffraction-limited imaging of anisotropy. A method using Fourier transform infrared spectroscopy was developed and transferred to the Infrared Microspectroscopy Beamline at the Australian Synchrotron, which demonstrated anisotropy recognition and mapping below the spatial resolution limit for the first time [4,5]. The method described—the 4-pol method—involves measuring the sample at four polarisation angles (0°, 45°, 90°, and 135° (−45°)). In addition to molecular anisotropy, the orientation of a circular grating with a 200 nm pitch and 100 nm line width was detectable with a system resolution of 5 μm, a 25 × difference [6]. To demonstrate the ability of the 4-pol method to move beyond the IR wavelengths and microscopy scale, a 4-pol visible light camera was used to show orientation detection in the visible wavelength range [7]. In this preliminary study, we extended the method even further and mounted the camera on a drone, which functioned as a remote sensing platform analogue and operated in reflection mode.

2. Materials and Methods

2.1. Drone Flight with Polarisation Camera

The CS505MUP monochrome polarisation camera (Thorlabs Inc.) has a 2448 × 2048 pixel sensor with micro-polarisers directly over each 3.5 μm pixel. Each 2 × 2 group of pixels provides 0°, 45°, 90°, and 135° intensity information for a total of 1224 × 1024 for each polarisation angle. A Navitar MVL8M23 f/1.4, 8 mm lens was attached to the camera, which was then mounted on a custom mounting rig which provided vibration damping, a battery, a LattePanda computer with a receiver, and a GoPro video camera (Figure 1a). This rig was then bolted under a Kraken 130V2 octocopter drone platform (Figure 1b). After takeoff, the acquisition was remotely triggered to conserve power and storage space. To account for different lighting conditions, the camera was programmed to record a bursts of 10 images at varying exposure times, every 30 s. The drone was flown up in steps of 20 m and hovered for 1 min to ensure a full set of images were acquired at each altitude. 3D-printed targets (Figure 2) consisting of 0.5 mm lines with a 1 mm pitch (one circular with an Archimedean spiral and one rectangular with different orientations in each quadrant) were set up on the ground as points of known orientation. The diameter of the circle was 20 cm and the dimensions of the rectangle were 24 × 20 cm.

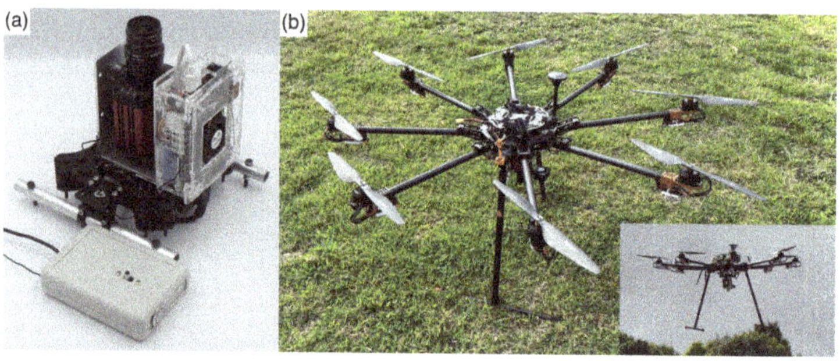

Figure 1. (**a**) Photograph of the polarisation camera in its mounting rig along with a remote control. (**b**) Preflight photo of the octocopter with camera module attached underneath. Inset shows the drone and camera just after takeoff.

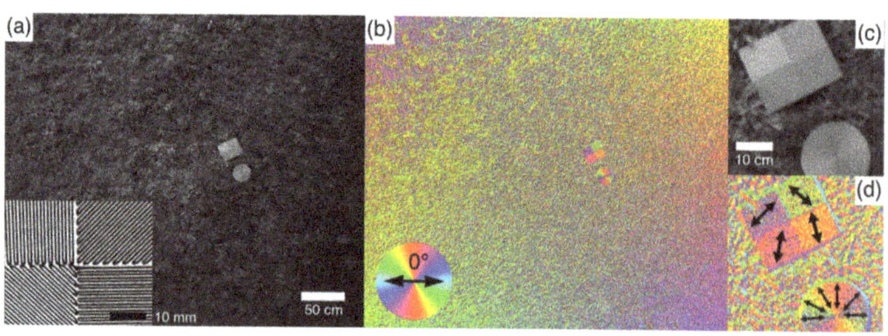

Figure 2. (**a**) Intensity image of 3D-printed targets on a grass field at an altitude of ∼4 m. Inset shows a photo of the centre of the rectangular target showing the orientation in each quadrant. (**b**) Corresponding azimuth map with the colours indicating azimuth angle. (**c**) Enlarged intensity image of the targets. (**d**) Enlarged azimuth image of the targets with black arrows indicating approximate calculated azimuth orientation.

2.2. Data Processing

The data were subsequently converted to intensity and orientation information using two methods—analytical and curve fitting [4,8]. For the analytical solution, the intensity is given by

$$I = \frac{I_{0^\circ} + I_{45^\circ} + I_{90^\circ} + I_{135^\circ}}{2}, \quad (1)$$

where I is the image intensity and I_{θ° is the intensity at a specific polarisation angle. The orientation (azimuth) is given by

$$\psi = \frac{1}{2}\text{arctan2}\left(\frac{I_{45^\circ} - I_{135^\circ}}{I_{0^\circ} - I_{90^\circ}}\right), \quad (2)$$

where arctan2 is the four-quadrant inverse tangent. The equation used for fitting was

$$I(\theta) = A\cos(2\theta - 2\psi) + c, \quad (3)$$

where $I(\theta)$ is the intensity at a specific polariser angle, A is a factor proportional to the degree of polarisation, θ is the polariser angle, ψ is the azimuth, and c is a factor proportional to I. The bounds of the fit were $0 < A < 5$, $-\frac{\pi}{2} < \psi < \frac{\pi}{2}$, and $-5 < c < 5$. The initial guesses were $A = 1$, $\psi = 0$, and $c = 1$. In all cases, these were the values used for all pixels.

3. Results and Discussion

3.1. Orientation Determination

Figure 2a shows an intensity image, calculated with Equation (1), at an altitude of approximately 4 m, and the inset shows a photo of 3D-printed lines on the rectangular target. This height is equivalent to 3 pixels/cm in the image or 6 pixels/cm over the full sensor (calculated using the 20 cm diameter circle as a reference). The highest theoretical resolution of a sensor is when two points are imaged onto separate neighbouring pixels. Naively for the above sensor with a pixel size of 3.5 µm, the sensor resolution R_{sensor} can be considered 3.5 µm, although physical limitations preclude this. The polarisation image (equivalent pixel size of 6.9 µm) has half the resolution of the full sensor. The magnification factor

$$m = \frac{\text{Sensor width}}{\text{Image width}}$$

between the object space and sensor can be used to calculate the maximum object resolution R_{object}. For the image in Figure 2a,

$$m = \frac{8.445 \text{ mm}}{4080 \text{ mm}} = 0.00207.$$

Hence,

$$R_{\text{object}} = \frac{3.45 \text{ µm}}{0.00207} = 1.67 \text{ mm}$$

for the full sensor and 3.33 mm for the polarisation intensity image.

Since the spaces between the lines were 0.5 mm, the lines could not be resolved in the image. Applying Equation (2) resulted in Figure 2b, which shows colours representing the azimuth angle with 0° defined as shown in the bottom left. It is evident that each quadrant of the rectangular target shows a different orientation, and the circular target shows radial changes. Figure 2c,d shows enlarged versions of the region of interest. Arrows in Figure 2d show approximate azimuth orientation. The azimuth is orthogonal to the orientation of the printed lines for both rectangular and circular targets. This is in contrast to transmission mode, where the calculated azimuth is parallel to the alignment [6]. However, the fact that the azimuth is consistent still indicates that the 4-pol method detects orientation that cannot be resolved by the optics in reflection. This system, with a theoretical 3.33 mm resolution, was able to determine the orientation of 0.5 mm lines spaced 0.5 mm apart, demonstrating detection of patterns over six times smaller than the resolution limit. It also shows that it works in cases where either the detected features are much smaller or much larger than the wavelength of the probing light. The difference in azimuth alignment might have been due to the illumination from the sky, since polarisation varies with angle from the sun [9].

3.2. Data Processing Method

There are cases wherein there is a low signal-to-noise ratio, the angle of a polariser cannot be set precisely, or the polarisation combinations are not 45° apart. The analytical method cannot accommodate these, and so curve fitting can be used to determine the azimuth and intensity. This comes at a time cost, however. While Equations (1) and (2) can be vectorised so the calculation can be performed on an entire image at once, curve fitting needs to be done for each pixel. As a single-thread process, the analytical method takes <1 s to process 1 frame. The same frame using curve fitting takes over 3 h.

While the fitting method is slow, it ultimately returns the very similar results to the analytical method, when the bounds are set correctly. Figure 3a shows an enlarged cropped section from Figure 2b, and Figure 3b shows the same area but calculated by fitting each pixel. At a glance, they are indistinguishable; however, Figure 3c shows that there are several pixels which are different. These differences are not very significant, as the number of pixels that do not agree are few. The time disparity demonstrates that the fitting method should only be used when a great number of polarisation measurements are needed—when the signal-to-noise ratio is low.

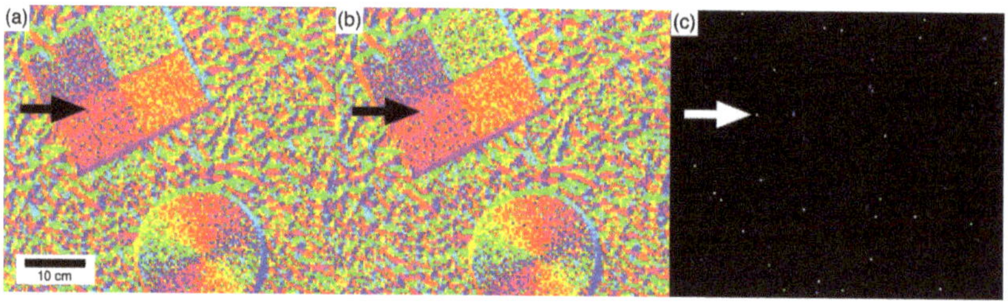

Figure 3. (**a**) Azimuth image calculated using Equation (2). (**b**) Azimuth image calculated by fitting Equation (3). (**c**) Images (**a**,**b**) overlaid with the difference image blend mode showing pixels that are different in colour and pixels which are the same as black. Arrows are eye guides to one pixel which is different.

4. Conclusions

The 4-pol method was successfully transferred from the Infrared Microspectroscopy Beamline to a drone platform. It was shown that orientation could be determined even when the aligned features (0.5 mm lines 0.5 mm apart) could not be spatially resolved (3.33 mm resolution). This demonstrated that patterns over six times smaller than the system's resolution limit could be extracted from the image. The calculated azimuth was orthogonal to the alignment direction, and the reason for this requires further study. This is a stepping stone towards satellite-based remote sensing using the same technique, and opens up possibilities for ocean monitoring, planetary observation, and astronomy, where orientation information is important.

Author Contributions: Conceptualisation, S.J. and S.H.N.; methodology, S.H.N.; software, S.H.N., V.A.; investigation, S.H.N., B.A., D.I., V.A., A.B. and S.J.; resources, B.A., D.I. and A.B.; data curation, S.H.N.; writing—original draft preparation, S.H.N. and S.J.; writing—review and editing, S.H.N., S.J., B.A., D.I. and V.A.; visualization, S.H.N.; supervision, D.I., A.B. and S.J. All authors have read and agreed to the published version of the manuscript.

Funding: This research was funded by the Australian Research Council, grant number LP190100505.

Data Availability Statement: The data presented in this study are available on request from the corresponding author.

Conflicts of Interest: The authors declare no conflict of interest.

References

1. Atkinson, G.A.; Ernst, J.D. High-sensitivity analysis of polarization by surface reflection. *Mach. Vis. Appl.* **2018**, *29*, 1171–1189. [CrossRef]
2. Schuler, D.L.; Lee, J.S.; Kasilingam, D.; Pottier, E. Measurement of ocean surface slopes and wave spectra using polarimetric SAR image data. *Remote. Sens. Environ.* **2004**, *91*, 198–211. [CrossRef]
3. Viana, R.D.; Lorenzzetti, J.A.; Carvalho, J.T.; Nunziata, F. Estimating energy dissipation rate from breaking waves using polarimetric SAR images. *Sensors* **2020**, *20*, 6540. [CrossRef] [PubMed]
4. Hikima, Y.; Morikawa, J.; Hashimoto, T. FT-IR Image Processing Algorithms for In-Plane Orientation Function and Azimuth Angle of Uniaxially Drawn Polyethylene Composite Film. *Macromolecules* **2011**, *44*, 3950–3957. [CrossRef]
5. Ryu, M.; Balčytis, A.; Wang, X.; Vongsvivut, J.; Hikima, Y.; Li, J.; Tobin, M.J.; Juodkazis, S.; Morikawa, J. Orientational Mapping Augmented Sub-Wavelength Hyper-Spectral Imaging of Silk. *Sci. Rep.* **2017**, *7*, 1–10. [CrossRef] [PubMed]
6. Honda, R.; Ryu, M.; Moritake, M.; Balčytis, A.; Mizeikis, V.; Vongsvivut, J.; Tobin, M.J.; Appadoo, D.; Li, J.L.; Ng, S.H.; et al. Infrared Polariscopy Imaging of Linear Polymeric Patterns with a Focal Plane Array. *Nanomaterials* **2019**, *9*, 732. [CrossRef] [PubMed]
7. Ng, S.H.; Anand, V.; Duffy, A.; Babanin, A.; Ryu, M.; Morikawa, J.; Juodkazis, S. Remote-sensing concept using polariscopy for orientation determination below the spatial resolution limit. In *Photonic Instrumentation Engineering VIII*; Soskind, Y., Busse, L.E., Eds.; SPIE: Bellingham, WA, USA, 2021; Volume 1169306, p. 4. [CrossRef]
8. Ryu, M.; Nishijima, Y.; Morimoto, S.; To, N.; Hashizume, T.; Matsubara, R.; Kubono, A.; Hu, J.; Ng, S.H.; Juodkazis, S.; et al. Hyperspectral Molecular Orientation Mapping in Metamaterials. *Appl. Sci.* **2021**, *11*, 1544. [CrossRef]
9. Coulson, K.L. *Polarization and Intensity of Light in the Atmosphere*; A. Deepak: Hampton, VA, USA, 1988.

engineering proceedings

MDPI

Proceeding Paper

Manufacturing and Testing of 3D-Printed Polymer Isogrid Lattice Cylindrical Shell Structures †

César M. A. Vasques *, Fernando C. Gonçalves and Adélio M. S. Cavadas

proMetheus, Escola Superior de Tecnologia e Gestão, Instituto Politécnico de Viana do Castelo, Rua Escola Industrial e Comercial de Nun'Álvares, 4900-347 Viana do Castelo, Portugal; fegs@ipvc.pt (F.C.G.); adelioc@estg.ipvc.pt (A.M.S.C.)

* Correspondence: cmavasques@gmail.com

† Presented at the 2nd International Electronic Conference on Applied Sciences, 15–31 October 2021; Available online: https://asec2021.sciforum.net.

Abstract: This article focuses on the use of fused deposition modeling (FDM) technology to manufacture and test polymer isogrid lattice cylindrical shell (LCS) structures with equilateral triangular unit-cells using non-professional and conventional 3D printing software and hardware. A parametric and automated 3D model for these structures is created in SolidWorks using the Visual Basic (VBA) programming language. Different configurations of the isogrid LCS structure are modeled, manufactured, and tested in order to determine the compressive structural strength and stiffness, as well as to investigate structural instability. The experimental results are used to deduce the inherent limitations of 3D printing, including the inhomogeneities, imperfections, and non-isotropic nature of FDM, as well as the effect of the configurations on local buckling behavior. The results suggest that coupling between local and global buckling has an impact on the compressive stiffness and strength of LCS structures, reducing the accuracy of structural designs neglecting these effects.

Keywords: isogrid; cylindrical shell; polymer structures; additive manufacturing; fused deposition modeling; 3D printing; local buckling

Citation: Vasques, C.M.A.; Gonçalves, F.C.; Cavadas, A.M.S. Manufacturing and Testing of 3D-Printed Polymer Isogrid Lattice Cylindrical Shell Structures. *Eng. Proc.* **2021**, *11*, 47. https://doi.org/10.3390/ASEC2021-11174

Academic Editor: Filippo Berto

Published: 15 October 2021

1. Introduction

Lattice cylindrical shell (LCS) structures have been applied in a wide variety of fields, from aerospace to medicine, and are now recognized as a viable design alternative for critical geodesic structural applications requiring lightweight, low-cost structures with high mechanical performance. Lattice structures are typically used as reinforcements for homogeneous shells, such as aerial aircraft fuselage sections, or as standalone structures that perform critical functions, such as stent devices [1–4]. In general, these structures combine axial, circumferential, and helical ribs to increase the strength-to-mass ratio of cylinder shells when subjected to axial, bending, and torsion stresses [5–7].

Different approaches to designing, developing, and manufacturing these lightweight structures have been pursued in different countries since McDonnell–Douglas and NASA first idealized and consolidated the isogrid concept for metallic aerospace structures in the 1970s in the United States [1,8]; these lightweight structures were indeed manufactured by first machining in the flat and then forming. Their design and manufacturing were extended in the 1980s by CRISM in the former Soviet Union to the so-called *anisogrid* (in relation with the anisotropy of composite materials) design concept, which, instead, considered composite carbon-fiber materials and LCS structures formed via filament winding. Due to the manufacturing constraints and disadvantages of metallic isogrid LCS structures, several studies on more general composite anisogrid geodesic designs have been conducted, covering design [9–11], manufacturing [12], mechanical properties [13], buckling [5], failure [14], and applications [2,15–18], to name only a few. Structural models of the lattice may be simplified by "smearing out", averaging, or taking mean values of the lattice properties and considering them as a solid continuous sheet of isotropic material with elastic properties [8,19–22]. Alternatively, more precise but more complex approaches may be used in conjunction with finite element technology, when using numerical computational analysis tools [5,6,21,23]. In the case of metallic lattices, the term *isogrid* has been used because the structure has been originally assumed to behave like an isotropic material [8], but recent studies appear to indicate otherwise [24].

Numerous LCS topologies can be created by varying the number of ribs and their arrangement, including the typical hexagonal, triangular, and mixed grid lattice shells [6,22]. Due to their complexity in terms of geometry definition, model generation, and physical realization, these structures pose significant design and manufacturing challenges, for which the increased creativity and design freedom enabled by additive manufacturing (AM) and 3D printing technologies are extremely beneficial, with potential to rapidly transform the engineering of lattice structures as we classically know it. Nowadays, AM is a widely used manufacturing process in which critical build-up parts are manufactured by layering or fusing materials in accordance with precise computerized 3D solid digital models. As such, new disruptive design and manufacturing paradigms are emerging, characterized by increased design flexibility and optimization, manufacturing simplicity, product customization, and degree of automation, where numerous remarkable technologies are converging at the moment: intelligent software, novel materials, novel manufacturing processes (particularly 3D printing and AM), robotics and automation, and a slew of web-based services. Additionally, freely available and widely disseminated computer-aided design (CAD) and manufacturing (CAM) software, as well as the well established computer numerical control (CNC) standards, combined with readily available and affordable commercial 3D printers and materials, have fostered a boom in the number of users and potential applications for these technologies. In the context of lattice structures, recent attention has been focused on both metal [25–30] and polymer [31–35] 3D printed parts for a variety of different applications and purposes (cf. also the review works in [36,37] and the references therein), most commonly limited to multipurpose planar designs or volumetric lightweight cellular structural infills.

This work is based on a preliminary investigation conducted as part of an undergraduate mechanical engineering student's final project motivated by Vasiliev and Totaro's advancements in composite LCS structures fabricated with the filament winding process [2,10,15,16,23]. While the latter studies are fascinating and instructive, they are limited to composite materials and filament winding processes on large-scale structures, rendering them inaccessible to the general public, where large-scale testing is also more challenging. With the advancement of 3D printing technologies and the availability of new materials and large-format printing underway, we are on the verge of witnessing a paradigm shift in lattice structure engineering that will enable greater design flexibility and physical realization, circumventing the primary limitations of filament winding composites.

As such, the overall goal of this work is to provide a first step and experimental contribution toward a better understanding of the feasibility of 3D-printed isogrid LCS structures at a small scale and to assess their local buckling mechanics through testing. In summary, this article starts by describing a 3D geometric model of isogrid LCS structures with a unit equilateral triangular cell that enables automatic generation of various configurations in SolidWorks. Next, polymer isogrid LCS structural test samples are fabricated using fused deposition modeling (FDM) and mechanically tested to determine the LCS structure's strength, stiffness, and mode of failure during buckling. While 3D printing by FDM introduces uncertainty regarding material properties and homogeneity by introducing some degree of anisotropy, it also simplifies analysis by utilizing smaller-scale, more affordable, and manageable polymer isogrid LCS structural samples. Furthermore, despite the focus being on the use of an accessible and affordable FDM technology and 3D printing software and hardware to manufacture cylindrical shell structures with equilateral triangular unit cells via 3D printing and on the study of local buckling, more refined and professional systems may be used to make the process more reliable, scalable, and applicable to large-formats and high-performance engineering materials. Lastly, the article ends with a summary of the most important findings and conclusions.

2. Geometry and 3D Modeling

An *isogrid* is an array of continuous equilateral triangles formed by a lattice of stiffening ribs; it is the simplest arrangement of bar elements with isotropic properties, hence the name. The intersecting ribs form a complete lattice (or grid) structure, regardless of whether they are attached to a single (or doubled) skin or used as an open lattice. The isogrid lattice's ability to control stresses via the ribs enables it to replace traditional solid structural elements with equivalent lattice shapes, thereby reducing weight and increasing mechanical strength. Despite their structural efficiency, these structures are still being investigated using novel materials and lattice morphologies. The morphology used in this study is the same as that used by Vasiliev and Totaro [16,23] for isogrid geodesic shells, which is defined by a cylindrical shell structure without skin (also typically ignored in design for load bearing), with ribs forming triangular unit cells patterned in the geodesic surface of the LCS structure. This morphology is comparable to that of an isotropic material shell structure and typically results in a combination of compression and tensile stresses in the ribs under general loading. As shown in Figure 1, seven parameters were considered to completely define the full LCS geometry, namely the LCS external diameter, D, and length, L; the constant width of the ribs on the tangent plane, a, and along the radial direction, e; the number of helical and circumferential ribs, n_h and n_c, respectively (no axial ribs are necessary for this unit cell orientation and morphology without vertical sides); and the height h of each stage in between two successive circumferential ribs. Of these, only five parameters are actually independent and required for its modeling; for example, we have considered the height h and length L as the dependent parameters and the remainder as the independent ones.

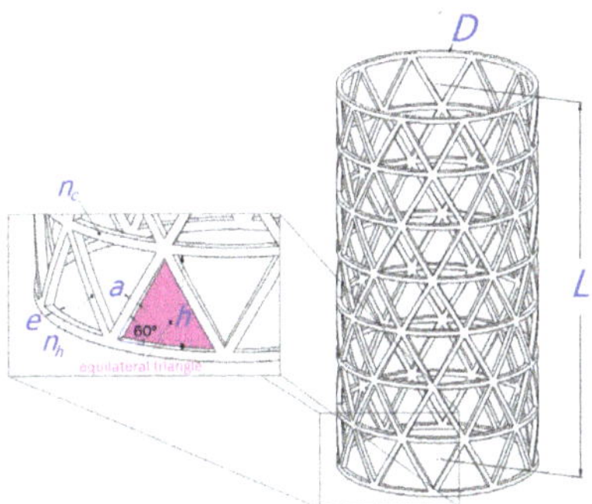

Figure 1. Geometric parameters required to fully define the 3D model of the isogrid LCS structure.

The geometry was chosen to favor local instability phenomena such as local buckling by having a low density of equilateral triangular unit cells and shells with a low total length-to-diameter ratio. Due to practical and manufacturing constraints, such as the printing volume and filament diameter of conventional polymer FDM 3D printing systems, small-scale samples can be printed with sufficient representativeness for a small number of unit cells only if the interrelated parameters of maximum diameter, number of circumferential ribs, and width are properly adjusted, thus precluding the analysis of global buckling. Considering previous works [15,23] and structural stability theory, the geometry was chosen to favor the appearance of local rib buckling along the tangential circumferential (as opposed to radial) axis of the cylinder (ensuring that $e > a$) and establishing a distance between circumferential ribs that ensures the application of Euler's column formula [38] for the determination of the critical rib buckling load, as well as an appropriate rib slenderness ratio that ensures the critical Euler stress is less than half the material's yield stress. As such, once the material properties and internal loads are known, Euler's buckling theory can be used to approximate the local buckling critical load by treating the ribs as beams. This approach, however, may be oversimplified, as complicating effects such as those caused by the inherent material anisotropy of FDM 3D printing, border effects at loaded extremity faces, border effects and stress distributions at rib junctions in the triangle vertices, and those caused by the LCS structure's global deformation behavior in compression complicate the analysis. These key aspects will be partially examined in better detail in the sections that follow and ongoing work.

The isogrid LCS structure's geometry is created parametrically and automatically in SolidWorks, allowing models to be created with minimal effort and in the shortest amount of time possible, thereby optimizing the design cycle. The required parameters input, operations, and graphical user interface (GUI) were programmed in Visual Basic (VBA) and made available as a button in the SolidWorks environment, as well as a GUI pop-up that appears when necessary as depicted in Figure 2.

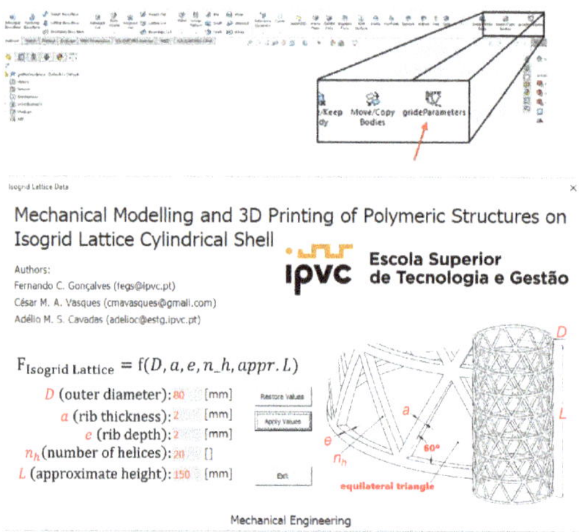

Figure 2. Graphical user interface used to introduce the parameters to automatically build the geometric 3D model of the isogrid LCS in SolidWorks and dedicated call-up button integrated into the SolidWorks environment workspace.

In a first stage, the geometry was designed in SolidWorks to ensure minimal operations and a clear understanding of how the geometry could be constructed. From there, complications arose when attempting to identify the internal variables that SolidWorks assigns to each operation and each parameter of each operation during the interface's VBA programming. This interface provides aids and easy visual identification for each parameter, as well as a friendly environment in which the user can enter the input data. Existing warnings and error messages assist in rectifying incorrectly entered values. To facilitate integration with SolidWorks, the VBA interface was inserted into the SolidWorks environment as illustrated. The VBA program's overall functional rationale is to automatically modify an existing geometry based on new desired parameters, where the user initially sees an example structure.

3. Manufacturing and 3D Printing

The FDM as a material extrusion AM process has gained popularity due to the limitless and simple design possibilities it provides in comparison to traditional manufacturing. To carry out this research and bring to life the various isogrid LCS structure configurations intended for testing, an Anet A6 3D printer was used. The main characteristics of the 3D printer are listed in the following Table 1. The PLA material from the filament manufacturer BeeVeryCreative was chosen because it is inexpensive, readily available, easy to print in great quality, and biodegradable; additionally, its mechanical behavior meets the requirements and purposes of this analysis (see Table 2).

Table 1. Main specifications of the Anet A6 3D printer used to print the isogrid LCS structure samples.

Layer thickness [mm]	0.1–0.3
Printing speed [mm/s]	10–120
XY axis position accuracy [mm]	0.012
Z axis position accuracy [mm]	0.004
Printing material	ABS, PLA, ...
Filament diameter [mm]	1.75
Nozzle diameter [mm]	0.4 (can be changed)
Build size [mm^3]	220 × 220 × 240
File format	STL, G-Code, OBJ

Table 2. Main specifications and properties of the BeeVeryCreative PLA 3D printer filament.

Diameter [mm]	1.75 ± 0.05
Printing temperature [°C]	205 ± 10
Specific weight [g/cm^3]	1.24
Tensile modulus [MPa]	3120
Yield strength [MPa]	70
Ultimate strength [MPa]	N/A
Strain at break [%]	20
Strain at yield [%]	5
Glass transition [°C]	57

Obviating the details of the initial process relative to the printer adjustments and setup, the printing (or slicing) direction was chosen with the cylinder in the upright position, so that the material anisotropy is kept simple to understand and all the successive layers are printed exactly the same way along the LCS structure with the axis of slicing aligned with the direction of the compressive loading. The Ultimaker Cura slicing software was initially used, but despite attempts to optimize printing performance without printing supports by adjusting the printing parameters, these efforts failed to resolve the existing issue of suspended material and unsatisfactory results caused by still-hot material not remaining in the proper filling by gravity. Another option was to print the LCS structure horizontally, but this resulted in even worse results. To address the issue, an attempt was made to model auxiliary geometry by introducing artificial bridging supports between the auxiliary and the real geometry, thereby improving material separation, but without success. Finally, we investigated the market for software slicers, and after experimenting with several programs that included support material, we obtained excellent results with Simplify3D, where material usage was minimized and the finished product was quite acceptable from a macroscopic standpoint. One example of a printing failure and a successful one with and without the support material are shown in Figure 3.

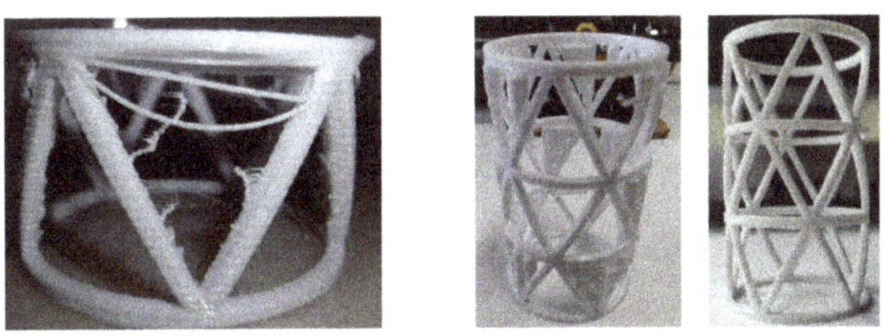

Figure 3. 3D printing failed attempt with suspended material (**left**); successful 3D printing solution using optimized supports generated in Simplify3D with (**middle**) and without (**right**) support material.

4. Compressive Testing and Results

As mentioned previously, one of the objectives of this research is to determine whether representative reduced-scale models of large isogrid LCS structures can accurately simulate their mechanical behavior under compressive loads and observe their local buckling behavior. As a result, it was decided to conduct compression tests using four distinct configurations, as illustrated in Figure 5, where the diverse morphologies were chosen to investigate the effect of the length, symmetry, proximity to the applied compressive load, and border effects on the local buckling behavior.

Three samples of each configuration were printed randomly, yielding a total of 12 samples in total; they were tested, also randomly, to avoid systematic errors inherent in the AM process and testing procedures. The final printer settings and configurations were as follows: nozzle diameter = 0.4; infill = 100%; layer height = 0.2 mm; nozzle temperature = 200 °C; build platform temperature off; printing speed = 45 mm/s; and cooling off. The total printing times for all the test samples was 12 h and 12 min; detailed information of this and the geometric description of the samples are listed in Table 3 by type of configuration. Due to time constraints, the printing head velocity was increased. As such, a slight vibration of the printer was observed in some of its displacement trajectories during printing, which may have exacerbated the samples' material and geometric inhomogeneity and reproducibility, possibly resulting in significant interference with the final results, as will be seen.

Table 3. Geometric description of the samples, post-processed characteristics, and printing time (simulation and real) for each configuration.

Configuration	a [mm]	e [mm]	D [mm]	n_h #	n_c #	L [mm]	Mass [g]	Rib Length [mm]	Rib Section [mm^2]	Printing Time Sim. [min]	Printing Time Real [min]
A	1.6	2.4	40	10	4	66.90	5.58	20.17	3.84	55	87
B					3	45.13	3.91			37	59
C					2	23.37	2.24			19	30
D					4 *	47.13	4.50			41	68

*: not generated automatically.

The compression tests were conducted using a compression test rig on a universal mechanical testing machine, model Shimadzu AG-X Plus 100 kN (Figure 4). The test speed was set to 0.25 mm/min, and data on both head displacement and applied force were collected at a 50 Hz rate. The tests were conducted in a random order to minimize systematic errors, and the results for the various tests are shown in Figure 5 for the 12 samples. As can be seen from the results, the four configurations support a range of critical compressive loads, suggesting that local buckling cannot be used to accurately predict strength and that a coupling effect exists between global compressive deformation and local buckling effects. As expected, given that all stages have equal stiffness, serial springs (stiffnesses) displacements sum up and the same force is applied to all the unit cell cylinder stages (i.e., with only one triangle along the longitudinal axis); as the stages between two circumferential ribs are piled up, we observe experimentally that the displacement increases almost proportionally, disregarding outliers, as we progress through configurations A, B, and C. The A configuration has the lowest strength, possibly because the local to global buckling coupling is more pronounced due to its length and higher global slenderness. The B configuration has the higher strength, most likely due to the longitudinal rib arrangement and reinforcing central circumference ring, whereas the D and C configurations have slightly lower strength than B. The results suggest that the local-to-global coupling has a greater effect on compressive strength than the local effect of reinforcement and different rib topologies at the load-applied border surfaces.

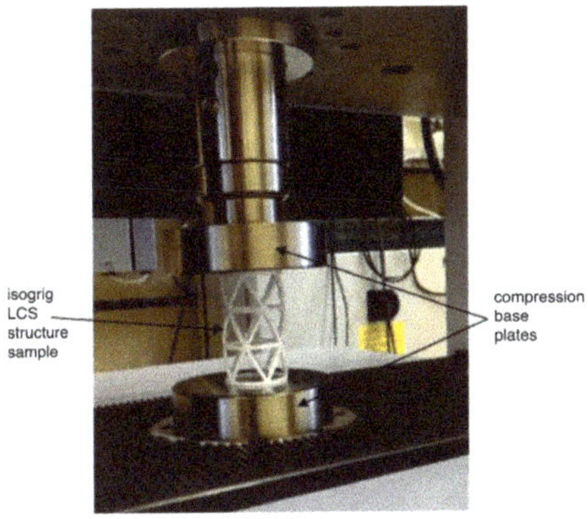

Figure 4. Compression testing apparatus.

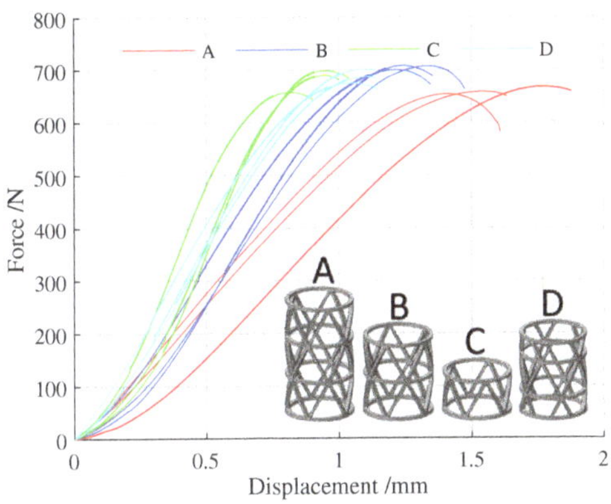

Figure 5. Compression test results for the 4 different configurations.

5. Conclusions

The purpose of this article is to demonstrate the feasibility of using fused deposition modeling (FDM) technology to fabricate polymer isogrid lattice cylindrical shell (LCS) structures with equilateral triangular unit cells using non-professional and conventional 3D printing software and hardware, and to infer experimentally about local buckling behavior. For these structures, a parametric and automated 3D model was created in SolidWorks using the Visual Basic (VBA) programming language. Due to the geometric complexity of isogrid LCS structures, they were realized through a progressive 3D printing trial-and-error improvement process that included progressive parameter adjustment and a better choice of slicing software to automatically generate 3D printing material supports. The printer's total run time has always been greater than the slicer program simulator's, typically by around 60%. As previously stated, the simulator does not account for extruder heating time, and the speed at which the lines of code execute on the printer varies, so speeds must be adjusted (printing assuming varied movements) to equalize these values and thus obtain a valid time simulation.

To improve control over the final mechanical properties and local buckling behavior repeatability, sensitivity analysis could have been conducted to examine printing parameters such as printing speed, layer height, extruder diameter, and printing temperature (statistical DoE approach). However, the trial and error method used, combined with the use of Simplify3D to generate trajectories and material supports, appears to be a successful strategy to produce isogrid LCS structures with satisfactory quality and compressive strength-to-mass performance. Concerning the compression testing of the various configurations, the print head speed was likely set too high, resulting in a significant discrepancy in results between the three samples of the same configuration. Although the maximum critical compression load and strength are less dispersed, the results indicate that additional research is necessary to determine the exogenous effects impacting local buckling behavior and printing quality. This preliminary study suggests that the order of dispersion of the obtained results precludes a detailed analysis of the sensitivity of various printing parameters using a DoE approach, at least until manufacturing and testing repeatability are improved, as the expected sensitivity of these parameters may be of the same order of magnitude as the obtained result's dispersion. Following this initial approach, more refined subsequent studies will focus on reducing result dispersion by ensuring reduced anisotropy and increased material homogeneity, conducting formal DoE sensitivity analyses to printing parameters, and validating the results using numerical and analytical structural modeling for better understanding of local buckling and global deformation. Nonetheless, this study demonstrates a previously unknown sensitivity of compressive strength to configuration, here denoted as local-to-global buckling coupling, implying that the analytic formulations for the critical local buckling load determination used by Vasiliev and Totaro should be used with caution, a behavior that will be further investigated in future studies, at least for structures with low-order number of helical ribs and cylindrical stages.

Author Contributions: Conceptualization, C.M.A.V.; methodology, C.M.A.V. and A.M.S.C.; software, F.C.G.; validation, C.M.A.V. and F.C.G.; formal analysis, C.M.A.V. and A.M.S.C.; investigation, C.M.A.V., F.C.G. and A.M.S.C.; resources, C.M.A.V., F.C.G. and A.M.S.C.; data curation, F.C.G.; writing—original draft preparation, C.M.A.V. and F.C.G.; writing—review and editing, C.M.A.V.; visualization, C.M.A.V. and A.M.S.C.; supervision, C.M.A.V. and A.M.S.C. All authors have read and agreed to the published version of the manuscript.

Funding: The authors gratefully acknowledge the support provided by the Foundation for Science and Technology (FCT) of Portugal, within the scope of the project of the Research Unit on Materials, Energy and Environment for Sustainability (proMetheus), Ref. UID/05975/2020, financed by national funds through the FCT/MCTES.

Data Availability Statement: Not applicable.

Conflicts of Interest: The authors declare no conflict of interest.

References

1. Huybrechts, S.; Hahn, S.E.; Meink, T. Grid stiffened structures: A survey of fabrication, analysis and design methods. In Proceedings of the 12th International Conference on Composite Materials (ICCM/I2), Paris, France, 5–9 July 1999; p. 10.
2. Vasiliev, V.; Barynin, V.; Razin, A. Anisogrid composite lattice structures—Development and aerospace applications. *Compos. Struct.* **2012**, *94*, 1117–1127. [CrossRef]
3. Abad, E.M.K.; Pasini, D.; Cecere, R. Shape optimization of stress concentration-free lattice for self-expandable Nitinol stent-grafts. *J. Biomech.* **2012**, *45*, 1028–1035. [CrossRef] [PubMed]
4. Beyi, A.F.M.; Ismail, A.E.; Taib, I.; Ibrahim, M.N. Stent classifications and effect of geometries on stent behaviour using finite element method. *Int. J. Mech. Eng. Robot. Res.* **2020**, *9*, 329–340. [CrossRef]
5. Morozov, E.; Lopatin, A.; Nesterov, V. Finite-element modelling and buckling analysis of anisogrid composite lattice cylindrical shells. *Compos. Struct.* **2011**, *93*, 308–323. [CrossRef]
6. Lai, C.; Wang, J.; Liu, C. Parameterized Finite Element Modeling and Buckling Analysis of Six Typical Composite Grid Cylindrical Shells. *Appl. Compos. Mater.* **2014**, *21*, 739–758. [CrossRef]
7. Belardi, V.; Fanelli, P.; Vivio, F. Structural analysis and optimization of anisogrid composite lattice cylindrical shells. *Compos. Part Eng.* **2018**, *139*, 203–215. [CrossRef]
8. Meyer, R.R.; Harwood, O.P.; Harmon, M.B.; Orlando, J.I. *Isogrid Design Handbook*; NASA Technical Report CR-124075; NASA Marshall Space Flight Center: Huntsville, AL, USA, 1973.
9. De Nicola, F.; Totaro, G.; Lenzi, F.; Ferrigno, A. Mechanical properties of composite anisogrid shell structures. In Proceedings of the 59th International Astronautical Congress, Glasgow, SCT, UK, 9 September–3 October 2008; Number IAC-08.C2.4.4.
10. Totaro, G.; Gürdal, Z. Optimal design of composite lattice shell structures for aerospace applications. *Aerosp. Sci. Technol.* **2009**, *13*, 157–164. [CrossRef]
11. Totaro, G. Multilevel Optimization of Anisogrid Lattice Structures for Aerospace Applications. Ph.D. Thesis, Delft Technical University, Delft, NL, USA, 2011.

12. Buragohain, M.; Velmurugan, R. Study of filament wound grid-stiffened composite cylindrical structures. *Compos. Struct.* **2011**, *93*, 1031–1038. [CrossRef]
13. Fan, H.; Fang, D.; Jin, F. Mechanical properties of lattice grid composites. *Acta Mech. Sin.* **2008**, *24*, 409–418. [CrossRef]
14. Zhang, Y.; Xue, Z.; Chen, L.; Fang, D. Deformation and failure mechanisms of lattice cylindrical shells under axial loading. *Int. J. Mech. Sci.* **2009**, *51*, 213–221. [CrossRef]
15. Vasiliev, V.; Barynin, V.; Rasin, A. Anisogrid lattice structures—Survey of development and application. *Compos. Struct.* **2001**, *54*, 361–370. [CrossRef]
16. Vasiliev, V.; Razin, A. Anisogrid composite lattice structures for spacecraft and aircraft applications. *Compos. Struct.* **2006**, *76*, 182–189. [CrossRef]
17. Del Olmo, E.; Grande, E.; Samartin, C.R.; Bezdenejnykh, M.; Torres, J.; Blanco, N.; Frovel, M.; Canas, J. Lattice structures for aerospace applications. In Proceedings of the 12th European Conference on Spacecraft Structures, Materials and Environmental Testing, Noordwijk, NL, USA , 20–23 March 2012; Ouwehand, L., Ed.; ESA SP-691; European Space Agency: Paris, France, 2012; p. 6.
18. Giusto, G.; Totaro, G.; Spena, P.; Nicola, F.D.; Caprio, F.D.; Zallo, A.; Grilli, A.; Mancini, V.; Kiryenko, S.; Das, S.; et al. Composite grid structure technology for space applications. *Mater. Today Proc.* **2021**, *34*, 332–340. [CrossRef]
19. Azarov, A.V. Theory of composite grid shells. *Mech. Solids* **2013**, *48*, 57–67. [CrossRef]
20. Vasiliev, V. *Advanced Mechanics of Composite Materials and Structures*; Elsevier: Amsterdam, NL, USA, 2018.
21. Beerhorst, M.; Hühne, C. Optimization of axially compressed cylindrical grid structures using analytical and numerical models. *Compos. Struct.* **2016**, *157*, 155–162. [CrossRef]
22. Abbasi, M.; Ghanbari, J. A comprehensive analytical model for global buckling analysis of general grid cylindrical structures with various cell geometries. *Int. J. Comput. Methods Eng. Sci. Mech.* **2021**, *22*, 477–499. [CrossRef]
23. Totaro, G. Local buckling modelling of isogrid and anisogrid lattice cylindrical shells with triangular cells. *Compos. Struct.* **2012**, *94*, 446–452. [CrossRef]
24. Delgado, X.M.; Merrett, C.G. Examination of the isotropy assumption in isogrid structures through analysis of nine isogrid variations. In *AIAA Scitech 2021 Forum*; American Institute of Aeronautics and Astronautics: Reston, VA, USA, 2021; p. 19. [CrossRef]
25. Challis, V.J.; Xu, X.; Zhang, L.C.; Roberts, A.P.; Grotowski, J.F.; Sercombe, T.B. High specific strength and stiffness structures produced using selective laser melting. *Mater. Des.* **2014**, *63*, 783–788. [CrossRef]
26. Maskery, I.; Aboulkhair, N.; Aremu, A.; Tuck, C.; Ashcroft, I. Compressive failure modes and energy absorption in additively manufactured double gyroid lattices. *Addit. Manuf.* **2017**, *16*, 24–29. [CrossRef]
27. Li, M.; Lai, C.; Zheng, Q.; Han, B.; Wu, H.; Fan, H. Design and mechanical properties of hierarchical isogrid structures validated by 3D printing technique. *Mater. Des.* **2019**, *168*, 107664. [CrossRef]
28. Ji, B.; Han, H.; Lin, R.; Li, H. Failure modes of lattice sandwich plate by additive-manufacturing and its imperfection sensitivity. *Acta Mech. Sin.* **2019**, *36*, 430–447. [CrossRef]
29. Guo, Y.; Yang, H.; Lin, G.; Jin, H.; Shen, X.; He, J.; Miao, J. Thermal performance of a 3D printed lattice-structure heat sink packaging phase change material. *Chin. J. Aeronaut.* **2021**, *34*, 373–385. [CrossRef]
30. Cao, X.; Ji, B.; Li, Y.; An, X.; Fan, H.; Ke, L. Multi-failure analyses of additively manufactured lattice truss sandwich cylinders. *Compos. Part Eng.* **2021**, *207*, 108561. [CrossRef]
31. Mancia, F.; Marchetti, M.; Regi, M.; Lionetti, S.; Marranzini, A.; Mazza, F.; Coluzzi, P.; Centro, C.; Materiali, S. Development of 3D advanced rapid prototyping multipurpose structures with micro and nano materials. In *Cost Effective Manufacture via Net-Shape Processing;* Meeting Proceedings RTO-MP-AVT-139; RTO: Neuilly-sur-Seine, France, 2006; Paper 20, pp. 20-1–20-24.
32. Ananth, S.; Whitney, T.; Toubia, E. Buckling stability of additively manufactured isogrid. In Proceedings of the 33rd Technical Conference of the American Society for Composites, Seattle, WA, USA, 24 September 2018; DEStech Publications, Inc.: Lancaster, PA, USA, 2018; Volume 5, pp. 3180–3193. [CrossRef]
33. MacDonald, E.; Espalin, D.; Doyle, D.; Muñoz, J.; Ambriz, S.; Coronel, J.; Williams, A.; Wicker, R. Fabricating patch antennas within complex dielectric structures through multi-process 3D printing. *J. Manuf. Process.* **2018**, *34*, 197–203. [CrossRef]
34. Forcellese, A.; Simoncini, M.; Vita, A.; Pompeo, V.D. 3D printing and testing of composite isogrid structures. *Int. J. Adv. Manuf. Technol.* **2020**, *109*, 1881–1893. [CrossRef]
35. Forcellese, A.; di Pompeo, V.; Simoncini, M.; Vita, A. Manufacturing of isogrid Compos. Struct. by 3D printing. *Procedia Manuf.* **2020**, *47*, 1096–1100. [CrossRef]
36. Velasco-Hogan, A.; Xu, J.; Meyers, M.A. Additive manufacturing as a method to design and optimize bioinspired structures. *Adv. Mater.* **2018**, *30*, 1800940. [CrossRef]
37. Reddy, A.H.; Davuluri, S.; Boyina, D. 3D printed lattice structures: A brief review. In Proceedings of the IEEE International Conference on Nanomaterials: Applications & Properties (NAP-2020), Sumy, Ukraine, 9–13 November 2020; IEEE: Piscataway, NJ, USA, 2020; p. 5. [CrossRef]
38. Budynas, R.G.; Nisbett, K.J. *Shigley's Mechanical Engineering Design*, 11th ed.; McGraw-Hill Education: New York, NY, USA, 2020; pp. 207–210.

engineering proceedings

MDPI

Proceeding Paper

Sensorized T-Shirt for Cardiological Patients in Telemonitoring †

Noemi D'Abbondanza [1,2], Martina Ferrazza [1,2], Leandro Lucangeli [2,3], Emanuele Piuzzi [4] and Antonio Pallotti [2,5,6,*]

1 Department of Basic and Applied Sciences for Engineering, Sapienza University of Rome, Via Eudossiana 18, 00184 Rome, Italy; noemidab@gmail.com (N.D.); ferrazza.martina@gmail.com (M.F.)
2 Technoscience—Consorzio Parco Scientifico e Tecnologico Pontino, Via di Val Cannuta 247, 00166 Rome, Italy; l.lucangeli@studenti.uniroma4.it
3 Department of Motor, Human and Health Sciences, "Foro Italico" University of Rome, Piazza Lauro de Bosis 15, 00135 Rome, Italy
4 Department of Information Engineering, Electronics and Telecommunications, Sapienza University of Rome, Via Eudossiana 18, 00184 Rome, Italy; emanuele.piuzzi@uniroma1.it
5 Department of Management and Law, University of Rome Tor Vergata, Via Columbia 2, 00133 Rome, Italy
6 Department of Human Sciences and Promotion of Life Quality, "San Raffaele" University of Rome, Via di Val Cannuta 247, 00166 Rome, Italy
* Correspondence: antonio.pallotti@uniroma2.it or antonio.pallotti@uniroma5.it

† Presented at the 2nd International Electronic Conference on Applied Sciences, 15–31 October 2021; Available online: https://asec2021.sciforum.net.

Abstract: Technological innovations in the development of wearable sensors have led to advancements in smart wearable devices targeted at health monitoring. In this work, a multisensor T-shirt for remote telemonitoring of vital signs related to cardiovascular diseases was developed. The prototype includes a single-lead electrocardiogram (ECG), a pulse oximeter, a temperature sensor, and a three-axis accelerometer. Data collected are sent by a Bluetooth module, then filtered and visualized thanks to a MATLAB script to provide information about heart rhythm, average temperature and oxygen saturation, and respiratory rate. The result is a simple, low-cost, and low-power system; an easily applicable solution within everyone's reach.

Keywords: wearable sensors; telemonitoring; electrocardiography; pulse oximetry; body temperature; respiratory monitoring

Citation: D'Abbondanza, N.; Ferrazza, M.; Lucangeli, L.; Piuzzi, E.; Pallotti, A. Sensorized T-Shirt for Cardiological Patients in Telemonitoring. *Eng. Proc.* **2021**, *11*, 48. https://doi.org/10.3390/ASEC2021-11130

Academic Editors: Nunzio Cennamo and Stefano Mariani

Published: 15 October 2021

1. Introduction

The term telemedicine, coined in the 1970s, means "healing at a distance" [1], and the World Health Organization (WHO) defines health telematics as "health-related activities, services and systems carried out over a distance by means of information and communications technologies" [2]. Smart wearable devices targeted at health monitoring can be used in telemonitoring systems to reduce healthcare system costs and to provide remote and personalized patient care that is accessible to all [3]. These devices are particularly useful for the follow-up of chronic diseases because they offer the possibility of continuous remote assistance [4–6]. There are several applications reported in literature, such as chronic obstructive pulmonary disease, diabetes, end-stage kidney disease, hypertension, or heart failure [6–8]. Evidence supports the use of wearable devices, particularly successful for exploring cardiac health and detecting arrhythmias [9–12]. Cardiovascular diseases (CVDs) are the major cause of death: in 2019, 32% of all global deaths were due to CVDs [13]. Detecting these diseases as early as possible leads to several benefits in prevention, diagnosis, and management.

In Ref. [14] the 12-lead electrocardiogram (ECG) is presented as the noninvasive gold standard for detecting several heart conditions. On the other hand, the use of the ECG machine does not offer the possibility of continuous and remote monitoring. When a longer recording during daily activities is needed, such as in arrhythmias detection, Holter monitoring or external cardiac event recorders can be used for noninvasive monitoring. The ambulatory ECG (Holter) can record multiple leads over 24–72 h, but it could have problems in terms of comfort because the patient must carry the device and keep sticky electrodes for all the recording duration [15]. The use of wearable devices can allow the overcoming of these limitations: in [16] the state of art of wearables in cardiovascular care is presented. Most of the common smart wearable devices on the market provide information about heart rate (HR) and heart rhythm through electrocardiography (ECG). In fact, common arrhythmias, such as atrial fibrillation (AF), can be detected using at least a single-lead ECG during their occurrences [17]. AF is associated to stroke and heart failure, so ECG monitoring through wearable devices can play an important role in early diagnosis [18].

Another interesting vital sign associated with adverse cardiac events is the Respiratory rate (RR): it can be used to identify heart failure or heart attack [19–21] because RR can increase even hours before the event, so frequent monitoring can be fundamental. Although respiratory-rate estimation is possible with ECG signal-processing algorithms [22], there are different ways to measure RR: Ref. [21] provides an overview of the available contact-based methods, while [20] presents a comprehensive overview of the design of respiratory-sensing system, describing the results of literature research.

Other vital signs can be provided to assess patient health condition: a review on the state of research and development in these wearable systems for health monitoring is presented in [21]. Concerning CVDs, it is important to monitor oxygenated hemoglobin in the blood through blood oxygen-saturation (SpO2)-monitoring systems because these diseases can provoke a reduction in oxygen levels in blood [23,24].

In Ref. [23], the state of the art of body-temperature-monitoring systems is also reported. It is difficult to establish a relation between body temperature and heart diseases: some researchers report a relationship with heart rate and respiration rate [25], others [26] consider temperature monitoring as a useful tool for detecting symptoms of medical stress related to stroke and heart attacks. However, there is an interesting correlation between temperature and strokes [27]: for example, in Ref. [28] it was found that monitoring—and thus preventing—high temperatures days after ischemic strokes can improve outcome.

The aim of this work is the development of a multisensor T-shirt for telemonitoring vital signs in cardiological patients. We developed a system including a single-lead ECG, a pulse oximeter, a temperature sensor, and a triaxial accelerometer. The more general goal concerns the development of an inexpensive and highly versatile wearable device, adaptable to the individual patient both from a firmware and hardware point of view. Its accessibility and low energy consumption will allow the collection of a large amount of data remotely, which can be directly accessible by the doctor for continuous and daily monitoring and used in the development of ML models for the automatic follow-up of the patient's health status.

2. Materials and Methods

In this section, we discuss the details of hardware, firmware, and software of the developed prototype T-shirt (Figure 1a) for telemonitoring of cardiological patients. The design developed presents an acquisition unit, a transmission unit, and an elaboration unit (Figure 1b).

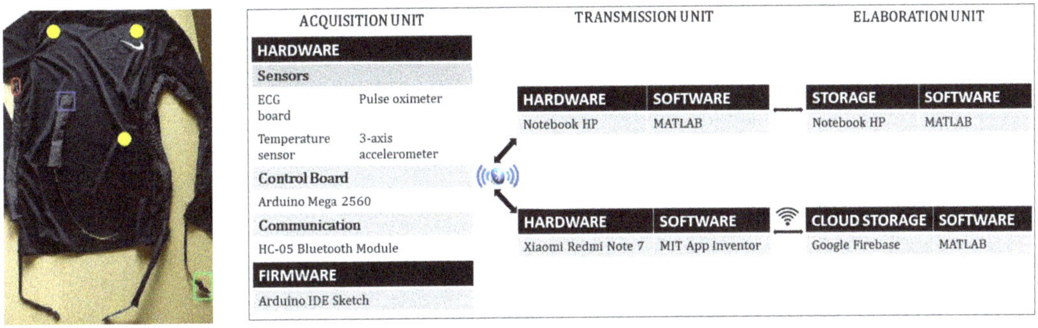

(a) (b)

Figure 1. (**a**) Sensorized T-shirt integrating temperature sensor (red), triaxial accelerometer (blue), pulse oximeter (green), and electrode placement (yellow). (**b**) Illustration of system design with both transmission units. In particular, the Android App has been created for telemonitoring applications.

2.1. Sensors

A single-lead ECG, a pulse oximeter, a temperature sensor, and a 3-axis accelerometer were used to monitor cardiac health.

The AD8232 SparkFun Single Lead Heart Rate Monitor [29–33] was chosen to measure the electrical activity of the heart. The AD8232 is an integrated signal-conditioning block for ECG that acts as an operational amplifier to extract, amplify, and filter biopotential signals in the presence of noisy conditions, such as those created by motion or remote electrode placement. The advantage of this sensor is that the electronic components needed for the ECG acquisition, such as amplifiers and filters, are contained in a 35.6 mm × 27.8 mm board, powered with a 3.3 V supply voltage. Data acquisition is made through three disposable foam solid hypoallergenic-gel electrodes of FIAB SpA. These electrodes are CE marked according to the directive 93/42/EEC and low cost (€0.20/electrode).

To monitor blood-oxygen saturation, the SparkFun Pulse Oximeter and Heart Rate Sensor, a 3.3 V sensor, were used. This board integrates two Maxim Integrated chips: the MAX32664 Biometric Sensor Hub and the MAX30101 Pulse Oximetry and Heart Rate Module [34]. The MAX30101 has an internal LED to light the finger and measures absorbed light through its photodetectors, while the MAX32664 processes the data and provides heart rate and blood-oxygen saturation (SpO2), reported as percentage of hemoglobin that is saturated with oxygen. An interesting feature of this sensor is the possibility of reporting measurements' confidence and information about correct finger position. The MAX30101 has an internal analog-to-digital converter (ADC) with 18-bit resolution; the device is easy to use and set, and it is very small (25.4 mm × 12.7 mm), so it is comfortable for the patient.

The DS18B20 digital thermometer [35,36] was chosen as temperature sensor because it provides up to 12-bit measurements and ±0.5 °C accuracy from −10 °C to +85 °C. Moreover, thanks to its size (5 mm × 4 mm × 6 mm.), it does not cause discomfort.

For an easy and fast evaluation of respiratory rate (RR), it was decided to use an accelerometer, like in more than 11% of works analyzed in Ref. [20]. The ADXL335, a 3-axis accelerometer [21,37] powered with 3.3 V, was chosen. It can measure acceleration with a minimum full-scale range of ±3 g, so it is used to measure the dynamic acceleration resulting from torso motion.

2.2. Control Board and Connectivity

The sensors were connected to an Arduino Mega 2560, a microcontroller board based on the ATmega2560. The choice was due to its 16 analog inputs, each of which provide 10 bits of resolution, useful for the future integration of other sensors. The Mega 2560 also supports two-wire interface (TWI) communication through the pins SDA (data line) and SCL (clock line), so it can communicate with I2C devices such as the pulse oximeter. The microcontroller board is powered by a rechargeable battery with a capacity of 10,000 mAh. A consumption test was conducted by measuring the current absorbed by the electronics. The resulting current was 100 mA and the estimated battery life in continuous acquisition was at least 48 h.

Data acquired from the sensors were transmitted through HC-05 Bluetooth Module, a Serial Port Protocol module designed for wireless communication. It supports different baud rates [38], which can be set in AT mode. The minimum baud rate sufficient to send all the data was chosen—that is, 115,200. This module was already used in other ECG applications [32,33].

2.3. Acquisition Unit

The prototype was tested using disposable electrodes: for this reason, they were not integrated in the T-shirt. Einthoven's lead II configuration was used for electrodes positioning, as also suggested in the Sparkfun Hookup Guide. Therefore, two electrodes were placed on the chest near the arms, while the right leg electrode was placed on the right lower abdomen (Figure 1a).

After electrode positioning, the patient can put on the sensorized T-shirt, in which pockets for the cables are sewn. The sensors can be connected after the fitting, so they cannot be damaged during the operation. Furthermore, to wash the T-shirt, every electronic component of the device can be easily removed and replaced.

The temperature sensor was fixed under the left armpit, while the pulse oximeter was in contact with the right index finger. For the accelerometer positioning, Ref. [39] reports that respiration frequency can be accurately measured from the placement "at the clavicular, pectoral and lateral sites on the chest as well the mid abdominal site". The accelerometer was fixed on the left costal margin in order to not interfere with ECG cables.

An Arduino sketch was written to acquire data from sensors at different frequencies. A different timer was activated for each sensor thanks to the use of the "timer.setInterval" function of the "SimpleTimer.h" library for Arduino Mega 2560. The temperature sensor and pulse oximeter were acquired every 1 s (1 Hz); the accelerometer every 200 ms (5 Hz); and the ECG every 4 ms (250 Hz). In this way, every second the firmware collected 270 samples: 1 start character "s", 1 temperature sample, 1 heart rate and 1 SpO2 samples from pulse oximeter, 15 samples from accelerometer (5 for each axis), 250 samples from ECG board, and 1 blank space as end character. Data were directly transmitted through HC-05 Bluetooth Module.

2.4. Transmission and Elaboration Unit

A serial connection between Arduino board and PC was created through the Bluetooth module. The transmission was managed by a MATLAB script, which allows for the selection of acquisition time and the saving of data in a .txt file at the end of the procedure.

As an alternative transmission mode, an Android application was developed, using MIT App Inventor, to have a transmission unit usable in remote monitoring. The app receives data from the Arduino board and saves them in completely anonymous form using an identification index into Google Firebase database. Using appropriate credentials (username and password), it is possible to access the database via MATLAB script and then process the data.

Depending on the modality of transmission, stored signals are processed with a MATLAB algorithm, allowing the filtering and the visualization of data and the extraction of parameters.

3. Results and Discussion

The system described in this work was tested on a participant and the obtained signals were compared with similar monitoring devices, including medical devices. The duration of the validation test was 30 s for ECG, temperature, and SpO2, while it continued for 60 s for breathing.

The signal obtained from the AD8232 SparkFun Single Lead Heart Rate Monitor was compared with the GIMA Cardio-B ECG, a medical device compliant with 93/42/CEE Directive for monitoring the heart signal. Both devices detected 33 cardiac cycles in 30 s, returning a heart rate of ~66 bpm. Furthermore, by comparing the time interval between the R peaks, it appears that the average of the errors was (6.125 ± 3.70) ms, while the maximum percentage error was 1.36%. The average of all the intervals between the peaks calculated with the GIMA ECG is (883.94 ± 41.17) ms, while for the SparkFun ECG it was (883.13 ± 40.12) ms. The amplitude of the two traces is of the same order: 0.4 mV for the GIMA ECG, and between 0.3 and 0.35 mV for the SparkFun one, which was slightly lower because the electrodes were positioned more externally.

For the validation of the SparkFun Pulse Oximeter sensor, the Oxy-2 Pulsoximeter by GIMA (medical device compliant with Directive 93/42/CEE) was used. The latter provides the measurement of saturation with an uncertainty of ±2%, and heart-rate measurement with an accuracy of ±2 bpm. The comparison of the sensors was carried out wearing the devices on the right- and left-hand indexes, respectively. Both devices showed the same SpO2 measurement for the entire acquisition. The heart rate measured by the SparkFun was within the range with an underestimation of the average rate by the sensor integrated in the T-shirt. The average frequencies calculated were in fact (64.39 ± 1.09) bpm for the GIMA and (62.81 ± 1.28) bpm for the SparkFun.

The DS18B20 temperature sensor was compared to a thermistor (with a resolution of 0.0001 °C) with both sensors positioned under the left armpit. The average temperatures obtained by the two sensors differed by 0.4 °C, with (36.32 ± 0.01) °C for the thermistor and (35.89 ± 0.03) °C for DS18B20.

The estimated respiratory rate obtained from the accelerations measured with the integrated ADXL335 was compared with the information provided by the UFI-1132 Pneumotrace II™, a piezoelectric breathing belt. Both devices recorded the same number of breaths (10 breaths in one minute), with similar power spectra and in particular with the same predominant frequency content. In fact, the respiratory rate was calculated as the frequency corresponding to the maximum value of the acceleration power spectrum. The prototype provided sufficient information to identify heart-rhythm disturbances, such as atrial fibrillation, by acquiring the single-lead ECG (Figure 2a). The acceleration perpendicular to the chest (Figure 2b) allowed for evaluation of the respiratory rate, that is, the number of breaths taken per minute. Finally, the temperature sensor and the pulse oximeter were integrated into the T-shirt to have additional parameters in the monitoring phase. After processing, the mean values of the acquired samples were calculated. In Figure 3, an extract of sensors output plotted with the mean values is reported.

Furthermore, the framework and the prototype developed in this work were used in the Ref. [40] study for the classification of healthy and arrhythmic subjects. The protocol provided for the acquisition of a population of 20 control participants, by means of a 2 min static test with the subject seated. The data obtained, together with signals from arrhythmic patients taken from public databases, were used for the development of classification models. The results obtained in Ref. [40] show the potential of the device developed and described in this work.

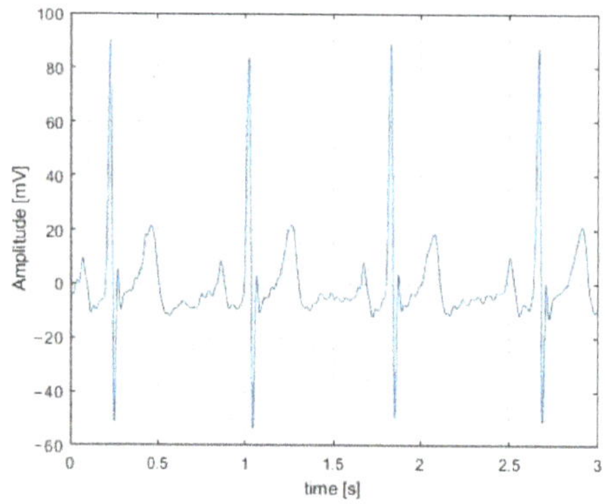

(a)

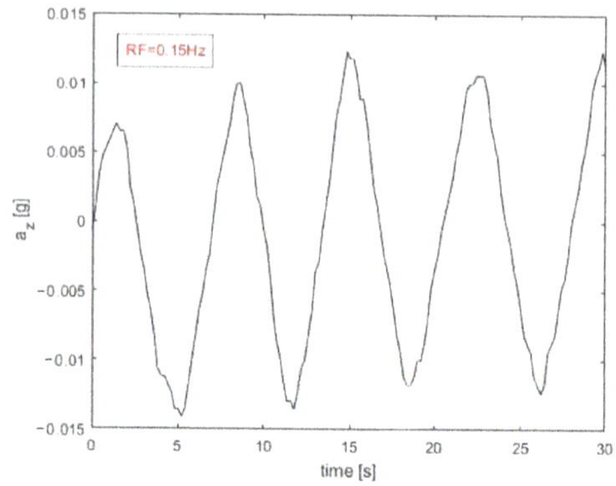

(b)

Figure 2. (**a**) Three cardiac cycles extracted from an ECG acquisition. (**b**) Accelerometer output along *z*-axis, perpendicular to the torso. In the first 5 s, the subject was not breathing.

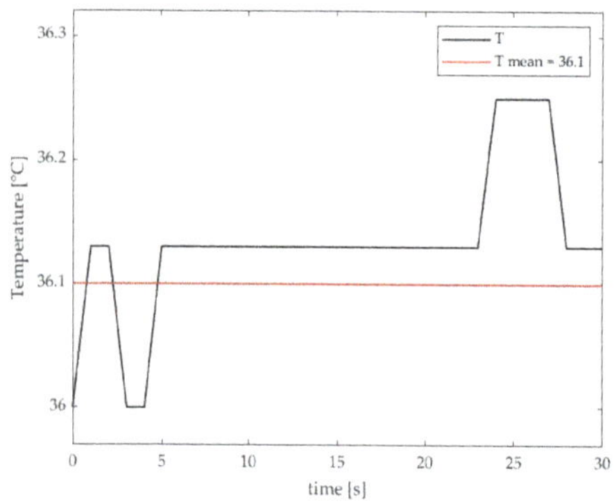

(a)

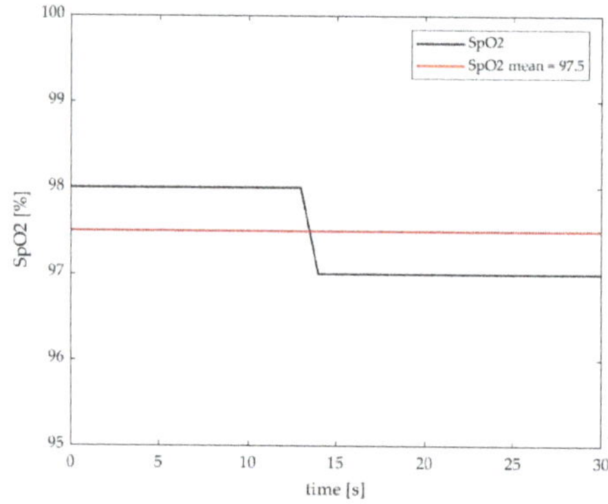

(b)

Figure 3. (**a**) Temperature sensor output and mean value; (**b**) Pulse oximeter output and SpO2 mean value.

4. Conclusions

In this work, a multisensor T-shirt was developed to measure vital signs important for heart monitoring: the prototype integrated a single-lead ECG, a pulse oximeter, a temperature sensor, and a triaxial accelerometer. The result was a simple, low-cost, and low-power telemonitoring system, able to collect enough data for machine-learning applications. The acquisition of a single-lead ECG signal is sufficient to detect heart rhythm diseases but it is also a limitation. Future developments will concern the measure of 12-lead ECG and the use of the prototype in dynamic tests, for the evaluation of health conditions during daily activities. The prototype described in this work has limitations in terms of wearability, but in the next versions particular attention will be paid to integrate noninvasive and flexible sensors and to reduce the size of the electronics used.

Author Contributions: Conceptualization, A.P. and E.P.; methodology, A.P.; software, N.D.; validation, N.D., E.P.; formal analysis, N.D. and M.F.; investigation, N.D., M.F. and L.L.; resources, N.D.; data curation, A.P.; writing—original draft preparation, N.D.; writing—review and editing, L.L.; visualization, N.D.; supervision, A.P. and E.P.; project administration, A.P.; funding acquisition, A.P. All authors have read and agreed to the published version of the manuscript.

Funding: This research received no external funding.

Institutional Review Board Statement: The protocol was approved by the Consorzio Parco Scientifico e Tecnologico Pontino Technoscience, with the code PST/205/P/140322.

Informed Consent Statement: Informed consent was obtained from the subject involved in the study.

Data Availability Statement: The datasets underpinning this work are available from the corresponding author upon request (agreement signing).

Acknowledgments: We thank Engineering Emanuele D'Angelantonio for his support and contribution in the research work.

Conflicts of Interest: The authors declare no conflict of interest.

References

1. Strehle, E.M.; Shabde, N. One hundred years of telemedicine: Does this new technology have a place in paediatrics? *Arch. Dis. Child.* **2006**, *91*, 956–959. [CrossRef] [PubMed]
2. World Health Organization. *A Health Telematics Policy in Support of WHO's Health-For-All Strategy for Global Health Development: Report of the WHO Group Consultation on Health Telematics, 11–16 December, Geneva, 1997*; World Health Organization: Geneva, Switzerland, 1998.
3. Deen, M.J. Information and communications technologies for elderly ubiquitous healthcare in a smart home. *Pers. Ubiquitous Comput.* **2015**, *19*, 573–599. [CrossRef]
4. Asua, J.; Orruño, E.; Reviriego, E.; Gagnon, M.P. Healthcare professional acceptance of telemonitoring for chronic care patients in primary care. *BMC Med. Inform. Decis. Mak.* **2012**, *12*, 1–10. [CrossRef] [PubMed]
5. Li, J.; Varnfield, M.; Jayasena, R.; Celler, B. Home telemonitoring for chronic disease management: Perceptions of users and factors influencing adoption. *Health Inform. J.* **2021**, *27*, 1460458221997893. [CrossRef]
6. Queirós, A.; Pereira, L.; Dias, A.; Rocha, N.P. Technologies for Ageing in Place to Support Home Monitoring of Patients with Chronic Diseases. In Proceedings of the 10th International Joint Conference on Biomedical Engineering Systems and Technologies, Porto, Portugal, 21–23 February 2017; pp. 66–76. [CrossRef]
7. Walker, R.C.; Tong, A.; Howard, K.; Palmer, S.C. Patient expectations and experiences of remote monitoring for chronic diseases: Systematic review and thematic synthesis of qualitative studies. *Int. J. Med. Inform.* **2019**, *124*, 78–85. [CrossRef] [PubMed]
8. Lucangeli, L.; D'Angelantonio, E.; Camomilla, V.; Pallotti, A. SISTINE: Sensorized Socks for Telemonitoring of Vascular Disease Patients. In Proceedings of the 2021 IEEE International Workshop on Metrology for Industry 4.0 & IoT (MetroInd4.0&IoT), Rome, Italy, 7–9 June 2021; pp. 192–197. [CrossRef]
9. Henriques, J.; Carvalho, P.; Paredes, S.; Rocha, T.; Habetha, J.; Antunes, M.; Morais, J. Prediction of Heart Failure Decompensation Events by Trend Analysis of Telemonitoring Data. *IEEE J. Biomed. Health Inform.* **2014**, *19*, 1757–1769. [CrossRef]
10. Dunn, J.; Runge, R.; Snyder, M. Wearables and the medical revolution. *Pers. Med.* **2018**, *15*, 429–448. [CrossRef]
11. Eletter, S.; Yasmin, T.; Elrefae, G.; Aliter, H.; Elrefae, A. Building an Intelligent Telemonitoring System for Heart Failure: The Use of the Internet of Things, Big Data, and Machine Learning. In Proceedings of the 2020 21st International Arab Conference on Information Technology (ACIT), Giza, Egypt, 28–30 November 2020; pp. 1–5. [CrossRef]
12. Kitsiou, S.; Paré, G.; Jaana, M. Effects of Home Telemonitoring Interventions on Patients with Chronic Heart Failure: An Overview of Systematic Reviews. *J. Med. Internet Res.* **2015**, *17*, e63. [CrossRef]
13. World Health Organization. Cardiovascular Diseases (CVDs). 2021. Available online: https://www.who.int/news-room/fact-sheets/detail/cardiovascular-diseases-(cvds) (accessed on 1 August 2021).
14. Khunti, K. Accurate interpretation of the 12-lead ECG electrode placement: A systematic review. *Health Educ. J.* **2013**, *73*, 610–623. [CrossRef]
15. Kwon, S.; Lee, S.-R.; Choi, E.-K.; Ahn, H.-J.; Song, H.-S.; Lee, Y.-S.; Oh, S. Validation of Adhesive Single-Lead ECG Device Compared with Holter Monitoring among Non-Atrial Fibrillation Patients. *Sensors* **2021**, *21*, 3122. [CrossRef]
16. Bayoumy, K.; Gaber, M.; Elshafeey, A.; Mhaimeed, O.; Dineen, E.H.; Marvel, F.A.; Martin, S.S.; Muse, E.D.; Turakhia, M.P.; Tarakji, K.G.; et al. Smart wearable devices in cardiovascular care: Where we are and how to move forward. *Nat. Rev. Cardiol.* **2021**, *18*, 581–599. [CrossRef] [PubMed]
17. Kirchhof, P.; Benussi, S.; Kotecha, D.; Ahlsson, A.; Atar, D.; Casadei, B.; Castella, M.; Diener, H.C.; Heidbuchel, H.; Hendriks, J.; et al. 2016 ESC Guidelines for the Management of Atrial Fibrillation Developed in Collaboration with EACTS. *Eur. Heart J.* **2016**, *37*, e636–e683. [CrossRef]
18. Duncker, D.; Ding, W.Y.; Etheridge, S.; Noseworthy, P.A.; Veltmann, C.; Yao, X.; Jared Bunch, T.; Gupta, D. Smart Wearables for Cardiac Monitoring—Real-World Use beyond Atrial Fibrillation. *Sensors* **2021**, *21*, 2539. [CrossRef] [PubMed]
19. Nicolò, A.; Massaroni, C.; Schena, E.; Sacchetti, M. The Importance of Respiratory Rate Monitoring: From Healthcare to Sport and Exercise. *Sensors* **2020**, *20*, 6396. [CrossRef] [PubMed]
20. Vanegas, E.; Igual, R.; Plaza, I. Sensing Systems for Respiration Monitoring: A Technical Systematic Review. *Sensors* **2020**, *20*, 5446. [CrossRef] [PubMed]

21. Massaroni, C.; Nicolò, A.; Presti, D.L.; Sacchetti, M.; Silvestri, S.; Schena, E. Contact-Based Methods for Measuring Respiratory Rate. *Sensors* **2019**, *19*, 908. [CrossRef]
22. Orphanidou, C.; Fleming, S.; Shah, S.A.; Tarassenko, L. Data Fusion for Estimating Respiratory Rate from a Single-Lead ECG. *Biomed. Signal Process. Control* **2013**, *8*, 98–105. [CrossRef]
23. Majumder, S.; Mondal, T.; Deen, M.J. Wearable Sensors for Remote Health Monitoring. *Sensors* **2017**, *17*, 130. [CrossRef]
24. Masip, J.; Gayà, M.; Páez, J.; Betbesé, A.; Vecilla, F.; Manresa, R.; Ruíz, P. Pulsioximetría en el diagnóstico de insuficiencia cardiaca aguda. *Rev. Esp. Cardiol.* **2012**, *65*, 879–884. [CrossRef]
25. Kirschen, G.W.; Singer, D.D.; Thode, H.C.; Singer, A.J. Relationship between body temperature and heart rate in adults and children: A local and national study. *Am. J. Emerg. Med.* **2020**, *38*, 929–933. [CrossRef]
26. Mukhopadhyay, S.C. Wearable Sensors for Human Activity Monitoring: A Review. *IEEE Sens. J.* **2014**, *15*, 1321–1330. [CrossRef]
27. Reith, J.; Jørgensen, H.; Pedersen, P.; Nakamaya, H.; Jeppesen, L.; Olsen, T.; Raaschou, H. Body temperature in acute stroke: Relation to stroke severity, infarct size, mortality, and outcome. *Lancet* **1996**, *347*, 422–425. [CrossRef]
28. Geurts, M.; on behalf of the DUST investigators; Scheijmans, F.E.V.; van Seeters, T.; Biessels, G.J.; Kappelle, L.J.; Velthuis, B.K.; van der Worp, H.B. Temporal profile of body temperature in acute ischemic stroke: Relation to infarct size and outcome. *BMC Neurol.* **2016**, *16*, 233. [CrossRef]
29. Chowdhury, M.H.; Hossain, Q.D.; Saha, P.; Rahaman, M. Design, fabrication and performance evaluation of a three electrode ECG recorder. In Proceedings of the 2016 International Conference on Innovations in Science, Engineering and Technology (ICISET), Dhaka, Bangladesh, 28–29 October 2016; pp. 1–4. [CrossRef]
30. Guvenc, H. Wireless ECG Device with Arduino. In Proceedings of the 2020 Medical Technologies Congress (TIPTEKNO), Antalya, Turkey, 19–20 November 2020; pp. 1–4. [CrossRef]
31. Gifari, M.W.; Zakaria, H.; Mengko, R. Design of ECG Homecare:12-lead ECG acquisition using single channel ECG device developed on AD8232 analog front end. In Proceedings of the 2015 International Conference on Electrical Engineering and Informatics (ICEEI), Denpasar, Indonesia, 10–11 August 2015; pp. 371–376. [CrossRef]
32. Prasad, A.S.; Kavanashree, N. ECG Monitoring System Using AD8232 Sensor. In Proceedings of the 2019 International Conference on Communication and Electronics Systems (ICCES), Coimbatore, India, 17–19 July 2019; pp. 976–980. [CrossRef]
33. Ghifari, A.F.; Perdana, R.S. Minimum System Design of The IoT-Based ECG Monitoring. In Proceedings of the 2020 International Conference on ICT for Smart Society (ICISS), Bandung, Indonesia, 19–20 November 2020; pp. 1–6. [CrossRef]
34. Penica, M.; Mohandas, R.; Bhattacharya, M.; Vancamp, K.; Hayes, M.; O'Connell, E. A Covid-19 viral transmission prevention system for embedded devices utilising deep learning. In Proceedings of the 2021 32nd Irish Signals and Systems Conference (ISSC), Athlone, Ireland, 10–11 June 2021; pp. 1–8. [CrossRef]
35. Saha, R.; Biswas, S.; Sarmah, S.; Karmakar, S.; Das, P. A Working Prototype Using DS18B20 Temperature Sensor and Arduino for Health Monitoring. *SN Comput. Sci.* **2021**, *2*, 33. [CrossRef]
36. Fajrin, H.R.; Ilahi, M.R.; Handoko, B.S.; Permatasari, I. Body temperature monitoring based on telemedicine. *J. Phys. Conf. Ser.* **2019**, *1381*, 012014. [CrossRef]
37. Bates, A.; Ling, M.J.; Mann, J.; Arvind, D. Respiratory Rate and Flow Waveform Estimation from Tri-axial Accelerometer Data. In Proceedings of the 2010 International Conference on Body Sensor Networks, Singapore, 7–9 June 2010; pp. 144–150. [CrossRef]
38. Ahsanuzzaman, S.; Ahmed, T.; Rahman, A. Low Cost, Portable ECG Monitoring and Alarming System Based on Deep Learning. In Proceedings of the 2020 IEEE Region 10 Symposium (TENSYMP), Dhaka, Bangladesh, 5–7 June 2020; pp. 316–319. [CrossRef]
39. Hughes, S.; Liu, H.; Zheng, D. Influences of Sensor Placement Site and Subject Posture on Measurement of Respiratory Frequency Using Triaxial Accelerometers. *Front. Physiol.* **2020**, *11*, 823. [CrossRef] [PubMed]
40. Ferrazza, M.; D'Abbondanza, N.; Piuzzi, E.; Pallotti, A. Statistical classification for the screening of cardiological patients by means of a sensorized T-shirt. In Proceedings of the 2021 International Electronic Conference on Applied Sciences, Online, 15–31 October 2021.

engineering proceedings

Proceedings

Classification-Based Screening of Parkinson's Disease Patients through Graph and Handwriting Signals †

Maria Fratello [1,2], Fulvio Cordella [1,2], Giovanni Albani [3], Giuseppe Veneziano [3], Giuseppe Marano [4,5], Alessandra Paffi [6,7] and Antonio Pallotti [2,8,9,*]

1 Dipartimento di Scienze di Base e Applicate per l'Ingegneria, Università degli Studi di Roma "La Sapienza"-Piazzale Aldo Moro, 5, 00185 Roma, Italy; maria.fratello97@gmail.com (M.F.); fulviocordella@yahoo.com (F.C.)
2 Consorzio Parco Scientifico e Tecnologico Pontino Technoscience, 15, 04100 Latina, Italy
3 Casa di Cura Le Terrazze, 10, 21100 Varese, Italy; g.albani@auxologico.it (G.A.); josip@inwind.it (G.V.)
4 AGIF Associazione Grafologica Italo-Francese, 8, 00153 Roma, Italy; giuseppemaranogm@gmail.com
5 Department of Geriatrics, Neuroscience and Orthopedics, Policlinico Universitario Agostino Gemelli IRCSS, Università Cattolica del Sacro Cuore, 1, 20123 Milano, Italy
6 Centro Interuniversitario sulle Interazioni tra Campi Elettromagnetici e Biosistemi (ICEmB), Università di Genova, 5, 16126 Genova, Italy; alessandra.paffi@uniroma1.it
7 Dipartimento di Ingegneria dell'Informazione, Elettronica e Telecomunicazioni (DIET), Sapienza University of Rome, 5, 00185 Roma, Italy
8 Dipartimento di Management e Diritto, Università degli Studi di Roma "Tor Vergata", 50, 00133 Roma, Italy
9 Dipartimento di Scienze Umane e Promozione della Qualità della Vita, Università Telematica San Raffaele Roma, 247, 00166 Roma, Italy
* Correspondence: antonio.pallotti@uniroma5.it or antonio.pallotti@technoscience.it
† Presented at the 2nd International Electronic Conference on Applied Sciences, 15–31 October 2021; Available online: https://asec2021.sciforum.net/.

Abstract: Parkinson's disease (PD) is one of the most common neurodegenerative diseases, affecting millions of people worldwide, especially among the elderly population. It has been demonstrated that handwriting impairment can be an important early marker for the detection of this disease. The aim of this study was to propose a simple and quick way to discriminate PD patients from controls through handwriting tasks using machine-learning techniques. We developed a telemonitoring system based on a user-friendly application for drawing tablets that enabled us to collect real-time information about position, pressure, and inclination of the digital pen during the experiment and, simultaneously, to supply visual feedback on the screen to the subject. We developed a protocol that includes drawing and writing tasks, including tasks in the Italian language, and we collected data from 22 healthy subjects and 9 PD patients. Using the collected signals and data from a preexisting database, we developed a machine-learning model to automatically discriminate PD patients from healthy control subjects with an accuracy of 77.5%.

Keywords: graph signal; handwriting signal; Parkinson's disease; machine learning; telemonitoring

Citation: Fratello, M.; Cordella, F.; Albani, G.; Veneziano, G.; Marano, G.; Paffi, A.; Pallotti, A. Classification-Based Screening of Parkinson's Disease Patients through Graph and Handwriting Signals. *Eng. Proc.* **2021**, *11*, 49. https://doi.org/10.3390/ASEC2021-11128

Academic Editor: Nunzio Cennamo

Published: 15 October 2021

1. Introduction

Parkinson's disease (PD) is one of the most common neurodegenerative disorders in the world, second only to Alzheimer's disease [1]. The differential diagnosis of PD is still an ongoing challenge for the scientific community: to this day, confirmation of the disease is available only postmortem and the rate of misdiagnosis is high; it has been estimated that 25% of diagnoses are incorrect [2]. The main cause of PD is the lack of dopamine production, and its main motor symptoms are bradykinesia, tremor, and rigidity [3]; neurologists rely on imaging techniques such as MRI (Magnetic Resonance Imaging), CT (Computed Tomography), or PET (Positron Emission Tomography), and patient clinical evaluations [3]. Machine-learning techniques have been studied to help the diagnosis of

PD and have shown promising results. Pereira et al. presented a review on recent studies concerning computer-assisted methods to aid PD recognition [4], which included speech, gait, and handwriting analysis. This work is part of a home-monitoring project that aims to aid in PD detection through a combined analysis of graphological and vocal signals [5]. An accuracy of 98.5% was reached through the analysis of vocal data from 55 subjects: 18 healthy control subjects and 33 drug-free and newly diagnosed PD patients.

In this study, we focused on handwriting of PD subjects; handwriting requires a complex coordination of consecutive movements, and the motor symptoms of PD can provoke handwriting impairments in the size of letters, which is referred to as micrographia, and in the pressure and kinematics of the pen [6,7], together with a general difficulty in writing, which involves different graphological patterns. Since "graphology is a discipline that deals with the dynamic study of the graphic gesture" [8], we based our analysis on computational graphology. Several studies have investigated the most relevant writing features and tasks for the diagnosis of PD. Reference [9] presents the state of the art of these studies. It is possible to collect relevant information from drawings (Archimedean spiral [10–15], circles [16], meanders [13,14], etc.) and from handwritten words and graphemes. The drawing of an Archimedean spiral (spirography) is a common task for tremor and other movement disorder analysis [10]. Thanks to the development of digitizing tablet technologies, it is possible to analyze not only the offline images, but also the kinematic characteristics of the graphic signal and the pressure applied to the tablet [17,18]. "Online" data are those collected while the user writes, while "offline" data are those available after the writing is completed [19].

In the past decade, important databases have been constructed in order to study handwriting impairments in PD: the PaHaW database [11], which includes real-time data (pen position, pen pressure, and pen inclination) collected from 38 PD patients and 37 control subjects, and the HandPD [13] and NewHandPD [14] databases, which include offline images collected by Pereira et al.

Dròtar et al., analyzing the PaHaW database, obtained an accuracy of 85.61% [20]; they demonstrated the relevance not only of the on-tablet movements, but also of the in-air movements, i.e., the variation of the pen position while the pen is not touching the table. Considering only the spiral task, they obtained an accuracy of 62.8% [12].

The aim of this work was to analyze handwriting signals from both PD patients and control subjects and to propose a way to automatically distinguish these two classes. In order to collect the necessary data, we developed a telemonitoring system based on a user-friendly application for drawing tablets that enabled us to collect real-time information about the digital pen during the experiment and, simultaneously, to supply visual feedback on the screen to the subject. Through this system, data can be collected remotely in order to allow patients to execute tasks in the comfort and safety of their home, reducing the demand on hospital services. We decided to propose a protocol to explore writing and drawing impairments, including specific tasks for subjects who declared Italian as their first language, since, to our knowledge, the literature is lacking in automatic handwriting classification studies with Parkinson's subjects whose first language is Italian.

2. Methods and Materials

In this study, we collected data from 22 healthy subjects and from 9 PD patients. The data from the PD subjects were collected thanks to a collaboration with the Casa di Cura Le Terrazze institute. All participants were right-handed except for one PD patient, with an age in the range of 60 ± 25 years. Information about subjects' age, gender, dimensions of the hand, and level of education are collected in Table 1. The educational level was classified according to UNESCO's ISCED 2011 (International Standard Classification of Education) [21]. This classification distinguishes nine levels of education, from early child education (level 0) to doctoral or equivalent level (level 8). These levels can be aggregated into three categories: low (0–2), medium (3–4), and high (5–8) [22]. The hand dimension was quantified by measuring the distance between the wrist and the top of the distal phalanx of the dominant hand's middle finger.

Table 1. Subject data. For each group (healthy control (C) or subject with Parkinson's disease (PD)), the age and the number of males and females are specified. Information about the anatomical measurements of the hand and the level of education is indicated only for the control group.

Group	Age (Mean ± SD)	Number Male/Female	Middle Finger–Wrist Distance (cm) (Mean ± SD)	Level of Education (ISCED 2011) (Mean ± SD)
C	55.8 ± 6.5	8/14	19.6 ± 1.8	5.1 ± 2.1
PD	69.3 ± 10	5/4	-	-

Information about the PD patients is collected in Table 2. For every patient, the Hoehn and Yahr Scale level is indicated. The Hoehn and Yahr Scale is a clinical scale used to describe the progressive motor impairment of subjects with PD [23]. This scale goes from 1 to 5 in order of severity of the motor symptoms; the first level corresponds to only unilateral involvement with minimal motor dysfunction, while the fifth and last stage corresponds to a more serious level of motor dysfunction wherein the subject is confined to bed or wheelchair unless aided. Table 2 also indicates the side of the body affected by the motor dysfunction (left, right or bilateral if both sides are involved), years since the diagnosis of PD, and the levodopa-equivalent daily dose (LEDD) corresponding to each PD patient.

Table 2. This table shows, for every PD patient, the Hoehn and Yahr Scale level (H&Y), the side of the body affected, the years since the PD diagnosis, and the levodopa-equivalent daily dose (LEDD) assumed by the subject.

Patient ID	H&Y	Side	Years since Diagnosis	LEDD
#1	2	Left	3	450
#2	3	Left	2	750
#3	2	Bilateral	1	200
#4	3	Right	6	625
#5	3	Left	11	400
#6	3	Left	4	250
#7	4	Bilateral	4	400
#8	1	Right	1	50
#9	3	Left	13	900

A commercial Wacom One drawing tablet with a screen was used for this test in order to be able to extract both "online" and "offline" features. Wacom tablets are widely used in handwriting movement analysis, as they offer high spatial and temporal resolution [9].

An application was developed by our team using the development platform Unity, which allowed us to collect information about pen position (x, y), pressure, and inclination with a frequency of 133 Hz and, simultaneously, to supply visual feedback on the tablet's screen to the subjects. For the protocol, the Wacom tablet was connected to a Lenovo Thinkpad T495 computer with Windows 10 as the operating system. The "Duplicate" modality was selected in order to have the same screen shown on the computer and the Wacom One tablet, as shown in Figure 1; Figure 1a shows the point of view of the operator, while Figure 1b shows the point of view of the participant.

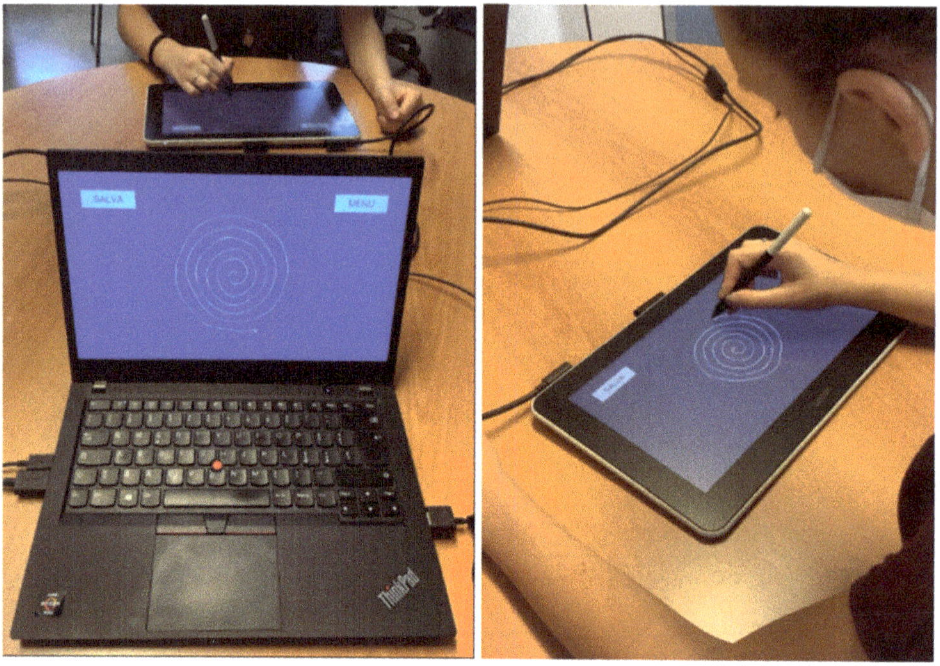

Figure 1. Experimental setup: (**a**) shows the operator's point of view, while (**b**) shows the participant's point of view. The subject uses the Wacom tablet to complete the protocol tasks, while the operator follows the experiment in real time from the monitor of his computer.

The application has a start page where the participant's ID can be entered and which includes a menu from which the user can choose which task to take. The data are saved locally in different.tsv files for every acquisition.

In order to analyze the data, we used the software MATLAB.

The protocol was divided into four parts: drawing an Archimedean spiral, writing the bigram "le" six times and two Italian sentences, drawing ten concentric circles, and writing seven lines of free text. For each part of the protocol, a different screen was shown to the subject: firstly, an image of an Archimedean spiral was shown and the subject was asked to trace it at a comfortable speed; secondly, a blank screen was shown and the subject was asked to write six times in cursive the bigram "le" and the two Italian phrases: "I fiori sono sul prato" and "Nel cielo ci sono le stelle". On the third screen, a circle was shown and the subject was asked to draw ten concentric circles inside it. Lastly, a blank screen was shown and the subject was asked to write seven lines of free text in cursive. The overall duration of the test varied between 10 and 15 min from subject to subject. The subjects were given the opportunity to try the tablet before the test. During the execution of the tasks, the subjects were seated in a comfortable position on a chair without armrests, and the tablet was positioned on a table in front of them.

Features were extracted separately from each task.

Data were separated into components, i.e., lines that are traced without lifting the pen from the tablet. In order to do that automatically, indices of the samples where pressure went from positive to zero and vice versa were saved in a vector of markers. Both in-air and on-tablet features were extracted.

Figure 2 shows an example of the "le" bigram task, where the different components, automatically detected, are represented in different colors and the "in-air" points of the pen position are represented as blue points. For each component, the velocity was calculated as follows:

$$v = \sqrt{{v_x}^2 + {v_y}^2}, \tag{1}$$

where $v_x = \frac{x_{i+1} - x_i}{t_{i+1} - t_i}$ and $v_y = \frac{y_{i+1} - y_i}{t_{i+1} - t_i}$,

Figure 2. Image of the "le" bigram task written by a healthy control subject from our dataset. The "in-air" points of the pen position are represented in blue. Different components of the "on-tablet" pen position are represented in different colors.

Where x_i, y_i, and t_i correspond to the pen position along x, pen position along y and time at a specific index i of the recorded data, respectively.

Acceleration and jerk of the components were also calculated. To analyze the spiral, the angular and radial velocity were calculated. Furthermore, the distance of the drawn spiral from the spiral guide was calculated using the following algorithm:

1. For each point of the drawn spiral (x_i, y_i), we found the spiral guide's closest point to it, (sx_i, sy_i).
2. We calculated the distance of each couple of points (x_i, y_i), (sx_i, sy_i), as

$$d_i = \sqrt{({x_i}^2 - {sx_i}^2) + ({y_i}^2 - {sy_i}^2)} \tag{2}$$

3. We found the parameter

$$p = \sum_i {d_i}^2, \tag{3}$$

that describes how much the drawn spiral is distant from the spiral guide. A smaller value of p meant a higher precision.

Furthermore, the power spectral density between 4 Hz and 9 Hz of the absolute velocity of the pen during the spiral task was calculated as a feature.

In order to develop a model for automatic classification of PD, we used data from 36 PD subjects and 35 healthy control subjects from the PaHaW dataset and data from the 22 healthy control subjects and 9 PD patients that we collected in our database. Two tasks that the two databases have in common were analyzed: the guided spiral and the bigram "le". The features that were considered are reported in Table 3.

After the feature extraction, we proceeded with the selection of the most significant features and the generation of the classification model. In Figure 3, a scheme of the workflow is presented.

Table 3. Features extracted: *1* if the feature was extracted from the spiral task analysis, *2* if the feature was extracted from the "le" bigram task analysis.

Features	Task
Velocity: absolute, vertical, and horizontal	*1,2*
Acceleration: absolute, vertical, and horizontal	*1,2*
Jerk: absolute, vertical, and horizontal	*1,2*
Radial velocity	*1*
Angular velocity	*1*
Variation of velocity, acceleration, and stroke between components	*2*
Number of changes of direction in velocity	*1,2*
Number of changes of direction in acceleration	*1,2*
Number of changes of direction in jerk	*1,2*
Normalized in-air time (time in air over total time)	*1,2*
In-air velocity: absolute, vertical, and horizontal	*2*
In-air acceleration: absolute, vertical, and horizontal	*2*
In-air jerk: absolute, vertical, and horizontal	*2*
Power spectral density of the absolute velocity	*1*

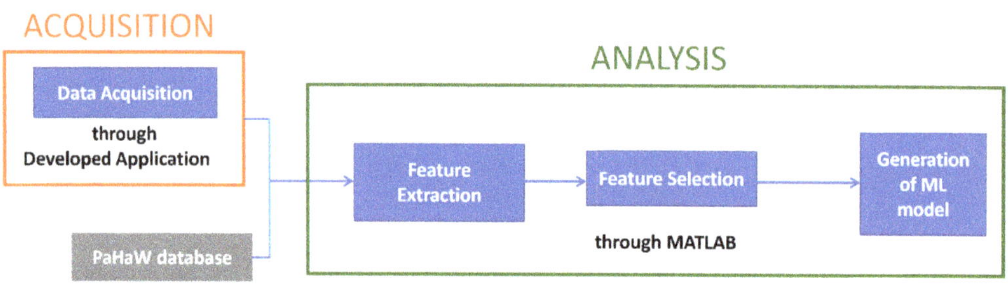

Figure 3. Workflow diagram of the process.

3. Results and Discussion

In order to discriminate between PD patients and healthy control subjects, three models were constructed: one using only data from the spiral task, one using only data from the "le" bigram task, and one using data from both of them.

We selected the most discriminant features using a Mann–Whitney test ($p < 0.05$). Table 4 presents the features with the lowest p-scores, divided per task.

Table 4. Feature ranking. The left column presents the most discriminant features for the spiral task, and the right column presents the most discriminant features for the "le" task.

Spiral	le
skewness of the pen acceleration on the x axis	mean of the maximum pen velocity on the y axis of the single le component in the air
kurtosis of the pen velocity on the x axis	mean of the range of the pen velocity on the y axis of the single le component in the air
25th percentile of the power spectral density of the absolute pen velocity	maximum of the acceleration on the x axis in the air

A 10-fold cross-validation was conducted. Results are reported in Table 5. Accuracy, specificity, sensitivity, F1 score, and precision were calculated in terms of *TP* (true positive), *FP* (false positive), *TN* (true negative), and *FN* (false negative), as follows:

$$Accuracy = \frac{TP + TN}{TP + TN + FP + FN} * 100 \quad (4)$$

$$Specificity = \frac{TN}{FP + TN} * 100 \quad (5)$$

$$Sensitivity = \frac{TP}{FN + TP} * 100 \quad (6)$$

$$F1\ score = \frac{TP}{TP + \frac{1}{2}(FP + FN)} * 100 \quad (7)$$

$$Precision = \frac{TP}{TP + FP} * 100 \quad (8)$$

Table 5. Model used and classification accuracy, specificity, sensitivity, F1 score, and precision of the two tasks analyzed separately and of the two tasks combined.

	Spiral	le	Spiral and le
Model	Linear SVM	Linear SVM	Medium KNN
Accuracy	71.6%	75.5%	77.5%
Specificity	79%	73.7%	77.1%
Sensitivity	62.2%	77.8%	77.8%
F1 Score	65.9%	73.7%	75.3%
Precision	70%	70%	72.9%

Considering the two tasks separately, we obtained a higher accuracy for the "le" bigram tasks than for the spiral tasks.

Moreover, considering the spiral and the "le" bigram task separately, the accuracy that we obtained for the spiral (71.6%) was higher than the accuracy obtained for the spiral by Dròtar et al. (62.8%) and, similarly, considering only the "le" bigram task, the accuracy that we obtained (75.5%) was higher than the accuracy obtained by Dròtar et al. for this task (71%).

The confusion matrix for each one of the three models (spiral, "le" bigram and combined tasks) are reported on the Figure 4.

	C'	PD'
C	45	12
PD	17	28

(a)

	C'	PD'
C	42	15
PD	10	35

(b)

	C'	PD'
C	44	13
PD	10	35

(c)

Figure 4. Confusion matrix of the three models generated. In (**a**), (**b**), and (**c**), PD′ and C′ indicate, respectively, subjects that were predicted to be subjects with PD and control subjects, while PD and C indicate the true categories of the subjects. PD corresponds to PD patients and C corresponds to healthy control subjects.

The highest accuracy (77.5%) and sensitivity (77.8%) were obtained by combining the two tasks, while the highest specificity was obtained when using only the data from the spiral task. The machine-learning algorithms that were employed were the support vector machine (SVM) for the spiral task and the "le" bigram task, and the medium k-nearest neighbors (Medium KNN) for the combined tasks.

4. Conclusions

In this study, an application is presented that allowed us to register data from tablets with a frequency of 133 Hz, in order to aid the recognition of PD through handwriting impairments. The tool that is proposed is simple and easy to use, allowing subjects to complete the test in the comfort of their home.

Data from 22 healthy subjects and 9 PD patients were collected and added to the PaHaW database [11,20], a pre-existing dataset that includes data from PD patients and healthy control subjects. Using only two of the eight tasks that the PaHaW database includes, an accuracy of 77.5%, was obtained, close to the 85.61% accuracy that Dròtar et al. obtained when considering all the eight tasks together [20]. We could not compare the other tasks because the first language declared by our subjects (Italian) was different from the first language of the PaHaW database's subjects (Czech).

The major limitations of this study are linked to the limited number of subjects involved and to the fact that we compared data from different databases, collected under different experimental conditions, such as the position of the subject during the tasks, and acquired with different devices, which could lead to a bias in the measurements. Moreover, subjects' characteristics such as age could lead to misclassification; for example, a control subject could present some tremor or bradykinesia not linked to Parkinson's disease, and for this reason could be misclassified as a PD subject.

However, the protocol that we developed can be used in future studies to collect more data from Italian PD subjects, in order to be able to create a model using only data from our protocol, using the combination of six tasks proposed here (drawing of an Archimedean spiral, writing the bigram "le", writing two Italian phrases, drawing ten concentric circles, and writing seven lines of free texts.

Moreover, this work is part of a home-monitoring project that aims to aid in PD detection through a combined analysis of graphological and vocal signals [5].. The sets of subjects tested for the vocal tasks and the graphological tasks were different from each other, so we could not create a classification model using combined vocal and graphological data, but the aim of this project is to continue to collect both vocal and graphological data in order to create a single, more complete, classification model.

Author Contributions: Conceptualization, A.P. (Antonio Pallotti), A.P. (Alessandra Paffi) and G.M.; methodology, A.P. (Antonio Pallotti); software, M.F.; validation, M.F., A.P. (Alessandra Paffi), G.M., G.A., G.V. and A.P. (Antonio Pallotti); formal analysis, M.F. and F.C.; investigation, M.F. and A.P. (Antonio Pallotti); resources, M.F., G.A., G.V. and A.P. (Antonio Pallotti); data curation, A.P. (Antonio Pallotti); writing—original draft preparation, M.F. and A.P.(Antonio Pallotti); writing—review and editing, M.F. and A.P. (Antonio Pallotti); visualization, M.F. and A.P. (Antonio Pallotti); supervision, A.P. (Antonio Pallotti), A.P. (Alessandra Paffi) and G.M.; project administration, A.P. (Antonio Pallotti); funding acquisition, A.P. (Antonio Pallotti). All authors have read and agreed to the published version of the manuscript.

Funding: This research received no external funding.

Institutional Review Board Statement: The protocol was approved by the Consorzio Parco Scientifico e Tecnologico Pontino Technoscience, with the code PST/342/P/130422.

Informed Consent Statement: Informed consent was obtained from the subject involved in the study.

Data Availability Statement: The datasets underpinning this work are available from the corresponding author upon request (agreement signing).

Acknowledgments: We thank Joint Research Laboratory "Computational Graphology"(Consorzio Parco Scientifico e Tecnologico Pontino Technoscience and San Raffaele University of Rome)for the support and contribution in the research work.

Conflicts of Interest: The authors declare no conflict of interest.

References

1. McKeith, I.G.; Burn, D. Spectrum of parkinson's disease, parkinson's dementia, and lewy body dementia. *Neurol. Clin.* **2000**, *18*, 865–883. [CrossRef]
2. Gupta, J.D.; Chanda, B. Novel Features for Diagnosis of Parkinson's Disease From off-Line Archimedean Spiral Images. In Proceedings of the IEEE 10th International Conference on Awareness Science and Technology (iCAST), Morioka, Japan, 23–25 October 2019. [CrossRef]
3. Tolosa, E.; Wenning, G.; Poewe, W. The diagnosis of Parkinson's disease. *Lancet Neurol.* **2006**, *5*, 75–86. [CrossRef]
4. Pereira, C.R.; Pereira, D.R.; Weber SA, T.; Hook, C.; de Albuquerque, V.H.C.; Papa, J.P. A survey on computer-assisted Parkinson's Disease diagnosis. *Artif. Intell. Med.* **2019**, *95*, 48–63. [CrossRef] [PubMed]
5. Cordella, F.; Paffi, A.; Pallotti, A. Classification-based screening of Parkinson's disease patients through voice signal. In Proceedings of the 2021 IEEE International Symposium on Medical Measurements and Applications (MeMeA), Lausanne, Switzerland, 23–25 June 2021; pp. 1–6. [CrossRef]
6. Emamzadeh, F.N.; Surguchov, A. Parkinson's Disease: Biomarkers, Treatment, and Risk Factors. *Front. Neurosci.* **2018**, *12*, 612. [CrossRef] [PubMed]
7. Zham, P.; Arjunan, S.P.; Raghav, S.; Kumar, D.K. Efficacy of Guided Spiral Drawing in the Classification of Parkinson's Disease. *IEEE J. Biomed. Health Inform.* **2018**, *22*, 1648–1652. [CrossRef] [PubMed]
8. Mazza, M.; Marano, G.; Traversi, G.; Gaetani, E.; Sani, G.; Mazza, S. Graphology: An Interface Between Biology, Psychology and Neuroscience. *Psychol. Disord. Res.* **2021**, *2020*, 1–13. [CrossRef]
9. Homas, M.; Lenka, A.; Pal, P.K. Handwriting Analysis in Parkinson's Disease: Current Status and Future Directions. *Mov. Disord. Clin. Pract.* **2017**, *4*, 806–818. [CrossRef]
10. Miralles, F.; Tarongí, S.; Espino, A. Quantification of the drawing of an Archimedes spiral through the analysis of its digitized picture. *J. Neurosci. Methods* **2006**, *152*, 18–31. [CrossRef] [PubMed]
11. Drotar, P.; Mekyska, J.; Rektorová, I.; Masarová, L.; Smekal, Z.; Faundez-Zanuy, M. Evaluation of handwriting kinematics and pressure for differential diagnosis of Parkinson's disease. *Artif. Intell. Med.* **2016**, *67*, 39–46. [CrossRef] [PubMed]
12. Pereira, C.R.; Pereira, D.R.; Silva, F.A.; Masieiro, J.P.; Weber, S.A.T.; Hook, C.; Papa, J.P. A new computer vision-based approach to aid the diagnosis of Parkinson's disease. *Comput. Methods Programs Biomed.* **2016**, *136*, 79–88. [CrossRef] [PubMed]
13. Pereira, C.R.; Weber SA, T.; Hook, C.; Rosa, G.H.; Papa, J. Deep Learning-aided Parkinson's Disease Diagnosis from Handwritten Dynamics. In Proceedings of the 2016 29th SIBGRAPI Conference on Graphics, Patterns and Images (SIBGRAPI), Sao Paulo, Brazil, 4–7 October 2016. [CrossRef]
14. Sadikov, A.; Zabkar, J.; Mozina, M.; Groznik, V.; Georgiev, D.; Bratko, I. ParkinsonCheck A Decision Support System for Tremor Detection. 2015. Available online: https://ailab.si/parkinsoncheck/pc-tr.pdf (accessed on 24 November 2021).
15. Lamba, R.; Gulati, T.; Al-Dhlan, K.A.; Jain, A. A systematic approach to diagnose Parkinson's disease through kinematic features extracted from handwritten drawings. *J. Reliab. Intell. Environ.* **2021**, *7*, 253–262. [CrossRef]
16. Eichhorn, T.E.; Gasser, T.; Marquardt, T.; Arnold, G.; Schwarz, J.; Oertel, W.H. Computational Analysis of Open Loop Handwriting Movements in Parkinson's Disease: A Rapid Method to Detect Dopamime tic Effects. *Mov. Disordcr Soc.* **1996**, *11*, 3. [CrossRef]
17. Taleb, C.; Likforman-Sulem, L.; Khachab, M.; Mokbel, C. Feature selection for an improved Parkinson's disease identification based on handwriting. In Proceedings of the 2017 1st International Work- shop on Arabic Script Analysis and Recognition (ASAR 2017), Nancy, France, 3–5 April 2017; pp. 52–56. [CrossRef]
18. Senatore, R.; Marcelli, A. A paradigm for emulating the early learning stage of handwriting: Performance comparison between healthy controls and Parkinson's disease patients in drawing loop shapes. *Hum. Mov. Sci.* **2019**, *65*, 89–101. [CrossRef]
19. Tappert, C.C.; Suen, C.Y.; Wakahara, T. The state of the art in online handwriting recognition. *IEEE Trans. Pattern Anal. Mach. Intell.* **1990**, *12*, 787–808. [CrossRef]
20. Drotar, P.; Mekyska, J.; Rektorová, I.; Masarová, L.; Smékal, Z.; Faundez-Zanuy, M. Analysis of in-air movement in handwriting: A novel marker for Parkinson's disease. *Comput. Methods Programs Biomed.* **2014**, *117*, 405–411. [CrossRef]
21. UNESCO Institute for Statistics. *International Standard Classification of Education: ISCED 2011*; UIS: Montreal, QC, Canada, 2021. [CrossRef]
22. International Standard Classification of Education (ISCED)—Statistics Explained (europa.eu). 2021. Available online: https://ec.europa.eu/eurostat/statistics-explained/index.php?title=International_Standard_Classification_of_Education_(ISCED) (accessed on 25 November 2021).
23. Bhidayasiri, R.; Tarsy, D. Parkinson's Disease: Hoehn and Yahr Scale. In *Movement Disorders: A Video Atlas*; Current Clinical Neurology; Humana: Totowa, NJ, USA, 2012. [CrossRef]

engineering proceedings

MDPI

Proceeding Paper
Optical Coatings: Applications and Metrology †

Paola Zuppella [1,*], Paolo Chioetto [1,2], Chiara Casini [1,2], Simone Nordera [1], Nunzio Cennamo [3], Luigi Zeni [3] and Vania Da Deppo [1]

1 Institute for Photonics and Nanotechnologies (IFN), National Research Council (CNR), Via Trasea 7, 35131 Padova, Italy; paolo.chioetto@pd.ifn.cnr.it (P.C.); chiara.casini@pd.ifn.cnr.it (C.C.); simone.nordera@pd.ifn.cnr.it (S.N.); vania.dadeppo@pd.ifn.cnr.it (V.D.D.)
2 Centro di Ateneo di Studi e Attività Spaziali "Giuseppe Colombo"—CISAS, Via Venezia 15, 35131 Padova, Italy
3 Department of Engineering, Università degli Studi della Campania "Luigi Vannvitelli", Via Roma 29, 81031 Aversa, Italy; nunzio.cennamo@unicampania.it (N.C.); luigi.zeni@unicampania.it (L.Z.)
* Correspondence: paola.zuppella@pd.ifn.cnr.it
† Presented at the 2nd International Electronic Conference on Applied Sciences, 15–31 October 2021; Available online: https://asec2021.sciforum.net/.

Abstract: The development of optical coatings has experienced rapid growth in the last few decades for a wide range of applications. The strong demand is motivated by the progress of new-generation sources, large-scale facilities, new lithography arrangements, innovative methods for materials science investigation, biosensors, and instruments for space and solar physics observations. The research activities carried out at the Padova branch of the Institute for Photonics and Nanotechnologies of the National Research Council range from the design and characterization of optical components for space activities to the development of nanostructured coatings for tools, such as biosensors and surface plasmon resonance devices. In recent years, we have dealt with the optical characterization of 2D materials in order to explore the feasibility of innovative optical elements designed and optimized to cover wide spectral ranges. In this manuscript, we show the results on the optical characterizations of MoS_2 and graphene samples, both monolayers, deposited on thick SiO_2 film. We present the preliminary and comparative analysis of the samples in question, showing a direct comparison with the optical performance of the pristine SiO_2 over the visible spectral range.

Keywords: optical coatings; metrology; thin films; MoS_2; graphene

Citation: Zuppella, P.; Chioetto, P.; Casini, C.; Nordera, S.; Cennamo, N.; Zeni, L.; Deppo, V.D. Optical Coatings: Applications and Metrology. *Eng. Proc.* **2021**, *11*, 50. https://doi.org/10.3390/ASEC2021-11137

Academic Editor: Saulius Juodkazis

Published: 15 October 2021

1. Introduction

The Institute for Photonics and Nanotechnologies (IFN) of Italy's National Research Council (CNR) carries out pioneering research in several fields of photonics. The Padova branch stands out in the technological activities related to the development of optical devices. Applications range from space instrumentation to sensors platforms, including optical metrology, and are strongly oriented to applied physics and technology transfer [1].

Thin films and optical coatings are transversal topics, common to all activities just mentioned. In the field of biosensors development, nanostructured films find a very interesting application in the use of innovative metals for surface plasmon resonance (SPR) platforms based on prism and fiber [2,3]. The scope is to improve sensitivity, detection accuracy, dynamic range, and application fields of this type of biodevice [4,5]. In space optics, high-performance optical coatings are optimized both to fulfill the scientific requirements of the instruments and to survive harsh operation environments [6]. Furthermore, in some spectral regions, such as vacuum ultraviolet and soft X-ray, the structures of the optical films become particularly complex, requiring design and fabrication of multilayer stacks [7].

Over the years, the CNR-IFN focused on the design and characterization of nanostructured thin films to be used as sensitive layers for biosensors [8,9], mirrors [10], filters, phase retarders [11], and polarizers. We have collaborated on the fabrication of a palladium/gold bilayer designed for an SPR sensor based on D-shaped optical fibre (POF). The novel SPR-POF platform was optimized to work in the 1.38–1.42 refractive index range, where it exhibits excellent performances in terms of sensitivity and signal to noise ratio [9].

Another interesting application we dealt with was the development of innovative biochips for Kretschmann SPR tools [8]. The new chips are based on palladium thin films deposited on plastic substrate. The plastic support is low cost and commercially appealing, and the palladium is interesting from the scientific point of view, showing inverted surface plasmon resonance response. The biochips were tested for the detection of DNA chains, selected as the target of the experiment, since they can be applied to several medical early-diagnosis tools, such as different biomarkers of cancers or cystic fibrosis [8].

With regard to characterization and metrology capability, an ellipsometry system dedicated to the study of optical properties, composition, and contamination of materials has been recently developed in the CNR-IFN laboratories [12]. One of the recent applications for this system was the study of the optical performance of a few layers graphene at hydrogen Lyman-alpha (121.6 nm) [13]. We determined the optical constants of such a material at this spectral line and observed the optical anisotropy and the effects induced on the substrate performances as a shift of the pseudo-Brewster angle [13].

Graphene is interesting for several reasons [14]: for example, its excellent chemical and thermal stability [15]. The arrangement of the carbon atoms makes it inert and impermeable to all atomic species, including helium [16]. It is an outstanding candidate as protective layer of optical coatings designed to operate in hostile conditions.

However, graphene is not the only 2D material under investigation. Among the others, MoS_2 is appealing for biosensing and optical sensors development [17]. The bio-applications, such as DNA, cancer, and COVID-19 detection, are the most relevant [18], and are supported by a study that shows MoS_2 is compatible with human bodies, while graphene is still under study for this aspect. Whatever the potential in optics, any feasibility assessment can be made after a careful evaluation of the optical performance of the materials. We therefore sought to measure the effects of monolayer MoS_2 deposited on the top of a widely used substrate. The substrate chosen for this comparative analysis was a thick SiO_2 film. The SiO_2 is well known due to excellent properties, such as anti-resistance, hardness, corrosion resistance, dielectric, and optical transparency [19–21].

Starting from this scenario, we characterized the optical response of MoS_2 in the visible spectral range under light polarization control and compared its optical throughput with that of graphene, the most popular of the 2D materials, and pristine SiO_2.

We describe the preliminary results in this manuscript, that includes a section dedicated to the experimental equipment and samples, and a section dedicated to the experimental achievements and discussion. The conclusion reports a summary of what is presented in the paper by envisioning a future experimental campaign.

2. Experimental Arrangement and Samples

The samples whose measurements we report in this paper are:

a. Monolayer graphene onto SiO_2 (300 nm)/Si and the corresponding bare sample of SiO_2 (300 nm)/Si (named "SiO_2 [R-G), both provided by Graphenea, Inc.;
b. Monolayer of MoS_2 onto SiO_2 (300 nm)/Si and the corresponding bare sample of SiO_2 (300 nm)/Si, (named "MoS_2 [R-MoS_2]"), both provided by 2D semiconductors.

The thicknesses of the graphene and MoS_2 were not experimentally determined at the time of this manuscript, but this is planned as part of the measurement campaign we have planned. We want to use Raman spectroscopy for this purpose, as we have already for the study of the three-layer graphene in [13]. Then, we are referring to "monolayer", according to what it is declared by the suppliers.

The nominal thickness of SiO_2 is 300 nm. The reflectivity measurements described hereafter make it possible to estimate the value from the experimental measurement test.

A proper thickness of the SiO_2 substrate allows the direct observation of graphene and MoS_2 by optical microscope. In mono layered graphene and MoS2 materials, few layers, discontinuities, and wrinkles are directly detectable in a very simple way [22–25]. The first measurement performed on the samples under examination was an observation by optical microscope. The images were acquired at 100× magnification.

We also investigated the morphology of the monolayer graphene sample by atomic force microscope (AFM) in no-contact mode operation (Park System XE-70).

After the observation of the surface, the samples were characterized in terms of optical performance.

The reflectometers equipment available at the CNR-IFN laboratories located in Padova cover a wide spectral range extending from the extreme ultraviolet (EUV) to visible wavelengths. An additional tool for infrared measurements has also recently been acquired.

The optical characterizations at shorter wavelengths (30–400 nm) require the use of normal incidence Johnson–Onaka reflectometer operating in vacuum. Figure 1 exhibits the sketch of the device [13].

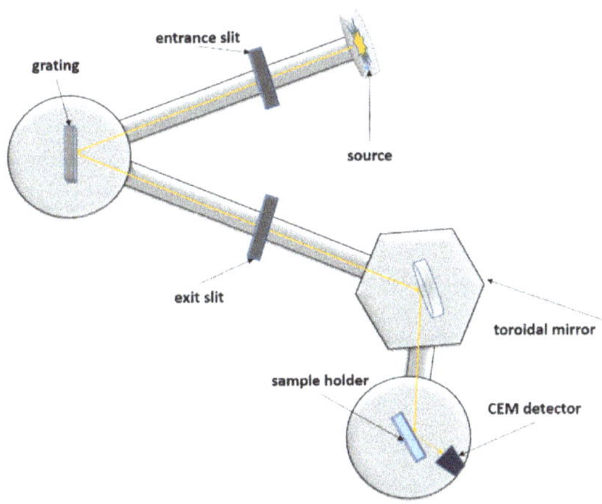

Figure 1. The figure used under CC-BY 4.0 depicts the EUV-VUV-UV Johnson–Onaka reflectometer.

The dispersion element mounted on the reflectometer is a Pt-coated toroidal grating with 600 lines/mm. The main radius is 0.5 m and the subtended angle between the entrance and the exit slits is 25°. A toroidal mirror working at 45° incidence angle focuses the monochromatic radiation on the sample. In the experimental chamber, the samples are hosted on a holder that can be rotated to the desired incidence angle. We have recently implemented this facility with a rotating linear polarizer optimized for the VUV. In this way, the reflectometer is suitable for ellipsometry measurements [12,13]. The upgraded system has recently been used for the characterization of phase retarders for EUV-VUV wavelengths [12,13].

The optical characterization in the visible spectral range is accomplished by using the VIS reflectometer depicted in Figure 2. The system designed for testing the optical response of samples at variable incidence angles, working in a θ-2θ configuration, was recently assembled. It consists of a compact, stabilized, broadband light source (360–2600 nm), a rotator stage to hold the sample at a desired working angle, and a spectrometer coupled with a cosine corrector for the detection.

The measurements, in reflection and transmission modes, can be performed for any type of sample with no restrictions in size. The optical response can be tested with polarized and unpolarized light in order to investigate the polarization response of the specimen of interest. A commercial polarizing filter was used in the measures described in this paper [25]. The filter has a p-polarized transmission close to 90% in the spectral range

450–900 nm. A calibrated sample of protected aluminium was used as a reference specimen (Filmetrics KLA Corporation).

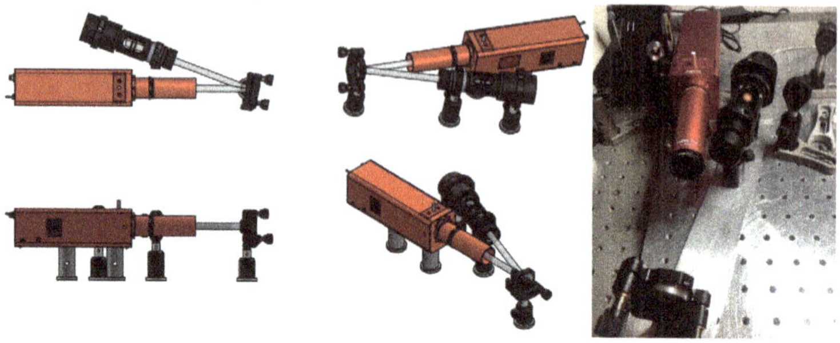

Figure 2. The design and the photograph exhibit the reflectometer available at CNR-IFN in Padova and optimized for the optical characterization in the visible range.

3. Results

In the sample under analysis, the nominal thickness (300 nm) of the pristine SiO_2 is tuned to enhance the optical visibility of the 2D materials on the top [22–25]. The top surface quality of graphene and MoS_2 can be easily determined by using an optical microscope, that allowed us to estimate the linear dimensions of the superficial defects. The presence of film discontinuities and wrinkles can be assessed in a qualitative and semi-quantitative way over a wide region. In case of regions of the same sample with a different number of layers, we usually observe areas with different contrast. An illustrative example is depicted in Figure 3 [24]. The image by Y. Stubrov et al. [24] shows a graphene plate on a Si substrate, covered with 300 nm-thick SiO_2 layer.

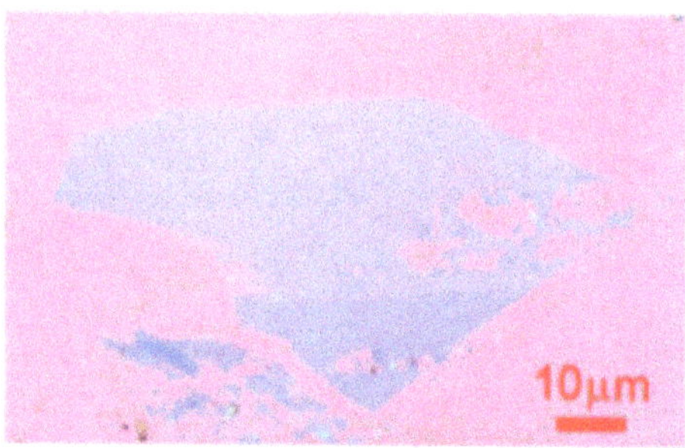

Figure 3. The figure used under CC-BY 4.0 depicts the optical microscopy image of an investigated graphene sample located on Si substrate, covered with 300 nm-thick SiO_2 layer.

For the specimens characterized in this paper, we cannot determine how many layers there are, but we can say that the number of layers is the same in all regions of the sample. The images (see Figure 4), acquired by a camera connected to the optical microscope, show a good quality of the samples surface, good homogeneity for both monolayers, and no relevant defects.

Several areas of the specimens were observed; we report the representative ones.

Figure 5 depicts the surface of the graphene sample (left) analysed by AFM over 3.5 µm × 3.5 µm area. At the time of writing, we have only analysed the AFM measurements of graphene, but we plan to perform the characterization on all samples under study. The surface quality of the graphene is good even at the nm scale, showing some wrinkles and defects mainly due to the transfer process of graphene, grown by chemical vapor deposition and then transferred onto SiO_2 substrate. The roughness estimated on the profiles (right) corresponding to the white cross in Figure 4 is Ra = 0.6 nm and Rq = 0.75 nm.

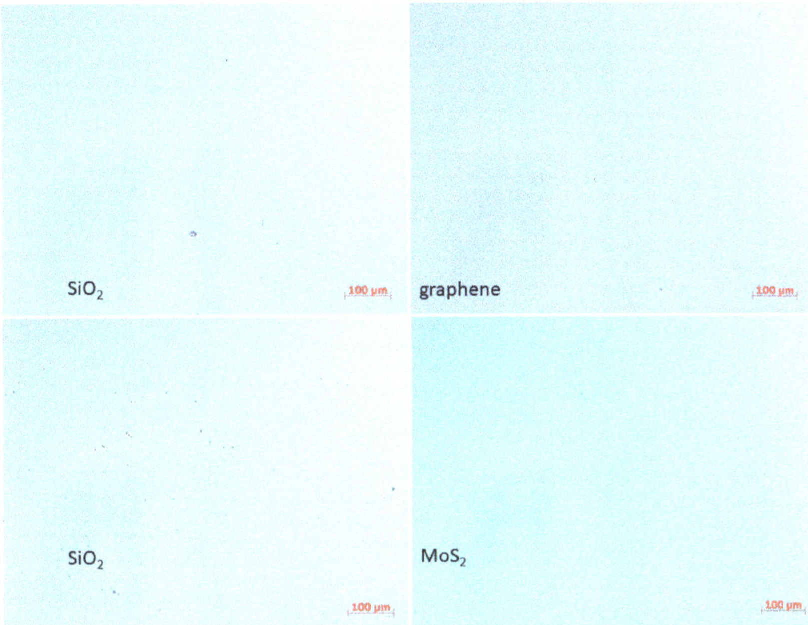

Figure 4. The figure depicts regions of the samples: SiO_2 provided by Graphenea (**left**, top); graphene provided by Graphenea (**right**, top); SiO_2 provided by 2D semiconductors (**left**, bottom); MoS_2 provided by 2D semiconductors (**right**, bottom).

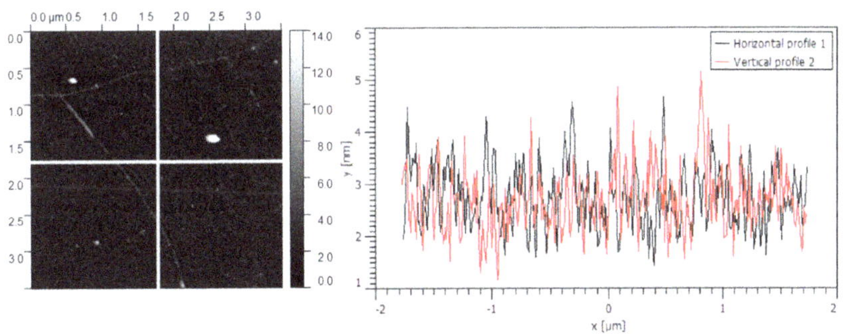

Figure 5. The figure depicts AFM measurement (**left**) of the graphene sample provided by Graphenea (**left**, top); the horizontal and vertical profiles (**right**) correspond to the white cross.

Once the samples were observed by optical microscope, their optical performance was measured. The experimental results are reported in Figure 6. All tests have been performed at 8° angle of incidence by using p-polarized light as probe.

In this test campaign, we used a known reference sample dealt by Filmetrics KLA Corporation for determining the experimental reflectance of any sample, without measuring the direct beam and without moving the experimental arrangement.

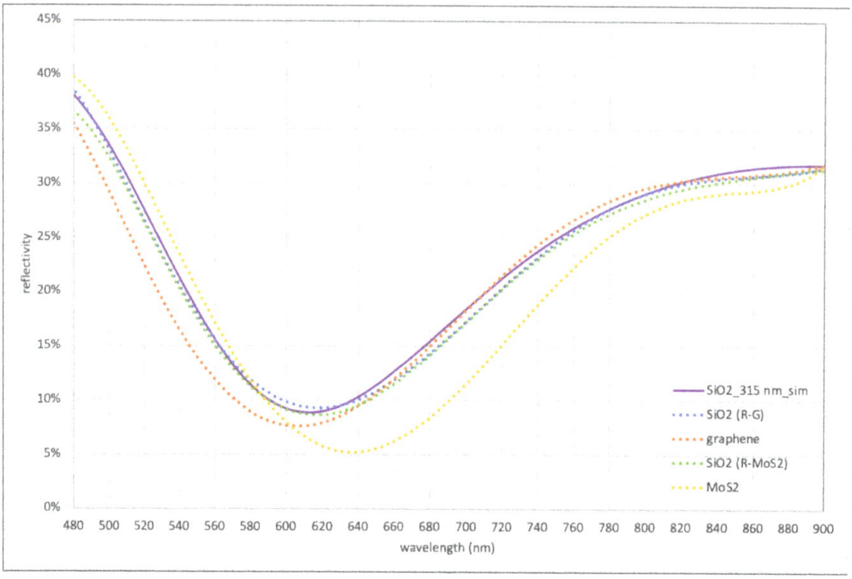

Figure 6. The plot depicts the experimental p-reflectance of the four samples under investigation together with the simulation of a SiO_2 (315 nm)/Si.

It is required to determine the factor F (Equation (1)):

$$F = \frac{R_{ref}}{R_{cal}} \quad (1)$$

which is the ratio between the value of the light reflected by the reference sample (R_{ref}) and the reflectance of the same sample measured and certified by the manufacturer (R_{ref}).

Given the experimental value of the light reflected by the specimen that we want to characterize (R_{exp}), the experimental reflectance, R, of such as sample is given by the following relationship:

$$R = \frac{R_{exp}}{F} \quad (2)$$

Figure 6 shows the measured reflectance according to Equation (2) for the four samples we are analysing. It is worth to note that the performances of the two SiO_2 provided by Graphenea, Inc. (see Figure 5, "SiO_2 [R-G] exp") and 2D semiconductors (see Figure 5, "SiO_2 [R-MoS_2] exp"), even if they come from two different suppliers, are both in good agreement with the simulation of the structure SiO_2 (315 nm)/Si. Then, the actual thickness of the SiO_2 that we can estimate from the reflectance measurements is 315 nm.

The reflectance of the SiO_2 is the reference with respect to we want to observe the optical effects of 2D materials. The sample with graphene on the top (see Figure 5, "graphene exp") reflects sensitively less and shows a blue shift of the minimum, that occurs around 605 nm against 610 nm of SiO_2. On the contrary, MoS_2 (see Figure 5, "MoS_2 exp") induces a red shift of the minimum, which is observed around 640 nm. The reflectance is higher than that of SiO_2 up to 580 nm, then becomes significantly lower at longer wavelengths.

The present study is qualitative and shows how reflectance measurements with polarization control are sensitive to the presence of 2D materials on the surface of a SiO_2/Si substrate, despite the sub-nanometric structures (nominal thickness of monolayer graphene is 0.34 nm and nominal thickness of monolayer MoS_2 is 0.72 nm). For a quantitative analysis, we plan to perform experimental thickness measurements and a full characterization on samples with different thicknesses of graphene and MoS_2.

Combined with experimental determination of thicknesses, the reflectance response can be used to retrieve the optical constants of such a material at the wavelength of interest. For graphene, this analysis has been addressed by several authors; for MoS_2, there are still interesting studies that could be assessed.

4. Conclusions

In this manuscript, we analysed the optical response of four samples based on the structure SiO_2/Si, two of them capped by MoS_2 and graphene, respectively. The experimental results show a good sensitivity of the reflectance to 2D materials, offering great potential for their characterization in view of many application scenarios.

Author Contributions: Conceptualization P.Z., methodology P.Z. and P.C., validation P.Z. and P.C., formal analysis P.Z., investigation, resource, data curation P.Z., P.C., C.C., S.N., N.C., L.Z., V.D.D., writing-original draft preparation P.Z., writing-review and editing P.Z., P.C., C.C., S.N., N.C., L.Z., V.D.D., visualization, supervision P.Z., project administration P.Z., funding acquisition V.D.D. All authors have read and agreed to the published version of the manuscript.

Funding: This activity has been financially supported by the contracts n. 2020-4-HH.0 and n. 2019-34-HH.0 between Agenzia Spaziale Italiana (ASI) and the Istituto Nazionale di Astrofisica (INAF).

Institutional Review Board Statement: Not applicable.

Informed Consent Statement: Not applicable.

Data Availability Statement: Not applicable.

Conflicts of Interest: The authors declare no conflict of interest.

References

1. Available online: http://www.pd.ifn.cnr.it (accessed on 1 April 2022).
2. Yesudasu, V.; Shekhar Pradhan, H.; Jasvanthbhai Pandyab, R. Recent progress in surface plasmon resonance based sensors: A comprehensive review. *Heliyon* **2021**, *7*, e06321. [CrossRef] [PubMed]
3. Zeng, Y.; Zhou, J.; Sang, W.; Kong, W.; Qu, J.; Ho, H.P.; Zhou, K.; Zhi Gao, B.; Chen, J.; Shao, Y. High-Sensitive Surface Plasmon Resonance Imaging Biosensor Based on Dual-Wavelength Differential Method. *Front. Chem.* **2021**, *9*, 801355. [CrossRef] [PubMed]
4. Cai, H.; Shan, S.; Wang, X. High Sensitivity Surface Plasmon Resonance Sensor Based on Periodic Multilayer Thin Films. *Nanomaterials* **2021**, *11*, 3399. [CrossRef]
5. Zhu, J.; Li, N. Novel high sensitivity SPR sensor based on surface plasmon resonance technology and IMI waveguide structure. *Results Phys.* **2020**, *17*, 103049. [CrossRef]
6. Garoli, D.; Rodriguez De Marcos, L.V.; Larruquert, J.I.; Corso, A.J.; Proietti Zaccaria, R.; Pelizzo, M.G. Mirrors for Space Telescopes: Degradation Issues. *Appl. Sci.* **2020**, *10*, 7538. [CrossRef]
7. Slaughter, J.M.; Schulze, D.W. Structure and performance of Si/Mo multilayer mirrors for the extreme ultraviolet. *J. Appl. Phys.* **1994**, *76*, 2144. [CrossRef]
8. Zuppella, P.; Pasqualotto, E.; Zuccon, S.; Gerlin, F.; Corso, A.J.; Scaramuzza, M.; De Toni, A.; Paccagnella, A.; Pelizzo, M.G. Palladium on Plastic Substrates for Plasmonic Devices. *Sensors* **2015**, *15*, 1138–1147. [CrossRef]
9. Cennamo, N.; Zuppella, P.; Bacco, D.; Corso, A.J.; Pelizzo, M.G.; Zeni, L. SPR Sensor Platform Based on a Novel Metal Bilayer Applied on D–Shaped Plastic Optical Fibers for Refractive Index Measurements in the Range 1.38–1.42. *IEEE Sens. J.* **2016**, *16*, 4822–4827. [CrossRef]
10. Zuppella, P.; Monaco, G.; Corso, A.J.; Nicolosi, P.; Windt, D.L.; Bello, V.; Mattei, G.; Pelizzo, M.G. Iridium/silicon multilayers for extreme ultraviolet applications in the 20–35 nm wavelength range. *Opt. Lett.* **2011**, *36*, 1203–1205. [CrossRef]
11. Gaballah, A.E.H.; Nicolosi, P.; Ahmed, N.; Jimenez, K.; Pettinari, G.; Gerardino, A.; Zuppella, P. Vacuum ultraviolet quarter wave plates based on SnTe/Al bilayer: Design, fabrication, optical and ellipsometric characterization. *App. Surf. Sci.* **2019**, *463*, 75–81. [CrossRef]
12. Gaballah, A.E.H.; Nicolosi, P.; Ahmed, N.; Jimenez, K.; Pettinari, G.; Gerardino, A.; Zuppella, P. EUV polarimetry for thin film and surface characterization and EUV phase retarder reflector development. *Rev. Sci. Instrum.* **2018**, *89*, 015108. [CrossRef] [PubMed]
13. Malik, N.A.; Nicolosi, P.; Jimenez, K.; Gaballah, A.; Giglia, A.; Lazzarino, M.; Zuppella, P. Experimental Study of Few-Layer Graphene: Optical Anisotropy and Pseudo-Brewster Angle Shift in Vacuum Ultraviolet Spectral Range. *Adv. Photonics Res.* **2021**, *2*, 2000207. [CrossRef]
14. Geim, A.K.; Novoselov, K.S. The rise of graphene. *Nat. Mater.* **2007**, *6*, 183–191. [CrossRef] [PubMed]
15. Galashev, A.; Rakhmanova, O. Mechanical and thermal stability of graphene and graphene-based materials. *Phys. Uspekhi* **2014**, *57*, 970–989. [CrossRef]

16. Su, Y.; Kravets, V.G.; Wong, S.L.; Waters, J.; Geim, A.K.; Nair, R.R. Impermeable barrier films and protective coatings based on reduced graphene oxide. *Nat. Commun.* **2014**, *5*, 5843. [CrossRef]
17. Zhang, W.; Zhang, P.; Su, Z.; Wei, G. Synthesis and sensor applications of MoS2-based nanocomposites. *Nanoscale* **2015**, *7*, 18364–18378. [CrossRef]
18. Zhao, L.; Kong, D.; Wu, Z.; Liu, G.; Yan, X.; Liu, F.; Liu, X.; Wang, C.; Cui, J.; Lu, G. Interface interaction of MoS2 nanosheets with DNA based aptameric biosensor for carbohydrate antigen 15–3 detection. *Microchem. J.* **2020**, *15*, 104675. [CrossRef]
19. Gitlin, D.; Karp, J.; Moyzhes, B. Graded tunnelling barrier and oxygen concentration in thermally grown ultrathin SiOx gate oxide. *J. Phys. D Appl. Phys.* **2007**, *40*, 2143–2149. [CrossRef]
20. Gravalidis, C.; Gioti, M.; Laskarakis, A.; Logothetidis, S. Real-time monitoring of silicon oxide deposition processes. *Surf. Coat. Technol.* **2004**, *180−181*, 655–658. [CrossRef]
21. Aharonovich, I.; Lifshitz, Y.; Tamir, S. Growth mechanisms of amorphous SiO[sub x] nanowires. *Appl. Phys. Lett.* **2007**, *90*, 263109. [CrossRef]
22. Blake, P.; Hill, E.W. Making graphene visible. *Appl. Phys. Lett.* **2007**, *91*, 063124. [CrossRef]
23. Zhang, X.; Kawai, H.; Yang, J.; Goh, K.E.J. Detecting MoS2 and MoSe2 with optical contrast simulation. *Prog. Nat.Sci. Mater. Int.* **2019**, *29*, 667–671. [CrossRef]
24. Stubrov, Y.; Nikolenko, A.; Strelchuk, V.; Nedilko, S.; Chornii, V. Structural Modification of Single-Layer Graphene Under Laser Irradiation Featured. *Nanoscale Res. Lett.* **2017**, *12*, 297. [CrossRef] [PubMed]
25. Xu, H.; Zhou, W.; Zheng, X.; Huang, J.; Feng, X.; Ye, L.; Xu, G.; Lin, F. Control of the Nucleation Density of Molybdenum Disulfide in Large-Scale Synthesis Using Chemical Vapor Deposition. *Materials* **2018**, *11*, 870. [CrossRef]

engineering proceedings

Proceeding Paper

ARC Knee Brace: Neoprene Knee Brace with Active Control Using Wearable Sensors †

Bilge Koyuncu *, Cevza Candan and Banu Nergis

Faculty of Textile Technologies and Design, Istanbul Technical University (ITU), Istanbul 34437, Turkey; candance@itu.edu.tr (C.C.); uygunf@itu.edu.tr (B.N.)

* Correspondence: koyuncubil@itu.edu.tr

† Presented at the 2nd International Electronic Conference on Applied Sciences, 15–31 October 2021; Available online: https://asec2021.sciforum.net/.

Abstract: Remote monitoring of a patient's physical rehabilitation process after knee surgery is crucial, especially in instances such as pandemics, where patients may not be able to acquire ongoing postoperative care owing to the precautions implemented. Wearable technology can be used to track a patient's development and ensure that they follow rehabilitation guidelines. An active rotation control (ARC) knee brace was developed to guide and facilitate physical therapy movements of patients with knee injuries in an actively controlled manner. The system can trigger a visual feedback mechanism when the subject performs various knee postures. Through the Internet, caregivers could obtain patients' overall knee-related rehabilitation metrics. The ARC knee brace employs inertial motion tracking technology which is based on low-cost inertial sensors and data processing algorithms to capture user's knee posture in real-time during rehabilitation process. The inertial measurement units (IMUs) containing a combination of accelerometer and gyroscope are used as motion sensors to measure accelerations and rotational rates of knee. The sensors track data include acceleration, rotation, and temperature. The processing system calculates various metrics from the posture of knee.

Keywords: active rotation control; knee brace; IMU; sensor; data processing; physiotherapy; health care; wearable technology

Citation: Koyuncu, B.; Candan, C.; Nergis, B. ARC Knee Brace: Neoprene Knee Brace with Active Control Using Wearable Sensors. *Eng. Proc.* **2021**, *11*, 51. https://doi.org/10.3390/ASEC2021-11115

Academic Editor: Nunzio Cennamo

Published: 15 October 2021

Publisher's Note: MDPI stays neutral with regard to jurisdictional claims in published maps and institutional affiliations.

1. Introduction

Knee injuries are frequent, irrespective of the gender, age, and demographic factors, and may necessitate costly and lengthy treatment methods [1–3]. A patient's recovery from a knee injury is generally reliant on their commitment to their allocated rehabilitation procedures. It is advantageous for sportsmen to do the motions prescribed for knee injuries at physical therapy facilities under the supervision of a professional. However, if an athlete is unable or does not choose to visit the physical therapy clinic, these motions are performed alone. When the intended knee stretching/opening cannot be accomplished during individual studies due to the fear of making a mistake or injuring oneself, the treatment procedure and, thus, the time to achieve knee mobility rises.

Wearable technologies are commonly utilized in the therapeutic field to examine mechanical analyses of movement in neurological and orthopedic disorders for helping patients' treatment programs inside and outside of a clinical setting [4–10]. The literature survey demonstrates that joint angle estimation/measurement is quite important when it comes to monitoring orthopedic injury recovery [11], e.g., knee for such wearable technologies. Consequently, many different types of sensors, including textile-based ones, have either been developed or used to accomplish this very task. The survey also points out that modelling wearable physical therapy systems without compromising user's comfort and evaluating these models on user study data whenever possible are the main concerns of the studies [12–16]. In addition to the products of the scientific works, there are also some commercially available wearable devices for injury rehabilitation. However, majority of these systems are based on "monitoring and/or providing continuous passive motion (CPM) with the wearer". Since these systems have isotonic movements, they are typically utilized as devices to ensure joint mobility in injury rehabilitation processes without an active muscle contraction. Although medical experts' overall assessment of the benefits of such devices is positive, they do not allow the muscle to contract voluntarily; thus, depending on the patient's condition, the expected effect in regaining "old" muscle strength may not be seen. Moreover, continuous passive motion (CPM) devices do occupy large areas and may be composed of many components and accessories; therefore, they may need assistance for use. Products such as patella open lateral support knee pads, which are prescribed for athletes with knee injuries, are, on the other hand, simply meant to assist the athlete protect himself and/or promote such motions during every day routine activities (such as walking), though such products may be employed together with CPM or isotonic movement devices [17–19].

In light of the literature, the study under discussion is conducted to develop a wearable knee brace that is not only light-weight, easy to use, and affordable, but also contributes to the development of sustainable e-textiles approach to some extent. The active rotation control (ARC) knee brace is composed of medical textile products (e.g., knee pads and knee cuffs), an electronic circuit that can be integrated into the textile substrate, and a code (software) for the circuit which is written to produce real-time data for monitoring one's knee mobility gaining performance in parallel to the physical therapy movements prescribed. Unlike continuous passive motion (CPM) devices, the active rotation control (ARC) knee brace offers a rehabilitation process with an active muscle contraction.

2. Material and Method

2.1. Material

A patella open lateral support knee pad is used as a substrate to integrate the electrical designs, such that it largely results in dependable and minimal patient obstruction. The electrical design that can be integrated to the textile-based lateral support is, however, composed of gyro sensors, an Arduino Wi-Fi serial module ESP8266 with built-in TCP/IP networking software, and connecting cables with microcontroller capability and a light diode. The Wi-Fi serial transceiver module ESP8266 and the sensors are covered using special cases created using SketchUp 3D Design Software modelling program [20–23]. Instead of location sensors which are very expensive and not easy to find, MPU6050gyro sensors are employed in the system to monitor the angular rotation rate of knee. The other component of the model is LED indication for a visual control system [24]. Finally, the complementary filters (CFs), connected to the high-frequency gyro output, are employed to monitor the knee's angular position by eliminating noise so that much clearer data reading can be realized.

2.2. Method

Modelling and Prototype Development

The ARC knee brace employs inertial motion tracking technology which is based on low-cost inertial sensors and data processing algorithms to capture the user's knee posture in real-time during rehabilitation process. The inertial measurement units (IMUs) containing a combination of accelerometer and gyroscope are used as motion sensors to measure accelerations and rotational rates of knee. The ARC knee brace prototype is divided into two sections. The first section is the control patch houses all of the electronic components and cannot be washed. The second section, however, is the textile-based neoprene brace (patella open lateral support knee pad) itself. The smart electronic parts can be integrated into the textile-based knee brace of small, medium, and large sizes. The ARC knee brace weighs approximately 350 g. So, the whole system does not put an unnecessary pressure on user's knee during therapy. The system is designed to be employed for short- and long-term physical therapy and/or recovery movements. Figure 1a depicts the electronic system as well as its interaction with the user. Figure 1b, on the other hand, presents the working principle of the prototype, namely the algorithm of the code.

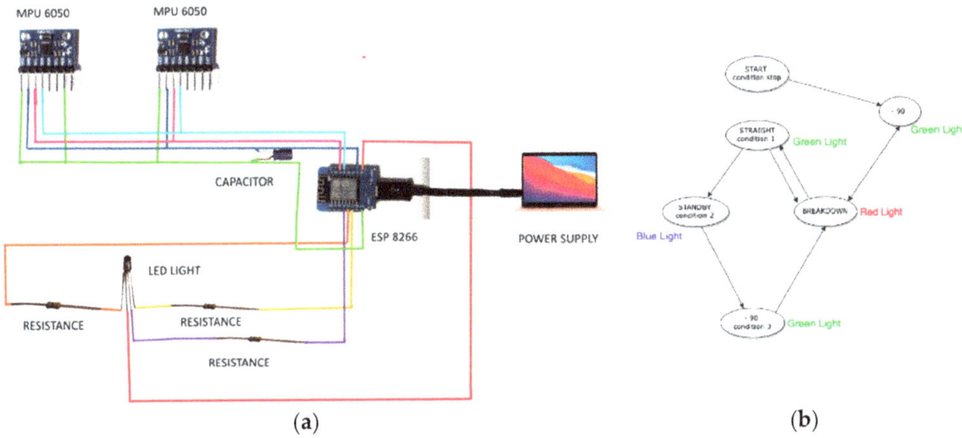

Figure 1. Details of the model: (**a**) design of electronic system and (**b**) schematic representation of working principle of the model.

The free-source Arduino library gives three metrics for the gyro sensors. These are acceleration (m/s^2 in X-Y-Z coordinates), rotation X (degree per second), and temperature (degree Celsius) [25]. Due to the incompatibility between the read values (degree/s) and the intended value (degree), the unit change is implemented in the code by adding the relevant mathematical computations so that the model can calculate knee angles based on the angular velocity and acceleration. The gyro sensor "MPU6050 0x68" collects and transmits data to IoT devices with eight pins. The VCC, GND, SCL, and SDA pins of the sensors are used in the circuit, whereas the gyro sensor "MPU6050 0x69" communicates through one extra pin, namely AD0. Since the gyro sensors contain noise, i.e., one of their most common output errors, an additional (complimentary) filter is also utilized for the circuit design. For data collection, the first gyro sensor MPU6050 0x68 is attached to the ESP8266 Arduino Wi-Fi module through a 0x68 channel, whereas the second gyro is connected to the very same Wi-Fi module via 0x69 channel [25]. Furthermore, 44 Hz is selected as the best acceptable frequency for our study from a range of 5–260 Hz intermittent values offered by the gyro. The minimum rotation value (degree/s) for the gyro is 250 degrees/s and is selected as the maximum limit for the design. The data collected comprises acceleration and rotation angle, and it is processed such that the graph of knee position against acceleration is obtained using the accelerometer's functioning principles. For the LED lighting of three colors, a connection between three distinct legs and three identically resistances are used. The goal of including a capacitor in the design is to protect IoT devices, in addition to extending the lifespan of the sensor. The pin and color codes are established in the developed code. For the work, three primary colors, namely red, green, and blue, are chosen, and depending on the voltage levels transmitted to the LEDs, the system activates the correct color.

The extremities of the knee movements are full extension and complete flexion [16,25]. As is well known from medicine, knee motion covers the maximum range of 140 degrees [25]. Accordingly, in the design of the ARC brace, extension of the knee refers to an angle increase in the bent knee joint. As previously explained, the bending angle information in the code is read in terms of rotation per second (degree/s), and the system architecture converts it to degrees.

3. Result and Discussion

The ARC knee brace can be integrated with the appropriate textile-based knee pads recommended by physicians. The brace, together with the relevant knee bending positions, is given in Figure 2, whereas the validation of the system running is demonstrated in Figures 3 and 4.

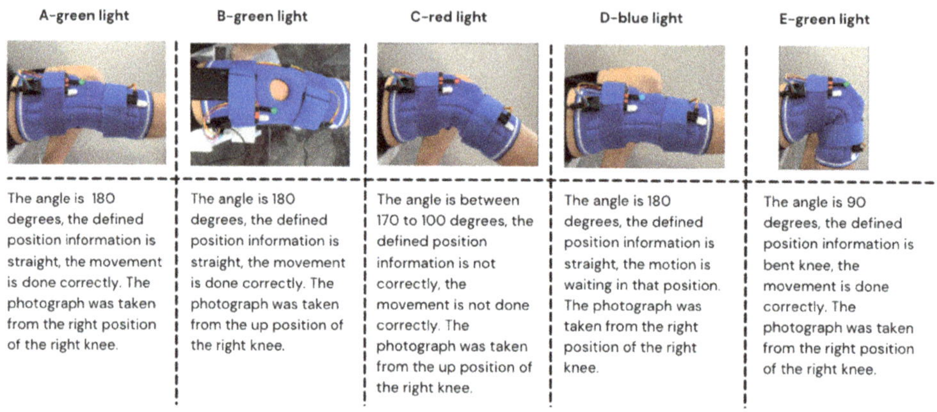

Figure 2. The knee bending test of the ARC knee brace.

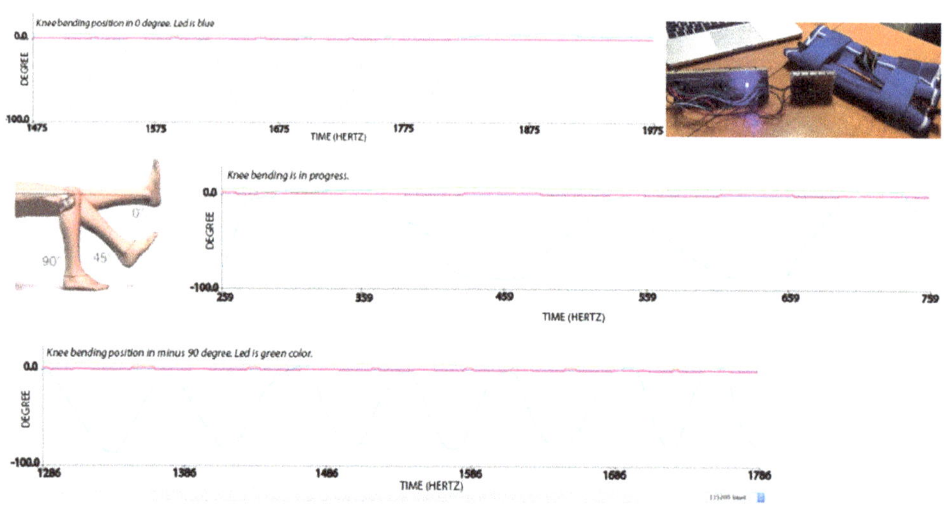

Figure 3. The prototype of the ARC knee brace and the momentary data taken from it for different knee bending positions from minus 90 and 0 degree (from down to up).

When the brace is run, it reads the real-time angular data for the current knee position. If the knee angle hits −90 degree, the LED turns into green, whilst the LED goes red for the incorrect knee positioning. Finally, the LED changes back to green at the angle of 0 degrees (Figure 1b, Figure 2A–E and Figure 3). During a treatment, the system waits for 20 s before it changes the green light to the blue one at the appropriate bending angle of the knee. The blue light indicates that the movement required for that particular phase of the treatment has been completed properly (Figure 2A–E and Figure 3). As seen from Figure 3, the knee improvement can be measured by the amount of knee extension (in the angular sense), which is more objective criteria, in comparison to the use of pain scales for knee mobility recovery testing, such as visual analogue scale (VAS) [26]. The data viewing, recording, and sharing options of the system are provided with users. The data can also be shared with the relevant specialist through the mobile application developed (Figure 4), if required. As is well known, the UI design process is a collection of interface animation, visual elements, screen layout, and content. The user experience and interface (UX/UI) design of the application allows the user to capture and monitor the observed data (Figure 4). For the brace under discussion, the IOS design color and the knee position codes are developed using the relevant public sources [27]. To do so, offering the user a more attentive interface has become the focal point of the main CSS codes (Figure 5), though there is no doubt that the application may be further improved to make it more user-friendly.

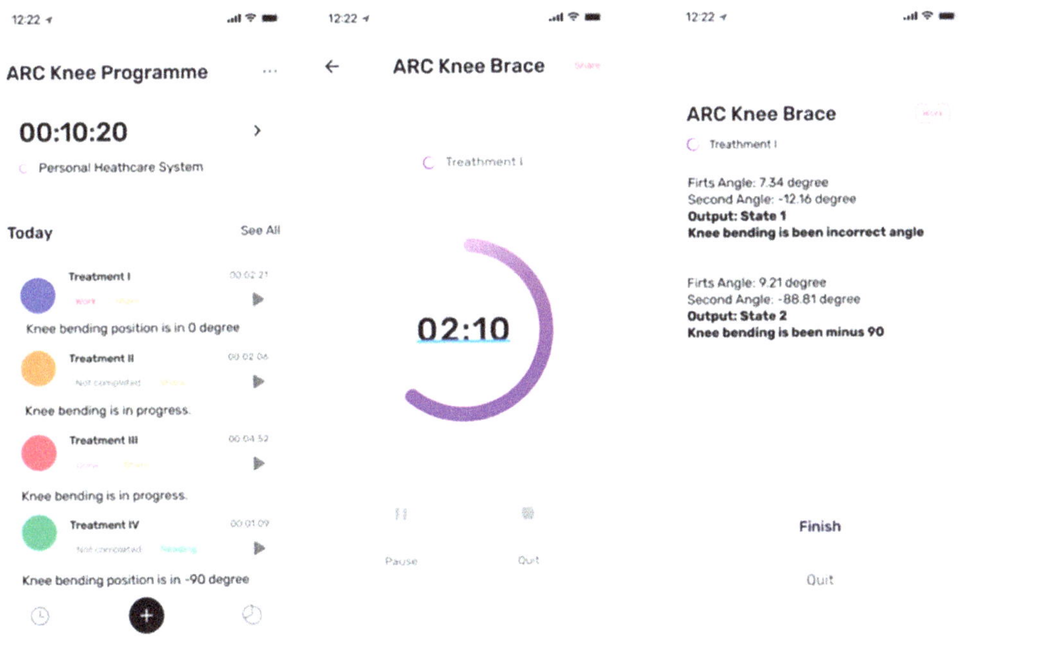

Figure 4. UX/UI design from the application interface for ARC knee brace: (**a**) exercise data collection; (**b**) timer motivation; and (**c**) ARC knee brace program outcome.

The power supply provided to the system can be connected to the power. It works in connection with a computer via USB so that it can monitor the real-time graphics produced corresponding to the relevant knee movements. External power banks, the phone with a converter patch cable, or some other computer-derived equipment are the other alternative sources that can be attached to the system using the appropriate intermediate cables.

As a final word, it should be noted that feedback from actual users is needed to make the system more user-friendly and more efficient, which is the scope of further study.

```
Cover Page:

self.view.backgroundColor = UIColor(red: 0.898, green: 0.898,
blue: 0.898, alpha: 1)

position: relative;
width: 375px;
height: 812px;

background: #FAFAFF;
Blue: #3D4ABA
Orange: #FFA656
Pink: #FD5B71
Green: #07E092
Black: #070417

Additions:
#FFEFF1, #F2F2F2, #E6FCF4, #FEF5ED, #FFFFFF, #7012CE
```

```
Design of the icons:

// Treatment I - Light

var view = UILabel()
view.frame = CGRect(x: 0, y: 0, width: 343, height: 84)
view.backgroundColor = .white

var parent = self.view!
parent.addSubview(view)
view.translatesAutoresizingMaskIntoConstraints = false
view.widthAnchor.constraint(equalToConstant: 343).isActive = true
view.heightAnchor.constraint(equalToConstant: 84).isActive = true
view.leadingAnchor.constraint(equalTo: parent.leadingAnchor,
constant: 200).isActive = true
view.topAnchor.constraint(equalTo: parent.topAnchor, constant:
1086).isActive = true
```

Figure 5. Some parts of the CSS codes.

4. Conclusions

The ARC knee brace discussed in this study is developed to contribute to physical therapy processes of athletes and/or ordinary individuals with knee injuries, in a relatively more self-dependent and affordable manner. The brace differentiates itself from the similar devices/products such that, unlike the other devices, it is designed for active knee motion control. Furthermore, the ARC knee brace is easy to use and affordable because it employs inertial motion tracking technology which is based on low-cost inertial sensors and data processing algorithms to capture the user's knee posture in real time during rehabilitation process. Moreover, it is an environmentally friendly system because it is integrated to a medical textile product (e.g., knee pad) that has already been available in the market. When a therapy period has ended, the knee pad itself can be used as it is meant by detaching the electronic components from the brace. Finally, the application (software) is developed for the brace which can be downloaded on the patient and/or physician's IoT device; thus, the progress of the treatment can be observed and recorded objectively by the patient and/or their physician.

Author Contributions: Conceptualization B.K., C.C. and B.N.; methodology, B.K., C.C. and B.N.; software, B.K.; validation, B.K.; formal analysis, C.C. and B.N.; investigation, B.K., C.C. and B.N.; writing—review and editing, B.K., C.C. and B.N. All authors have read and agreed to the published version of the manuscript.

Funding: This research received no external funding.

Institutional Review Board Statement: The study was conducted in accordance with the National Utility Model Application of Türk Patent "Smart Knee Brace for active motion control", TPE 2020/12076, Document Registration Number: 2020-GE-333942 (pending).

Informed Consent Statement: Not applicable.

Data Availability Statement: Not applicable.

Conflicts of Interest: The authors declare no conflict of interest.

References

1. Gage, B.E.; McIlvain, N.M.; Collins, C.L.; Fields, S.K.; Dawn Comstock, R. Epidemiology of 6.6 million knee injuries presenting to United States emergency departments from 1999 through 2008. *Acad. Emerg. Med.* **2012**, *19*, 378–385. [CrossRef] [PubMed]
2. Losina, E.; Paltiel, A.D.; Weinstein, A.M.; Yelin, E.; Hunter, D.J.; Chen, S.P.; Klara, K.; Suter, L.G.; Solomon, D.H.; Burbine, S.A.; et al. Lifetime medical costs of knee osteoarthritis management in the United States: Impact of extending indications for total knee arthroplasty. *Arthritis Care Res.* **2015**, *67*, 203–215. [CrossRef] [PubMed]
3. Deshpande, B.R.; Katz, J.N.; Solomon, D.H.; Yelin, E.H.; Hunter, D.J.; Messier, S.P.; Suter, L.G.; Losina, E. Number of persons with symptomatic knee osteoarthritis in the us: Impact of race and ethnicity, age, sex, and obesity. *Arthritis Care Res.* **2016**, *68*, 1743–1750. [CrossRef] [PubMed]

4. Van den Bogert, A.J.; Geijtenbeek, T.; Even-Zohar, O.; Steenbrink, F.; Hardin, E.C. A real-time system for biomechanical analysis of human movement and muscle function. *Med. Biol. Eng. Comput.* **2013**, *51*, 1069–1077. [CrossRef] [PubMed]
5. Shinbane, J.S.; Saxon, L.A. Digital monitoring and care: Virtual medicine. *Trends Cardiovasc. Med.* **2016**, *26*, 722–730. [CrossRef] [PubMed]
6. Jonas, S.; Hannig, A.; Spreckelsen, C.; Thomas, M.D. Wearable Technology as a Booster of Clinical Care. In Proceedings of the SPIE—The International Society for Optical Engineering, San Diego, CA, USA, 15–20 February 2014; Volume 9039. [CrossRef]
7. Lu, L.; Zhang, J.; Xie, Y.; Gao, F.; Xu, S.; Wu, X.; Ye, Z. Wearable Health Devices in Health Care: Narrative Systematic Review. *JMIR mHealth uHealth* **2020**, *8*, e18907. [CrossRef] [PubMed]
8. Geijtenbeek, T.; Steenbrink, F.; Otten, B.; Even-Zohar, O. D-flow: Immersive virtual reality and real-time feedback for rehabilitation. In Proceedings of the 10th International Conference on Virtual Reality Continuum and Its Applications in Industry (VRCAI '11), Hong Kong, China, 11–12 December 2011; ACM: New York, NY, USA, 2011; pp. 201–208.
9. Barrios, J.A.; Crossley, K.M.; Davis, I.S. Gait retraining to reduce the knee adduction moment through real-time visual feedback of dynamic knee alignment. *J. Biomech.* **2011**, *43*, 2208–2213. [CrossRef] [PubMed]
10. Teran-Yengle, P.; Birkhofer, R.; Weber, M.A.; Patton, K.; Thatcher, E.; Yack, H.J. Efficacy of gait training with real-time biofeedback in correcting knee hyperextension patterns in young women. *J. Orthop. Sports Phys. Ther.* **2011**, *41*, 948–952. [CrossRef] [PubMed]
11. Hewett, T.E.; Myer, G.D.; Ford, K.R.; Heidt, R.S., Jr.; Colosimo, A.J.; McLean, S.G.; Van Den Bogert, A.J.; Paterno, M.V.; Succop, P. Biomechanical Measures of Neuromuscular Control and Valgus Loading of the Knee Predict Anterior Cruciate Ligament Injury Risk in Female Athletes: A Prospective Study. *Am. J. Sports Med.* **2005**, *33*, 492–501. [CrossRef] [PubMed]
12. Shyr, T.-W.; Shie, J.-W.; Jiang, C.-H.; Li, J.-J. A Textile-Based Wearable Sensing Device Designed for Monitoring the Flexion Angle of Elbow and Knee Movements. *Sensors* **2014**, *14*, 4050–4059. [CrossRef] [PubMed]
13. Bergmann, J.H.M.; Anastasova-Ivanova, S.; Spulber, I.; Gulati, V.; Georgiou, P.; McGregor, A. An Attachable Clothing Sensor System for Measuring Knee Joint Angles. *IEEE Sens. J.* **2013**, *13*, 4090–4097. [CrossRef]
14. Papi, E.; Spulber, I.; Kotti, M.; Georgiou, P.; McGregor, A.H. Smart Sensing System for Combined Activity Classification and Estimation of Knee Range of Motion. *IEEE Sens. J.* **2015**, *15*, 5535–5544. [CrossRef]
15. Gholami, M.; Ejupi, A.; Rezaei, A.; Ferrone, A.; Menon, C. Estimation of Knee Joint Angle Using a Fabric-Based Strain Sensor and Machine Learning: A Preliminary Investigation. In Proceedings of the 2018 7th IEEE International Conference on Biomedical Robotics and Biomechatronics, Enschede, The Netherlands, 26–29 August 2018; pp. 589–594. [CrossRef]
16. Watson, A.; Sun, M.; Pendyal, S.; Zhou, G. TracKnee: Knee angle measurement using stretchable conductive fabric sensors. *Smart Health* **2019**, *15*, 100092. [CrossRef]
17. Patent US2009024065A1. Available online: https://patents.google.com/patent/US20090024065A1/en?oq=US2009024065A1 (accessed on 14 September 2021).
18. Patent US201117844A1. Available online: https://patents.google.com/patent/US201117844A1/en?oq=US201117844A1 (accessed on 14 September 2021).
19. Takeda, R.T.; Shigeru, T.; Masahiro, M.; Manabu, N.; Minoru, Y.S. Gait analysis using gravitational acceleration measured by wearable sensors. *J. Biomech.* **2009**, *42*, 223–233. [CrossRef] [PubMed]
20. Sketch up 3D Design Sofware. Available online: https://www.sketchup.com (accessed on 10 September 2021).
21. Rengier, F.; Mehndiratta, A.; Tengg-Kobligk, H.; Zechmann, C.M.; Unterhinninghofen, R.; Kaucozr, H.U.; Giesel, F.L. 3D printing based on imaging data: Review of medical applications. *Int. J. Comput. Assist. Radiol. Surg.* **2010**, *5*, 336–341. [CrossRef] [PubMed]
22. Allen, D. History and Physical Exam of the Knee (10 2016). Available online: https://www.orthobullets.com/knee-and-sports/3003/history-and-physical-exam-of-the-knee (accessed on 10 September 2021).
23. Karagöl, B. 3D Printing: What Does It Offer and for Whom? TEKPOL Working Paper Series. 2015, pp. 1–17. Available online: https://open.metu.edu.tr/bitstream/handle/11511/92358/stps_wp_1502.pdf (accessed on 10 September 2021).
24. Liao, X.Z.; Zheng, K.; Zhuo, G.; Fangfang, L.; Qingliang, Z.Y. Ultrasensitive and stretchable resistive strain sensors designed for wearable electronics. *Mater. Horiz.* **2017**, *4*, 502–510. [CrossRef]
25. Arduino. Available online: https://www.arduino.cc/en/software (accessed on 10 September 2021).
26. Visual Analogue Scale. Available online: https://www.physio-pedia.com/Visual_Analogue_Scale (accessed on 10 September 2021).
27. GitHub. Retrieved from UI for IOS. 2021. Available online: https://github.com/search?q=UI+for+IOS (accessed on 10 September 2021).

engineering proceedings

MDPI

Proceeding Paper

The Relationship between the Degree of Conversion in Dental Dimethacrylate Polymers Determined by Infrared Spectroscopy and Polymerization Shrinkage †

Marta W. Chrószcz *, Izabela M. Barszczewska-Rybarek and Promise Wori

Department of Physical Chemistry and Technology of Polymers, Silesian University of Technology, Strzody 9, 44-100 Gliwice, Poland; izabela.barszczewska-rybarek@polsl.pl (I.M.B.-R.); woripromise.wp@gmail.com (P.W.)
* Correspondence: marta.chroszcz@polsl.pl
† Presented at the 2nd International Electronic Conference on Applied Sciences, 15–31 October 2021; Available online: https://asec2021.sciforum.net/.

Abstract: In this work, the relationship between the degree of conversion (DC) in the Bis-GMA/TEGDMA polymer networks, determined by two methods—Fourier transform infrared spectroscopy (DC_{IR}) and polymerization shrinkage (DC_S)—was studied. The DC_{IR} was calculated by using the internal standard method, whereas the DC_S was calculated by measuring differences in the monomer and polymer densities, resulting in the polymerization shrinkage. Both methods revealed the same trend in the DC changes with alterations in the Bis-GMA/TEGDMA ratio. However, the DC_S values were lower, in comparison to the DC_{IR} values on average by 18%.

Keywords: dental resin composite; dimethacrylate composite matrix; degree of conversion; Fourier transform infrared spectroscopy; polymerization shrinkage

Citation: Chrószcz, M.W.; Barszczewska-Rybarek, I.M.; Wori, P. The Relationship between the Degree of Conversion in Dental Dimethacrylate Polymers Determined by Infrared Spectroscopy and Polymerization Shrinkage. *Eng. Proc.* **2021**, *11*, 52. https://doi.org/10.3390/ASEC2021-11151

Academic Editor: Mike Barbeck

Published: 15 October 2021

1. Introduction

The degree of conversion (DC) is the most useful parameter in the characterization of dental composite restorative materials based on dimethacrylates. Its value informs about curing efficiency. If the curing level is insufficient, the physicochemical and mechanical properties of the composite can significantly deteriorate [1]. Inadequate curing also decreases the biocompatibility of the material, due to an increase in free monomer content, which tends to leach from the composite matrix [2,3].

It is known from the literature that the DC depends on the monomer chemical structure [1,3,4]. Monomers of highly elastic molecules polymerize to higher DCs, whereas those of stiff molecules polymerize to lower DCs. In particular, bisphenol A glycerolate dimethacrylate (Bis-GMA), the most important and commonly used dental monomer, has a stiff and spacious molecule, willing to form hydrogen bonds, which causes a serious DC limitation. Therefore, it can achieve a limiting DC of 39% when homopolymerized [4]. It is the major drawback of Bis-GMA because the DC below 55% is not acceptable in practical applications [5]. To achieve a sufficiently high DC, Bis-GMA has to be copolymerized with monomers of more elastic and smaller molecules [6,7]. Triethylene glycol dimethacrylate (TEGDMA) is the most commonly used in this field. In addition to the monomer chemical structure, several other factors influence the DC. They include (i) initiation method and initiator type [8,9], (ii) irradiation time [7], (iii) sample thickness [10], (iv) irradiation lamp [11], and (v) filler type and content [12].

The most popular techniques used for the DC determination include (i) Fourier transform infrared spectroscopy (FTIR), (ii) Raman spectroscopy (RS), (iii) solid-state nuclear magnetic resonance (ssNMR), and (iv) differential scanning calorimetry (DSC).

The DC determination method that uses FTIR is the most common, simple, and gives the most reliable results [1]. It is based on the monitoring of changes in the absorption intensity of several bands resulting from vibrations of the C=C double bond, present in the methacrylate group. They include: (i) twisting vibrations (816 cm^{-1}), (ii) wagging vibrations (948 cm^{-1}), and (iii) stretching vibrations (1637 cm^{-1}) [1,13]. The intensity of these bands decreases due to polymerization. As the band located at 1637 cm^{-1} is the most resolved and intense, it is the most commonly used for DC determination [13]. However, to comply with the Beer–Lambert law, the C=C absorption intensity has to be related to the absorption intensity of an internal standard—the band whose absorption intensity does not alter due to the polymerization [14]. The band corresponding to the skeletal stretching vibrations of the C–C bonds in the aromatic ring, located at 1608 cm^{-1}, usually serves as an internal standard [4]. If the system lacks an aromatic ring, the band corresponding to the C=O stretching vibrations, located within 1715 and 1720 cm^{-1}, can be used [4]. However, this method produces underestimated DC values and therefore it is recognized as less reliable compared to that using aromatic internal standard [15,16].

An alternative solution for the DC determination in dimethacrylate composites might be a method based on measuring differences in the monomer and polymer densities resulting from the polymerization shrinkage and calculating its theoretical value (DC_S). The calculation of the latter uses the literature information that the molar volume of one mole of the methacrylate group decreases by 22.5 cm^3 [17] due to polymerization, in the course of which van der Waals forces that occur between monomer molecules turn into covalent bonds that constitute crosslinks of hardened composite matrix [1]. This method has already been applied in the survey on the DC of dimethacrylate systems, however, its reliability has not yet been established.

In this study, a series of dimethacrylate copolymers consisting of various Bis-GMA/TEGDMA various ratios were subjected to FTIR and polymerization shrinkage analyses to determine the DC and relationship between the results from both methods.

2. Materials and Methods

2.1. Materials

Bis-GMA, camphorquinone (CQ), dimethylaminoethyl methacrylate (DMAEMA), and TEGDMA were purchased from Sigma Aldrich (St. Louis, MO, USA).

2.2. Sample Preparation

Seven Bis-GMA/TEGDMA mixtures were prepared. The Bis-GMA weight fraction ranged from 20 to 80% and increased by 10% each time. TEGDMA content decreased proportionally. The mixtures were admixed with the 0.4 wt.% CQ and 1 wt.% DMAEMA that was a photoinitiating system, composed of, respectively, initiator/and accelerator. Thus, prepared compositions were introduced into silicon molds with a diameter of 15 cm and 5 cm thick, covered with PET film to prevent oxygen inhibition, and irradiated with the UV-VIS lamp (Ultra Vitalux 300, Osram, Munich, Germany, 280 to 750 nm, 2400 mW/cm^2) for 1h, from a distance of 15 cm.

2.3. Fourier Transform Infrared Spectroscopy (FTIR)

2.3.1. Instrumentation

FTIR spectra were recorded with the use of a Spectrum Two (Perkin-Elmer, Waltham, MA, USA) spectrometer, with 128 scans at a resolutions of 1cm^{-1}. Monomers and polymers were tested as KBr pellets. A thin layer of a monomer was placed between two KBr pellets, whereas a polymer was ground into a fine powder with a grain size smaller than 25 μm, mixed with KBr powder, and pressed into a pellet.

2.3.2. Calculation of the Degree of Conversion (DC_{IR})

The DC_{IR} was calculated with the use of the following equation:

$$DC_{IR}(\%) = \left(\frac{\left(\frac{A_{C=C}}{A_{Ar}}\right)_{polymer}}{\left(\frac{A_{C=C}}{A_{Ar}}\right)_{monomer}} \right) \times 100, \quad (1)$$

where $A_{C=C}$ is the absorption intensity of the band resulting from the carbon–carbon double bond stretching vibrations, located at 1637 cm^{-1}, and A_{Ar} is the absorption intensity of the internal standard—the band resulting from the skeletal stretching vibrations of the carbon–carbon bonds in the aromatic rings, located at 1608 cm^{-1}.

2.4. Polymerization Shrinkage

2.4.1. Density Measurements

Densities of monomers (d_m) were measured with the use of a 1 mL pyknometer, according to the ISO 1675 standard [18]. Densities of polymers (d_p) were determined with the use of an analytical balance (XP balance, Mettler Toledo, Greifensee, Switzerland), equipped with a density determination kit that uses the Archimedes' principle.

2.4.2. Calculation of the Polymerization Shrinkage

The experimental polymerization shrinkage (S) was calculated with the use of the following equation:

$$S(\%) = \left(1 - \frac{d_m}{d_p}\right) \times 100, \quad (2)$$

where d_m is the density of a monomer mixture, and d_p is the density of the corresponding polymer.

The theoretical polymerization shrinkage (S_{theor}) was calculated on the assumption that the volumetric shrinkage of one mole of the methacrylate double bonds is equal to 22.5 cm^3 [17], according to the following equation:

$$S_{theor}(\%) = \left(1 - \frac{\frac{MW}{d_m} - 2 \times 22.5}{\frac{MW}{d_m}}\right) \times 100, \quad (3)$$

where MW is the molecular weight of a monomer mixture, d_m is the density of a monomer mixture, 2 is the number of double bonds in the monomer molecule, and 22.5 is the volumetric contraction of one mole of the methacrylate group due to its polymerization [17].

2.4.3. Calculation of the Degree of Conversion (DC_S)

The DC_S was calculated according to the following equation:

$$DC_S(\%) = \frac{S}{S_{theor}} \times 100, \quad (4)$$

where S is the experimental polymerization shrinkage, and S_{theor} is the theoretical polymerization shrinkage.

3. Results and Discussion

In this study, seven compositions of Bis-GMA and TEGDMA monomers were prepared, polymerized, and characterized for the density, polymerization shrinkage, and degree of conversion. The latter was determined by two methods—Fourier transform infrared spectroscopy and polymerization shrinkage.

The weight ratios of prepared compositions, their molecular weights (MW), concentrations of double bonds (x_{DB}), and densities of the samples, before and after curing, are summarized in Table 1. The x_{DB} values ranged from 4.52 to 6.37 mol/kg. Its value decreased as the Bis-GMA content increased. The percentage difference between the x_{DB} of the B20:T80 and B80:T20 samples was 41%. As the x_{DB} value informs about the theoretical crosslink density of the dimethacrylate polymer network, it can be said that the greater the Bis-GMA concentration in the monomer mixture, the lower the crosslink density in the resulting polymer network [1,4,8].

Table 1. Sample names, their chemical compositions, and properties: MW, molecular weight, x_{DB}, concentration of double bonds, d_m, density of uncured samples, and d_p, density of cured samples.

Sample Name	Weight Ratios		MW (g/mol)	x_{DB} (mol/kg)	d_m (g/cm^3)		d_p (g/cm^3)	
	Bis-GMA	TEGDMA			avg.	SD	avg.	SD
B20:T80	20	80	314.07	6.37	1.106	0.024	1.194	0.029
B30:T70	30	70	330.09	6.06	1.113	0.052	1.204	0.025
B40:T60	40	60	347.78	5.75	1.121	0.072	1.207	0.061
B50:T50	50	50	367.51	5.44	1.127	0.046	1.210	0.014
B60:T40	60	40	389.60	5.13	1.133	0.062	1.219	0.016
B70:T30	70	30	414.50	4.83	1.141	0.094	1.233	0.033
B80:T20	80	20	442.81	4.52	1.147	0.135	1.216	0.019

As can be seen from Table 1, the densities of monomer mixtures (d_m) ranged from 1.106 to 1.147 g/cm^3. The d_m value increased as the Bis-GMA content increased. The percentage difference between the d_m values determined for the B20:T80 and B80:T20 samples was 4%. As expected, polymerization resulted in tighter packing, which was reflected in higher densities of cured materials compared to the densities of their uncured counterparts (on average by 7.5%). The polymer densities (d_p) ranged from 1.194 to 1.233 g/cm^3. Its value increased with the increase of the Bis-GMA content up to 70 wt.% and then decreased. The percentage difference between the largest and the smallest d_p values was 3%. The density values of cured and uncured samples were used to determine the experimental and theoretical polymerization shrinkages (respectively, S and S_{theor}). These results are summarized in Table 2.

Table 2. The theoretical (S_{theor}) and experimental polymerization shrinkage (S) as well as the degree of conversion, calculated from the polymerization shrinkage (DC_S) and absorption intensity (DC_{IR}).

Sample Name	S_{theor} (%)	S (%)		DC_S (%)		DC_{IR} (%)	
		avg.	SD	avg.	SD	avg.	SD
B20:T80	15.85	7.39	0.45	46.64	2.32	55.30	5.23
B30:T70	15.17	7.55	0.52	49.81	2.56	60.47	4.78
B40:T60	14.49	7.21	0.41	49.75	2.31	61.01	5.23
B50:T50	13.80	6.89	0.46	49.88	3.04	63.97	3.56
B60:T40	13.09	7.11	0.39	54.33	4.34	64.87	4.87
B70:T30	12.38	7.57	0.55	61.23	3.45	72.83	5.44
B80:T20	11.66	5.44	0.25	46.68	2.98	61.34	5.34

The S values ranged from 5.44 to 7.57%. They showed no clear trend throughout the studied series. The highest S value was determined for the B70:T30 sample, whereas the lowest S value was determined for the B80:T20 sample. The percentage difference between the largest and smallest S values was 40%. The S_{theor} values ranged from 11.66 to 15.85% and decreased with the increase in the Bis-GMA content. This relationship resulted from a decreasing concentration of double bonds in this order, because the lower the concentration of double bonds, the lower the volumetric contraction can occur. The percentage difference between the lowest and highest S_{theor} values was 34%. S_{theor} and S were further used to determine the DC_S (Table 2). The DC_S values ranged from 46.68 to 61.23%. It can be seen that the DC_S increased with the increasing Bis-GMA content up to 70 wt.% and then decreased.

The degree of conversion was also examined by using the FTIR technique (DC_{IR}). Representative FTIR spectra of the B80:T20 uncured and cured samples are shown in Figure 1. The FT IR spectra of the remaining compositions in their cured and uncured forms can be found in the Supplementary Materials (Figures S1–S6). The results obtained for the DC_{IR} are summarized in Table 2. DC_{IR} values ranged from 55.30 to 61.34%. As can be seen from Table 2, the DC_{IR} values were higher than the corresponding DC_S values. Nevertheless, it is worth noting that a similar tendency was observed in both methods. The DC_{IR} values increased as the Bis-GMA content increased up to 70% and then decreased when the Bis-GMA content exceeded 70%. The difference between the DC_S and DC_{IR} values ranged from 15.66 to 23.90% (Figure 2).

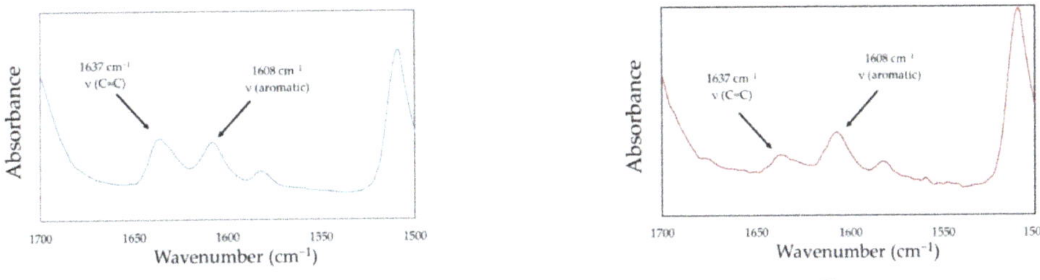

(**a**) (**b**)

Figure 1. The representative FTIR spectra of the B80:T20 composition in its uncured (**a**) and cured (**b**) forms.

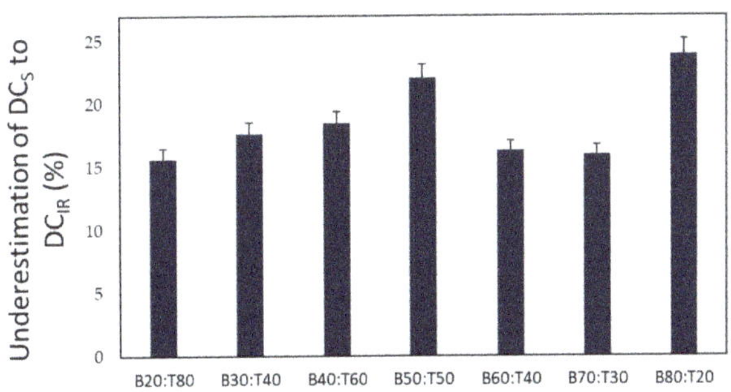

Figure 2. The percentage of underestimation of the DC_S values to the DC_{IR} values.

The underestimation of the DC_S values to the DC_{IR} values can be explained by the fact that the volumetric contraction does not only depend on the concentration of double bonds, but that other factors, too, play an essential role in this phenomenon. They include dimensions, shapes, and hydrophilicity of monomer molecules [19]. The Bis-GMA molecule is large, stiff, and hydrophilic. On the contrary, the TEGDMA molecule is small, elastic, and much less hydrophilic. By increasing the Bis-GMA content, its molecular features increase their impact on the ability to pack tightly. This probably resulted in lower d_p values than expected. In addition, the increase in the Bis-GMA content caused an increase in the viscosity of the monomer mixture, which might increase the inaccuracy of the d_p measurement (the higher the viscosity, the higher the air trapping probability that decreases density). However, the literature shows that the DC_{IR} values calculated with the use of the carbonyl internal standard (instead of the aromatic one) are also underestimated. Collares et al. [16] found that the DC_{IR} values of the Bis-GMA/TEGDMA compositions, determined with the use of carbonyl internal standard, were lower by 23% in comparison to the DC_{IR} determined with the use of the aromatic internal standard.

4. Conclusions

The methodology of the degree of conversion determination in dimethacrylate polymers based on the measurements of the polymerization shrinkage can be a valuable alternative to the methodology based on the FTIR measurements. It produces underestimated results compared to the methodology by using an aromatic internal standard. However, the results achieved with the use of the carbonyl internal standard can be less consistent.

Supplementary Materials: The following supporting information can be downloaded at: https://www.mdpi.com/article/10.3390/ASEC2021-11151/s1, Figure S1: The FTIR spectra of the B20:T80 composition in its uncured (a) and cured (b) forms; Figure S2: The FTIR spectra of the B30:T70 composition in its uncured (a) and cured (b) forms; Figure S3: The FTIR spectra of the B40:T60 composition in its uncured (a) and cured (b) forms; Figure S4: The FTIR spectra of the B50:T50 composition in its uncured (a) and cured (b) forms; Figure S5: The FTIR spectra of the B60:T40 composition in its uncured (a) and cured (b) forms; Figure S6: The FTIR spectra of the B70:T30 composition in its uncured (a) and cured (b) forms.

Author Contributions: Conceptualization, I.M.B.-R.; methodology, I.M.B.-R. and M.W.C.; investigation, M.W.C. and P.W.; resources, I.M.B.-R. and M.W.C.; data curation, I.M.B.-R., M.W.C. and P.W.; writing—original draft preparation, I.M.B.-R. and M.W.C.; writing—review and editing, I.M.B.-R.; visualization, M.W.C.; supervision, I.M.B.-R.; project administration, I.M.B.-R. All authors have read and agreed to the published version of the manuscript.

Funding: This research was funded by the Rector's grant for the scientific research and development activities in the Silesian University of Technology, grant number: 32/014/RGJ22/2004.

Institutional Review Board Statement: Not applicable.

Informed Consent Statement: Not applicable.

Data Availability Statement: Data supporting reported results are available from the authors.

Conflicts of Interest: The authors declare no conflict of interest. The funders had no role in the design of the study; in the collection, analyses, or interpretation of data; in the writing of the manuscript, or in the decision to publish the results.

References

1. Barszczewska-Rybarek, I.M. A Guide through the Dental Dimethacrylate Polymer Network Structural Characterization and Interpretation of Physico-Mechanical Properties. *Materials* **2019**, *12*, 4057. [CrossRef] [PubMed]
2. Gajewski, V.E.S.; Pfeifer, C.S.; Fróes-Salgado, N.R.G.; Boaro, L.C.C.; Braga, R.R. Monomers used in resin composites: Degree of conversion, mechanical properties and water sorption/solubility. *Braz. Dent. J.* **2012**, *23*, 508–514. [CrossRef] [PubMed]
3. Kannurpatti, A.; Anseth, J.; Bowman, C.H.N. A study of the evolution of mechanical properties and structural heterogeneity of polymer networks formed by photo-polymerizations of multifunctional (meth)acrylates. *Polymer* **1998**, *39*, 2507–2513. [CrossRef]
4. Sideridou, I.; Tserki, V.; Papanastasiou, G. Effect of chemical structure on degree of conversion in light-cured dimethacrylate-based dental resins. *Biomaterials* **2002**, *23*, 1819–1829. [CrossRef]
5. Alshali, R.Z.; Silikas, N.; Satterthwaite, J.D. Degree of conversion of bulk-fill compared to conventional resin-composites at two time intervals. *Dent. Mater.* **2013**, *29*, e213–e217. [CrossRef] [PubMed]
6. Dickens, H.; Stansbury, J.W.; Choi, K.M.; Floyd, C.J.E. Photopolymerization kinetics of methacrylate dental resins. *Macromolecules* **2003**, *36*, 6043–6053. [CrossRef]
7. Pfeifer, C.S.; Shelton, Z.R.; Braga, R.R.; Windmoller, D.; Machado, J.C.; Stansbury, J.W. Characterization of dimethacrylate polymeric networks: A study of the crosslinked structure formed by monomers used in dental composites. *Eur. Polym. J.* **2011**, *47*, 162–170. [CrossRef] [PubMed]
8. Stansbury, J.W. Dimethacrylate network formation and polymer property evolution as determined by the selection of monomers and curing conditions. *Dent. Mater.* **2012**, *28*, 13–22. [CrossRef] [PubMed]
9. Barszczewska-Rybarek, I.; Chladek, G. Studies on the curing efficiency and mechanical properties of Bis-GMA and TEGDMA nanocomposites containing silver nanoparticles. *Int. J. Mol. Sci.* **2018**, *19*, 3937. [CrossRef] [PubMed]
10. Par, M.; Gamulin, O.; Marovic, D.; Klaric, E.; Tarle, Z. Raman spectroscopic assessment of degree of conversion of bulk-fill resin composites—Changes at 24 hours post cure. *Open. Dent.* **2015**, *40*, e92–e101. [CrossRef] [PubMed]
11. Randolph, L.D.; Palin, W.M.; Bebelman, S.; Devaux, J.; Gallez, B.; Leloup, G.; Leprince, J.G. Ultra-fast light-curing resin composite with increased conversion and reduced monomer elution. *Dent. Mater.* **2014**, *30*, 594–604. [CrossRef] [PubMed]

12. Halvorson, R.H.; Erickson, R.L.; Davidson, C.L. The effect of filler and silane content on conversion of resin-based composite. *Dent. Mater.* **2003**, *19*, 327–333. [CrossRef]
13. Lin-Vien, D.; Colthup, N.B.; Fateley, W.G.; Grasselli, J.G. *The Handbook of Infrared and Raman Characteristic Frequencies of Organic Molecules*; Academic Press: London, UK, 1991; pp. 137–281.
14. Moraes, L.G.; Rocha, R.S.; Menegazzo, L.M.; de Araújo, E.B.; Yukimito, K.; Moraes, J.C. Infrared spectroscopy: A tool for determination of the degree of conversion in dental composites. *J. Appl. Oral. Sci.* **2008**, *16*, 145–149. [CrossRef] [PubMed]
15. Barszczewska-Rybarek, I.M. Quantitative determination of degree of conversion in photocured poly(urethane-dimethacrylate)s by FTIR spectroscopy. *J. Appl. Polym. Sci.* **2012**, *123*, 1604–1611. [CrossRef]
16. Collares, F.M.; Portella, F.F.; Leitune, V.C.B.; Samuel, S.M.W. Discrepancies in degree of conversion measurements by FTIR. *Braz. Oral. Res.* **2014**, *28*, 9–15.
17. Patel, M.P.; Braden, M.; Davy, K.W.M. Polymerization shrinkage of methacrylate esters. *Biomaterials* **1987**, *8*, 53–56. [CrossRef]
18. *ISO 1675:1985*; Plastics—Liquid Resins—Determination of Density by the Pyknometer Method. International Standard Organisation: London, UK, 1985.
19. Oliveira, K.M.; Lancellotti, A.C.; Ccahuana-Vásquez, R.A.; Consani, S. Shrinkage stress and degree of conversion of dental composite submitted to different photoactivation protocols. *Acta. Odontol. Latinoam.* **2012**, *25*, 115–122. [PubMed]

www.ingramcontent.com/pod-product-compliance
Lightning Source LLC
Chambersburg PA
CBHW061553220926
51850CB00005B/327
9783036548173

MDPI
St. Alban-Anlage 66
4052 Basel
Switzerland
Tel. +41 61 683 77 34
Fax +41 61 302 89 18
www.mdpi.com

Engineering Proceedings Editorial Office
E-mail: engproc@mdpi.com
www.mdpi.com/journal/engproc